PRINTING & DESIGN

Dr. Sabrie Soloman

Ph.D., Sc.D., MBA, PE

KHANNA BOOK PUBLISHING CO. (P) LTD.

Publisher of Engineering and Computer Books

4C/4344, Ansari Road, Darya Ganj, New Delhi-110002

Phone : 011-23244447-48 **Mobile:** +91-9910909320

E-mail : contact@khannabooks.com

Website : www.khannabooks.com

Dear Readers,

To prevent the piracy, this book is secured with HIGH SECURITY HOLOGRAM on the front title cover. In case you don't find the hologram on the front cover title, please write us to at contact@khannabooks.com or whatsapp us at +91-99109 09320 and avail special gift voucher for yourself.

Specimen of Hologram on front Cover title:

Moreover, there is a SPECIAL DISCOUNT COUPON for you with EVERY HOLOGRAM.

How to avail this SPECIAL DISCOUNT:

Step 1: Scratch the hologram
Step 2: Under the scratch area, your "coupon code" is available
Step 3: Logon to www.khannabooks.com
Step 4: Use your "coupon code" in the shopping cart and get your copy at a special discount
Step 5: Enjoy your reading!

Indian Price: ₹ 2995.00
International Price: $ 146.99

ISBN: 978-93-89139-17-4

3D Printing & Design *by*
Dr. Sabrie Soloman

First Edition: 2018
Reprint: 2021, 2022 **2023**

To view complete list of books, Please scan the QR Code:

Published by:
Khanna Book Publishing Co. (P) Ltd.
CIN: U22110DL1998PTC095547

Visit us at: **www.khannabooks.com**
Write us at: **contact@khannabooks.com**

Printed in India by: Vaibhav Products, Delhi.

DEDICATION

To the one ...

Who is Clothed in Perfection,
Entwined the golden rays by the warmth of her illuminating Soul,
Weaved the roads to glory by the strides of her unwavering Feet,
Soared higher than eagles, carrying her offspring on her back to climb even higher,
Living loving example unfailingly she is maintaining ...

For these and more - I dedicate the humble work of this book!

To My Sister in Law - Anne

Preface

"Rising costs. Shorter lead time. Complex customer specifications. Intricate product design. Competition from across the street—and around the world."

Business today faces an ever-increasing number of challenges. The manufactures that develop more effective and efficient forms of production, development, and faster time to marketing, will be the ones who meet these challenges.

The use of Additive Manufacturing—3D printing makes a fundamental commitment to rapid manufacturing solutions based on simple and affordable 3D printing technology. With this technology, one can integrate Additive Manufacturing—3D printing processes, react to rapidly changing product performance conditions, help management to react to rapidly changing production conditions, help personnel to react more effectively to complex qualitative decisions, and lower the cost of and improve product quality throughout the manufacturing enterprise.

The first step in achieving such flexibility is to establish a rapid entity system that can be reshaped whenever necessary, thus enabling it to quickly respond to the changing requirements of the enterprise-and the environment. This reshaping must be accomplished with minimal cost and disruption to the ongoing operation.

Additive Manufacturing—3D printing will play a key role in achieving flexibility in the product performance and manufacturing system. However, this technology alone cannot shorten lead time, reduce inventories, and minimize excess capacity to the extent required by today's manufacturing operation. This can be accomplished only by integrating various types of materials presented in different innovative forms with appropriate control means throughout the manufacturing process within computer integrated digital manufacturing strategy. The result is that individual product design and manufacturing processes will be able to flow—communicate —and respond together as a unified cell, well-structured to face rapid efficient product to market demands.

In order to develop a Computer Aided Design & Manufacturing CAD/CAM and control information system that will achieve these objectives, the enterprise must start with a specific long-range strategy, one providing a foundation that accommodates today's needs as well as taking those of tomorrow's —including the support of new manufacturing processes, incorporating new data functions, and establishing new data bases and distributed channels-into account. The tools for this control and integration are available today through the implementation of Additive Manufacturing Technology.

The United States leads the world in inventing new products; however, many of these new products, ultimately, are manufactured by other countries. The inability of U.S. manufacturers to compete globally cannot be only blamed on low-cost labor in other countries; more than one- half the trade deficit comes from foreign industries that pay higher wages. The inability to apply rapid affordable design and manufacturing systems for production can be a contributing factor to this dilemma. Additive Manufacturing—Rapid Prototyping—3D Printing is the game changer.

A rapid and affordable design and manufacturing system is, simply, a system that contains a variety of reliable parts, harmoniously joined together to generate a specific product that will achieve a particular manufacturing operation, directed and controlled by simple and effective sensors and control systems. Modern rapid design and manufacturing technology is prevalent in the rapid change in CAD, and in applying the sensory fast-feedback and control technology in the rapid iteration of manufacturing improvement systems to simplify product fabrication, assembly, and performance. This is greatly enabled through the fast iterations in 3D printing technology.

When considering the Computer Aided Design aspect of a system the following must be taken into account: dynamics, kinematics, statics, and even styling of parts. All of these play a vital role in forming optimum manufacturing parameters in product design. The generation and control through various iterations of rapid prototyping provides a review of manufacturing engineering concepts and performance from a system's point of view, directed towards solving the manufacturing and mass production problems.

Additive Manufacturing—3D printing plays a key role in translating new product specifications from design engineering into manufacturing process plans which are then used to manufacture the final part. As the part is being deigned and reiterated, manufacturing technical evaluators work with design engineers to ascertain if material strength and part specifications for manufacturing can be integrated into the design of a total product, and at what cost. Tolerances, materials, clearances, appropriate handling of parts, acceptable types and orientation of part and product assembly times are particularly important factors in this evaluation because they directly affect productivity, the guidelines of which are essential to this analysis.

Cost estimates are equally important. If a new part or process is needed because, for example, existing parts or processes are deemed too expensive or incapable of producing the desired product, process engineers would be asked to either develop a new economical part or process or to change product design. Choosing the best alternative could be very difficult because such a decision is based on many conflicting objectives, e.g., cost, feasibility, timeliness, market share, part's available material, standard parts, availability of part tooling, sensors locations, and the like. A well CAD for a part incorporating simple strategy for the rapid generation of a new part to improve the performance of a manufacturing system plays a fundamental role in the new manufacturing thinking.

Advanced 3D printing technology is more than an implementation of new technologies. It is a long-range strategy that allows components of the entire manufacturing operation to work together to achieve the qualitative and quantitative goals of business. It must have top management commitment and may entail changing the mind—set of people in the organization and managing that change. However, the rewards are great since successful implementation of this technology is, in large measure, responsible for the success of rapid prototyping through Additive Manufacturing—3D printing strategy today.

In the last few years, Additive Manufacturing—3D printing technology has taken the world by storm, revolutionizing applications in every field from education to astrophysics. The manufacturing and medical world has grasped numerous benefits as 3D printing becomes more affordable and more versatile. Everything from microelectronics to personalized organs and from implants to medical prostheses can be 3D printed. The future of medicine certainly contains 3D printers. Whether medical professionals are using them to help students practice and research new treatments and procedures, or patients are receiving new organs and prosthetics, 3D printing has hundreds of possible applications.

The future of industries and medicine certainly contains countless 3D printers. Whether entrepreneurs creating novel products or medical professionals using them to help students practice and research new treatments and procedures, or patients are receiving new organs and prosthetics, Additive Manufacturing—3D printing technology has countless of applications. It is a game changer!

Sabrie Soloman

Foreword

The ongoing revolution in Additive Manufacturing– 3D Printing Technology, well into its third decade, has now (albeit somewhat belatedly) become recognized and joined by government, industry, health and academe in USA and around the world. It is based on a number of concepts which have made their way into the professional jargon and have been brought to the public's awareness by technical and business writers: concurrent engineering, Additive Manufacturing, digital manufacturing, flexible manufacturing, just-in-time production and inventory, automation, and manufacturing quality. Each of these are ingredients that contribute to the ultimate goal, which, simply stated, is to achieve the highest quality products at the lowest possible cost, and to do so in a timely fashion. A tall order, this, but one on which depend the welfare of a host of individual companies and, even more importantly, the economic health of entire countries, with political and social implications beyond overstatement.

A principal ingredient in the process, perhaps the most important one, is the achievement and implementation of error-free 3D printing production, at one and the same time a guarantor of quality and a minimizer of waste of materials and labor. At first impression, the term "error-free" will sound like a pious ideal, to be striven for but impossible to attain. A moment of reflection will persuade, however, that the aim need not be a philosophical abstraction. In the final analysis, it is the end 3D printing product alone that must fall within the range of prescribed tolerances, not each of the many steps in the printing process. That is to say, given within the context of computer-aided-design integrated 3D printing a sufficient array of monitors distributed throughout the printing envelope-i.e., sensors measuring layer thickness, temperature, speed, materials (and appropriate means to feedback and respond to, in real time, the information gathered by them) and control systems which can identify, rectify, or remove defects in the course of part printing- every item that reaches the end of the part printing will be, ipso facto, acceptable.

The book before the reader contains not only an exposition of 3D printing systems and controls, but a host of invaluable asides, comments, and extended discourses on key topics of modern Additive Manufacturing. The author, Dr. Sabrie Soloman is an active practitioner of advanced manufacturing techniques and a highly regarded teacher of the subject. In this volume, he brings error-free 3D printing a step nearer to realization; the world of additive manufacturing – 3D printing, digital manufacturing and design engineering owes him a debt of gratitude.

H. Deresiewicz
Chairman
Applied Science and Engineering
Columbia University

About the Author

Dr. Sabrie Soloman, Ph.D., Sc.D., MBA, PE – He is the Chairman & CEO of American SensoRx, In, USA; Founder of Advanced Manufacturing Post Graduate Studies at Columbia University, USA; Professor of Advanced Technology at Columbia, where he lectures on Sensors & Control Systems in Manufacturing, Affordable Automation, Computer Integrated Manufacturing (CIM), Flexible Manufacturing Systems (FMS), Design for Manufacturability, Introduction to Electromechanical Engineering, Modern Welding Technology, and 3D Printing/Bioprinting Technology. Dr. Soloman authored numbers of technical books published and translated worldwide: Sensors Handbook (2 editions), Sensors and Control Systems in Manufacturing (2 editions), Affordable Automation, Introduction to Electromechanical Engineering, Modern Welding Technology, to name a few. Dr. Soloman holds numerous Patents, Technical Awards, and several US Product Registrations. He is a Fellow of the Society of Manufacturing Engineers, USA, The Royal Society of Manufacturing Engineers (England), and L'Ordes Des Ingenieurs Du Quebec (Canada). He received several awards from the American Society of Mechanical Engineers (ASME), the Society of Manufacturing Engineers (SME), and the American Management Association (AMA). Dr. Soloman is considered an international authority on advanced manufacturing technology, robotics, biomedical engineering, pharmaceuticals, and automation in the microelectronic, automotive, beef, pork, poultry industries. He has been and continues to be instrumental in developing and implementing several industrial and modernization programs through the United Nations to European, Asian, and African countries. He is the first to introduce and implement unmanned flexible synchronous/asynchronous manufacturing systems the microelectronic and meat industries, and the first to incorporate advanced vision technology in a wide array of robot/micro-robot manipulators. Dr. Soloman was selected to deliver the US Presidential closing address, "Innovative Remote Sensors Technologies," at the Universal Design Conference, New York, USA.

Contents

1 Digital Manufacturing Digital Revolution

INTRODUCTION

This chapter is specifically tailored to provide the reader with the necessary knowledge of Additive Manufacturing (AM) Technology, which is often referred to as "3D Printing Technology." Also, the following chapters will provide in-depth details of the fundamental Design and Implementations of the Printing Technology.

While each consecutive chapter is a complementary to the preceding one, it will stand as a unique entity irrespective of its location amid the other chapters in the book. The history, and the causes of advancing the 3D Printing Technology is not only detailed as foundational knowledge, but also to equip the reader with distinctive abilities to discern the unique values of each 3D printing system to merit innovative implementations in many manufacturing fields as deemed appropriate

The reader may find the last two chapters to be vital for our health, and physical welfare, as well as the means to combating diseases, and remedy the human organs, if needed. The last chapter may encourage, and enthuse distinctive readers, or innovative enterprises to consider undertaking "3D Printing & Design" Technology as a lucrative future business worldwide, irrespective the geographical location around the globe.

School Students, as well as undergraduates, and post graduate university students will find the book of immense value to equip them not only with the fundamental in design and implementation but also will encourage them to acquire a system and practice creating their own innovative samples. Furthermore, professionals and educators will be well prepared to use the knowledge and the expertise to practice and advance the technology for the ultimate good of their respective organizations.

The rest of the chapters are educational in nature, assisting the reader to gradually acquire the knowledge, the understanding, and the insight, steps which are eminent to lead you gaining the necessary wisdom to adopt, and acquire this technology and justify its benefits qualitatively, quantitatively, as well as in particularly, financially.

Moreover, the details of each type of 3D printing technology are discussed at length in the subsequent chapters.

Small-Scale Innovative Science

Try to envisage yourself beamed down to the 14th century, traveling back in time through a normal traditional village in Western Europe, without losing the memories of the 21st century. Imagine you were able to wind back the clock to 8:00AM to the year 1350AD. You would find yourself in a perpetual shock. Preparing your food would be a challenging adventure. You would not find a microwave, an oven or a proper fire to cook your food. You would not find your favorite glossy decorative plate, nor your elegant silver wear or even your throw away carton of orange juice with its plastic straw.

You may be able to walk to your work destination, using a donkey, a horse, or even a mule! You may find a steam locomotive train to take to a next city. There is no possibility to convey a message to your loved ones of the possible delay the train may encounter.

Obviously, you would not be able to send a quick message to your acquaintances, friends, and family. You would be wearing cumbersome clothes that are hardly convenient for work. You would not find a close by mall to shop for convenient clothes with certain style, material, or color. There would be no internet, Mobil or a landline phone to place an order of house hold items, or even food.

Imagine your life for a moment if the 21st century technology were instantly abated or how different your life would be if you were living in the 14th century.

TECHNOLOGY

Technology impacts all items of your daily use. Airplanes, high speed trains, automobiles, hybrid automobiles, electric automobiles, cell phones, microwave ovens, washing machines, carton containers, plastic milk bottles, plastic cups are all the result of scientific discoveries combined with engineering innovation, which enabled people to discover new life-long applications and invent new products directly improved the way of life to all people.

Our health has substantially improved due to technological advances. Technology has impacted the food we eat, the clothes we wear, the cities we travel, and how we communicate with one another.

There are a few downsides to some facets of technology, such as pollution, obesity, cancerous substances, and less social acquaintances with new friends and even with family member. Nonetheless, overall technology has greatly improved many aspects of living for most people. The average life span of a male in Western Europe in the 14th century was 31.3 years of age, while the life span of a male in Western Europe in the 21st century is 71.2 years of age contributed to advancement in technological health improvements.

Today, technology advancement has enabled parts and devices to be manufactured instantly and used immediately in machinery through 3D Printing. Technology innovation has experienced a giant leap forward — even to manufacture precision parts in microns.

The word technology comes from the Greek words "techne," which means craft, and logy, which means scientific study of. Accordingly, technology means the scientific study of craft. In this case, craft means any method or invention that allows humans to control or adapt to their environment.

ARCHIMEDES: THE FIRST INVENTOR

How Did Technology Start?

Inventions and tool making have been around for as long as mankind has walked the earth. After the 15th century modern technology began to rise, once scientific concepts, philosophy, astronomy, and mathematics blended together in harmony.

The first inventor who harmonized engineering and science applications was Archimedes of Syracuse (287 B.C.–212 B.C.). The Archimedean screw was credited to Archimedes for raising water, which is still used in Egypt. Also, Archimedes is credited with inventing the cross-staff utilized in astronomy, and the odometer, measuring the distance traveled.

The most famous story about Archimedes is when he was told by the king to find out if the king's crown was made of solid gold. He had to figure out a way to test if the king's crown was made of gold without melting it. This was a puzzle for Archimedes. It is said that one day he noticed that his body would displace the bath water. Seeing this, he realized he could use the displacement of water as a way to measure the crown's volume, and thus its density upon weighing it. The story goes that at this point Archimedes jumped out of the tub and ran through the streets naked shouting "Eureka I have found it!" No one is certain if the story is true, but it does give an idea how exciting new discoveries in technology can be!

Science Sparks Technology

The broad disciplines in science were represented by astronomy, physics, chemistry, and biology. The early inventors experimented with objects combining these disciplines attempting to discover ways to improve human lives.

More often inventions and discoveries happen by mere accident. The Phoenician sailors in 4000 B.C for example have discovered the glass while cooking their pots on nitrate blocks. The sand below mixed with and the nitrate blocks and were melted by the fire creating a rudimentary glass. This pleasant accident at the time, with the help of the sailors paved the way for Galileo and others to manufacture telescopes for observing the planets and the stars.

The early star gazers and astronomers had no means to see beyond naked eye observation in the sky. The early astronomers have ever longed for miraculous technology to seeing beyond our solar system and galaxy.

Heavens were observed through the eyes of Galileo Galilei in the early 1600s, through the first telescope. The craft of glass making paved the way to the scientific concepts that glass could magnify far distant objects, thus for the first telescope was invented. Sir Isaac Newton's accelerated the improvements in how far a telescope could magnify by using a curved mirror instead of flat glass for the lens. Thanks to Sir Isaac Newton understanding the science of light optics, he was able to contribute to the technological advance of the telescope.

Technological advances in many applications would be impossible if basic science principles are not clearly understood, diligently applied, and carefully observed.

Technology Sparks Science

Galileo Galilei and Isaac Newton have changed our understanding of not only our own solar system, but also the whole universe through the invention of their telescope. The telescope steered a novel and captivating scientific discipline of "astronomy" —observing and studying the planets, stars, and other objects in the universe. The Nebula, for example, could not be observed prior to the invention of the telescope. No one could have estimated how many planets were in our solar system. Thanks to the technology of the telescope, the knowledge of universe was revealed.

Cosmology perceptions were revealed through the invention of the telescope, which changed the understanding of the universe. It would be practically impossible to study the universe in the absence of means to look beyond our own planet. The telescope equipped scientists to estimate the span of the universe, predict planets and stars trajectories, and quantify galaxies' stars, and black holes of the universe.

Atoms, can be seen through the use of scanning tunneling microscopes (STM), atomic force microscopes (AFM)— as well as the discovery of nanotechnology paving the road to a very small-scale science and metrology.

Scanning Tunnel Microscopes (STM)

Objects molecules and human cells are made of billions of Atoms, and trillions of cells. Atoms are too small to see by the naked eye. No one could imagine atoms existed before the invention of the Scanning Tunnel Microscope (STM).

In 1980, an innovative magnification technology was discovered, through the invention of the STM. STM made it possible to see the smallest object known to mankind exhibited in the structure of the atom. Although several credible scientific experiments had pin pointed to the existence of atoms, however, atoms could not be seen prior to the invention of the STM technology, which made it possible to "see" atoms.

A scanning tunneling microscope is not a typical microscope. It does not work with light or lenses, and one cannot look through it. In fact, when using an STM, one does not actually "see" the atoms, at least not in the way as looking at this page of this book.

The STM scans the surface of an object and then projecting its surface image on a very high resolution computer monitor, or other screen. The STM has an exceptionally metal probe called a stylus. The stylus actually performs the scanning process. The stylus point is exceedingly sharp; it measures one atom wide, which is 0.1 to 0.5 nanometers (1 × 10–10 m to 5 × 10–10 m). This stylus moves very close to the surface of the object being scanned. The gap between the tip of the stylus and the object is only about as wide as an atom, or even closer.

The STM works by passing the stylus back and forth over the surface of the object being scanned. The moving stylus is computer controlled. Human beings are not precise enough to keep the stylus the right distance away from the scanning surface. As the stylus moves, it "picks up" electrons from the surface of the object. The electrons show where the atoms in the object are placed. The signals created by these electrons are strengthened and then projected on to the monitor to create an image.

The STM can produce distinctive images of a surface, but it has another intriguing function. An STM can be used to "capture" individual atoms! The computer controlling the STM can then arrange the atoms into specific locations.

In 1990, researchers at IBM used the STM to grab individual xenon atoms. It took over 20 hours, but they were able to arrange 35 atoms into the letters I, B, M to make the smallest company logo ever.

Since then, researchers have been working on ways to move atoms around more quickly. They are discovering ways to make incredibly tiny structures, one atom at a time.

Atomic Force Microscope (AFM)

One of the drawbacks to the early scanning tunneling microscopes was that they could only be used to scan objects that conduct electricity easily, like metals. Therefore, in matters that do not conduct electricity, they could not be used to create images of many substances, such as plastics, for 3D Additive Printing (as it will be demonstrated in this book), or living tissues, as in 3D Bio-Printing (in a separate book).

In the years since STMs were invented, several other types of probe microscopes have been developed. They work in slightly different ways, but the basic principal remains similar. The microscope allows scientists to get an extremely close-up image of the object to generate scaffolding useful for 3D Additive Printing or 3D Bio-Printing. One type is called an Atomic Force Microscope, or AFM.

The AFM can scan many different types of surface, including metals and nonmetallic images of DNA.

Like the STM, the AFM has a very short tip. However, instead of picking up electrons like an STM, the AFM can "see" atoms just by bumping into them (that is, by measuring the force between an atom and the tip).

Because everything is made of atoms, the AFM can see all kinds of objects, not just conductors.

The Science of Nanotechnology and 3D Nano Printing

The science of designing and working with extremely small objects is called nanotechnology. The nano in nanotechnology means nanometer and so nanotechnology means building things that are nanometers in size. A nanometer is one billionth of a meter (10-9m).

Some nanotechnology researchers are developing new ways of using various types of STMs and AFMs to work at these incredibly small sizes. Biologists can use a special type of AFM to look at proteins and other compounds one molecule at a time, useful for 3D additive manufacturing printing. Through 3D additive printing, computer makers can produce super-miniature computer circuits. Similarly, through 3D bio-printing, medical researchers can examine how a particular medicine affects specific cells or they can create a scaffold to build certain organ.

Manufacturers, biologist, and material scientists are using nanotechnology techniques to develop completely new substances that have never been made before.

Nano-3D Printing - Nanobots

One of the challenges with nanotechnology is the difficulty of using large machines to make such micro-structures. Therefore, some engineers are designing nano-machines. These exceptionally small machines might be able to build miniscule robots called "nanobots" "Figure 1.1."

Fig.1.1: 3D Printing – Nanonbot

Courtesy: Seoul National University – S. Korea

3D additive manufacturing may be the only known method for constructing nanobots, which will overcome the speed disadvantage of 3D additive printing, thereby enabling the technology to be widely deployed in every manufacturing aspect. If millions of nanobots worked together, they might be able to do significant manufacturing innovations.

Scientists and researchers constructed teams of nanobots able to perform microscopic surgery inside a patient's body. Some groups of nanobots have been programmed to build objects by arranging atoms precisely so there would be no waste. Other nanobots might even be designed to build more nanobots to replace ones that wear out!

Compared to other areas of science like manufacturing and biology, nanotechnology is a very new area of 3D printing research. Working with microns and nanometers is still a very slow and difficult task.

However, as the science of nanotechnology continues to develop, researchers will certainly find faster and easier ways to manipulate very small structures, including individual atoms in manufacturing and cell biology.

Manufacturing of the Future Today

Instead of bashing, bending, and cutting material — the way manufacturing always has been — 3D printers build objects by depositing material, layer by layer. The process is more properly described as additive manufacturing. For example, a 3D system can be used to print a hammer complete with a natty wood-effect handle and a metalized head. This is what manufacturing will be like in the future.

Economies of Scale Matter Less

Today, if you ask a factory to make you a single hammer to your own design, you will be presented with a bill for thousands of dollars. The makers would have to produce a mold, cast the head, machine it to a suitable finish, turn a wooden handle, and then assemble the parts. To do that for one hammer would be prohibitively expensive. If thousands of hammers are needed, each one of them will be much cheaper, thanks to economies of scale.

For a 3D printer, however, economies of scale matter much less. Its software can be endlessly fine-tuned to reduce just about anything. The cost of setting up the machine is the same, whether it makes one item or as many items as can fit inside the machine. Like a 2D (two-dimensional) office printer that pushes out one letter or many different ones until the ink cartridge and paper need replacing, a 3D printer will keep going, at about the same cost for each item.

Additive manufacturing is not yet good enough to make a car or an iPhone, but it is already being used to make specialist parts for cars and customized covers for iPhones. Although it is still a relatively young technology, most people probably already own something that was made with the help of a 3D printer. It might be a pair of shoes, printed in solid form as a design prototype before being produced in bulk. It could be a hearing aid, individually tailored to the shape of the user's ear. Or it could be a piece of jewelry; cast from a mold made by a 3D printer or produced directly using a growing number of printable materials.

3D Printing May lead to the factory of the future

Additive manufacturing is only one of a number of breakthroughs leading to the factory of the future. Conventional production equipment is becoming smarter and more flexible too. For example, Volkswagen has a new production strategy called "Modularer Querbaukasten," or MQB. By standardizing the parameters of certain components, such as the mounting points of engines, the German car maker hopes in time to produce all its models on the same production line. The process is being introduced this year, but will gather pace as new models are launched over the next decade. Eventually the process should allow its factories in America, Europe, and Asia to produce locally whatever vehicle each market requires, Figure 1.2.

Fig.1.2: Revolutionary Modular Using 3D Printing
Courtesy: Volx-Wagon Der Modulare QuerbaukastenVW revolutioniert den Autobau

Efficient Factories

Factories are becoming vastly more efficient, thanks to automated milling machines that can swap their own tools and cut in multiple directions, together with robots equipped with vision and other sensing systems. The days of gigantic factories filled with massive numbers of people may be numbered.

Relocating Back to Rich Countries

As the number of employed personnel declines, the cost of labor as a proportion of the total cost of production will diminish too. This will encourage makers to move some of the work back to rich countries. This movement can be attributed to highly affordable manufacturing techniques, making domestic factories faster to respond to changing local demands, as demonstrated by the USA manufacturing entrepreneurs relocating their fabrication operations back to USA.

Also, material scientists and metallurgists are constantly providing engineers, and manufacturers with new and superior materials to make parts in the most economical and effective means. Carbon-fiber composites, for instance, are replacing steel and aluminum in products ranging from simple mountain bikes to sophisticated airliners. Sometimes the materials are farmed, cultivated and may be grown from biological substances and from micro-organisms that have been genetically engineered for the task of fabricating useful parts.

ONE-OFF PROTOTYPES

3D printing was originally conceived as a way to make one-off or very few prototypes, but as the technology has become worthy of implementation, more applications have been printed as finished goods. 3D printing is an additive manufacturing process opposite to traditional subtractive manufacturing.

One-off prototypes can be hideously expensive to produce, but a 3D printer can bring down the cost by a sizable margin. Many consumer goods, mechanical parts, shoes, and architects' models now appear in a 3D-printed form for appraisal by engineers, stylists, and clients before obtaining the final approval. Any changes can be swiftly reprinted in a few hours or overnight, whereas waiting for a new prototype to emerge from a machine shop could take weeks, and sometimes months. Some designers are already printing ready-to-wear shoes and dresses from plastic and nylon materials.

Prominent fashion designers have lately produced striking 3D-printed collections for the catwalks. Print leather has not been achieved yet; scientists are searching for such an application.

As there are barely any economies of scale in additive manufacturing, the technology is currently ideally suited to low-volume production. Also, it allows the mass customization of finished parts. Millions of dental crowns and shells for hearing aids are already being made individually with 3D printers.

Constraints-Free of Traditional Factories

Freed of the constraints of traditional factories, additive manufacturing allows designers to produce parts that were previously considered far too complex to make economically. Engineers are finding practical applications to use 3D additive manufacturing. Fluids flow more efficiently through rounded channels than they do around sharp corners. It is difficult to make such channels inside a solid metal structure by conventional means, whereas a 3D printer can do this easily. Through additive manufacturing, a gearbox was manufactured for a racing car with smooth internal pathways for hydraulic oil instead of drilled-out, right-angle bends. The gearbox not only allows faster gear changes, but also is 30% lighter. A Boeing F-18 fighter contains a number of printed parts such as air ducts, reducing part weight by at least 30%.

Weight Savings

Weight savings are part of the attraction of 3D-printed parts. With objects being built up layer by layer, it is possible to use just enough material to make the part work efficiently. Building parts in a traditional factory requires adding flanges and brackets so that objects can be handled, milled, and molded by machine tools, and provide surfaces for the parts to be bolted or welded together. A 3D printer is likely to print the item as a complete part that requires no assembly. It can even manufacture in a single attempt final mechanical objects with moving parts.

3D additive manufacturing promises sizable savings in material costs. In the aerospace industry, metal parts are often machined from a solid billet of costly high-grade titanium. This constitutes 90% of material that is wasted. However, titanium powder can be used to print parts such as a bracket for an aircraft door or part of a satellite. These can be as strong as a machined part, but use only 10% of the raw material.

Efficient Logistics

Vital tangible benefit of 3D printing technology is extended to the simplification of supply chain due to the following statements:

1. Shorter lead times
2. Reduced supply chain risk
3. Reduction of transportation costs

Facing the benefits of the current evolution of 3D printing technology, companies from all parts in the supply chain are experiencing the opportunities and threatens it may bring. First, to traditional logistic companies, 3D printing is causing a decline in the cargo industry, reducing the demand for long-distance transportation such as air, sea and rail freight industries. The logistic companies which did not realize the current evolution may not adapt rapidly enough to the new situation. As every coin has two sides, with 3D Printing, logistics companies could also become able to act as the manufacturers.

Some constraints of 3D printing are; it currently only focus on small products manufacturing and the relatively high cost. And changing traditional manufacturing machines to 3D printing may also induce a high cost, this could also be a reason manufacturing companies hesitate to use 3D printing. How to better switch a traditional manufacturing flow to a new model using 3D printing without inducing high costs? What is the way out for logistic firms? And is there a possibility that when everyone has a 3D printing machine at home which may lead to the extinction of manufacturing companies someday in future?

3D Printing Complex Design

The ability to produce highly complex designs with powerful computer software and turn them into real objects with 3D printing is creating a new design language. 3D-printed items often have an organic, natural look. "Nature has come up with some very efficient designs, Figure 1.3. Often it is prudent to mimic them," particularly in medical devices. By incorporating the fine, lattice-like internal structure of natural bone into a metal implant, for instance, the implant can be made lighter than a machined one without any loss of strength. It can integrate more easily with the patient's own bones and be crafted precisely to fit the intended patient. Surgeons in the Netherlands printed a new titanium jaw for a woman suffering from a chronic bone infection.

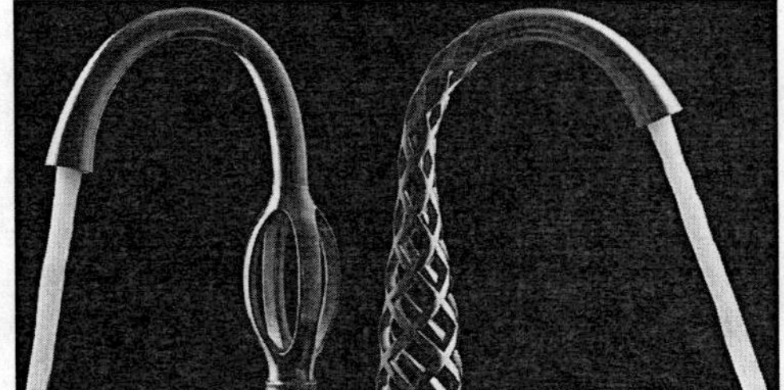

Fig.1.3: 3D Printing Intricate Design

Courtesy: American Standard – Twisted Faucets

Many companies are now wondering about the effect that additive manufacturing will have on their business. General Electric is exploring how it might use 3D printing in all its operations. It already has one product in the pipeline, in the form of a small ultrasound scanner. Such scanners are used by doctors to produce an image of features inside the body, such as unborn babies. The size, weight, and cost of the imaging consoles has shrunk, but the transducer probe which is placed on the body has remained largely unchanged and is now the most costly part of the system. The probe transmits pulses of high-frequency sound and receives signals back, using the reflections to produce images. It contains miniscule piezoelectric structures that are made by painstakingly micro-machining a brittle block of ceramic material.

Now General Electric has developed an additive system to print the transducer. This will greatly reduce production costs and allow new, inexpensive portable scanners to be developed, not only for medical use but also to inspect critical aerospace and industrial structures for crack flaws.

A number of worldwide manufacturers, Including GE, and Rolls-Royce, believe that producing complex structures utilizing 100% 3D additive printing is impossible; however, some form of hybrid printing system will emerge. This would produce the outline of a shape, thus saving on material. The shape can then be machined for precision.

The Replicator

A robotic rapid-manufacturing system is known as the Replicator. It is the same size as a large refrigerator and is capable of both subtractive and additive manufacturing, Figure 1.4. It uses a laser-based deposition system to build a basic shape that is finished by machining. As befits its name, the Replicator is capable of reverse engineering by digitally scanning an object placed inside it to produce the data needed to build an exact replica.

Fig.1.4: The Replicator - 3D Printer Machine - Courtesy of "MakerBot"

The Replicator can scan an object in one place while simultaneously communicating to another machine, locally or globally, developed to build a replica object. For example, urgently needed spares could be produced in remote places without having to ship the original object. Even parts that are no longer available could be replicated by scanning a broken item, repairing it virtually, and then printing a new one. It is likely digital libraries will appear online for parts and products that are no longer available. Just as the emergence of e-books means books may never go out of print, components could always remain available. Service mechanics could have portable 3D printers in their vans and hardware stores could offer part-printing services.

3D printers would be invaluable in remote areas. Rather than waiting days for the correct tool to be delivered, you could instantly print the tool on the job.

Some entrepreneurs already have desktop 3D printers at home. Industrial desktop 3D printing machines are creating an entirely new market. This market is made up of hobbyists, do-it-yourself enthusiasts, tinkerers, inventors, researchers, and entrepreneurs. Some 3D-printing systems can be built from kits and use open-source software. Machinists may be replaced someday by software technicians who service production machines.

"3D Systems," corporation, which produces a variety of prototyping and industrial machines, has launched a consumer range of relatively small 3D printers named the "Cube," Figure 1.5. These printers manufacture several objects such as ornaments, and children's toys. The 3D Systems have been developed along with an online platform called "Cubify" to deliver services for a community of users. The printer deposits thin film of layers from different cartridges of different color at a low cost, The deposited materials are cured and hardened quickly. The produced can reach up to 140mm cubed, but does not possess the high quality as several other larger printers. The typical cost of materials is about $3.50. While, the quality is not up to that of industrial printers, nonetheless, it is good enough for developers, designers, and entrepreneurs.

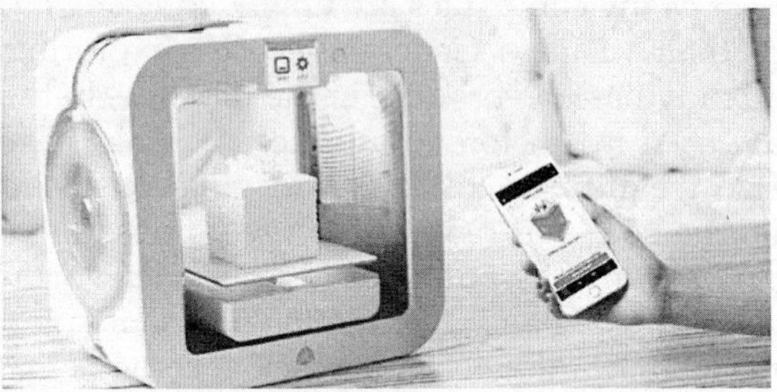

Fig.1.5: "Cubify" 3D Printer Fused Deposition Modeling (FDM)

Courtesy: 3D Systems

DEFINITION OF 3D PRINTING

3D additive manufacturing is a process tailored for making three-dimensional objects of varieties of different shapes created from digital models. The objects are produced using an additive process, where successive layers of materials are deposited down in different shapes. The 3D Additive Manufacturing is considered diverse from traditional machining techniques, which depends primarily on the removal of material by cutting or drilling. The removal of material is referred to as a "subtractive process."

3D additive manufacturing is dependent on limited types of industrial mechanisms and robots capable of carrying out successive additive process directed by computer program. In the early 2010s printer devices became widely available commercially, although the 3D additive manufacturing technology has been known since the 1980s. At the advent of the 21st century there has been a spur of growth in the sales of this additive manufacturing machines, and resulted in sharp drop in price.

In 2018 the market for 3D additive manufacturing printers and services reached $7.3billion worldwide. The 3D additive manufacturing technology is used for both prototyping and spread manufacturing with applications in automotive, construction, architecture, industrial design, military, aerospace, engineering, civil engineering, biotech, dental and medical industries, fashion, food, human tissue replacement, jewelry, footwear, eyewear, education, geographic information systems, and many other fields.

It is evident that open source 3D printing has become a mass market entity because domestic 3D printers has offset their capital costs by enabling consumers to avoid the escalated costs of purchasing common household objects.

TERMINOLOGIES

Additive

The term additive manufacturing refers to technologies that create objects through sequential layering. Objects that are manufactured additively can be used anywhere throughout the product life cycle, from pre-production or prototyping to full-scale production or rapid manufacturing, in addition to tooling applications and post-production customization.

Subtractive

In manufacturing, and machining in particular, subtractive methods refer to more traditional methods. The term subtractive manufacturing is a retronym developed in recent years to distinguish it from newer additive manufacturing techniques. Although fabrication has included methods that are essentially additive for centuries, such as joining plates, sheets, forgings, and rolled work via riveting, screwing, forge welding, or newer kinds of welding, it did not include the information technology component of model-based definition. Machining, or generating exact shapes with high precision, has typically been subtractive, from filing and turning to milling, drilling, and grinding.

Stereolithography

In a 1984 patent's General Principles, the term stereolithography was defined as a "system for generating three-dimensional objects by creating a cross-sectional pattern of the object to be formed."

3D Printable Models

3D printable models may be created with a computer-aided design package or via 3D scanner. The manual modeling process of preparing geometric data for 3D computer graphics is similar to plastic arts such as sculpting.

3D Scanning

3D scanning is a process of analyzing and collecting data of actual objects, their shapes, and appearances to digitally build corresponding three-dimensional models (Figure 1.6). Average consumes would find it difficult to create 3D printable models. Therefore, several 3D printing entrepreneurs have begun in the past few years.

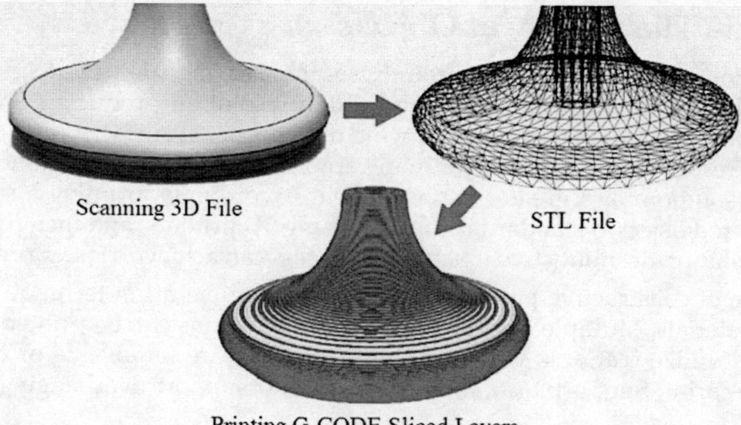

Fig.1.6: 3D Scanning and Slicing layer by layer

Courtesy: The 3D Bros

Digital Printing of Scanned Object

To obtain a 3D print, a digital scanner is used to create precise digital information of the actual physical object's shape. Alternatively, the developed CAD design of the object is stored in a printable file. Special slicing software divides the object into layers suitable to be printed layer by layer. Also, the 3D printable file deposits down successive layers of liquid, powder, paper, or sheet material to build the model in layers from a series of cross sections, Figure 1.6. These layers, which correspond to the virtual cross sections from the CAD model, are amalgamated or automatically fused to create the final shape. The crucial advantage of this technique is its ability to create almost any shape or geometric feature.

The Value of Printer Resolution

Machine resolution in general is defined by the least dimension the instrument can sense or deliver. Also, the 3D printer resolution is correlated to the dimension of the drop deposited from the nozzle of the printer. Also the printer resolution is the least stepping height, and the least horizontally stepping increment in the X-Y directions. The 3D resolution printer is the ability of the nozzle printer to compose the layer thickness.

Printer resolution defines the layer thickness, and the X-Y resolution in dots per inch, or micrometers. Typical layer dimension is approximately 100 µm, 250 DPI. Some 3D printers such as the Objet Connex , and 3D Systems' ProJet can print layers within 16 µm, 1,600 DPI (dot per inch) X-Y, which is comparable to that of laser printers. The particles, 3D dots, are around 50 to 100 µm in height, 510 to 250 DPI in diameter.

Construction of a model may take as long as several days using contemporary methods and as little as several hours, depending on the 3D printing method used, and the size and complexity of the model. Also, the printing time may be reduced further to a few hours depending on the number of models being produced simultaneously.

Injection molding techniques is traditionally affordable for manufacturing polymer products relatively in high quantities. However, additive manufacturing can be faster, more flexible, and less expensive when producing relatively small quantities of parts. Undoubtedly, 3D printers will give designers and developers the ability to produce parts and concept models using a desktop size printer.

Surface Finish- The Quality of Objects

A greater precision can be achieved through subtractive machining processes. Subtractive machining often begins by a fașt process using rough resolution producing low surface finish, followed by a greater resolution process producing high precision surface finish. Although the 3D printer-produced resolution is sufficient for many applications, yet some demanding applications dictate a better resolution for a greater surface finish. Accordingly, printing a slightly oversized version of the desired object in standard resolution using 3D printers, and then removing material with a higher-resolution of subtractive machining process can achieve greater precision.

In the course of constructive parts, some additive manufacturing techniques are capable of using multiple materials. Multiple colors and color combinations can be printed simultaneously. Some 3D printers utilize scaffolds when printing. Scaffolds are removable or dissolvable upon completion of the print. Supporting scaffolds are used to support overhanging features during construction.

TYPES ADDITIVE MANUFACTURING OPERATIONS

In the late 1970s several different 3D printers have been invented, Figure 1.7. Originally, the printers were initially large, expensive, and limited to produce several restricted configurations. A variety of large number of additive processes is now available. They differ in the technique layers are deposited to create parts, and in the substrate materials that can be used. Some approaches melt or soften material to produce the layers.

However, each method has its own benefits and downsides. Some 3D printer manufacturers consequently offer a choice between powder and polymer for the material from which the object is built. Some manufacturer use standard, off-the-shelf business paper as the build material to produce a durable prototype. Speed, cost of the 3D printer, cost of the printed prototype, and the cost of choice materials and color capabilities are the main considerations in selecting a 3D printing machine.

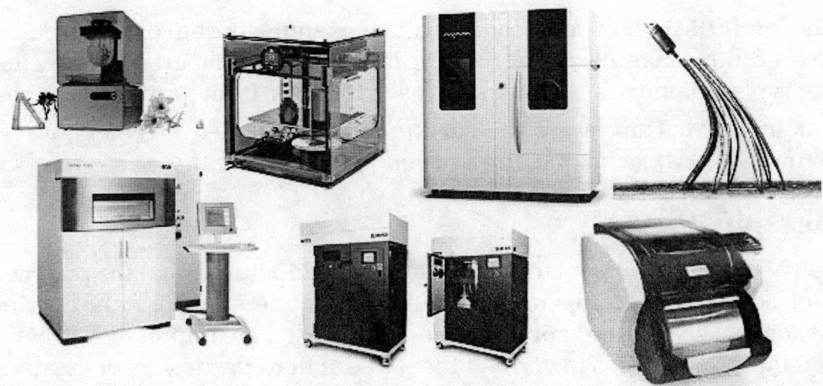

Fig.1.7: Several Types of Additive Manufacturing 3D Printing Machines

The expansive world of 3D printing machines has become a confusing place for beginners and professionals alike. The most well-known 3D printing techniques and types of 3D printing machines are stated below. The 3D printing technology is categorized according to the type of technology utilized. The categories are stated as follows:

- Stereolithography (SLA)
- Digital Light Processing (DLP)
- Fused deposition modeling (FDM)
- Selective Laser Sintering (SLS)
- Selective laser melting (SLM)
- Electronic Beam Melting (EBM)
- Laminated object manufacturing (LOM)

The focus in this guide is to provide each of the stated 3D printing technology the basic understanding of its operation and the similarity as well as the dissimilarity functions of each printer.

Definitions of SLA: Different sophisticated technologies are used to cure liquid materials, e.g., stereolithography (SLA).

Definitions of (DLP): Digital Light Processing, DLP is a technology developed by Texas Instruments used for projecting images from a monitor onto a large screen for presentations. Prior to the development of DLP, most computer projection systems were based on LCD technology which tends to produce faded and blurry images.

Definitions of (FDM): Fused filament fabrication (FFF), also known under the trademarked term fused deposition modeling (FDM), sometimes also called filament freeform fabrication, is a 3D printing process that uses a continuous filament of a thermoplastic material. ... The print head is moved under computer control to define the printed shape.

Definitions of (SLS): In the mid-1980s Selective Laser Sintering (SLS) was developed. The Selective Laser Melting (SLM) does not use sintering for the fusion of powder granules. However, it will completely melt the powder using a high-energy laser geared to create dense materials in a layer construction technique with desirable mechanical properties comparable to conventional manufactured metals.

Definitions of (SLM): The Selective Laser Melting (SLM) does not use sintering for the fusion of powder granules.

Definitions of (EBM): Electron-beam additive manufacturing or electron-beam melting (EBM) is a type of additive manufacturing, or 3D printing, for metal parts. The raw material (metal powder or wire) is placed under a vacuum and fused together from heating by an electron beam.

Definitions of LOM: Thin layers are cut to shape and joined together through laminated object manufacturing (LOM) technique, (e.g., paper, polymer, metal).

Stereolithography (SLA)

Stereolithography is a 3D printing method that can be used to implement the projects that involve 3D printing of objects. Although this method is the oldest one in history of 3D printing it is still being used nowadays. The idea and application of this method are impressive. Whether the user is a mechanical engineer, who needs to verify if the part can fit to the design, or creative person who wants to make a plastic prototype of new coming project, Stereolithography can help to turn the models into a real 3D printed object.

This method was patented in 1986. The process of printing involves a uniquely designed 3D printing machine called a stereo lithograph apparatus (SLA), which converts liquid plastic into solid 3D objects.

Most printing techniques require computer aid design (CAD) file to process the object. This file contains information about dimensional representation of an object. CAD file must be converted into a format that a printing machine can understand. There is Standard Tessellation Language (STL) format that is commonly used for stereolithography, as well as for other additive manufacturing processes. The whole process consists of consequent printing of layer by layer hence STL file that printing machine uses should have the information for each layer.

SLA printing machines do not work as usual desktop printers, which extrude some amount of ink to the surface. SLA 3D printers work with excess of liquid plastic that after some time hardens and forms into solid object. Parts built with 3D printers' type like this usually have smooth surfaces but its quality varies depending on the quality of SLA machine used.

The process of printing includes several steps. It starts from creation of 3D model in CAD program, special piece of software processes the CAD model and generates STL file that contains information for each layer. There could be up to ten layers per each millimeter. Then SLA machine exposes the liquid plastic and laser starts to form the layer of the item.

After plastic hardens a platform of the printer drops down in the tank a fraction of a millimeter and laser forms the next layer until printing is completed. Once all layers are printed the object needs to be rinsed with a solvent and then placed in an ultraviolet oven to finish processing.

The time required to print an object depends on size of SLA 3D printers used. Small items can be printed within 6-8 hours with small printing machine, big items can be several meters in three dimensions and printing time can be up to several days long.

Stereolithography is widely used in prototyping as it does not require too much time to produce an object and cost is relatively cheap comparing to other means of prototyping. This 3D printing method is rarely used for printing of the final product, Figure 1.8.

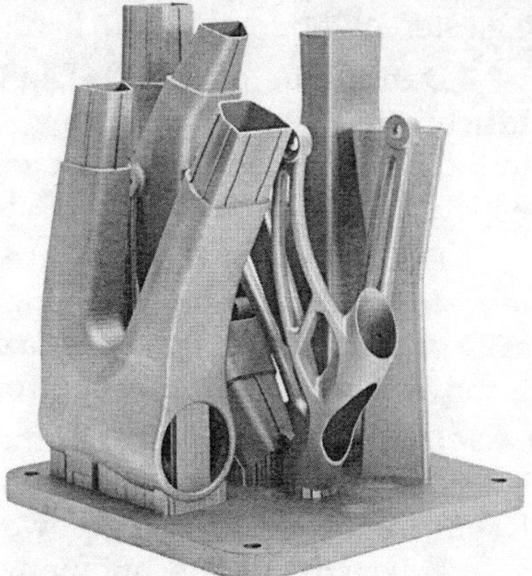

Fig.1.8: Metallic Bicycle Frame Part 3D Printed Using SLM Technology - Courtesy: Renishaw and Empire Cycles

Although Stereolithography is considered to be the oldest 3D printing technology many companies still use it to create prototypes of their projects. 3D Systems Inc., the company that started to use this method for prototyping, sells SLA machines to businesses and manufactures.

Home 3D printers usually use additive manufacturing (AM) processes and not stereolithography to produce 3D objects. But anyone interested in this technology can purchase SLA machine for home use and try this process on-their-own. One of examples of home use SLA 3D printer was created by "Formlabs."

Stereolithography or SLA is one of the earliest 3D printing methods and one that remains in in the forefront to this day.

The one thing that stands out the most about SLA 3D printers is the range of application that this printing method offers. This type of printer provides excellent representation of the necessary model of a creative design.

SLA uses a laser to cure liquid resin into hardened plastic in a process called Photopolymerisation. An SLA-based 3D printer runs on excess liquid polymer that can solidify. The end product of the material used in this type of SLA printers, usually has a smooth finish to it.

The 3D Model is imported into a slicing program like. A tank is filled with liquid photopolymer resin. A build platform is lowered into the tank and one layer of the design is traced by a UV laser. The laser is positioned using galvanometers which are sets of mirrors that rotate and reflect the laser. The liquid resin hardens into a solid creating a single layer of the object. This process is repeated and the build platform is raised until the object is completed.

The object is rinsed with the special solvent once all the layers have been printed. The object is then passed through an ultraviolet oven to complete the printing process.

The size of the design, and the specifications of the printer determine the time required to get a 3D print ready with an SLA printer. Smaller objects may take few hours, while large or complex object may take several days.

Digital Light Processing (DLP)

In Digital Light Processing (DLP), a tank of liquid polymer is intermittently exposed safely to light from a DLP projector. The exposed liquid polymer is hardened (Figure 1-9). The build base plate steps down in small increments allowing the liquid polymer to be again exposed to light. The process is repeated until the model is built. The liquid polymer is then drained from the tank, leaving behind the solid model. The "Envision TEC Perfactory" is an excellent example of a DLP rapid prototyping system.

The Objet PolyJet system is an Inkjet printer systems developed to progressively spray photopolymer materials onto a base-build tray in ultra-thin layers of a thickness vary between 16 and 30 μm until the part is finalized. The photopolymer layer thickness is cured with UV light after it is sprayed, producing completely cured models that can be used immediately, without post-curing. The gel-like unwanted supporting material, is designed to support intricate geometries. This gel-like material is removed manually or through water jetting. The 3D microfabrication technique can be used to make Ultra-small features. This technique traces the desired 3D object in a block of gel using a focused laser. As a result of the nonlinear nature of photo-excitation, the gel is cured to a solid only in the places where the laser was focused. The remaining gel is then washed away (Figure 1-9). Parts dimensions below 100 nm can be easily produced, as well as complex structures with moving and interlocked parts.

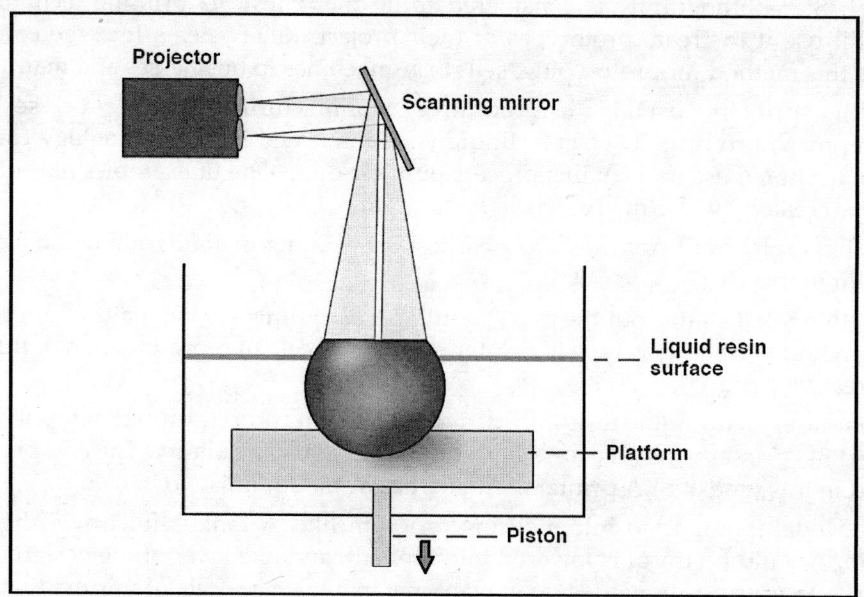

Fig.1.9: Digital Light Processing (DLP), Tank of Liquid Polymer Intermittently Exposed to Laser light

The LEDs produced light energy can be utilized as a new technique in solidifying a solid resin. A 3D digital model can be typically sliced in successive horizontal planes to effectively apply stereolithographic technique as "Mask-Image Projection-Base. Consequently, each slice is transformed into a digital two-dimensional mask image, Figure 1.9. The inherit property of the stereolithographic technique is to mask image and then project it onto a photo-curable liquid resin surface. LED's and light is projected onto the layer resin substance to cure it in the desired shape of the layer.

Several applications require light projection from below, enabling the resin to spread evenly in a very short time into uniform thin layers. This technique is tailored to reduce production time from hours to minutes. Additionally, this technique is used to generate objects consisted of multiple materials, which cure at different times. However, there are commercially available systems such as "Objet Connex," which applies the resin substance through small nozzles.

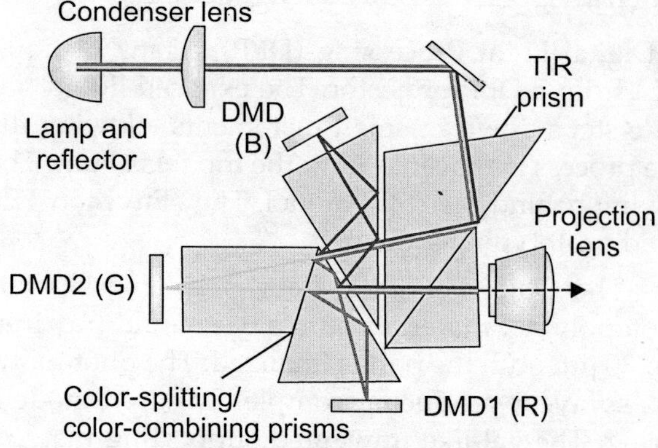

Fig.1.10: Digital Light Processing (DLP) – cutesy - photos: videoeffectsprod

Digital Light Processing is another 3D Printing process very similar to stereolithography. The DLP technology was created in 1987 by Texas Instruments and became very popular in Projectors production. It uses digital micro mirrors laid out on a semiconductor chip, Figure 1.10. Also, the technology is applicable for movie projectors, cell phones and 3D printing.

Both the DLP 3D printing, as well as SLA works with photopolymers. However, what makes SLA and DLP processes unalike is a different source of light. The DLP 3D amateurs generally use more conventional sources of lights such as arc lamps. The other important element of process is a liquid crystal display panel that is being applied to the whole surface of building material during single run of the DLP process. The material to be used for printing is liquid plastic resin that is placed in the transparent resin container. The resin hardens quickly when affected by large amount of light. Also, the printing speed is impressive. The layer of hardened material can be created with such printer in few seconds. When the layer is finished, it is moved up and the next layer is started to be worked on.

The results of such printing are robust and have excellent resolution. There is a greater advantage of the DLP technology over the SL technology, as there is much less materials to be used for a fine and detailed object, which results in lower material cost and less waste.

A DLP printer uses digital micro-mirrors fitted on a semiconductor chip for the printing process. The DLP Printers are similar to their SLA counterparts in as they both function on photopolymers. The DLA printers require additional light source, while the SLA printers can well function without the light.

The DLP printers use LCD panels, which is applied to the surface of every 3D printed layer during the printing process. DLP printers utilize a liquid polymer resin as raw material, which is typically placed in a transparent container. When the liquid polymer is exposed to light, or a high concentration of photons, the resin immediately solidifies. Accordingly, the DLP is similar to SLA. The DLP printer's speeds are very slow.

Fused Deposition Modeling FDM

In 1990 the fused deposition modeling (FDM), was commercialized, the part is produced by extruding small beads of material which harden immediately to form layers, Figure 1-11. A metal wire, often referred to thermoplastic filament is looped on a coil and then unreeled to supply material to an extrusion nozzle head. As the nozzle head heats the material allowing it to flow forming intended layer. The extrusion head moves by stepper motors or servo motors to adjust and control the flow. The nozzle head can be moved in the horizontal and vertical plane of motion. This 3D printer head nozzle mechanism is typically controlled by a computer-aided manufacturing (CAM) software package running on a high speed microcontroller.

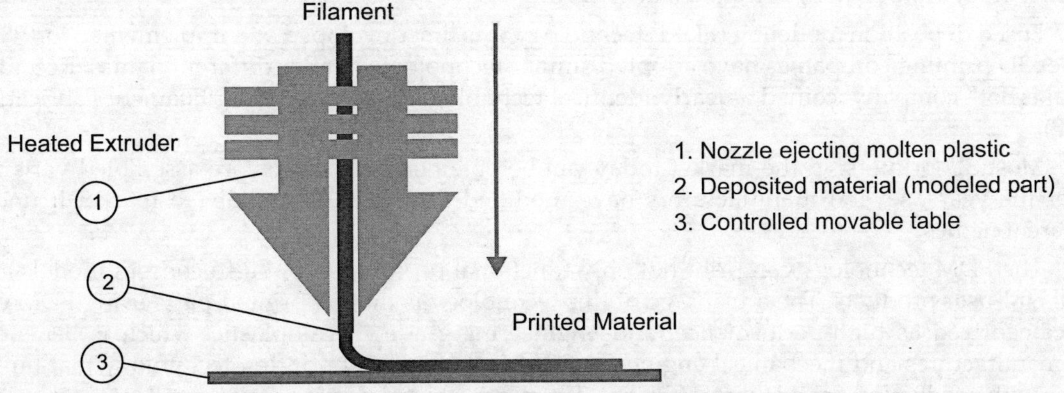

Fig.1.11: FDM-Material Hardens Immediately as Deposited

Various polymers are used, including acrylonitrile butadiene styrene (ABS), polycarbonate (PC), polylactic acid (PLA), high density polyethylene (HDPE), PC/ABS, and polyphenylsulfone (PPSU). In general, the polymer is fabricated from virgin resins, and shaped as a filament.

Recycling and processing post-consumer plastic waste turning the waste into filaments is feasible. Recycling the waste into filaments may involve machines to shred and extrude the plastic material, Figure 1.11.

FDM has some limitations to the shapes that can a fabricated. For instance, FDM usually cannot produce stalactite-like structures because they would be hanging and unsupported during the build. A thin support may be designed into the structure that can be broken away or dissolved during the finishing process.

Most 3D printers, in the market today can be obtained as a kits or pre-assembled variants. Over the years, several manufacturers have modified this technology only to market it under different names.

A popular application of FDM printers is the development of Fused Filament Fabrication (FFF) printers. FDM printers are widely used in just for printing operations prototypes. However, they are also used in part production of several applications such as plastic gears, Lego, jewelry, and much more.

The FDM printer type is building parts from the bottom up layer-by-layer, with a heated bed using an extruding thermoplastic filament.

The 3D printer heats up the filament until it reaches its melting point and then deposits it through the extrusion nozzle on the printing bed. FDM printers are developed to utilize a wide range of printing materials in addition to the thermoplastic.

The printer is controlled by Slicer software that controls the motions of its X, Y and Z axes, in addition to the extrusion nozzle.

The printing speed of this FDM type printer is slower speed in comparison to other printer types. However, it enjoys higher quality products. FDM printers are widely used by architects, automotive corporations, toy manufacturers, food industry, and medical sector.

Example of the available FDM printers in the market is:

Anct A8, Anct E10, AnyCubic Kossel, AnyCubic I3 Mega, Creality CR-10S, Creality CR-10, Creality CR-10 Mini, Easythreed E3D Nano, Tevo Tarantula, Tevo Tornado, Tevo Michelangelo, TronXY X3S, TronXY X5S, and Zonestar Z5F 90

Fused deposition modeling (FDM) technology was first developed and implemented in 1980s. Other 3D printing companies have adopted similar technologies under different names. Recently, "MakerBot" company coined a nearly identical technology known as Fused Filament Fabrication (FFF).

Most 3D printers, in the market today can be obtained as a kits or pre-assembled variants. Over the years, several manufacturers have modified this technology only to market it under different names.

The FDM technology can print not only functional prototypes, but also concept models and final end-user products. The advantage of this technology is that all printed parts using FDM can be categorized as high-performance, and engineering-grade thermoplastic, which is beneficial for manufactures and mechanical engineers. FDM is the only 3D printing technology that builds parts with production-grade thermoplastics. Therefore, objects printed are of excellent mechanical, thermal and chemical qualities.

A popular application of FDM printers is the development of Fused Filament Fabrication (FFF) printers. FDM printers are widely used in just for printing operations prototypes. However, they are also used in part production of several applications such as plastic gears, Lego, jewelry, and much more.

The FDM printer type is building parts from the bottom up layer-by-layer, with a heated bed using an extruding thermoplastic filament.

The 3D printer heats up the filament until it reaches its melting point and then deposits it through the extrusion nozzle on the printing bed. FDM printers are developed to utilize a wide range of printing materials in addition to the thermoplastic.

The printer is controlled by Slicer software that controls the motions of its X, Y and Z axes, in addition to the extrusion nozzle.

3D printing machines that use FDM Technology build objects layer by layer from the very bottom up by heating and extruding thermoplastic filament, Figure 1.12. The whole process is similar to stereolithography. Special software "slices" the CAD model into layers, and then calculates the manner printer's extruder builds each thermoplastic layer. The printer heats thermoplastic till its melting point, and then extrudes it throughout the nozzle onto a base that is called a build platform or a table, along the calculated path. The 3D printer controller (a computer) translates the dimensions of an object into X, Y and Z coordinates to control both the nozzle and the base to follow specific calculated path during printing. In order to support the upper layers of hanging walls the printer may place deposited support shapes underneath the hanging portion of the design, which can be dissolved after printing is completed.

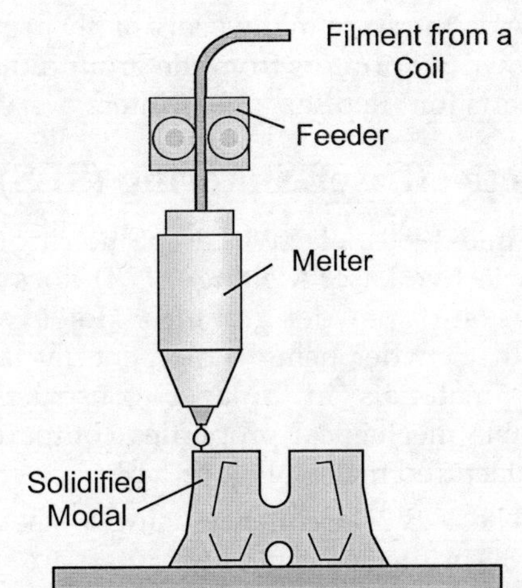

Fig.1.12: Fused Deposition Modeling FDM

Courtesy: 3DMT

When the thin layer of plastic binds to the layer beneath it, it cools down and hardens. Once the layer is finished, the base is lowered to start building of the next layer. Printing time depends on size and complexity of an object printed. Small objects can be completed relatively quickly while bigger or more complex parts require more time. Comparing to stereolithography this technique is slower in processing. When printing is completed support materials can easily be removed either by placing an object into a water and detergent solution or snapping the support material off. Then printed objects can also be milled, painted or plated afterwards.

Printing Complex Geometries

FDM technology is widely spread to a variety of industries such as automobile industries, such as Hyundai and BMW, or food industries like Nestle and Dial. FDM is used for new product development, model concept and prototyping and even in manufacturing development. This technology is considered to be simple-to-use and environment-friendly. With the use of this type of 3D printing method it became possible to build objects with complex geometries and cavities.

Different kind of thermoplastic can be used to print parts. The most common of those are ABS (acrylonitrile butadiene styrene) and PC (polycarbonate) filaments. There are also several types of support materials including water-soluble wax or PPSF, polyphenylsulfone.

Printed Objects using 3D FDM technology have very good quality of heat and mechanical resistance that allows the use of the printed objects in testing various prototypes. FDM is widely used to manufacture user products, particularly small, detailed parts and specialized manufacturing tools. Some thermoplastics materials can even be used in food and drug packaging, making FDM a popular 3D printing method within the food, medical and pharmaceuticals industry.

FDM Prices

The price for those 3D printers depends on size and model. Professional printers usually cost from $10,000 and more. 3D Printers designed for home use are not as expensive. There are several models such as the "Replicator" of "MakerBot," "Mojo" of Stratasys and "Cube" of 3D Systems. The price for these models varies from $1,200 to $10,000. However, the new starting cost offer more and more affordable versions of FDM 3D printers, the price of which can be just about $300-$400.

Also there are many users of 3D printing, as Do-It-Yourself community, who prefer to create their own 3D printers from the ground up. There are websites that offer large varieties of DIY kits and parts for "RepRap" 3D printers.

Selective Laser Sintering (SLS)

In the mid-1980s Selective Laser Sintering (SLS) was developed. The Selective Laser Melting (SLM) does not use sintering for the fusion of powder granules. However, it will completely melt the powder using a high-energy laser geared to create dense materials in a layer construction technique with desirable mechanical properties comparable to conventional manufactured metals, Figure 1.13.

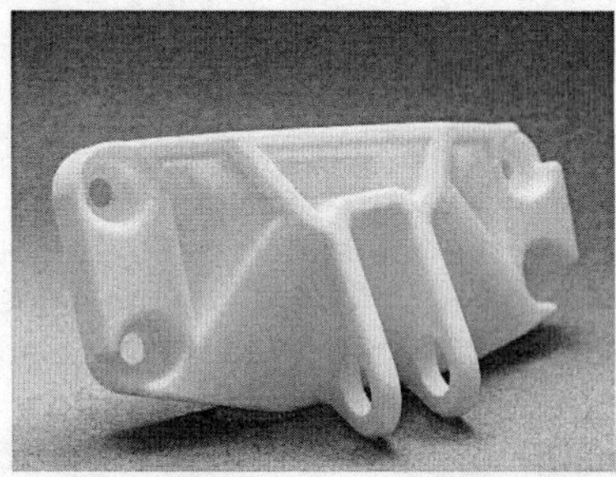

Fig.1.13: A functional Bracket Made from SLS Nylon

The SLS 3D printer is using a laser as a light source to build 3D printed objects. SLS printers are substantially different than FDM and SLA printers. The SLA and FDM depend on supporting structures unlike the SLS printers, which use non-sintered powder.

A CAD design for the SLS printers must be converted into an ".stl" format for printing with an SLS 3D printer. Several printing materials SLS printers can support a wide range of 3D printing, ranging from nylon to glass, ceramics, and even metals such as steel, silver or aluminum.

The required customized goods with 3D printing are favored the use of SLS printers due to their versatility of printing different materials.

However, business establishments and manufacturers are using these printers more often than 3D enthusiasts and amateurs due to the requirement of high-powered lasers and the associated high cost of these printers. .

Selective Laser Sintering (SLS), Figure 1.13, is a technique that uses laser as power source to form solid 3D objects. This technique was developed by Texas University in 1980s. Later the development extended to create the "Desk Top Manufacturing" (DTM) in 2001. "3D systems Inc." developed stereolithography, which is in much way similar to Selective Laser Sintering. The main difference between SLS and SLA is the use of powdered material in the vat instead of liquid resin as stereolithography does.

Unlike some other additive manufacturing processes, such as stereolithography (SLA) and fused deposition modeling (FDM), SLS does not need to use support structures because the object being printed is constantly surrounded by un-sintered supporting powder.

The initial process of the 3D printing (SLS) starts with the creation of computer-aided design of an object (CAD) file. This file must be converted to an ".STL" format by employing special software. The printing material employs wide varieties of nylon, ceramics and glass, including some metal-like aluminum, steel or silver. Due to wide variety of materials that can be used with this type of 3D printer, the technology is widely used in various industries for 3D printing customized products.

SLS is more spread among manufactures rather than 3D home amateurs as this technology requires the use of high-powered lasers, which makes the printer expensive. Though, currently there are several attempts to develop low-cost SLS 3D printing machines. For example, details about a newly developed SLS printer that uses carbon and wax for printing are poised to decrease the cost of the SLS system. Another encouraging example is the "Focus SLS" printer that can be easily used at home conditions and initially was presented at "Thingiverse."

Selective Laser Melting (SLM)

Selective laser melting (SLM) is a technique that also uses 3D CAD data file as a source to forms 3D object by means of a high-power laser beam that fuses and melts metallic powders together, Figure 1.14. The power source of (SLM) is a subcategory of selective laser sintering (SLS). Nonetheless, this is not totally true similarity because the (SLM) process fully melts the metal powder into solid 3D part unlike selective laser sintering. The SLM technology was conceived by German Research Institute "Fraunhofer Institute ILT" in 1995.

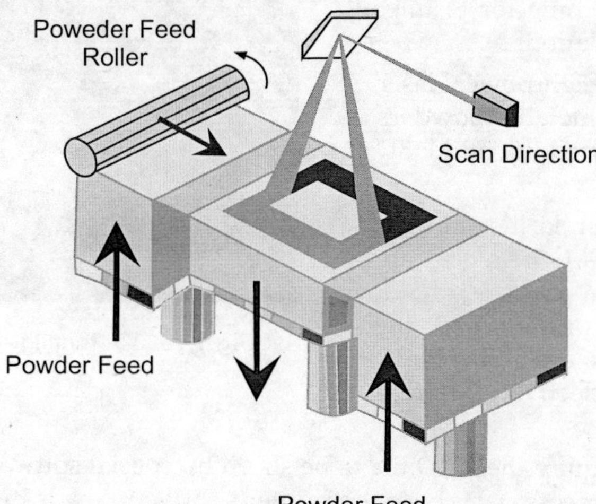

Fig.1.14: SLM - 3D Printing Schematic Diagram

SLM 3D printing technology also uses the object CAD file with special software to slice the CAD file information into 2D layers. Accordingly, the utilized software integrated with the object CAD file format creates a standard ".STL" file. Immediately, once the file is loaded the printing machine's software assigns parameters and values for object "construction and the laser path, Figure 1.14.

The fine metal powder is evenly distributed onto a plate. Each slice of the 2D layer image is intensively fused by applying high laser energy that is directed to the powdered plate. The energy of the laser is very intense that the metal powder is fully melts to form a solid object. After each layer is completed the process starts again on the next layer. The powder metals to use with the SLM technology include stainless steel, titanium, cobalt chrome and aluminum.

SLM technique of 3D printing is widely applied to print object with complex geometries and structures, in particularly with thin walls, and hidden voids, cavities or channels. Much of the pioneering SLM projects were dedicated to aerospace application of different lightweight components. SLM is suitable for objects with complex structure as in intricate tooling and difficulty in their physical access for various machining. The technology is not widely spread for home users, but mostly among manufactures of aerospace and medical orthopedics. Pharmaceutical and medical Objects constructed by the SLM 3D printing technique will face detailed protocol to validate the constructed parts. Any pharmaceutical and medical components manufactured through SLM technology may not be integrated with main operation unless the stringent validation procedure is successfully completed, Figure 1.15. A certification is usually issued upon successful completion of the validation process. This may involve in require scheduling for additional time for testing, qualification, and for validation.

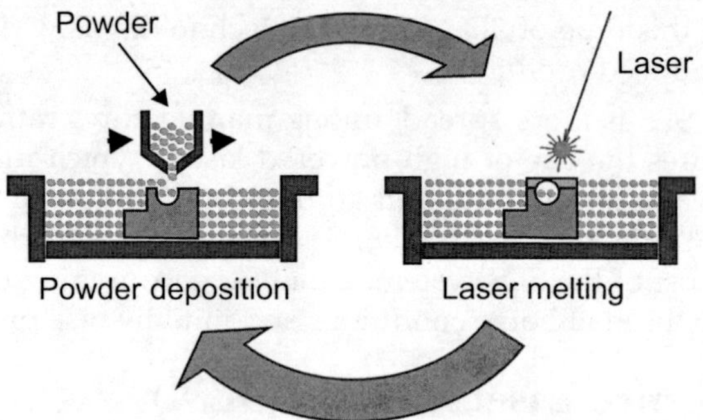

Fig.1.15: Schematic Diagram of 3D Printing SLM Technology

SLM printers use high-power laser beams to fuse and melt metallic powders together. CAD date is used to build 3D objects.

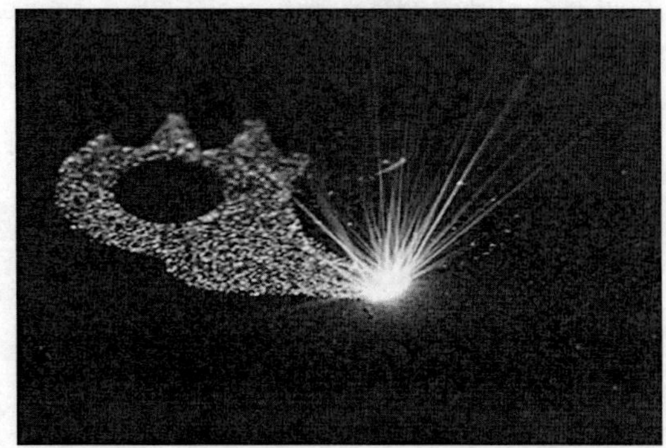

Fig.1.16: 3D Printing – SLM in Operation

Given the striking similarities in their printing techniques, SLM printers are often considered a natural progress of SLS printers. However, SLM printers melt the printing material to create solid 3D objects instead of utilizing selective sintering process used in SLS printers.

The SLM printers require the CAD file to be sliced by special software converting CAD into 2D layers. The printing process starts once the printer's software reads the file and assigns values for construction. The 3D printer evenly spreads out fine metal powder onto a plate, Figure 1.16.

A solid structure is formed using the 3D printer to print each 2D layer with the use of high laser energy, which melts the powder. The same process is repeated layer after layer until the entire design comes together. 3D object made of aluminum, cobalt chrome, stainless steel, and titanium materials can use SLM printers.

Electron Beam Melting (EBM)

Another innovative 3D printing innovation is the Electron beam melting (EBM) invention, Figure 1.17. It is a similar type of additive manufacturing technology for producing high strength metal parts (e.g. titanium alloys). EBM uses an electron beam in a high vacuum to manufacture parts by melting metal powder layer by layer. Parts produced through EBM technique are fully dense, void-free, and possess high tensile strength property, Figure 1.18. This is unlike the metal sintering techniques, which operates below the melting point. .

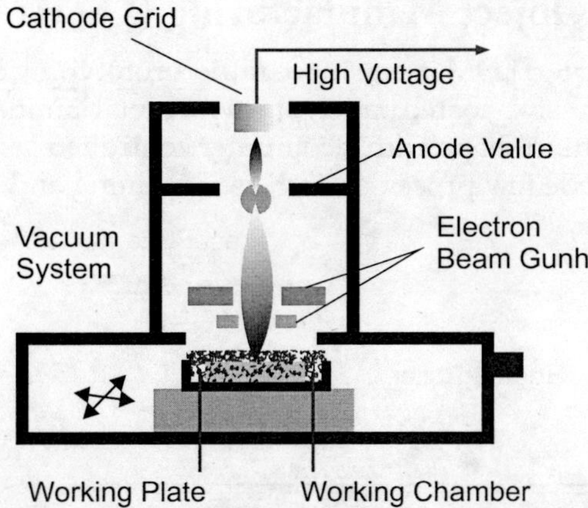

Fig.1.17: Schematic Diagram of EBM System– 3D Printing System

Additional 3D inkjet printing technique creates the model, or the object one layer at a time by printing a binder in the cross-section of the part through spreading a layer of powder, plaster, or resins, using an inkjet-like process Figure 1-19. This process is repeated until every layer has been printed. This technology enables the printing of full-color model, overhangs, and elastomer parts. The strength of bonded powder prints can be improved with wax or thermoset polymer impregnation.

EBM printers are 3D printers, which are primarily used to create 3D objects with metal parts. The EBM printers are almost identical to SLM printers. However, the EBM utilizes an electron beam instead of a high-power laser beam used by the SLM printers. The melting temperature of the metal powder can reach 1000°C.

EBM printers do not enjoy an enviable popularity on account of being slower and costlier, besides limited material support. For now, the application of EBM printers is primarily confined to aerospace and medical implants sectors.

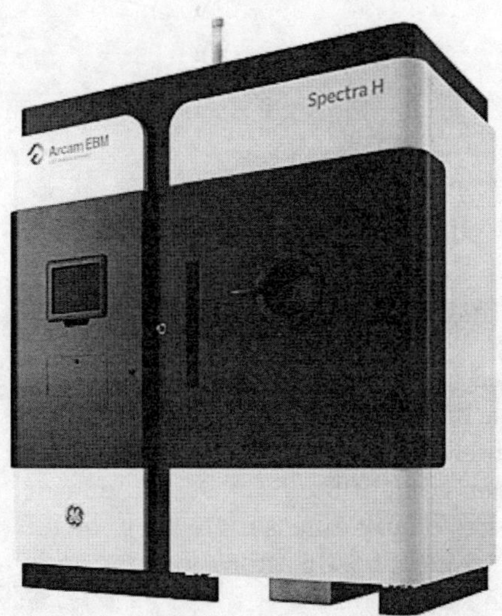

Fig.1.18: GE Additive's latest electron beam melting (EBM)

Courtesy: GE

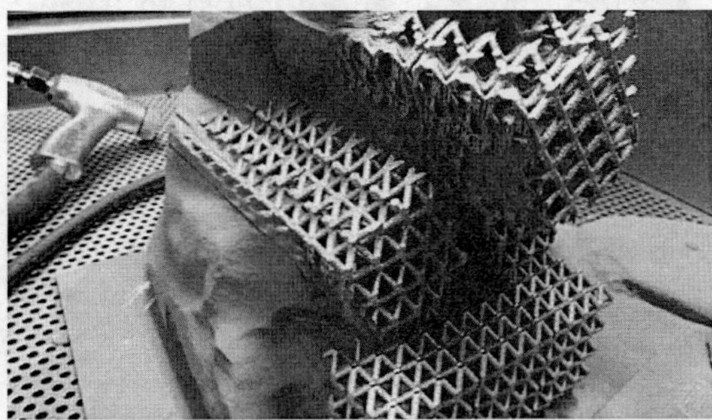

Fig.1.19: 3D Printing Complex Metal Part Using EBM

Details of Laminated Object Manufacturing (LOM)

Laminated object manufacturing, LOM is one more rapid prototyping system, Figure 1.20. During the LOM process, layers of adhesive-coated paper, plastic or metal laminates are fused together using heat and pressure and then cut to shape with a computer controlled laser or knife. Post-processing of 3D printed parts may include few processes such as machining and drilling, Figure 1.20.

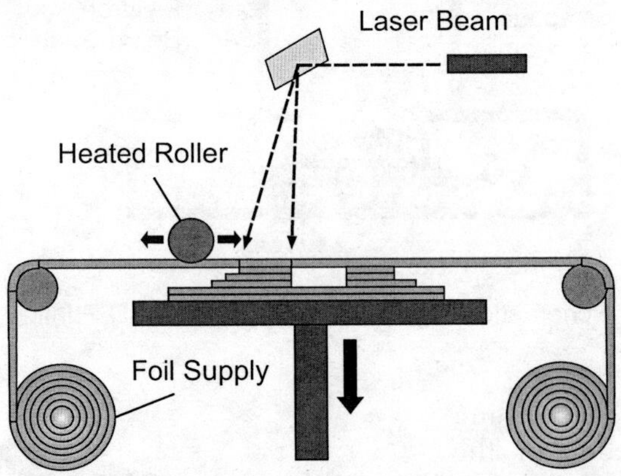

Fig.1.20: Schematic Diagram of LOM 3D Printing Technology

The LOM process includes several steps. Firstly, CAD file is transformed to computer format, which are usually STL or 3DS. LOM printers use continuous sheet coated with an adhesive, which is laid down across substrate with a heated roller. The heated roller that is passed over the material sheet on substrate melts its adhesive. Then laser or knife traces desired dimensions of the part. Also the laser crosses hatches of any excess material in order to help to remove it easily after the printing is done, Figure 1.21.

After the each layer is processed, the platform is moved down by about one-sixteenth of an inch. A new sheet of the material is pulled across substrate and adhered to it with a heated roller. The process is repeated again and again until 3D part if fully printed. When any excess material has been cut the part can be sanded or sealed with paint. If paper materials were used during printing, the object would have wood-like properties, which shows that may need protection from moisture. Thus, covering the 3D object with a lacquer or paint renders the object usable.

Fig.1.21: 3D Printer - LOM by "Mcor Technologies"

Courtesy: Mcor Techhnologies

Fig.1.22: 3D Printed Model Using LMO

LOM technology is not the most popular 3D printing technique. However, it is one of the most affordable and fastest techniques available. Furthermore, the cost of printing is advantageously low due low cost of raw materials. Additionally, objects printed with LOM technique can be relatively large, and requires no chemical reaction to print large objects, 1.22.

Currently, Cubic Technologies, Inc., is the main manufacturer of LOM printers. There are few companies use LOM technology. "Mcor Technologies Ltd." Produce LOM 3D printers. The LOM systems are widely being used among artists, architects and product developers to create affordable projects, usually, from letter papers.

Laminated Object Manufacturing (LOM) is one of the simplest methods of additive manufacturing. The process involves creating a model with a combination of layers consisted of paper and the polyester laminate. Contours are cut by a laser, knife or milling tool in a special paper which is coated with a thin layer of adhesive, Figure 1.22. The finished model has properties similar for models made of wood. The geometry of the model can be complicated keeping the dimensional accuracy of 0.1 mm.

The advantages of the LOM technology should be a very low cost and non-toxic material that is paper. Another advantage is the simplicity of the process concept, and the low cost of implementation and relatively high operating speed. A significant disadvantage of this rapid prototyping method is a large amount of waste, which, depending on the geometry of the model, can be as high as 90%. A negative trend is also significant anisotropy of mechanical properties made elements and the lack of feasibility model with a very thin wall (minimum thickness is about 1mm.

Laminated object manufacturing (LOM) printers fuse layers of adhesive-laced paper, metal laminates or plastic together using pressure and heat and cut using a computer-controlled knife or laser.

The LOM is the post-processing of 3D printed parts to give it the desired solid shape and structure.

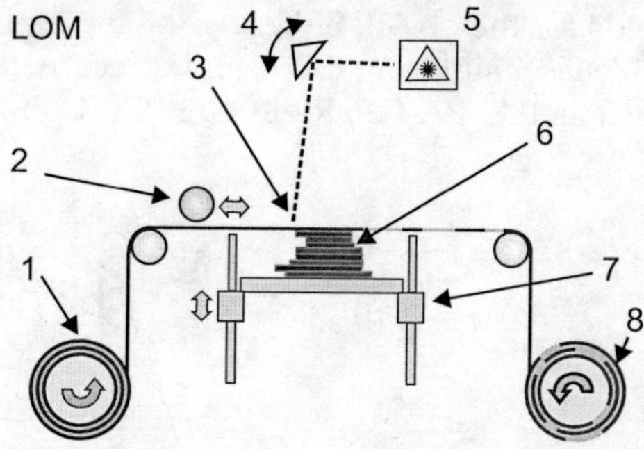

Fig.1.23: Schematic Diagram of LOM 3D Printing Technology

The LOM 3D printers require converting a CAD file .STL or 3DS format. A laminate sheet coated with adhesive is laid down on the substrate, and a heated roller is moved over the materials to melt the adhesive.

The layer is cut off by a knife or laser into the desired dimensions and shape. The excess material is removed, Figure 1.23. A new sheet is added to the substrate as the platform is moved. The process is repeated until the object is fully printed. The LOM printers that are sold by Cubic Technologies for home use

It is relatively expensive comparing to "MakerBot Replicator" or "3D System's Cube" devices. However printers from "Mcor Technologies" will become more popular in the nearest future, as they have partnered with office supplies retailer "Staples." They offer in-store printing services including printing of predefined objects like phone cases, pens, jewelry and many other useful objects. Also there can be custom objects ordered that would be based on customers own design.

The expansive world of 3D printing machines has become a confusing place for beginners and professionals alike. The most well-known 3D printing techniques and types of 3D printing machines are stated below:

Affordable printers can often be used to make a mold, which, in turn, can then be used to make metal parts. However, printers that work directly with metals are expensive, (Table 1.1).

Table 1.1: Cost of Different Materials Used in Different Additive Manufacturing Printing

Type of 3D Printing	Average Cost of Print Material
Stereolithography SLA	$200/L
Multi Jet Modeling MJM	$300/Kg
Selective Laser Sintering SLS	$500/Kg
Binder Jet Technique BJT	$100/Kg
Fused Deposition Modeling FDM	$ 50/Kg

GRANULAR MATERIALS BINDING

The selective fusing of granular-bed materials is also an innovative technique in 3D printing approach. The 3D selective fusing technique fuses portions of the layer, and then transports the working area downwards, adding another layer of granules, Figure 1.24. The process is repeating until the object is built up. The advantage of this technique is to support overhangs and thin walls using the unfused media in the object being produced Figure 1.25. This technique reduces the need for temporary auxiliary scaffolds for the object. Sintering the media into a solid object is typically accomplished by Laser. Selective laser sintering (SLS), and direct metal laser sintering (DMLS) use both metals and polymers, such as PA, PA-GF, Rigid GF, PEEK, PS, Alumide, Carbonmide, and elastomers, Figure 1.26.

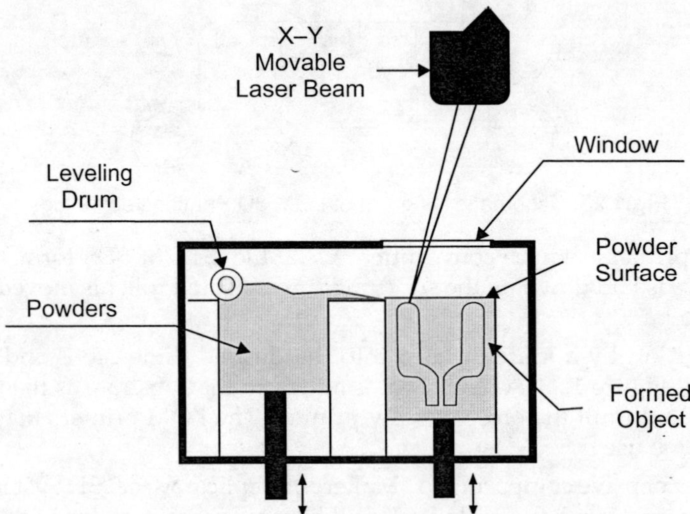

Fig.1.24: SLS and DMLS Techniques Fuses Portions of the Layer - working area Moves Downward

Fig.1.25: 3D Printing (DMLS) Technology

Courtesy: DMG Mori

Fig.1.26: 3D Printed Metal Part Using DLMS Technology

Courtesy: 3D Hub

Object Manufacturing

In the 1990s some 3D additive printers, have utilized paper as the build material, causing a lower cost to print. Some companies used a carbon dioxide laser to cut cross sections out of special adhesive coated paper, and then laminated them together forming the desired part shape, Figure 1.27.

In 2005, a different process was introduced using regular sheets of office paper, cut to shape by the use of a Tungsten carbide blade, utilizing selective deposition of adhesive and pressure to bond the prototype. Also, there are a number of companies developed printers to print laminated objects using thin plastic and metal sheets.

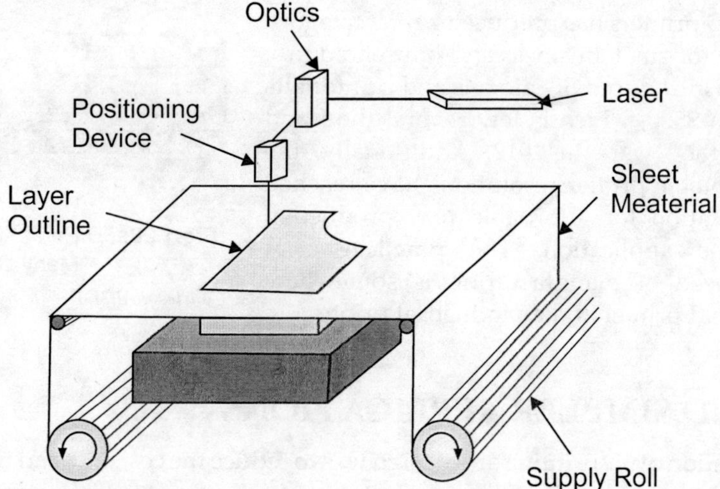

Fig.1.27: Laser Cutout Special Adhesive Coated Paper

Stereolithography (SLA) produces a solid part from a liquid substance. This process dramatically redefined previous efforts of creating a prototype from sheets of paper.

3D INNOVATIVE PRINTERS

Various start-up companies have developed printers for general use with substances that can be sprayed and/or dispensed through an orifice, varying from honey to silicone sealant and chemical respondent. "3D Printers Kits" provided by various suppliers have become available following the project's design in pre-assembled form since 2012 at prices in the US$2000 range.

General Electric uses the high-end 3D additive manufacturing printer classical model to construct parts for engine turbines. In 2014-2020, GE is committed an in-house investment of more than $2 billion on the 3D Additive Printing Technology. Numerous projects and corporations are committed to develop affordable 3D printers for household desktop use.

3D Printers Down-Trend Cost

The cost of 3D printers has decreased dramatically since 2010, with machines that used to cost $25,000 costing less than $1,000. For instance, as of 2014, several companies and entrepreneurs have taken promising steps since 2014 marketing components to build various 3D printer designs, at a price starting at about US $300.

The costs of 3D are becoming more appealing financially as printer's price substantially come down, to use for either personal use or manufactured products. The positive impact of the personal and household use of the 3D printer's products at home can improve environment by reducing manufacturing materials use and distribution.

Forming Waste Plastic into Inexpensive Filament

Waste-plastics, such as medical containers and milk jugs, can be converted into inexpensive filaments to be used in several 3D additive manufacturing printers. Also, this technique has provided positive impact on environment by such distributed recycling method, Figure 1.28. The development and hyper-customization of 3D printers has produced a new category of printers suitable for small business and household use. Several developers introduced 3D models and printer kits at a price less than $500, which is few several thousand dollars less than they were in 2012. Additionally, the practical and acceptable printer resolutions have reached reliable implementation for affordable part production depending on the application. The practical print resolution and speed of manufacturing is somewhere between a household printer and an industrial printer.

Fig.1.28: 3D Printer Filament Spools of Many Colors

Courtesy: All3DP

EARLIER AND SIMPLER APPLICATIONS

The earliest application of 3D printing and design was to reduce the cost and lead time of developing prototypes for new parts in tool room manufacturing environment. Rapid prototyping was one of the original additive alternatives to subtractive tool room methods. Traditional subtractive fabrication is painfully slow and often expensive. However, with 3D innovative manufacturing technological advancements, and the abilities to propagate and disseminate of those advancements into the business world, additive methods are rapidly moving deeper into the production end of manufacturing in creative methods. Consequently, additive 3D printers have become a viable alternative to manufacture parts instead of using subtractive methods that were formerly the sole authority of part manufacture. In some cases parts are made more profitably via additive 3D manufacturing. The widely spread applications include but not limited to design visualization, prototyping/CAD, metal casting, architecture, education, geo-spatial, healthcare, medicine, and entertainment/retail.

Industrial Rapid Prototyping RP

Industrial 3D printers have existed for almost four decades. 3D Rapid Prototyping has been used extensively for producing fast prototypes, in particularly in research work. Rapid Prototyping Machines are generally larger machines equipped with unique proprietary powdered metals technology using, casting molds, such as sand molds, plastics molds, paper, or cartridges. Universities and Commercial Enterprises are often use for rapid prototyping techniques.

Rapid Manufacturing

While Rapid Prototyping method was developed mainly to produce prototypes useful for further study and engineering analysis, it lead to a surprising implementation in producing reliable finished and useful manufactured products. Also, Rapid Prototyping has introduced reliable appropriate materials for final manufacture. This has also lead to introducing several possibilities to directly manufacturing finished parts. The relatively inexpensive production of small numbers of parts has made the 3D Rapid Prototyping a signature identity to this technology.

The technology of Rapid manufacturing is rapidly gaining ground in part manufacturing technique although it is still in the infant stage to prove its processes worthiness in manufacturing reliability. In general, the 3D printing technology is here to stay in the field of rapid manufacturing. Several experts considered 3D printing as a revolutionary technology, in particularly when employing selective laser sintering (SLS) as one of its most promising technology for part manufacturing at an affordable cost. The positive adaptation of Selective Laser Sintering (SLS), and it closely knit technology of Direct Metal Laser Sintering (DMLS) have given experts some assurance to use this technology of rapid prototyping methods for accurate part manufacturing. As of 2019, these techniques have matured substantially overcoming the infant mortality stage of new technology. Nonetheless, there are many obstacles to be overcome before rapid manufacturing could be considered a realistic manufacturing method.

Customization

Entrepreneurs and commercial Enterprises have created corresponding manufacturing and design services to empower consumers to customize design and manufacture of objects using highly intelligent but simplified web-based customization software. Customer is then able to order their unique 3D printed parts according to their own manipulation of the given flexibilities of software and part design configurations. This service allows consumers to create custom parts for their personal use, including household goods, and personal items.

Mass Production

Parts Mass Production using the current slow print speed of 3D printers is the major hurdle in utilizing this technology. To overcome this limitation, several fused filament machines have been developed offering simultaneous printing using multiple extruder heads. Additionally, printing in multiple colors became feasible using different polymers, making multiple prints simultaneously. The overall print speed thus increased, when utilizing multiple print heads during production. Accordingly, duplicate machines are not necessary as they can share a single controller. Multiple printing heads in a single machine requires less capital cost. Also, multiple heads can share a single controller.

Multiple Colors

Providing multi-color and multi-material features when needed using multi heads in one machine have substantially increased the speed of part production, distinct from the use of multiple machines. Also, multi-material machines are limited to produce identical replica of the same part. Additionally, the print speed increases proportionately to the number of heads. Furthermore, the overall cost is reduced by sharing the same heated print volume, and the same printing energy used. Recent printers recently offer twin print heads. However, these twin heads are used to manufacture single sets of parts in multiple colors.

Several studies have been conducted to evaluate this 3D Multi Color Printing technique concluded that this method colors 3D printing technique is bypassing the field of conventional subtractive methods.

Printer Replicator

Within the past several years 3D printing domestic and practical household applications did not gain adequate recognition and lacked behind 3D printing parts for enthusiastic hobbyists. However, many hobbyists have printed working-clocks made of various materials for home woodworking machines. Also, 3D printing was also used for printing ornamental objects.

Recently, 3D printers had been used to aid animals, as in the case of a 3D printed foot for a crippled duckling to walk again. A hermit crab 3D printed shell was able to inhabit a new style home. Attractive necklaces, rings, and handbags bags were 3D printed as decorative parts for ladies.

Substantial 3D printings were used in chemical research environments. Various chemical mixtures were mixed using 3D printing technology. The 3D printer ink can print anything as long as the printer nozzle can be dispensed from its orifice as liquid or paste. The inventors of the chemical application predicted that this technology could be used for both industrial and domestic use. Medicine today can be easily dispensed from remote locations to produce the necessary medicine or household chemicals.

Fashion and Clothing

The world of fashion and clothing did not escape the new world of 3D printing. The world famous fashion designers of clothing successfully experimented with 3D-printed bikinis, shoes, and dresses, Figure 1.29. Additionally, stylist shoe makers used 3D printing in commercial production to prototype and manufacture shoes for athletes. 3D printing can quickly re-manufacture custom players' shoes, and balance-fit them for different field environments, Figure 1.30.

Fig.1.29: 3D Printing Fashion Clothing Appeared Eccentric and Futuristic

Fig.1.30: 3D Printed Shoe Collection "Myth" Made on Desktop Machines

Courtesy: Continuum.

3D BIO-PRINTING

Biotechnology firms, academic, medical and pharmaceuticals institutions in USA have pioneered the research and development 3D Printing Technology in the field of Tissue Engineering. The aim of the innovative development was to utilize Tissue Engineering in Organ manufacturing techniques. 3D ink jet printing technique is appropriate to use in printing tissues leading to organ manufacturing. A three dimensional structure of an organ can be formed to generate successive deposition of layers of tissues of living cells. The tissue layers are deposited onto a gel-like mainly consisted of glucose matrix medium gradually built up to compose three-dimensional structures. Vascular systems can be generated in the same manner through 3D ink jet printing. Well designed and simplified Computer-aided tissue engineering was the initial success of organ printing technology.

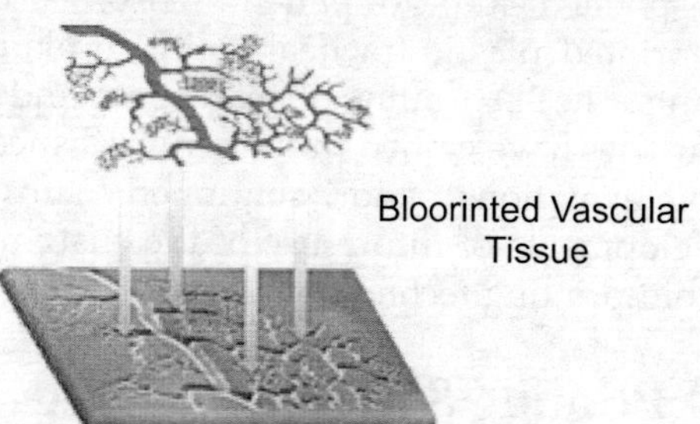

Fig.1.31: 3D Printing of Functional Biomaterials for Tissue Engineering

3D Human Tissue Accelerate Preclinical Drug Testing

The process of early stage of preclinical drug testing and discovery can be achieved through developing functional, three-dimensional human tissue for medical research and therapeutic applications, Figure 1.31. Medical Corporations have successfully utilized the 3D bio-printing of human tissues to accelerate and enable treatments including surgical therapy and transplantation at lower cost.

There are three fundamental steps in drug discovery:

1. Discovery
2. Preclinical Development, and
3. Clinical Trial.

The role of 3D ink jet tissue printing also enhances fast transition from discovery to preclinical development. The preclinical development is a continuum process resulting in preliminary pharmacology and toxicology testing, which often contributes to drug candidate selection.

3D PRINTING FOR IMPLANT AND MEDICAL DEVICE

Simple organs would not be feasible to print without the use of 3D printed biodegradable scaffolds. The scaffolds are constructed using bio-inks, which is made from the patient's own stem cells. The 3D printed scaffolds are dissolved in a timely fashion once the entire organ structure is printed. Usually the stem cells are taken and stored from a patient's own fat and bone to generate blood vessels and bone cells form the desired organ, such as 3D printed heart, Figure 1.32. Also, successful 3D printing of organs such as vaginas, windpipes, bladders and bones has been implanted in a number of patients, Figure 1.33.

Fig.1.32: Israeli Scientists Unveil World's First 3D Printed Heart

Courtesy: Tel Aviv University

Successful 3D printed titanium for making specific pelvic implanted into patients. Other 3D printing implants has been used successfully to print patient-specific implants and devices for medical use, Figure 1.32. A titanium lower jaw transplanted into to patient. Also, successful 3D printed plastic tracheal splint implanted in infants. The 3D printing of hearing aid and dental industries have gained prominent and successful implementations and continued substantial developments of future use of the custom made 3D bio-printing technology, Figure 1.33.

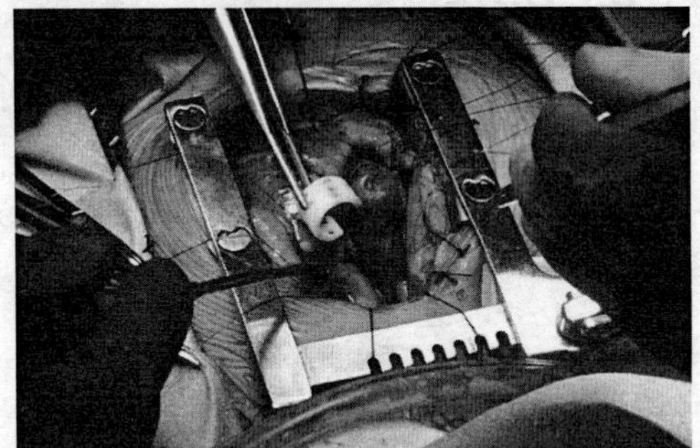

Fig.1.33: Implanting 3D-Printed Tracheal Splints

Courtesy: Children's Healthcare of Atlanta, USA.

3D Printing Services

Both consumers and industries recently enjoy the custom services offered by entrepreneurs and organizations offering on-line 3D printing services. The custom services require customers to upload their unique 3D designs through developers' service website. High speed industrial 3D printers create the required custom design.

OPEN SOURCE SCIENTIFIC EQUIPMENT

Open-source scientific equipment can be structured to create open source labs for numerous present and future 3D printing applications. The 3D printing applications are extended to print scientific-based applications to create, e.g., replicating ancient and priceless artifacts in archaeology, reconstructing fossils in paleontology, reconstructing bones and body parts in forensic pathology, and in crime scene investigations to reconstructing complicated damaged evidence assimilated during investigations.

Astronomical Very Large Telescope (VLT)

Utilizing 3D printing technology is being implemented to create innovative molds geared to cast new astronomical telescopes. The manufacture of molds is used to cast new telescope components. Chile requires telescopic components to be used in the MUSE instrument of the Very Large Telescope (VLT). The telescope is part of the Adaptive Optics Facility project. The 3D printing technique facilitated the manufacture of such complex astronomical custom instruments. The use of 3D priming technology delivers the components faster and more economically, and with greater flexibility.

Sensor Arm

Germany has manufactured the first sensor arm to be implemented in Chile astronomical telescope the MUSE, Figure 1.34. It is the structural part of a new sensor arm fitted in the telescope. The manufactured part for the MUSE is required to house the GALACSI adaptive optics module. The manufactured sensor arm is a metallic assembly used to grip three flat mirrors developed for:

Fig.1.34: The Multi Unit Spectroscopic Explorer "MUSE" is installed on the Nasmyth at the Very Large Telescope (VLT) of European Southern Observatory (ESO)

Courtesy: *R. Bacon*

- to supply light beams into sensors, which controls the active optics systems of the Very Large Telescope
- to guide the telescope during observations.

The sensor arm was manufactured using a traditional technique recognized as investment casting. The investment casting has been used to manufacture components with very complex profiles, e.g., blades for gas turbine engines. A model was generated using this casting method. The mechanical properties of the model were analyzed. A virtual model was created, and then a 3D pattern is printed —a type of thermoplastic polymethylmethacrylate material was used.

A negative of the original shape was created. A heat-proof ceramic is used to coat the pattern of the investment process. The ceramic heat coat is treated and hardened. The wax-infiltrated pattern is melted away from the new ceramic shell, resulting in a seamless die into which the metal for the final cast is poured.

The previous older sensor arm was manufactured using beryllium substance. It is a lightweight metal that gratified the necessities for the sensor arm. Nonetheless, beryllium is unsafe as it is highly toxic, and may not be fabricated or even modified safely. It is almost impossible to modify the item once it has been installed. Accordingly, more appropriate high grade aluminum was carefully chosen to make the final cast for the replacement.

Other means to manufacture the spacer for the VLT test camera. Also, Germany manufactured the spacer part from ductile cast iron and used a similar investment casting process. The pattern was 3D printed using laser sintered technique.

As acknowledged, 3D printing technology is an additive manufacturing promises to reduce manufacturing time, minimize waste materials at affordable cost. Traditional milling machining and lathe turning machines are subtractive manufacturing processes, as material is removed from a larger work-piece until the final configurations are reached. These traditional manufacturing processes have generated substantial waste of material in the form of chips. Fortunately, 3D Additive processes circumvent this wastage by constructing the form layer by layer to accomplish the final shape. 3D printing, in particularly 3D printing for casts permits intricate internal geometries to be fashioned with ease.

The purpose of the new sensor arm is to manipulate or tricking the telescope active optics system to focus 250 millimeters from the base line of the original focus position, resulting that the focal plane now lies 500 millimeters from its flange. This extra distance enabled the space required to fit GALACSI, the adaptive optics module for MUSE, Figure 1.34.

These mirrors are made from carbon-fiber reinforced silicon carbide "HB-CESIC®". This carbon-fiber reinforced silicon carbide is an exceptional material with exceptional hardness, high stiffness and enjoys low thermal expansion coefficient.

3D PRINTING PROOF OF PRINCIPLE

3D printing techniques has been proven to be a valuable tool for "A proof-of-Principle" stage in the life cycle of developing a new parts and product, in particularly, when a new material or chemical compound is sought to employ. "Chemical Ink" can be dispensed to print a new chemical material to form the object under investigation. A 3D printer is used to print reactants into the object to observe the result. The process may be repeated until satisfactory results are obtained.

3D scanning technologies

It is often imperative to duplicate the replication of an object. Thanks to the use of 3D scanning technologies an object replica became a common technique to precisely create exact replicate to an object. Such object may be implicated in the same scale or scaling it up or down as deemed suitable for the purpose of the object and its design function. Traditionally, molding techniques are often associated with costly endeavor, difficulties in preparing the mold process, and most likely excessively invasive to be performed, while the scanning technique associated with 3D print technology renders a highly cost effective accomplished in a relative short time, Figure 1.35. 3D printing associated with scanning techniques is very valuable in particularly for precious objects and delicate cultural heritage artifacts. The 3D printing associated with 3D scanning became fundamental technique to use, where direct contact with the molding substances may harm the original artifacts.

Fig.1.35: iReal 2S - Professional Handheld Low Cost - Color 3D Scanner

Courtesy: 3D Scanner Company

Architecture and Building Construction

The use of large scale 3D printing technology extended to the construction and building industry. The main objects in employing the 3D printing technology in construction and building is stated as follows:

1. Faster turn in construction and building projects
2. Lower cost
3. Creating construction and buildings with ease in harsh environments

Using 3D printing in construction of "off-Earth habitats" has made it feasible and highly economical without risking human lives and exorbitant overhead, Figure 1.36.

Fig.1.36: Australia's Construction Plan Using the World's Largest 3D Printer

Lunar Base Construction

A lunar base construction using 3D printing may employ a special base lunar material The base lunar material is a layer of loose, heterogeneous material covering solid rocks. The base lunar material contains dust, soil, broken rocks, and other materials present on the moon surface. Scientists have been successful in employing microwave sintering process replacing the additive binding agent to the base. This, in turn has yielded solid blocks from the raw material.

Terahertz Devices

Terahertz devices are complex shaped of wave guides, couplers, and many bends. The use of innovative 3D printing technology manufacturing of these devices have become highly economical and considered a common practice. The delicate and complex configurations of these devices could not be economically obtained employing traditional conventional fabrication techniques. The current professional and commercially available grade printers can reach a resolution of 16 µm. Accordingly, 3 D printing Terahertz Devices of a resolution of 100 µm can be achieved. Once the Terahertz structure devices is printed, they can be sputter-coated with gold, or any other electrically conductive metal, to create a Terahertz Device.

Recreation Coral-Shaped Structures

Coral damaged reefs can be remedied by the use of a large-scale additive 3D printing utilizing a special material similar of the shape, color and texture of coral reefs, Figure 1.37. The creation of unique coral-shaped structures will stimulate coral polyps to populate generating new healthy coral reefs. The structures created by 3D printing process have a much more natural shapes and colors than other structures created by other artificial reefs. Such 3D printed new coral reefs have a neutral pH, while other materials do not.

Fig.1.37: Australian Architect Restored Coral Reefs Using the World's Largest 3D Printer

Effects of 3D Printing

Manufacturers have realized that it is essential to remain competitive in this global manufacturing competitive environment. As a result, many world manufacturers have recently adopted healthy flexibility in their manufacturing strategy and adaptability to employ new technology. Additive manufacturing relatively, is no longer in the infant mortality stage. It has proven its worthiness in many manufacturing fields from various types of industries and research institutions. Educating consumers and producer alike of all available technologies becomes a bed-rock in new technology use. Additive manufacturing technology and strategy has made vertical integration economically feasible in many instances. The technological development of 3D printers is an industrial revolution countering globalization. End users manufacturing developers will execute their own manufacturing instead of engage in trade to buy products from manufacturers or corporations.

The considerable challenge in 3D implementation is integrating the 3D additive technologies into commercial production. Implementing the 3D additive technology is likely to succeed if it complements traditional subtractive technology rather than displacing it entirely.

Outer Space Exploration

Zero and or low gravity environments of space application development were the motivation of 3D printing initiation. The primary application dictated means to create remote, "On Demand" basic devices, and manipulating instruments of compound mechanical motions, to achieve important tasks in outer space contrasted with using valuable resources such sending a crew using fuel and cargo space to carry the items into space.

NASA has conducted extensive research with experienced manpower to evaluate "Made in Space" potential of 3D printing. The assessment intended to explore innovative means to more economical and highly efficient means to manufacture parts in space, such as rocket parts and articulated gripping tools.

Rocket engine injectors have achieved their functions as good as traditionally constructed parts, if not better, during stressed simulation test runs, which subjected them to temperatures approaching 6000° F - 3316° C with extreme pressures.

As astronauts need not to carry heavy loads into space NASA has prepared 3D printer to launch into space. NASA successfully demonstrated making spare parts using 3D printers on the fly.

Firearms

A working plastic gun has been successfully designed by a US based group and can be downloaded. The model can be reproduced by anybody with a 3D printer. The same US based group designed a 3D printable rifle lower receiver capable of lasting more than 650 rounds and a 30 round magazine.

SOCIAL AND ECONOMIC CHANGE

The advent of commercially-affordable 3D additive manufacturing technology has enthused many writers and commentators to speculate the seriousness of the social and cultural changes impact on the different societies. As more and more 3D implantations an all levels of industrial, commercial, and household applications, the conservative relationship between the home and the workplace might be swept away, in particularly, when the 3D printers begin to enter people's homes. Similarly, it is predicted that, as the 3D additive manufacturing becomes simpler for businesses to convey new objects designs everywhere around the world, so the need for affordable high-speed cargo services might also become significantly cheaper. Additionally, assumed the easiness of replicating objects, the changes of the current copyright legislation will remain to be seen if changes can be made to protect intellectual property (IP) rights as the new technology is widely spread.

SETTING THE STRATEGY FOR ADDITIVE MANUFACTURING ADVANCED 3D PRINTING

3D Additive Manufacturing technology, discussed in this book, allows the entire manufacturing and research operation to harmoniously work together to accomplish the business qualitative and quantitative goals. In fact, 3D Printing is more than an implementation of new manufacturing technologies. It is a long-range strategy. This technology must have the top management commitment. It must involve changing the attitude and the mint-set of people and the organization structure and management flexibility. The major success of this manufacturing strategy is largely credited to the success of implementing the advanced technology of additive manufacturing and 3D printing systems.

ADDITIVE MANUFACTURING - 3D BIO-PRINTING TECHNOLOGY

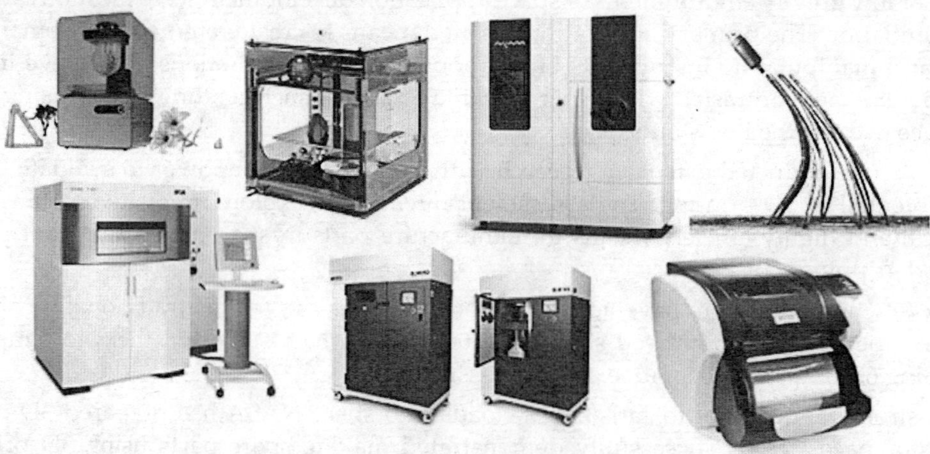

Fig.1.38: Additive Manufacturing 3D Printing Machines

Additive Manufacturing (AM) is a term to describe set of technologies that create 3D objects by adding layer-upon-layer of material, Figure 1.38. Materials can vary from technology to technology. But there are some common features for all additive manufacturing, such as usage of computer together with special 3D modeling software. First thing to start this process is to create CAD sketch. Then additive manufacturing device reads data from CAD file and builds a structure layer by layer from printing material, which can be plastic, liquid, powder filaments or even sheet of paper.

The term Additive Manufacturing holds within such technologies like Rapid Prototyping (RP), Direct Digital Manufacturing (DDM), Layered Manufacturing and 3D Printing. There are different 3d printing methods that were developed to build 3D structures and objects. Some of them are very popular nowadays; others have been dominated by competitors.

Electronic Beam Melting (EBM)

EBM 3D printing technology is another type of additive manufacturing for metal parts, Figure 1.38. It is the same as SLM 3D printing, utilizes powder bed fusion technique. While SLM uses high-power laser beam as a power source, EBM uses an electron beam instead. This represents the main difference between these two methods. The rest of the printing operations are similar.

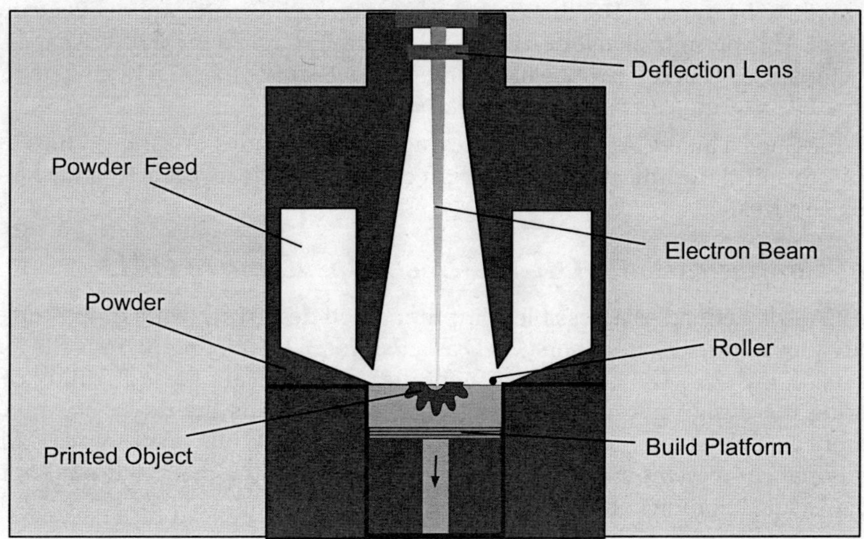

Fig.1.39: Schematic Diagram of Electronic Beam Melting (EBM)

Courtesy: Additive Blog

The material used in EBM is metallic powder, which melts and then solidifies to forms a solid 3D object build layer by layer through a computer controlled routines tailors specifically for each unique object, Figure 1.39. The computer controls electron beam in high vacuum. Contrary to SLS, the task of the EBM completely melts the metallic powder. The process is usually conducted under high temperature that may reach up to 1000 °C.

The operation of the Electron Beam Melting (EBM) is much slower and more expensive than the Selective Laser Melting operation. Also, the materials used in EBM are not as accessible as the materials used in the SLM. Accordingly, EBM methodology is not as popular as SLM. Though, EBM is still used in many applications of different manufacturing processes. Currently, the most well-known commercial materials used in EBM are pure "Titanium," "Inconel 718," and "Inconel 625." The application of EBM is mainly focused on medical implants and aerospace area.

PREPARING YOUR 3D CAD-DESIGN, 3D-SCANNING & PRINTING

3D Printing – Digital Operation

3D Digital Configuration – Off-the-Shelf Slash Software

3D software "Slash" is developed to deliver the designer an exclusive technique to interface with design features to offer a high level of 3D printing accuracy. This software empowers the designer to generate various logos and 3D texts. This 3D softer is free to use. Moreover, it is ideal for first time user, nonetheless add-on-features are accessible for complex structure of commercial applications as necessary by the consume. Naturally, the Off-the-shelf free versions have retractions in functions, such as the requirements for higher resolutions, and colors. Nonetheless, the interface is imputative and easy generating a block cutting style to creäte various figures makes used by beginners.

Furthermore, the beginner may find available multiple files of software ready for download to import. Beginners may then use a portion to generate a new design. This kind of software has a exceptional advantage to rapidly produce a design from start to finish.

3D Digital Structure – Off-the-Shelf Tinker CAD Also, This Off-The-Shelf on line 3D design app is tailor-made to beginners. The distinctive advantage of this software deliver intuitive block-building concept. This permits developed to model the design from a set of basic shapes. Additional unique characteristic of "TinkerCAD" is its sizable tutorials guide to assist beginners to acquire the desired design. File sharing can be done with ease.

The Off-the-Shelf TinkerCAD encompasses significant amount of hand-on files that can offer the best shapes to suit the application. The design can be easily incorporated and shared with 3rd party printing services.

3D Digital Configuration – Off-the-Shelf Open Source CAD

Open source CAD, is defined as accessible computer aided design software individuals may freely practice, modify and share it with others. CAD as a tool for 3D modeling permits the user to design real-life objects of any size. The CAD constituent makes modifications and editing of any design simple. Altering the design parameter through the CAD permits the designer's model to freely grasp a different model. No payment is required. Nevertheless, while open source CAD is an ideal tool for traditional component and an outstanding training for beginners, it may not be the best tool for specialists developing complex and proprietary design.

3D Digital Configuration – Off-the-Shelf SolidWorks

SolidWorks is a 3D CAD program often used by knowledgeable designers. This SolidWorks software encompasses a substantial amount of advanced features, such as design validation tools, and reverse engineering. "Solidworks" is best suited industrial implementations that stress extensive details. Though most software, mimic curves through gently inclining flat structures, SolidWorks is perfect to generate highly precise curves, evading the use of polygonal modelling, while other software utilizes flat surfaces averaged and joined together.

3D Digital Configuration - Inventor

"Inventor 3D CAD software" is a 3D mechanical design tool prepared with automation and simulation. Also, free form parametric modeling choices are integrated with this software. This Inventor software is custom-made for various level of users' proficiency. The visual data representation of this kind of software delivers the intended efficient deign.

3D Digital Configuration- Slicers & 3D Printer Hosts

Slicers are the easiest and most inventive techniques to execute the transformations from a 3D model to a printed object. The slicer routine acquires a CAD model, and then slices it into expressive layers to convert the 3D model into "G-code." Fortunately, the slicer routine integrates "3D Printer Settings" such as, temperature, layer thickness, and print speed, to the "G-Code." With much ease, the "3D Printer" can then read the "G-Code" to generate the model layer by layer as it follows the exact instructions set in the "G-Code."

Slicers & 3D Printer Host – "Slic3r"

This is open-source software, which embraces real-time incremental slicing, in addition to previewing 3D prints. This kind of slicing software is the most extensively used amongst users in 3D printing. This software enjoys real time incremental slicing ensuring flawless change and settings. This software has the capability to resume executing the routines where it stopped instead of starting from the beginning. However, for accuracy the G-Code needs to be recalculated for the parts in question. This features lead to greater flexibility, faster response and precise slicing program.

What Does Slicing Software Do?

3D slicing programs, every now and then referred to as slicers, convert digital 3D files into G-code which 3D printers understand. They calculate a path for the nozzle to follow based on the geometry of the part and user-specified settings.

3D printers are surprisingly simple machines. They progressively build objects layer-by-layer, and this process takes anywhere from a few minutes to many hours. These machines seem quite complex at first, but they operate by following a long list of very simple instructions to print out parts. Slicing software permits you to tune settings and adjust or chance how your part is printed.

Your 3D printer understands how to print layers but STL files on your computer are designated by a cluster of connected triangles. Your 3D printer would not know what to do if you directly send it a 3D object or an STL file --so first, you must import your file into a slicing program to turn it into lines of G-code.

Before turning your 3D file into G-code, slicing software permits you to modify printing parameters so you can adjust features such as layer height, wall thickness, and printing temperature. Slicers can create support structures so you can print overhangs, and they also can create material to help with construct plate adhesion such as skirts, brims, and rafts.

Slicers permit you to preview the layers of your model before you print it. This is useful so you can guarantee the part will be printed the way you intended. Furthermore, most slicers can repair and slice STL files that have errors or defects such as holes in the mesh, non-manifold faces, or overlapping geometry.

How Does Slicing Work?

Slicing programs cut a 3D object into numerous 2D cross-sections. Overhanging faces are detected; and, if the option for support material is enabled, it will create a support structure underneath these areas. From each of these cross-sections, a toolpath for the nozzle is calculated based on printing parameters. Startup and cooldown routines are added at the beginning and end of the print.

Some slicing programs are capable to directly connect to and control your 3D printer via USB or WiFi so you do not require leaving your desk to start a print!

Print Settings

When utilizing slicing software, you can select from a variety of print settings to adjust how your part is printed. Adjusting these settings can change the quality, strength, and total print time. Here are some basic print settings you should be familiar with.

Nozzle and Build Plate Temperature

These temperatures will depend on the material you are printing with. Typical printing temperatures for PLA plastic are 190-215 °C at the nozzle and 20-60 °C for the build plate.

Layer Height

This setting is quite self-explanatory. It adjusts how thick each layer is. Typical layer heights vary from 0.1 – 0.3 mm with 0.2 mm being the standard for most printers. Smaller layer heights result in smoother parts, but due to the increased number of layers, the print takes much more time.

Wall Thickness

Also referred to as "shell thickness", this setting is significant in determining the strength of your part. Again, it's quite self-explanatory. Each shell is corresponding to the width of your nozzle (0.4 mm is a standard nozzle thickness) in most applications, 2-3 shells is enough, nevertheless you might want to use more if you need your part to be stronger.

Infill Percentage

Most 3D printed parts are not hollow, and most are not solid either. The interior volume of most 3D prints consists of a pattern called infill which increases strength and supports the layers above it. Most parts are often printed have around 10% infill, but sometimes it is given a slightly highly percentage if needed the extra strength.

Flow

This is a percentage which modifies the rate at which material is fed into the extruder. Characteristically this value is set at 100%, but this can be modified based on the material you use. Most users found that different brands of PLA print best at different flow percentages, but this setting usually has to be tuned to your specific printer.

Classically, 3D printing filament is 1.75 mm in diameter (sometimes it is 2.85 mm, but most printers' support 1.75 mm). If you find that a certain brand of filament is under or over extruding, it could be that the diameter is not exactly 1.75 mm. Most brands are quite consistent, but if you have a spool of inexpensive material, you may have to adjust the flow percentage to account for this inconsistency.

Print Speed

Print speed is simply how fast the nozzle moves while it is printing. You can change the overall speed (30-60 mm/s is typical for desktop printers), but you can also change the speed depending on what's being printed.

For example, you can specify that you want your outermost wall printed slower than the inner walls, or that you want your first few layers to print slower than the rest of the print.

You must also take into account the material being printed. If you are printing a stiff plastic such as PLA, 60 mm/s should be no problem for most printers, but if you are printing flexible filament like TPE or TPU, you will perhaps want to slow down your print to 30 mm/s or even less.

For travel moves where no material is being extruded, speeds of 120-150 mm/s are quite characteristic. These faster travel movements can be considerably louder though, so if you wish to reduce the noise coming from the printer, keep these travel moves at around 80 mm/s.

Retraction

Sometimes during printing, the nozzle has to travel from one place to another without extruding any material. During this travel movement, excess material can leak from the nozzle leaving unsightly strings and blobs on your part.

The "retraction" reverses the extrusion mòtor slightly; relieving pressure on the molten material while the travel movement takes place. Retraction can be easy to overlook, but it is a vital setting, which can greatly improve the quality of your parts.

Enabling retraction increases print time slightly, but it improves the quality of your prints so much that it is certainly worth the extra time.

Retraction Distance

This setting determines how much material to retract back into the nozzle. For "Bowden-style" extruders (where filament is fed through a long PTFE tube before reaching the nozzle), this distance should be somewhere between 3— and 5 mm. If your extruder is direct-drive, 1 mm should be sufficient. If there is still stringing, gradually increase this distance until it stops.

Retraction Speed

This one is not too complicated to figure out. This determines how rapidly the material is retracted. This value should not be too high, because if the retraction is too quick, material which is not melted yet can separate from the molten material in the nozzle causing bubbles and defects. 30 mm/s is a good value to start with.

SUPPORT STRUCTURES

Overhangs are a challenge for 3D printers since plastic will droop downwards after exiting the nozzle. Enabling support material creates an added supporting structure that is printed at the same time as the part. This mesh is created just below overhanging features so that the plastic does not droop after being printed, Figure 1.40.

The process of removing support material can end up leaving flaws on the surface it was removed from, and it can also be difficult to remove, so it is highly recommend taking the time to tune your support settings.

Fig.1.40: 3D Printing Structure Support

SUPPORT PATTERN

Your support pattern governs how rigid your support structure is. Some patterns like 'concentric' and 'lines' are very easy to remove, but they can be quite weak. Interconnected patterns such as 'grid' tend to be quite sturdy, but they can be difficult to remove. If your support structures are tall, or prone to fall over mid-print, try using a more rigid pattern.

Support Density

This setting is similar to infill. Increase this percentage to make a denser mesh that is harder to remove, but results in better overhangs. It is recommended to use 15% for most of the prints.

Cooling

Your 3D printer has part-cooling fans which significantly improve print quality while they are active. Once material is extruded from the nozzle, it should be cooled as quickly as possible so that it stays in place. This is especially important when printing bridges and overhanging features.

The exception to this is printing with high-temperature materials such as ABS, polycarbonate, and nylon. They should not be printed with active cooling because they suffer problems with warping and layer adhesion. For most other materials though, proper cooling is highly recommended.

Minimum Layer Time

As an alternative of using fans, another way to improve cooling is increasing the minimum time for each layer. When printing very small parts, your 3D printer can go too fast, and extrude molten material on top of a layer that has not cooled down yet. This results in drooping and low-quality surface finish. A good value to start with is 10 seconds. If, for example, a layer would take 6 seconds to print normally, this setting will tell your printer to slow down so that the layer takes 10 seconds instead.

Lift Head

If enabled, this setting lifts the nozzle away from the print if the minimum layer time has not been reached. This setting should be utilized if the layer being printed is exceptionally small. Simply, slowing down the print speed will not permit the material enough time to cool, so lifting the nozzle completely away from the print can further help with cooling.

Build Plate Adhesion

Occasionally, your part will have issues sticking to the build plate, Figure 1.41. You can try to attempt different ways to give your parts the ability to stick, such as utilizing a PEI sheet or applying an adhesive, but your slicer also has choices to select from which can help solve the issue. There are typically 3 options in your slicer to build plate adhesion: skirt, brim, and raft.

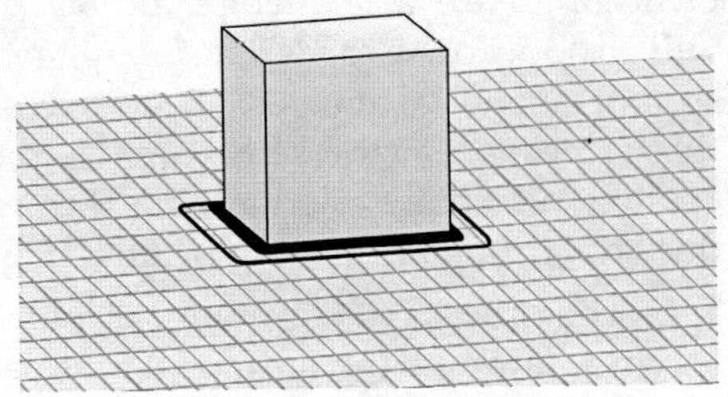

Fig.1.41: Build Plate Adhesions

Skirt

A skirt is a small length of extra material extruded at the beginning of a print. This is achieved to guarantee that plastic is flowing smoothly before the print begins. This setting improves build plate adhesion to some extent by keeping the bottom surface consistent.

Brim

A brim can be created if your part is having trouble sticking to the build plate. A brim modifies the first layer of your object to increase the surface area, improving build plate adhesion. Brims can be somewhat bothersome to remove so only use them if necessary.

Raft

When you enable rafts, a platform is created at which your object is then printed on top. This feature can be useful if you need the bottom surface of your print to be as smooth as possible. Rafts considerably add to print time and use too much material. Only use this option if you must. Usually, rafts reduce the effectiveness of a heated bed.

Other Settings

Here are a few settings which are important to discuss, but do not fit into any specific category.

Horizontal expansion

Once molten material leaves the nozzle of your printer, it tends to expand slightly. This effect can be bothersome if you have two parts which need to fit together. A common example of this is when printing holes for screws or rods. For instance, if a hole is 10 mm in diameter on your 3D model, it might end up being 9.6 mm when it is actually printed. Adding a horizontal expansion factor of -0.2 mm will negate this effect.

Combing

Travel moves which do not extrude material can cause defects if the nozzle travels across a line of material that has already been printed. By enabling combing, these intersections are made less visible because the extruder will move overtop of infill as much as possible. This setting will not really increase your print time by a significant amount, and it will improve the quality of your parts slightly as well.

G-code

G-code essentially instructs your 3D printer's nozzle where to go, and how much plastic to extrude along the way. It takes many instructions to print an object so these files can be very long. A small part that takes 1 hour to print can have over 100,000 lines of G-code.

Once you have sliced your object, you send the G-code file to your printer and start the print.

Other machines such as CNC routers, water jet cutters, and laser cutters all use G-code to control their motion. Understanding G-code is not strictly necessary if you want to use slicing software, but it is quite interesting to learn, and it can be beneficial if you want to fine-tune and modify and customize your printing routine.

Startup G-code

Before you start a print, there are a few things you can tell your printer to do so that you can improve your first layer:

1. First – All prints must start by homing all the axes. This ensures the print begins in the correct location.
2. Then the nozzle and build plate should heat up.
3. It is also a good idea to extrude a small amount of plastic before you begin printing so your nozzle is primed with fresh material.
4. Sometimes while heating, some liquid plastic will ooze out of the nozzle before the print, so extruding a small blob before your print will avoid any defects in your first layer.
5. Typically, this routine is already the default on most slicers, it found to be helpful to adjust this startup G-code to suit your particular setup.

Ending G-code

When a print finishes, you can specify a routine for your printer to follow. There should, at the very least, be some lines of G-code telling the printer to start cooling down the nozzle and build plate, as well as turning off the motors and any cooling fans. Also, it is also recommend lifting the nozzle a few millimeters and then homing in X and Y, so that the nozzle does not stick on the print once it cools down.

With more practice with the of "G-code know-how," you can program elaborate routines such as changing the color on RGB LED strips, or making your printer literarily, play a melody, so that you know it is finished printing.

Which Slicing Program to Use?

Using slicing software is an indispensable step in the 3D printing process. There are several slicing programs to choose from, and the one you utilize is up to you. Most programs are free to download, are open-source, and will work with almost any desktop 3D printer.

Some programs are very user-friendly, and will take most of the guesswork out of choosing the correct settings. More advanced users will desire to use more powerful and complex slicers to fine-tune every setting. The user should now be familiar with how slicing software operates. There is only so much knowledge the user may acquire through reading. Downloading some slicing software and attempting slicing some objects would complement the reader and equip the user with greater insight.

Digital Configurations of 3D Printing

The implementation of 3D Printing, Additive Manufacturing, cannot take place deprived of fundamental specific software programs. While Additive Manufacturing is dealing with tangible mechanical fabrication it is becoming revolutionary innovative technology to boldly depart from a pure mechanical engineering practice to a pure digital part fabrication. Accordingly, 3D Printing – Additive Manufacturing Technology is founded upon three major advancements in Digital Manufacturing Technology, stated as follows:

1. Digitally Deign Part - Using Computer Aided Design "CAD" – such as "Solid Works"
    (a) Alternatively Conduct 3D Scan on a Physical Existing Object
    (b) Transform three 3D Scan into Traditional 3D CAD form
2. Transform The Digital Program CAD to Slice the Mechanical Design into manageable Layers – such as ".STL," and then transform the ".STL" to printable "G-Code"
3. Transform the Layers' Slicing Program into Printable Layer by atop of Layer Using 3D Printer

Configuration of 3D model is tailored to create practical tool to build reliable structure printed for day to day consumption. Most of the tools stated in the digital manufacturing tools are simple to create, while other are highly sophisticated to satisfy complex structure, which may require experienced software developer.

3D DIGITAL CONFIGURATION - USE OFF-THE-SHELF "SLASH" SOFTWARE

New Beginners

The beginners may find available multiple files of software ready for download to import. Beginners may then use a portion to create a new 3D CAD design. This type of software has a unique advantage to quickly generate a CAD design from start to finish.

Samples of available CAD programs are listed below:

3D Digital Configuration: Off-the-Shelf Tinker "CAD"

Also, this Off-The-Shelf on line 3D design app is tailored to beginners. The unique advantage of this software provide intuitive block-building concept. This allows the developer to model the design from a set of basic shapes. Additional unique feature of "TinkerCAD" is its sizable tutorials to guide and assist beginners to obtain the desired design. File sharing can be accomplished with ease.

The "Off-the-Shelf TinkerCAD" contains considerable amount of hands-on files that can provide the best shapes to suit the application. The design can be easily integrated and shared with 3rd party printing services.

3D Digital Configuration: Off-the-Shelf Open Source "CAD"

Open source CAD, is defined as accessible computer aided design software individuals may freely use, modify and share it with others. CAD as a tool for 3D modeling enables the user to design real-life objects of any size. The CAD component makes modifications and editing of any design simple. Changing the design parameter through the CAD allows the designer's model to freely reach a different model. No payment is required. However, while open source CAD is an ideal tool for traditional component, and an excellent training for beginners, it may not be the best tool for professionals developing complicated and proprietary design.

3D Digital Configuration – Off-the-Shelf SolidWorks "CAD"

"SolidWorks" is a 3D CAD program frequently used by experienced designers. This "SolidWorks" software contains a considerable amount of advanced features, such as design validation tools, and reverse engineering. "Solidworks" is best suited industrial application that demands extensive details. While most software, mimic curves through gently inclining flat structures, "Solidworks" is ideal to create highly accurate curves, avoiding the use of polygonal modelling, while other software uses flat surfaces averaged and joined together.

3D Digital Configuration – Inventor "CAD"

"Inventor" 3D CAD software is a 3D mechanical design tool equipped with automation and simulation. Also, free form parametric modeling choices are incorporated with this software. This Inventor software is tailored for various level of users' proficiency. The visual data representation of this type of software provides the ability to reference 3rd party designs without the need to convert file formats.

Samples of available 3D Slicer and G-Code programs are listed below:

Executing 3D Slicing and Printing – The "G-Code"

The following software entities are tailored to provide the most favorable best "3D Printing Software Tools" to execute "3D Print."

3D Digital Configuration- Slicers & 3D Printer Hosts

Slicers are the easiest and most innovative techniques to execute the transfigurations from a 3D model to a printed component. The slicer routine acquires a CAD model, and then slices it into meaningful layers to convert the 3D model into "G-code." Fortunately, the slicer routine incorporates "3D Printer Settings" such as, temperature, layer thickness, and print speed, to the "G-Code." With much ease, the "3D Printer" can then read the "G-Code" to create the model layer by layer as it follows the precise instructions set in the "G-Code."

Slicers & 3D Printer Host – "Slicer"

This is open-source software, which includes real-time incremental slicing, in addition to previewing 3D prints. This type of slicing software is the most widely used among users in 3D printing. This software enjoys real time incremental slicing ensuring flawless change and settings. This software has the capability to resume executing the routines where it stopped instead of starting from the beginning. However, for accuracy the "G-Code" needs to be recalculated for the parts in question. This features lead to greater flexibility, faster response and precise slicing program.

REFERENCES

1. Pahole, I. Drstvenšek, I., Ficko, M. & Balič, J. (2005). Rapid prototyping processes give new possibilities to numerical copying techniques, Journal of Materials Processing Technology, 164-165, (1416 – 1422), ISSN 0924-0136
2. "Warehouse Management System: Cost Justification." http://www.idii.com/wp/kom_wms_justification.pdf. Retrieved July 24, 2011.
3. Arnett, Stephen. "Project Cost Justification." http://www.qualitydigest.com/apr99/html/body_teams.html. Retrieved July 24, 2011.
4. H Paul Barringer, P.E. "A Life Cycle Cost Summary." http://www.barringer1.com/pdf/LifeCycleCostSummary.pdf. Retrieved June 24, 2011.
5. Carnegie Mellon University. "Life-Cycle Cost Analysis Primer." http://www.ce.cmu.edu/~hsm/im2004/readings/DOT-LCCA-Primer.pdf. Retrieved June 24, 2011.
6. Sieglinde Fuller. "Life-Cycle Cost Analysis (LCCA)." http://www.wbdg.org/resources/lcca.php. Retrieved June 24, 2011
7. Soloman, Sabrie Sensors Handbook, 2nd Edition, McGraw-Hill Publishing Co., NY – ISBN 978-0-07-160570-0
8. Soloman, Sabrie "Sensors and Control Systems in Manufacturing," McGraw-Hill Publishing Co., NY – ISBN 0-07-059626-3
9. Soloman, Sabrie "Affordable Automation," McGraw-Hill Publishing Co., NY – ISBN 0-07-059633-6
10. Soloman, Sabrie Sensors Handbook, 1st Edition, McGraw-Hill Publishing Co., NY – ISBN 0-07-059630-1
11. Soloman, Sabrie "Sensors and Control Systems in Manufacturing," McGraw-Hill Publishing Co., NY – ISBN 978-07-10572-4

2 Digital Design for 3D Printing

INTRODUCTION

Rapid prototyping methodology is defined as a unique technique developed specifically to combine several manufacturing methods used to rapidly fabricate either a scaled model, a true part, or a group of parts referred to as an assembly of specific configurations using 3D computer-aided design (CAD) as input data. The construction of part configurations or group of parts as assembly is achieved utilizing 3D additive layer manufacturing technology.

In the late 1980 the inception of "Rapid Prototyping" began, and to the amazements of the manufacturing practitioners and technologists it gained a respectable reputation in the manufacturing arena. Various models of various design and numerous prototypes were fabricated using the "Rapid Prototyping" technique. Wide range of manufacturing applications has sprung forward utilizing 3D additive manufacturing technology known as 3D Printing. 3D additive manufacturing – 3D printing - can fabricate production-quality parts in relatively small volume, and short time. This technology can provide a viable alternative to typical traditional manufacturing technology of CNC for short-runs.

Equally impressive is the springing of "On-Line Service Application" to provide low cost rapid prototyping. Rapid prototyping technology was initially sparked in the 19th century sculptors discussing how to find new techniques to produce identical multiples sculptures. Recently, various modern sculptors are using the progeny technology to produce exhibitions. Recently, it is possible to interpolate volumetric data from several one-dimensional images. Consequently, it is feasible to reproduce viable designs from a set of data to create three-dimensional images leading to 3D designs.

3D design and 3D printing came to the scene as synonymous technology to fill the void for rapid prototyping and fabrication in the modern world technology. Since the inception of additive manufacturing - 3D printing the demand for 3D printers in professional and educational institutions explodes to the scene. 3D Traditional printers paved the way to the invention of the 3D printers. 3D printers provide a variety of advantages over traditional manufacturing. Also, in medical application 3D models have become crucial to the applications prior to production, as an excellent diagnostic tool for professionals and designer. 3D model help to save considerable development time and cost. 3D models help determining the components validity, and the accurate dimensions of the printed part to achieve optimum performance, with minimum cost, affordable product and with user friendly interface.

Additive Manufacturing has well served product testing function when a physical prototype is produced .The 3D conceptual designs combined with the rapid prototyping technologies provide a platform for design optimization as well as a foundation for further understanding of 3D printing.

TRANSFERRING PROTOTYPING TO 3D PRINTING

Additive manufacturing is the ability to uniquely fabricate a part through depositing a single dot at a time in a certain plane. When the collective dots are deposited in the same plane they form a unique layer. As further dots are deposited sequentially in all layers, a unique part is formed. The deposition of dots in all the layers are dispensed through 3D printer computer controlled. There are numbers of different types of 3D printers developed according different types of inventions. Each type of printing invention has its advantages and disadvantages. However, all printers are tailored to 3D print parts using different materials, speeds, quality, and cost. Each type of printer inventions has its set of rules to operate in a most efficient manner. 3D printing can:

Advantage

- Fabricate any geometry of parts
- Does not require extensive tooling cost
- Can rapidly verify part design and performance
- Low volume of part production

Limitations

- Material strength of 3D printing is less than that of subtractive traditional technology
- Variation of surface quality and tolerance due to temperature variations and warpage
- Variations of parts repeatability
- Cannot produce highly dense parts compared to subtractive technology

Design through Reverse Engineering

Transforming prototypes to 3D prints requires few several steps involving part design through computer aided design (CAD), reverse engineering though part (Scanning Technique), and ability to convert CAD into readable file.

Designing parts to employ varieties of 3D printing technologies is a fundamental process to ensure efficient part fabrication and simplicity in turning the conceptual design into real functional part. The following fundamental steps are common throughout the varieties of 3D printing technologies. These steps highlight the general process from design to final part. While each printing technology is different, the following steps are the same across all technologies.

Creating 3D File

The first step in the 3D printing is creating a digital model. Computer Aided Design (CAD) is the most shared method for creating a digital model (Figures 2.1(A), 2.1(B), and 2.1(C). In the absence of CAD reverse engineering can be applied to create a digital model through 3D scanning. Thus, CAD modeling and reverse engineering are synonymous to create 3D digital model. When designing part for 3D printing there are several considerations that must be considered. These considerations highlight the limitations of geometry. Also, these considerations provide clear choice of materials, and the methodology to clear a hole- construction in 3D printing technology.

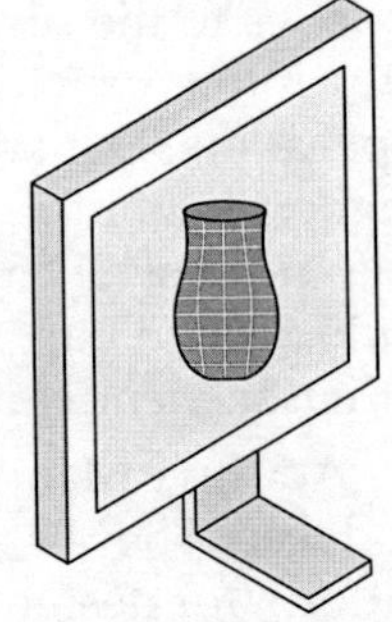

Fig.2.1: (A) The Image of the Printed Model is displayed

1. 3D Image is created utilizing CAD software.

2. The CAD file is forwarded to the Printer
3. The motor advances the polymer filament, melts it, and pushes the melted polymer through the nozzle.
4. The printer deposits successive layers, (liquid, powder, paper, or metal) and then constructs the model from a series of cross- sections.

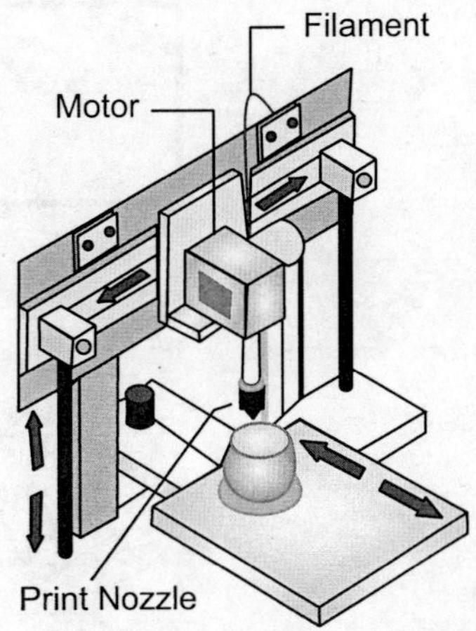

Fig 2.1: (B) The Printer deposits layers of according to the STL Slicing Routine

STL FILE FOR 3D PRINTING

Printing a part utilizing 3D printing technology, a CAD model must be converted into a format that any 3D printer can read and interpret its procedures, Figure 2.2 (1). Therefore, a procedure must begin to convert the CAD model into a Stereolithography (STL) file, Figure 2.2 (2). This type of file is referred to as "Standard Triangle Language file," Figure 2.2 (3). Also, a file named " OBJ" and a file named "3DP" are acceptable types of 3D printing files. While "OBJ" and "3DP" types are less common, they are simple and acceptable to use. Nonetheless, "STL" file type utilizes numerous polygons or triangles to describe intricate surfaces of an object as it substantially simplifies complex CAD model. CAD programs have the capability to export a model as an STL file.

Fig 2.1: (C) The Printer Equipment are Enclosed

Once the "CAD" program has created the "STL" file, it is imported into a unique slicer program, Figure 2.2 (3). It slices the part design into the corresponding layers to be used to construct the part. Following this step the slicer program takes the STL file and converts it into G-code. The "G-code" is defined as a numerical control programming language. It is primarily used in Computer aided manufacturing "CAM" to control traditional automated machines, such as the CNC machines, as well as the 3D printers, Figure 2.2 (4).

The 3D build parameters including the 3D support location, layer height, and part orientation are then defined by the printer operator. Each 3D printer has its unique Slicer programs, which is often considered a proprietary to each type of 3D printer. However, there are some universal slicer programs, such as "Netfabb," "Simplify3D," and "Slic3r." The designer is primarily concerned about providing a 3D printer operator with an "STL file." Accordingly, the operator will then prepare the required parameters for best print and produce the "G-code file." The constructed model may incorporate further processes to deliver a desired finished model, Figure 2.2 (5).

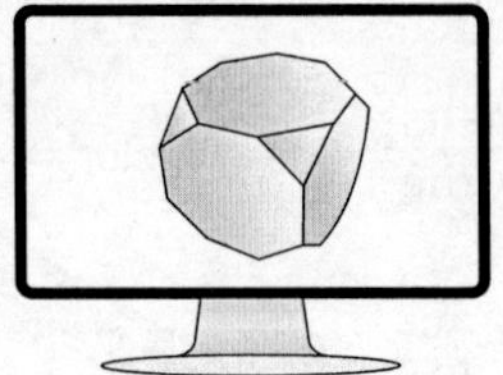

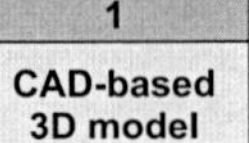

Fig.2.2: (1) CAD Based Model

Courtesy: Graphic: Deloitte University Press Du | DUPress

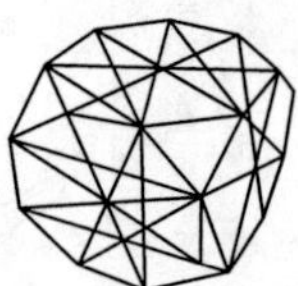

Fig 2.2: (2) ".STL" File

Courtesy: Graphic: Deloitte University Press Du | DUPress

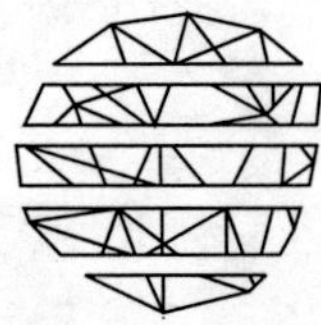

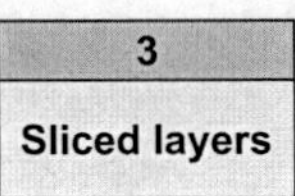

Fig 2.2: (3) Slicer Layers

Courtesy: Graphic: Deloitte University Press Du | DUPress

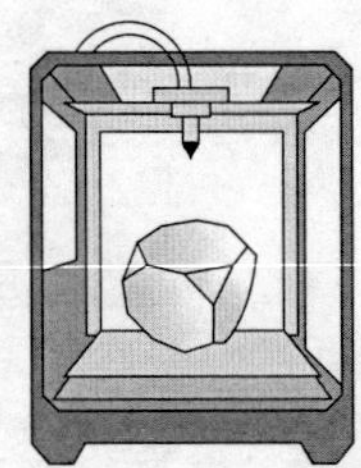

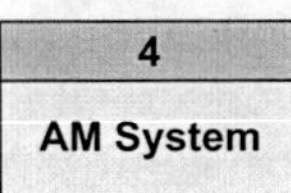

Fig 2.2: (4) Additive Manufacturing

Courtesy: Graphic: Deloitte University Press Du | DUPress

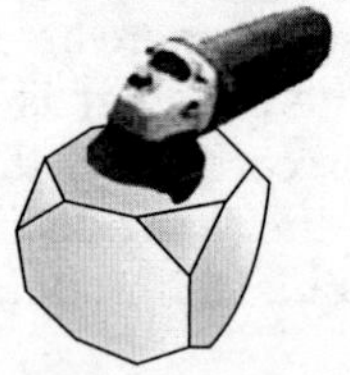

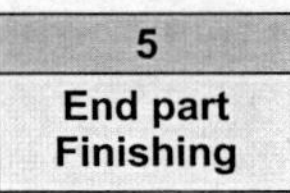

Fig 2.2: (5) Finished Model

Courtesy: Graphic: Deloitte University Press Du | DUPress

3D PRINTING – THE ADDITIVE MANUFACTURING – RAPID PROTOTYPING

3D additive manufacturing is the printing technologies used in fabricating parts in different printing techniques. Determining the ideal 3D printing process for a certain design can be problematic.

The variations in dimensional accuracy, surface finish and post processing requirements plays a fundamental role in 3D printing methods. Also, selecting the materials used in printing would influence parts function and reliability, Figure 2.3.

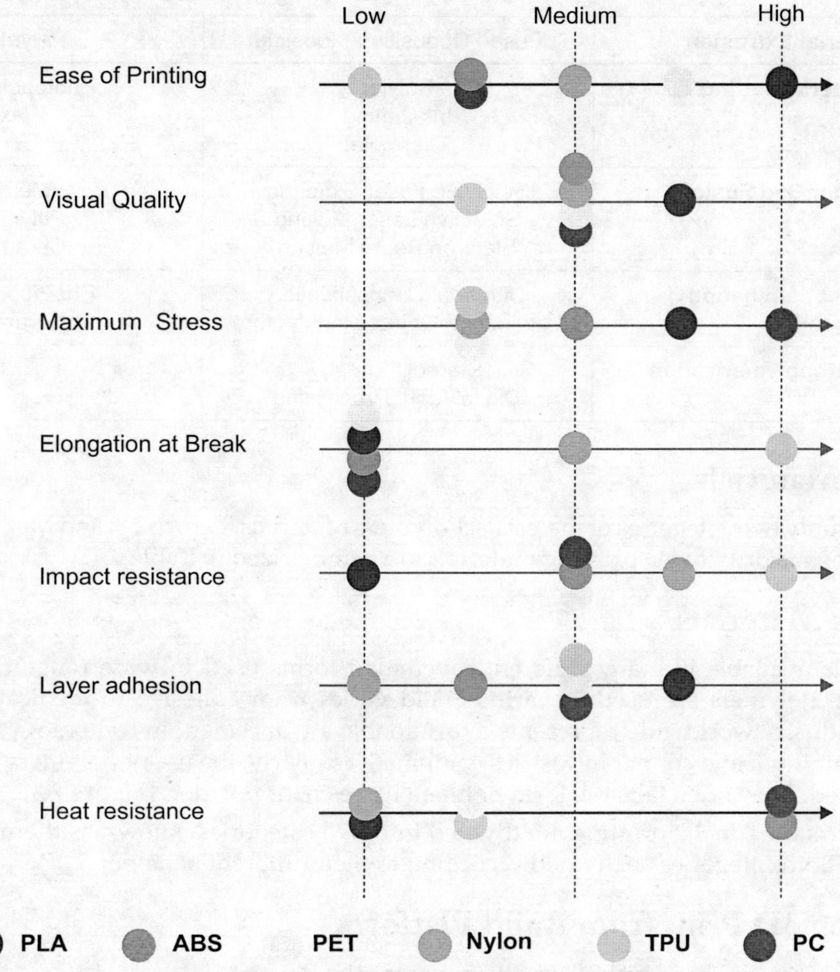

Fig.2.3: The Polymers' Rank at Different Criteria

CLASSIFICATION OF 3D PRINTING TECHNOLOGIES

Terminologies and classification of each of the different 3D printing methods were standardized in 2015 through The "ISO/ASTM 52900." Several categories of various processes were established.

Table 2.1 Additive Manufacturing Polymer and Metallic Materials

Process Category	Technology	Materials
Binder Jet	3D Printing Ink-Jetting	Metal Polymer Ceramics

Direct Energy Deposition	Direct Metal Deposition Laser Deposition Laser Consolidation Electron Beam Melting	Metal Powder Wire
Material Extrusion	Fused Deposition Modeling	Polymer
Material Jetting	Polyjet Ink-Jetting Thermojet	Photopolymer Wax
Powder Bed Fusion	Selective Laser Sintering Selective Laser Melting Electron Beam Melting	Metal Polymer Ceramic
Sheet Lamination	Ultrasonic Consolidation Lamination Object Manufacture	Photopolymer Ceramic
Vat Photopolymerization	Stereolithogrphy Digital Light Processing	

Printing Materials

There are mainly two categories of materials: Polymers Materials, Figure 2.3, and Metallic Materials, Table 2.1. The majority of 3D printing materials are summarized in Table 2.1.

Polymers Materials

Polymers are available in many different diversified forms used in wide range of 3D printing applications. Polymers are used in various field varies from adhesive to medical applications. Polymers industry worldwide represents a formidable advancement in technology that bypassed the steel, aluminum and copper industries combined. Similarly, the use of polymers in 3D printing is widely used. They often take the form of filament, resin, or powder, (Figure 2.3).

Polymers used in 3D printing are divided into two categories, known as thermoplastics and thermosets. Both categories ae divided according to their different behavior.

Separating 3D Print from Build Platform

A manual process is the simplest way to separate the 3D printed part from the build platform, Figure 2.4. Other ways may be more difficult, which requires more precise means to extract the 3D printed part while it is still encased within the build material. Other are more complicated to remove 3D printed part while it is still attached to the build plate. Complicated industrial 3D printed parts require high skilled operators and strict removal procedures combined with safety requirements and complete control of the environments.

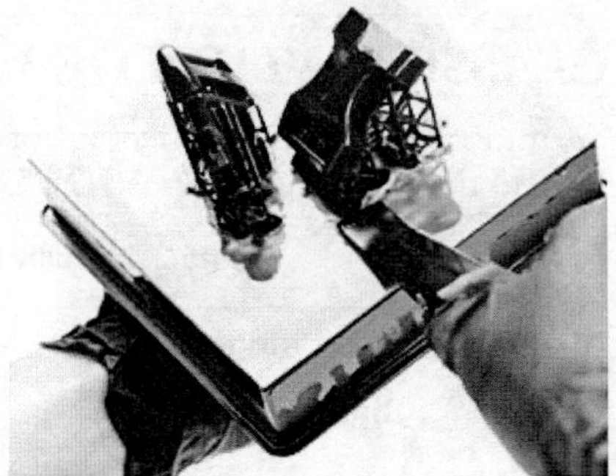

Fig.2.4: Removing the Model from the Build Plate

Post Processing

Printer type and the type of technology determine the procedure of post processing. Some printer types and technologies necessitate using UV curing agent prior handling. Few other printer technologies allow parts to be extracted immediately after the printing task. Other printing technologies, which use structural supporting post may extract the printed part at the time of removing the structural supporting post, Figure 2.4.

Determining the most efficient method to extract the printed part is by understanding how the specific 3D printing technology produces the part to find an optimum way to extract the 3D printed part.

CORRELATING 3D PRINTING TECHNOLOGIES

Navigating through vast numbers of 3D printing technologies may require effort to acquire the imperative knowledge to correlate the best 3D printed part with the optimum 3D printing technology. Equally imperative is the choice of the materials used to obtain optimum part function, and the suitable strength for the intended application purposed specifically for the part, Figure 2.5. Having the necessary knowledge of material and 3D printing technologies it is possible to determine the specific technique that is best suited for a specific intended design. However, specific design rules for each technology must also be followed to obtain reliable 3D printed part.

Thermoplastic Material Extrusion — FFF

Thermoplastic Material extrusion in 3D prints is utilizing a string of solid thermoplastic filament materials. The thermoplastic extruded material is inserted through a heated nozzle causing the filament to melt. Using a built platform the printer deposits the material on a build platform in a well predetermined path. The filament is then cooled and takes a solid form.

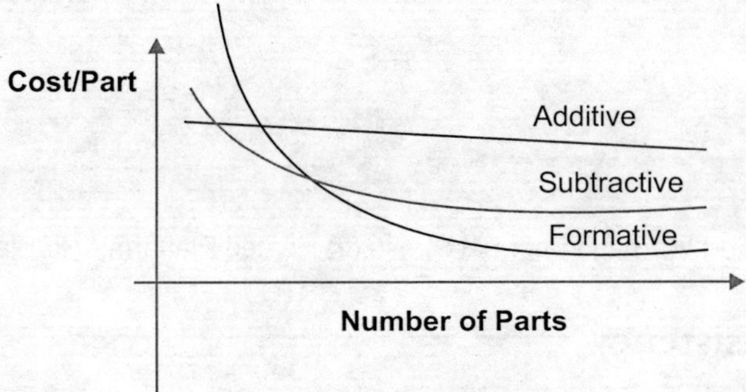

Fig.2.5: Cost of 3D Printing/Part

Fused Filament Fabrication - 3D Printing FFF

FFF is defined as "Fused Filament Fabrication." It is the most common 3D printer using polymer material in extrusion technology. This technology is often referred as "Fused Deposition Modeling" or FDM. FDM is a trademark of Stratasys. This technology enables the 3D printer extrusion head to feed a spool of filament loaded into the 3D printer. Once the temperature of the printer nozzle is elevated to a certain level, the filament is stepped forward by printer motor. The printer control directs the extrusion head to move in X-Y-Z directions. It deposits molten material at a precise location. The material is then cooled down and solidifies. Once a full layer is deposited, the build platform moves down, and the process repeats itself to continue building up the part layer-by-layer. The same technology may be used the construction industry, Figure 2.6.

Fig.2.6: 3D Printer Deposit Layers of Concrete Wall

Printer Characteristics

Most Fused Filament Fabrication (FFF) machines have the capabilities to adjust and monitor several parameters fundamental to produce highly reliable and accurate 3D prints. The consistency of the extruded filament is controlled by the build speed, extrusion speed, and nozzle temperature. These parameters are established by the operator, however, based on the type of materials used in the 3D print some machines use automatic presets parameters. The print resolution is determined the nozzle diameter and layer height of a printed part using FFF machines. High quality surface, dimensional accuracy, and a higher level of details of a 3D print are determined by a smaller nozzle diameter, and a small layer height. A fundamental parameter that must be considered is the available build envelope when using FFF printing machines. The standard site of the desktop printers is 200 x 200 x 200 mm as a build chamber. Industrial 3D printing machines are available in 1000 x 1000 x 1000 mm. Nonetheless, it is prudent to consider redesigning large part to a few smaller parts that can be assembled or fit together after printing.

Warping

When utilizing FFF 3D prints, it is likely to experience internal stress caused by thermal degradation of inner printed part compared to the outer printed part. This, in turn creates warping. Warping is resulted from differential cooling of different sections. Some sections of the print are cooled at different rates causing contraction and shrinkage. This phenomenon, Figure 2.7 (1), creates internal stresses leading to distortion. In order to limit the harmful effect of warpage, a heated bed as well as good bed adhesion is implemented to limit the harmful effect of warpage, Figure 2.7 (2).

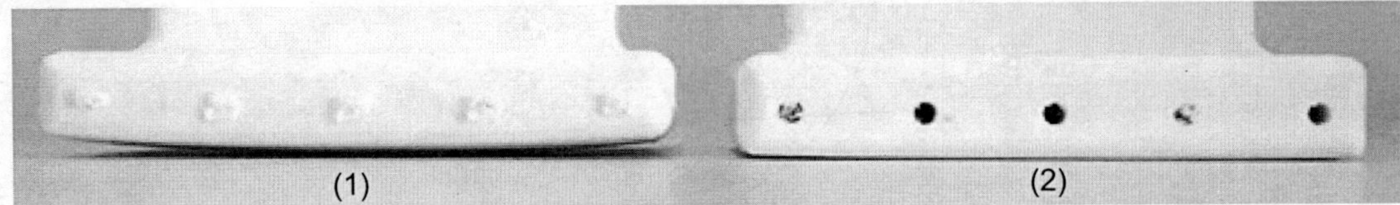

Fig.2.7: (1) Warpage Thermal Stress (2) Heated Platform - No Warpage

Dimensional Consistency

As thermoplastic is extruded and depositing one layer at a time resting onto a build plate causing each layer to cool down at a rate slightly different each time. Accordingly, internal stresses is generated resulting in a slight distortion in particularly the thin section of a printed part, while a larger thicker sections are subject to a greater warpage due to large temperature variations. Incorporating a large printing base layer on the bed for constructing the 3D print would limit the amount of thermal distortion. Also, heated beds and adding fillets at sharp edges and corners in the design will help to reduce the distortion.

Materials

The thermoplastics of the Fused Filament Fabrication (FFF) are provided on a spool. The filament diameter is typically 1.75 mm or 3 mm. The FFF filaments are the most versatile technique in 3D printing, while enjoys the lowest materials cost, Figure 2.5. The spool cost varies from ($20 - $40/ kg. High performance filaments, such as the "PEEK" material may cost $500/kg. Also, FFF printing filaments can be obtained in a variety of colors.

The likelihood of distortions occurs when the printing temperature increases during the printing process while parts are cooling at an increased rate. The thermal cooling generates intense internal stresses. As illustrated in Figure 2.8, the higher the thermoplastic lies in the pyramid, the higher the temperature it requires to print, and the better the engineering properties.

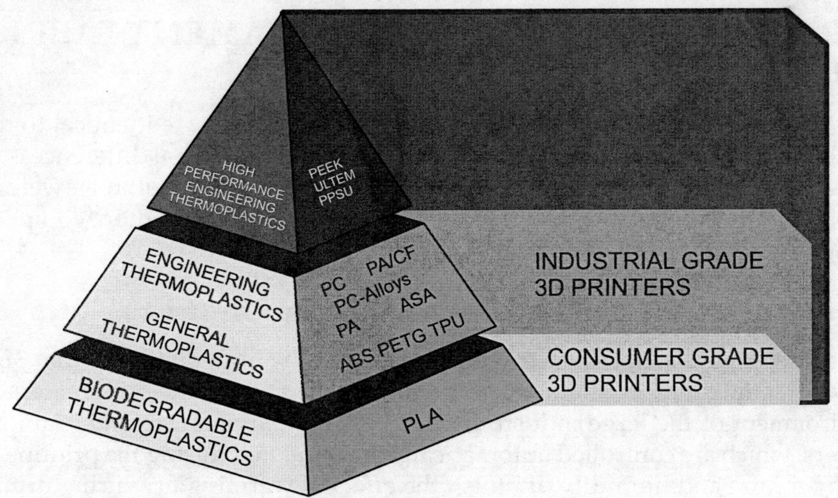

Fig.2.8: The Thermoplastic Polymer Pyramid

Courtesy: INTAMSYS Corporation

ABS and PLA polymers are generally are easy to use in 3D print. .Thermoplastics such as PEEK and PEI enjoy excellent engineering properties; however, it is often used in 3D print using industrial machines that are able to provide greater control over the print environment.

Post Processing

If a standing or inclined surface of a part is required to be 3D printed using "post processing supporting structure," such post will impact on the overall dimensions of the part. The sanding surface undoubtedly will impact the assembly of parts if a tight fit is imposed.

Benefits and limitations of Fused Filament Fabrication FFF

FFF is considered the most common 3D printing machine due to its Low cost of printing spool of materials as well as actual printing machine itself. Additionally, the FFF printing machines are simple to operate. Consequently, FFF machines are considered the most cost competitive in producing custom thermoplastic parts. FFF is well known globally as it was the first system ever introduced to the public. Also, it represents the largest install-base of 3D printers worldwide. The ease of available printing materials FFF also has become the most popular choice for rapid prototyping, in particularly, for functional applications that are not used commercially; However, some of the limitations are stated as follows:

- The anisotropic nature of parts is caused by the main limitations of FFF.
- Parts produced by FFF method through layer by layer are in general weaker in one direction.
- As the build direction will impact part performance, it forces the designer to highly consider the build direction.
- The percentage of the infill, curves, and fillets has an effect on the strength of a printed part. Usually, printers produce parts with 20% infill. Undoubtedly, Higher percentage of infill will result in a stronger part
- Build time will increase, thus the cost.

- Visible layer lines will appear as the part is built layer-by-layer method of printing
- Some form of post processing is needed, if a smooth surface finish is desired

INDUSTRIAL AND PERSONAL FUSED FILAMENT FABRICATION OF 3D PRINTING MACHINES

The technical capabilities of the low-end FFF 3D printing machines are identical to the industrial high end machines, but differ in output product scale. Also, an additional difference is that the high end industrial 3D printing machine uses standard thermoplastics material as well as advanced engineering thermoplastic materials to produce high quality parts with more complicated design configurations, as illustrated in Figure 2.8.

3D Printer Setting of Fused Filament Fabrication (FFF)

The FFF printer settings can be adjusted with ease within the high-end industrial 3D printers by automatically tightening the environment control parameters to suit the print material. Within the enclosed environment of the large industrial scale FFF, the temperature and humidity are other vital parameters, which are controlled automatically are regulated during the printing process. The control of temperature and humidity diminish the effect of warming up during printing process, as it reduces the cooling rate allowing the part to cool down successfully, eliminating the internal thermal stress and distortion.

Multiple Extrusion Heads

The FFF industrial 3D printing machines can use multiple extrusion heads enabling the support structures, if any, to be printed by using dissolvable materials, Figure 2.9. These lead to produce printing parts of a greater accuracy with a high level of repeatability. The high-end industrial machines contain a large build volume to print large parts of greater numbers, in a single build. Industrial FFF machines are used particularly for high quality printed parts for low to medium production runs.

Fig.2.9: 3D Printer Equipped with Dual Printing Heads

Courtesy: AXIOM Corporation

Whereby, desktop FFF 3D printing machines relish low cost as well as rapid method of producing 3D printing. The allowed tolerances of the low-end desktop FFF machines are capable to produce adequate products for most applications.

The gap between high-end industrial FFF 3D printing machines and low-end desktop FFF 3D printing machines is minimized over time due to continuous improvements of hardware, materials and software. Incorporating dual extrusions heads in modern low-end desktop FFF 3D printing machines would result in highly accurate printing process for low end machines. Also, bringing modernization will allow parts of industrial quality to migrate to the desktop low end less expensive machines. Additionally, advancements in desktop 3D printers will allow more control over the print parameters with respect to industrial machines. Accordingly, low-end machines would become ideal for high quality prototyping.

Testing Part Form, Fit, and Function

Fused Filament Fabrication FFF, 3D printing technology has served the community well. It is, in particularly used as powerful tool-design verification, quick prototyping, and cost analysis.

Hobbyists and professionals alike have used this technology to create non-commercial parts to test and verify part's function, part's fit, and part's form.

Replacing Investment Casting

Investment casting technique is relatively cumbersome and often costly. The advent of FFF 3D printing technology has waved the path to substantial material savings contrary to investment casting technique. Less investment cost was benefited the FFF new 3D printing technology. Additionally, the cost of FFF 3D printing machines is less than the cost of the investment casting machines.

Housing Verifications

Device housings and enclosures are the most popular applications for FFF 3D printing, Figure 2.10. Device housing designer can, in a matter of hours develop a viable housing or enclosure printed and even tested. Also, modifications can be easily incorporated to ensure enclosure strength and aesthetics. A final design with appropriate material and enclosure configurations as well as a rapid prototype can be viewed by numbers of decision makers. Substantial cost savings when using this type FFF technology compared to traditional technology in creating same prototypes.

Fig.2.10: 3D Printed Housing Enclosures

Courtesy: Sculpteo Corporation

Special Tools

Complex and specialized tools can be custom-made through FFF 3D printing machines, Figure 2.11. On the field manufacturing and assembly activities may demand instantaneous fabrication of special tools. The special part or tools can be designed and fabricated in a very short time to face the difficulty of operating at a remote location.

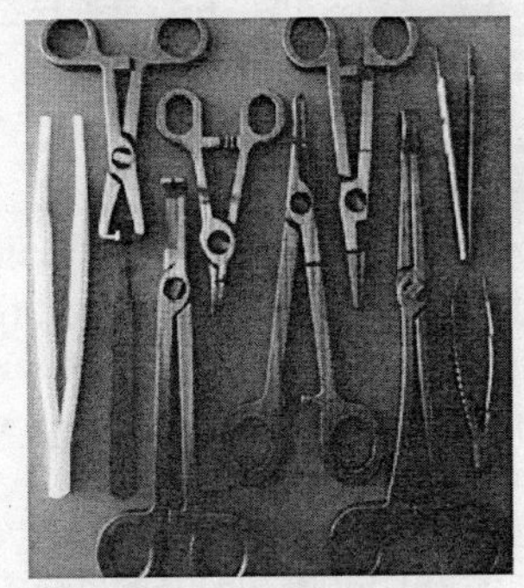

Fig.2.11: 3D Printed Surgical Tools in Space Flight

Courtesy: Louisville Education

CONTINUOUS FILAMENT FABRICATION CFF

Continuous Filament Fabrication (CFF) is a modified system of FFF technology where an additional print head is incorporated the same as FFF technology. The only differentiates is the integration of a second printing head. The second print head usually prints with a continuous strand of carbon fiber, known as (Kevlar) to reinforce the printed layers of polymers.

The continuous strand of carbon fiber, constructed throughout the entire build layers, provides the part with additional practical toughness to function better. Part designers can accurately imbed the carbon strand in specific section in the model to enhance material strength and part performance.

METAL -FUSED FILAMENT FABRICATION (MFFF)

Metal Fused Filament Fabrication (MFFF) is another breakthrough within the FFF technology applicable to industrial and desk top printing. Recently, 3D metal printing has a high worldwide demand. MFFF has already provided substantial cost saving in industrial and commercial applications.

ATOMIC DIFFUSION ADDITIVE MANUFACTURING (ADAM)

The Metal X machine, "Markforged," provides 3D print of a part made of polymer filament, and then permeated with metal powder. The part is then sintered in a furnace, causing the polymer and the metal powder to bond together. The outcome is a robust metal part. This process is referred to as Atomic Diffusion Additive Manufacturing (ADAM).

Similar to Metal X machine is the "Metal DM Studio" desktop. In a like manner, the "Metal DM Studio" (Figure 2.12) 3D prints layer-by-layer green state parts through heating specially extruded and formulated metal rods. The printed parts are then sintered with the polymer. The outcome is a solid metal part.

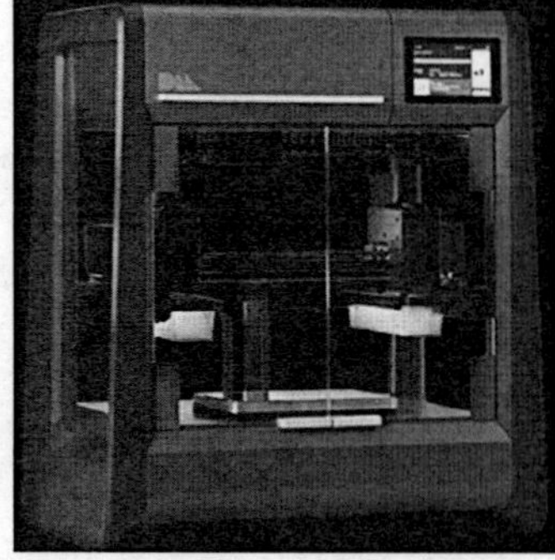

Fig.2.12: Atomic Diffusion Additive Manufacturing (ADAM) - Metal Studio Printer

Courtesy: Proto3000

Integrated Circuitry – Voxel8 DK 3D Printer

The type of multi material 3D printer "Voxel8 DK" deposits conductive metal-filled paste combined with thermoplastic polymer. This allows the 3D fabricated parts to be integrated with electronics. This unique method of 3D multi-material printing enables the "Voxel8" to pint parts with embedded conductive pathways and circuits.

Printing Simplifications

There is an increasing demand for 3D printer technologies in the market. The current printers available are expensive and relatively complicated to use. Apart from being expensive, previous means of producing a prototype took a long time, many specialized tools and skilled labor. Developing a printer which uses rapid prototyping technology would drastically reduce the time, money and man-hours for the production of a prototype. The main motivation is to develop a cost effective solution for printer requirements for laboratories as well as in machine shop. Another simplification factor is to select the most effective materials, and study subjected heat transfer while part is subject to dynamic environment.

Also, developing effective user interface can assist personnel who may not be very familiar with computer software. Since 3D printing is relatively new in printer technologies, it has a greater scope for development in the areas of design and marketing. A successful completion of printing part can lead to the creation of a new product with a promising potential when launched in the market.

3D Physical Models

Design of three dimensional printers is well established by many corporations.

Physical models of early printing machines have played important role in the history of 3D engineering development. Many of these models have progressed to form today's computational printer using new rapid-prototyping technology, which allowed reintroduction of physical models as an intuitive way to demonstrate new mechanical concepts of 3D Additive Manufacturing.

The use of physical models in engineering has had, until the last two decades, a long and useful history. This is especially true in machine design and engineering. Filippo Brunelleschi (1377-1436), the architect and engineer of the Duomo in Florence is known to have created construction models, including machines. Although physical models of machines were prevalent in early exhibitions and universities, their use has been largely replaced today by CAD models and simulations. These computational models are more versatile and of lower cost, but they lose the physical embodiment that is essential for an intuitive appreciation of many critical concepts of motion and force, such as friction, hysteresis, compliance, geometric tolerances, and dynamics

However, new rapid-prototyping technology allows reintroduction of physical models as an intuitive and simple way to demonstrate these fundamental mechanical concepts. People now use the web to integrate both these textual and artifact collections on the history of machines and mechanisms. Besides reading textual descriptions, viewing pictures and videos, and interacting with simulations, individuals may also download, 3D-print and interact with their own fully functional physical replicas. It is expected that as rapid-prototyping becomes more commonly available such forms of documentation will become increasingly prevalent.

Priorities of Building Part

A 3D Printing compatible application can be effectively achieved by evaluating and analyzing a working model of a three dimensional printer, to reach desired tolerance. Two main technologies have been extensively utilized to print parts as close as possible to the desired quality. A 3D Printer utilizing a technology of a molten plastic extruded from an independent extruder. Also, another technology is utilizing powder polymers to 3D print the desired parts. The sequence of selecting 3D printing system and the most adequate materials to print parts is described as follows:

1. Select the technology best suited to obtain the desired tolerance and surface texture
2. Find the best suited material to achieve part quality and fitness
3. Rapid Fabrication of parts and provide continuous improvements

Innovative 3D printing following Design Specifications

An internal mechanism can be incorporated into a 3D printer. A Cartesian robotic arm can be fitted 3D printer. The robot arm can precisely deposit each layer according to the programmed part. The object is firmly resting on a platform that is lowered a certain increment allowing the robotic arm to draw the next layer. A part can be configured to fit the 3D printer envelop of 25″x20″x20″.

The engineering design is a breed of design specifications. The optimum and effective specifications may also be improved by:

(a) Iterative process
(b) Feasibility study,
(c) Conceptual design,
(d) Preliminary design,
(e) Detail design.

The stage of the conceptual design, the customers' requirements plays the fundamental role in developing successful design. Therefore, validating the customers' needs may demand creating several solutions. Each solution is evaluated to ensure optimum part form, fit, and function, at a reasonable cost.

According to customers' demands a feasible solution is developed at the preliminary design stage. The requirements are defined in greater details. At this stage is likely to experience the followings:

(a) Extensive information is collected
(b) Optimal solution can be quantified to deal with the critical parameters
(c) Evaluating product function, strength, spatial compatibility, and the financial impact of the product.

Consequently, the expected product will result from the detail design of a 3D printer of a tested and producible product.

Testing and evaluating Parameters

The final stage is rapid sample testing of the developed product through optimum 3D printer. The test includes the form, dimensions, tolerances, surface texture of the product and all its associated individual parts. Also, test must be conducted to validate specific selected printer materials, as well as determining the valid manufacturing processes.

Nonetheless, alternative solutions to design and fabrication problems, and often lead to several trade-offs associated with each solution. However, a viable solution must be simple, easy to produce with excellent quality at an affordable low cost.

Through the engineering process evaluation the 3D printer developing and fabrication can be considered as a primary tool for evaluation. However, phase development constitutes a fundamental role in product innovation. Phase development assists in developing a 3D printer as a good tool for problem solving. It can easily validate a new design concept. Additionally, a phase development of 3D printer can be used as a visual tool to explain effectively the product's purpose. The collective developments by design teams will be of immense value as a stepping stone to further development by another design team to study the initial constructed unit to accomplish new products.

Team Participation

Intensive care must be taken when organizations adopt 3D printing - rapid development methodologies to avoid role confusion and scattered responsibility. Also, eliminate communication breakdown within the original development team, and client team. As the client participation is vital, it is imperative to request client's representation with equal authority in the development process. Moreover, no increment of the system may be developed without a formal thorough documentation of the final design phase.

The most sensitive issues in innovation are directly dealing with cost, quality, and time-to-market. Accordingly, without designing for affordable and reliable production the rise and fall of the product will rest upon the effective utilization of a 3D printer.

Product specification is the key to the success in the 3D printing product development process. However, the longest consuming activities, is the prototyping task. Design the 3D printer is the most effective time saving iterative process in which multiple iterations may be needed. It is therefore very critical to reduce prototyping and or achieve very rapid prototyping to shorten the entire product development cycle.

Utilizing additive manufacturing printer to construct physical objects automatically is the most effective way to develop varieties of products in short time. Rapid prototyping techniques had blossomed in the late 1980s, as it brought the simple means for producing prototype parts and models. Additive Manufacturing as means of rapid prototyping accelerated the product development process, and increased the product acceptance due to simplification of the early phases of conceptual design. Additionally, 3D additive manufacturing can test and improve design before the manufacture of expensive fabrication and assembly tooling.

Motion Trajectories

Motion trajectories are defined as the controlled path where the tool must precisely follow to form a new space taken out of an object in traditional subtractive fabrication technology or a new material added in 3D additive manufacturing. To create the necessary motion control trajectories to generate the actual additive form rapid prototyping is used in 3D printing, employing additive manufacturing mechanism. Therefore, a 3D geometric model is prepared to be sliced into layers in additive manufacturing. The slices are then scanned into lines. The lines are always produced in 2D drawing. This is the initial generation of the "trajectory" in subtractive technology as in CNC`s tool path. However, in additive manufacturing, the trajectory is identical yet reversed.

During the 1980's the era of the machine tool crisis swept through the USA. American dominance diminished in the field of Machine Tool Manufacturing. Industrial entrepreneurs and United States policy makers were encouraged to have an active role in regaining America's dominance! Substantial efforts had begun to counter the falling trends in traditional CNC and Computer Aided Manufacturing arena. U.S. rapid prototyping corporations and entrepreneurs embraced the worldwide leadership role in this technology. The National Science Foundation (NSF), and the National Aeronautics and Space Administration (NASA), the U.S. Department of Energy (DOE), the U.S. Department of Commerce (DOC) – National Institute of Standards and Technology (NIST), the U.S. Department of Defense (DOF), Defense Advanced Research Projects Agency (DARPA), and the Office of Naval Research (ONR) collaborated to gain the support of the US government.

Measures for 3D Additive Manufacturing

Executives and corporate administrators must consider many vital measures when evaluating the use of a 3D additive manufacturing system. Data correlated to its use and implementation must be acquired and carefully weighted. Often the speed, cost, and quality of 3D additive manufacturing systems are not sufficient to evaluate and adequately justify the implementation of such systems. The following parameters should be considered in creating a matrix of data comparison to reach a prudent decision, Figure 2.13.

Time

Although the speed of the 3D printing plays a vital role in justifying such a system implementation, it is not the only parameter that should be considered. Speed is only one element of the time to produce a ready-for-use part. Instead, investigate the total process time. The actual and unique parts pertinent to the intended user must be evaluated. Good representative samples of the object must be provided and analyzed by creating a simulation routine to envisage the actual time to process the fabrication and the productivity of the part. The following three steps — preparation, fabrication, and mechanization —highlight the protocol of creating the actual simulation routine.

Preparation

Consider both the time to prepare files for fabrication or building and the time to set up the system.

For file preparation, consider how long it takes to manipulate all the build settings, how long it takes to slice and support a file, and how long it takes to prepare the build file. For a low-scale, individual implementation of a prototype, the system may not necessarily require extensive quality and elaborate expenses. For a high quality, production-grade system, an extensive production-grade machine is necessary.

For system setup, determine how long it takes to load materials, complete per-build house cleaning, and warm up the system from a cold start. This process of valuation must not be ignored. Ignored parameters often prove to be vital to the project's success or failure.

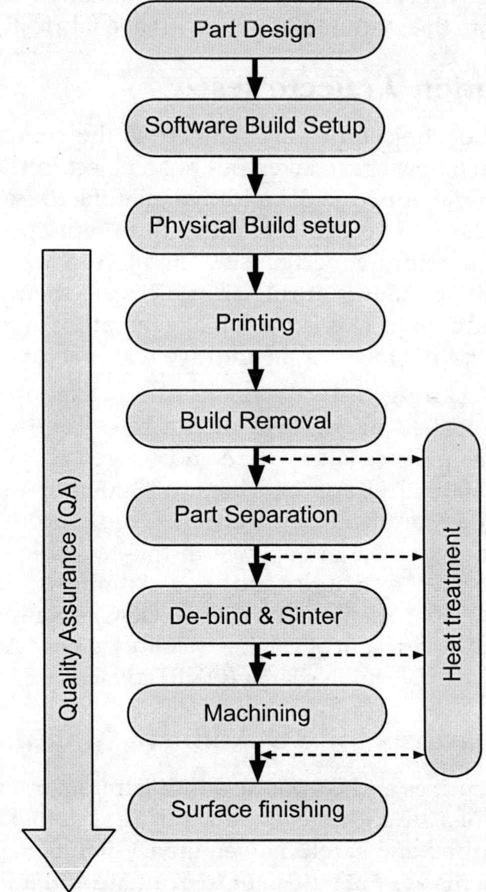

Fig.2.13: Additive Manufacturing Workflow

Courtsey: Digital Alloys

Fabrication

This fabrication step may become cumbersome because it varies with changes in manufacturing parameters and by part size and its complex geometry. Accordingly, the fabrication time must be measured for the typical parts to fabricate with the exact environment, Figure 2.14.

Consider the difference in time when parts fabricated individually, or fabricated simultaneously in groups. Some technologies are fastest when many parts are built in a single run.

Every 3D additive manufacturing printer requires a few steps to prepare a part for use. Evaluate the steps necessary for preparing distinctive parts, and the time required for the task. Examples include support removal, cleaning, curing, and deburring.

The surface texture and the aesthetic factors must be considered to reach the required quality and part appearance for the intended application. Therefore, calculating the time required for sanding, filling, priming, coating, or painting must be considered.

Mechanization

It is well known that 100% automation may not be fully implemented without exorbitant cost. Fabrication of an item undoubtedly would involve a manual involvement process as well as an unmanned process. An accurate measure of the time involved in each process must carefully take place. Attention must be paid to the processes most important to the fabrication operations. A fast process may need hours of direct labor whereas a slower process may need only several or a few minutes. If manpower is unavailable, a slower process that requires less labor may be employed.

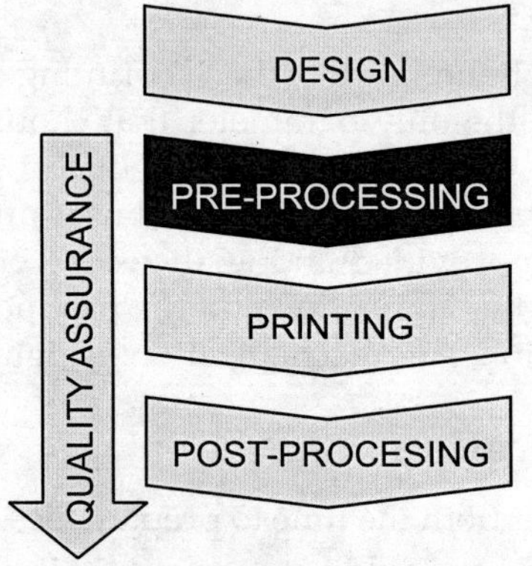

Fig.2.14: The Build Cycle Time–Preprocessing and Post Processing

Courtesy: Digital Alloys

Cost

An evaluation must be based on an unbiased system price. The total cost of start up as well as the ongoing cost of operation must be considered as part of the total cost of producing the model. The cost of tools, peripherals, and incidentals may also be considered in the cost evaluation parameter.

Total Cost of 3D Additive Manufacturing Systems

The principal cost is the base price for the 3D additive manufacturing printer. The additional costs that need to be added are categorized as follows:

1. Cost of hardware and PCs
2. Cost of third-party software
3. Cost of licenses, upgrades, and add-ons
4. Cost of cleaning, curing, de-powdering, de-burring, support removal devices
5. Cost of workstations, hand tools, supplies, vent hoods
6. Cost of training
7. Cost of warranty and periodical maintenance

Cost of Facilities

Although a small 3D additive manufacturing system may require only a 110VAC outlet, a higher-level system usually requires facility modifications to 220VAC or even 550VAC. Additional costs include:

1. Cost of utilities
2. Cost of HVAC, in particular the Cost of humidifiers, temperature controls, and ventilation
3. Cost of noise, vibration, dampening, and isolation
4. Cost of enclosed work area

Cost of Maintenance

There are two types of maintenance cost:

- annual programmable contract maintenance and
- routine maintenance.

The cost of service programmable contract maintenance often includes repairs, preventative maintenance, remote support, and on-line support.

Routine maintenance may be performed on a daily, weekly, and monthly basis. The cost of the tools required to achieve such maintenance tasks must be incorporated.

Cost of Materials

There are several ways to calculate cost of materials. The easiest method is to calculate the actual cost that raw materials can be economically purchased to produce a certain batch. This cost is measure by either $/weight or $/volume. Either way, the cost can be accurately calculated.

Cost of Consumables

There are many different types of consumables that must be used in conjunction with the actual 3D additive manufacturing printers. Some examples are trays, lasers, print heads, light bulbs, and nozzles.

Quality

A fast and affordable 3D printer may not perform according to expectations if the demand of quality and surface finish exceeds the 3D additive manufacturing printer resolution capabilities. Select the optimum printer with the optimum quality requirements. The cost of inexpensive printer often produces very low quality parts. However, it may be satisfactory to utilize such a low cost printer for a part that does not need the demand of a higher quality.

Surface Finish

The thickness of a layer in the 3D additive layer manufacturing system will determine the quality of the part. Traditionally, the resolution of depositing or fusing the material to compose each layer will also contribute to the quality of the layers. Repetitive and combined layers will form the desired surface texture. The desired part must not be achieved with the aid of sanding, grinding, or secondary machining process. 3D printers offer surface finishes from smooth and glossy to rough and textured. Higher surface quality often demands longer processing time to achieve.

Resolution

Resolution is the smallest drop that nozzle can deposit. Over the height of the part, the resolution is a function of the layer thickness as well as the space between each deposited or fused element of material. The resolution value varies according to the types of the 3D printer. In some cases, X-Y-Z resolution is user-definable; in other cases, it is a function of the resolution of motor driving the motions. The smaller the resolution is, the more costly the driving motor will be. The time to deposit each layer as well as the required resolution will ultimately affect the cost of the system.

Accuracy

Accuracy is a function of material, fabrication parameters, part geometry, and part size. In order to evaluate accuracy, typical parts and the intended fabrication configuration must be utilized. 3D scanning equipment may help build accurate configurations. However, the CMMs (Coordinate Measuring Machines) may not reflect the true intended quality. Therefore, when evaluating accuracy, a 3D scanning system may be employed to show best-fit and color-mapping a visual display of the tolerance across the entire part.

Repeatability

Repeatability is an accuracy parameter. It measures the ability of the 3D additive manufacturing system to be used many times to achieve the same expected accuracy each time. Failing to achieve the repeated accuracy will render the system invalid.

Dimensional tolerance must be considered for three conditions stated as follows:

1. Part-to-part — same fabrication, but in different locations
2. Different fabrication methods — same system, but over multiple means of fabrication
3. System-to-system — same job, but run on different machines

Stability

If the parts have a service life of more than a few days, one must consider the effects of time and environmental exposure. Dimensional accuracy and material properties may change over time when parts are exposed to UV light, heat, and moisture.

Operation

Effective implementation of 3D additive manufacturing demands a harmony between the proposed 3D printer capabilities with the company's work flow requirements, procedures, staffing resources, and safety codes.

Staffing

3D printer operations can be decentralized, with self-serve production, or centralized, with technician-run processes. Although there are some 3D printers that can work in either mode, most of them favor one method over the other. If staffing with a technician is required, an evaluation process is needed to determine if the technician should be a dedicated operator or a multi-tasking resource. The budget availability determines hiring the labor that demands for the post-processing efforts.

Environment

Determining the optimum location for the 3D printer is important. It can be an office environment, a lab, or the shop floor. Each one of these locations must be in harmony with the technology used in the 3D system. Independent of the 3D printer's location, shop space is also needed to produce parts to a certain level of quality or to gear up to a production-grade quality.

Harmful Materials

The organization's safety standards must be considered; including the safety of storage, handling and disposal of materials used in 3D printing, post-processing, and part prototyping or scaled up production. Resins and powders may require special procedures. The same safely measures must be applied for liquid agents to clean parts and remove debris.

Administrators and engineers may emerge to create a weighted matrix that emphasizes the importance of each parameter in the 3D additive manufacturing selection process.

Disadvantages of 3D Additive Manufacturing Printers

3D printing is considered a fast and highly efficient means of creating prototype parts to test for form, fit, and function before parts advance to full-scale manufacturing. It is a viable technology in terms of validating parts and ensuring the design and engineering aspects are fully adapted before product advances to production.

However, there are disadvantages in using the 3D additive manufacturing technology. These range from a limited use of materials to the uncertainty of the uses of technology for short-run and long-run manufacturing.

Materials Resources

The 3D printed parts are often built in through additive fashion, layer-by-layer from the ground up. Although the technology is a major process breakthrough, the materials that can be used are still limited. For instance, the 3D printing material of choice is polymers (plastic), because they can be deposited down in melted layers to form the final part. The types of polymers vary in areas such as strength and temperature; therefore, part strength cannot accurately be tested in many cases. Some developers may offer metal as a material, but final parts may not be fully dense. There are several more specialized materials that companies are using to print, such as glass and gold, but such technologies have yet to be commercialized.

Sustainable Accuracy

3D printing is primarily a prototyping technology, meaning that parts created via the technology are mainly test parts. As with any viable test part, the dimensions have to be precise and sustained during the life of the product in order for engineers to obtain accurate information of parts feasibility. Although 3D printers have made advances in accuracy in recent years, many of the plastic materials still come with an accuracy disclaimer. For instance, many materials print to ±0.1 mm in accuracy. This value indicates that there is a possibility for error.

Manufacturing Limitations

3D Additive Manufacturing is highly acceptable technology for creating prototype parts due to its economic benefits. It is an economical and inexpensive means of creating one-run parts for which hard expensive tools are not necessary required. Parts typically are created in hours and changes to the design and engineering of the part can be made in a CAD (computer-aided design) file after the part is analyzed. However, in terms of a manufacturing process, 3D Additive Manufacturing is not a fully recognized as a realistic option in every manufacturing process such as thermoforming and stamping, as several parts are typically made in few seconds instead of few hours as the case in 3D additive manufacturing.

Size Constraints

Parts created additively through 3D printing are also limited in size. For instance, the most affordable, common 3D printing machines typically are small enough to fit on desktop. 3D machines often have built-in chamber sizes of similar proportions. There are 3D printers that are able to create larger parts, but they're much more expens;kive and, therefore, an unrealistic option for many companies. In general, larger the part that needs to be fabricated, takes longer time to create.

3D Printing — Barriers to Adoption

The major limitation of 3D printing that may create barriers to more widespread uptake by industrial users is the capital cost. However, materials, accuracy requirements, and surface finish may play also a fundamental role.

Materials Limitation in Design Engineering

The materials issue continues to be a limitation for design engineers and manufacturing engineers for 3D printing of prototyping applications.

3D additive manufacturing could grow exponentially if engineering grade materials were available more readily. Testing the functionality of new products is a key application for 3D manufacturing implementations.

Higher performing materials would be a significant step forward if low cost materials can be controlled. Typical 3D printing materials such as ABS and PLA suffer from lack of tensile strength, which limits their usefulness for functional prototyping.

3D additive manufacturing applications have to be carefully selected to suit the materials, when surely it should be the other way around. The functionality of products produced using metallic materials has offered satisfactory results for some users. However, they remain cost prohibitive, which keeps the technology limited to low or prototyping requirements.

Accuracy and Surface Finish

Accuracy and surface finish have been the major factors in selecting material and method of fabrication. High accuracy is possible with some of the 3D printing processes. Nonetheless, although very tight tolerances are achievable, the parts that these processes produce are extremely fragile and next to useless for functional testing. When strength and robustness are built into the parts, surface finish is sacrificed and requires specific skills and significant amount of man-hours to finish the part to acceptable levels.

Magical parts cannot happen. Display parts behind attractive glass windows do not provide a true representation of parts with acceptable tensile strength at a reasonable cost. Seeing bicycle chains, figurines, skeletons, or magic gear trains are all attractive, but may be irrelevant to a true high-volume production application.

Industrial manufacturers desire to provide adequate products in form, fit, and function at a reasonable cost and with structural integrity. Also, the products must have a high quality surface finish. The product must accept heat and surface treatment to obtain acceptable material properties, such as hardness and shear strength. Additionally, it should be possible to machine the product to create acceptable screw threads if needed. Until these particular requirements can be met, the applications for 3D printing will remain relatively limited.

DESIGNING MECHANICAL PARTS FOR 3D ADDITIVE MANUFACTURING PRINTING

Minimum Specification for Parts

There are some specifications to bear in mind in order to avoid issues in parts, and to achieve the best quality.

The minimum printable features in planes X, Y, and Z are as follows:

- Minimum hole diameter at 1 mm thickness 0.5 mm
- Minimum shaft diameter at 10 mm height 0.5 mm
- Minimum printable font size for embossed or debossed letters or numbers
- Figure 2.15 an 2.16 6 pt
- Minimum clearance at 1 mm thickness 0.5 mm
- Minimum slit between walls 0.5 mm
- The dimensional accuracy that can be achieved by most 3D printer is ±0.2 mm up to 100 mm and 0.2% above that value, measured after sand blasting.
- Solid part or structural fill allows to print topology-optimized, generative designs or even small lattice structures. This kind of design helps to reduce the weight of the part and the quantity of material used, which not only reduces the cost of the part but also helps to reduce the operating cost in applications that are very weight-sensitive.

Fig.2.15: Emboss 1 mm

Fig.2.16: Deboss 1 mm

- The minimum gap recommended in a lattice structure to ensure that all the material inside the part can be removed is 1 mm, Figure 2.17.
- When printing a cantilever, the minimum wall thickness depends on the aspect ratio, which is the length divided by the width, Figure 2.18.
- For a cantilever with a width less than 1 mm, the aspect ratio should be less than 1.
- There are no specific recommendations for widths greater than 1 mm.
- In general, the minimum recommended wall thickness is 0.3 mm for short walls oriented in the XY plane, and 0.4 mm for short walls oriented in the Z plane.

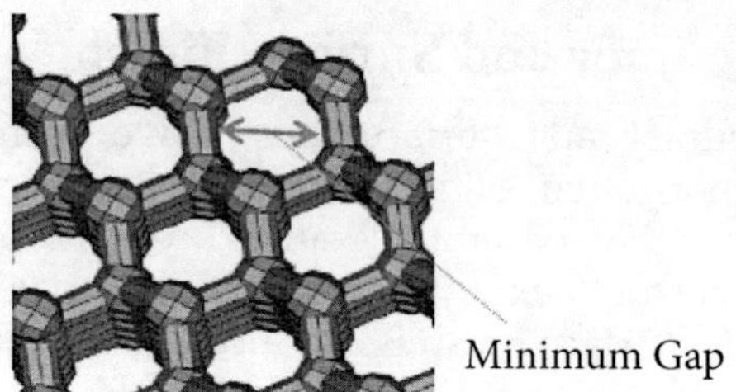

Fig.2.17: 1 mm Minimum Gap

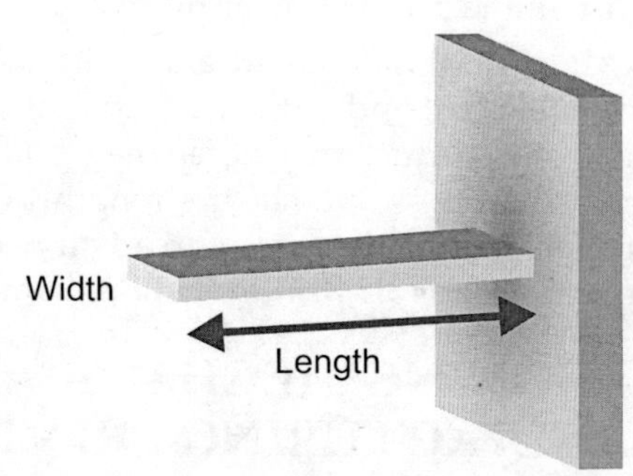

Fig.2.18: Aspect Ratio Equals Length/Width

For parts with a high aspect ratio, you are recommended to increase the wall thickness, or add ribs or fillets to reinforce the part, Figure 2.19.

Very thick walls can accumulate heat and cause spot shrinkage in dense areas with an accumulation of material.

- Minimum gap between parts to be assembled after printing. Sometimes a pair of printed parts needs to fit together for the final application. In these cases, you are recommended to have gaps of at least 0.4 mm (±0.2 mm of tolerance of each part) between the interface areas that should fit together, in order to ensure correct assembly, Figure 2.20.
- Minimum spacing and clearance between parts printed as assemblies. Assembly parts that are printed together should have a minimum clearance of 0.7 mm. Parts with very thick walls above 50 mm should have a higher gap in order to ensure proper performance.
- Hollow closed geometries. Multi Jet Fusion is a process in which the parts are built by selectively fusing the desired areas of a material-based layer. If the parts printed are hollow, drain holes need to be added to the design to remove the material. The minimum recommended diameter of the holes is 2 mm. You are recommended to include at least two holes.

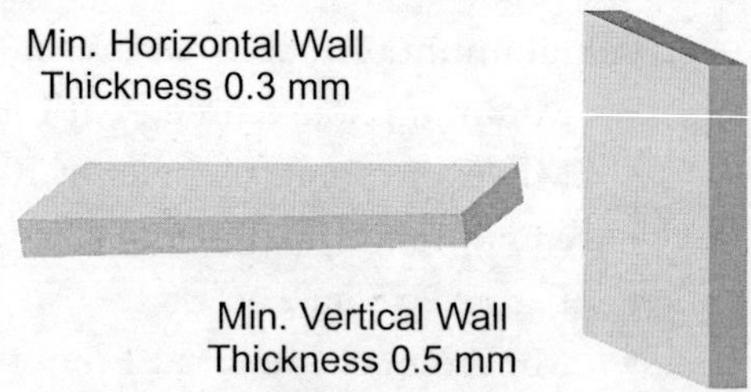

Fig.2.19: Minimum Vertical and Horizontal Wall Thicknesses

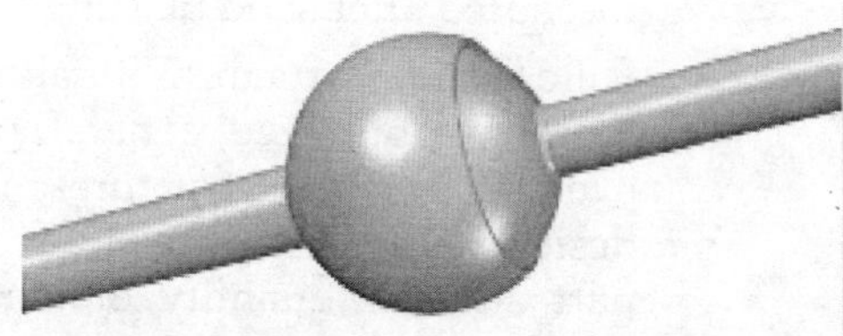

Fig.2.20: Minimum Spacing between Parts

- Glue lines. Parts larger than the maximum build size can be printed with Multi Jet Fusion by splitting them into different parts. They can then be joined together by gluing, welding, or by pin inserts.

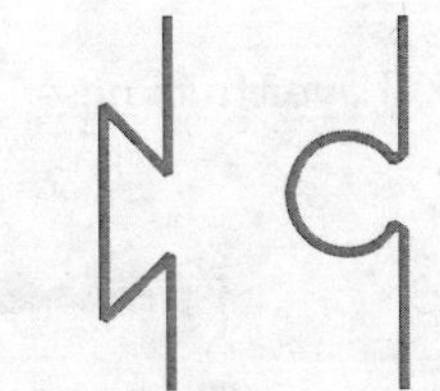

Fig.2.21: Interlocking Design Dovetailing Components together

If you plan to glue parts together, you are recommended to include interlocking features such as those shown in the pictures below: as a guide to position the parts, to help them to bond together, and to facilitate the gluing process. Remember to leave an additional space of 0.1–0.2 mm between parts for the glue, in addition to the minimum spacing between parts printed as assemblies, Figure 2.21.

- Ducts. To remove material from narrow ducts, consider designing and printing a strip or a chain through the duct. When the parts have been printed, you can pull out the chain to dislodge most of the material. Any remaining material can be removed by the normal cleaning process, Figure 2.22.

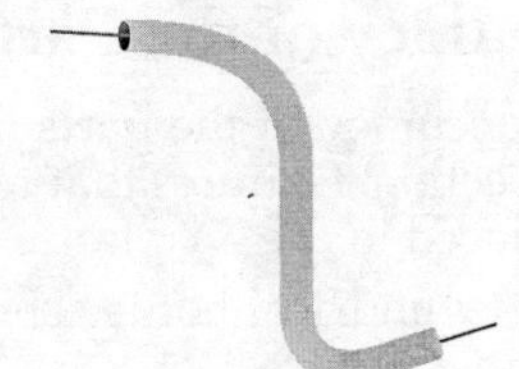

Fig.2.22: Duct Incorporating a Strip or a Chain

OTHER CONSIDERATIONS

Contraction coefficient

The technology of the process produces a small contraction of the fused area in each layer. For the best possible accuracy, the parts should compensate for this contraction to be as close as possible to the correct size, and this is already done by the "HP SmartStream," "3D Build Manager," and by plug-ins for "Magics" and "Netfabb."

Further adjustments of the material contraction coefficient can be done through the "HP Smart Stream" "3D Build Manager" or any other similar commercial software.

Orientation

There are several considerations to be taken into account when the orientation of the part is defined:

- Aesthetics
- Part curl
- Part strength
- Surfaces that need a higher level of detail
- Accuracy of the part

Aesthetics

Stepping is a phenomenon that appears when printing in Multi Jet Fusion. Multi Jet Fusion is a layer-by-layer process and, in some cases, the layers that form the part can be detected in the finished part.

In order to avoid stepping with curved or sloped surfaces, you are recommended to avoid whenever possible positioning those surfaces with angles less than certain values with the horizontal (XY plane), depending on the thickness of the layer. The minimum and maximum thickness values recommended for each layer are shown below, Figure 2.23.

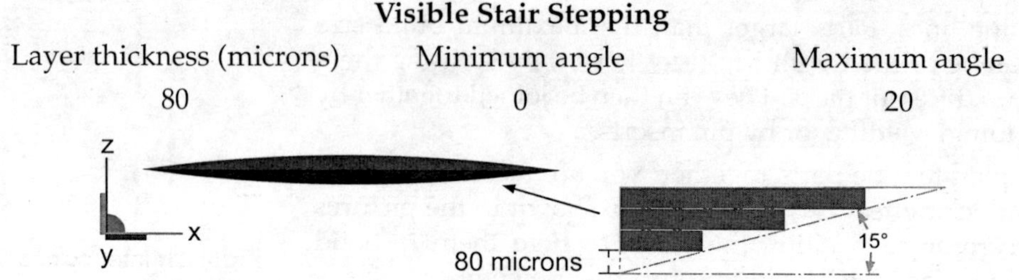

Fig.2.23: Avoid Stepping with Curved or Sloped Surfaces

Accuracy of the Part

The accuracy of the parts is better in the XY plane than in the Z direction. Features that require the highest resolution should be placed in the XY plane.

Cylindrical bodies should be oriented vertically as far as possible, Figure 2.24.

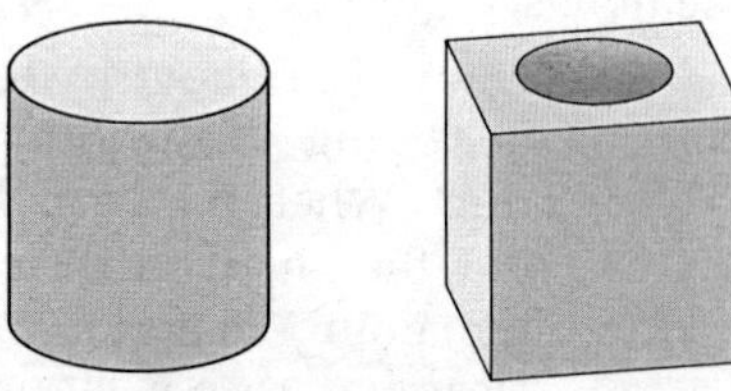

Fig.2.24: Cylindrical Objects should be Oriented Vertically

Engraving

The best orientation for embossing letters is to place them upside down in the build chamber, while for debossing letters it is better to place them face up to achieve the best resolution.

Warp Minimization

For best print quality and to minimize warping, long flat surfaces should be positioned flat in the horizontal plane. This is especially recommended if the parts are long and thin. Place the parts as low as possible in the build chamber, Figure 2.25.

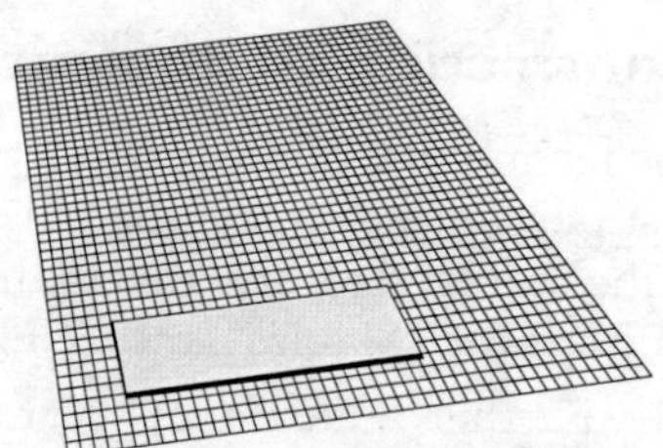

Fig.2.25: Position Object on Flat Surface to Avoid Warping

For critical parts, you are recommended to place them in the center of the bed to get the most uniform cooling. Minimization of the height in the Z plane will also help to avoid warping.

There are other design considerations that can be used to reduce warping, such as increasing the thickness of the part, or adding ribs in the areas that may be affected.

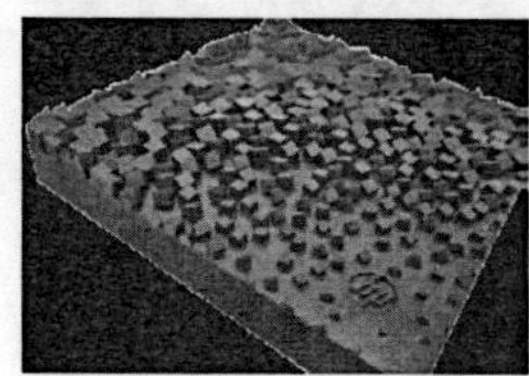

Fig.2.26: Sharp Surfaces Facing Upward

Curved surfaces

Smoothness is best achieved with surfaces that are upside down in the build chamber. If a part needs a particularly smooth surface, it should be positioned upside down.

Sharp surfaces

Sharp surfaces should be printed facing up, Figure 2.26.

Mechanical properties vary with orientation

The best mechanical properties of a part are achieved in the XY plane (horizontal). Any feature such as a pin or clip that needs to hold a load should be positioned horizontally whenever feasible.

Positioning of parts in the build chamber

If the part needs to be as accurate as possible, avoid positioning it near the sides of the chamber. Solid and massive parts are best placed close to the center of the print bed.

Repeatability

For maximum repeatability, try to keep the same printing conditions.

Distance between parts in the build chamber

The distance between parts should be at least 2 mm for the best printer stability.

Printing a part for final production

If a part is being made for final production, first try a small job with the part printed in different orientations, so that you can select the best one(s) to use for volume production.

Orientation flowcharts

The best orientation of a part in the build chamber depends to a great extent on the final use of the part and the properties that you want to optimize. The main attributes that can be optimized are mechanical properties of certain areas, accuracy of specific features, and the look and feel of the part.

Depending on the objectives, the following workflows are recommended, Figure 2.27.

Optimize Mechanical Properties

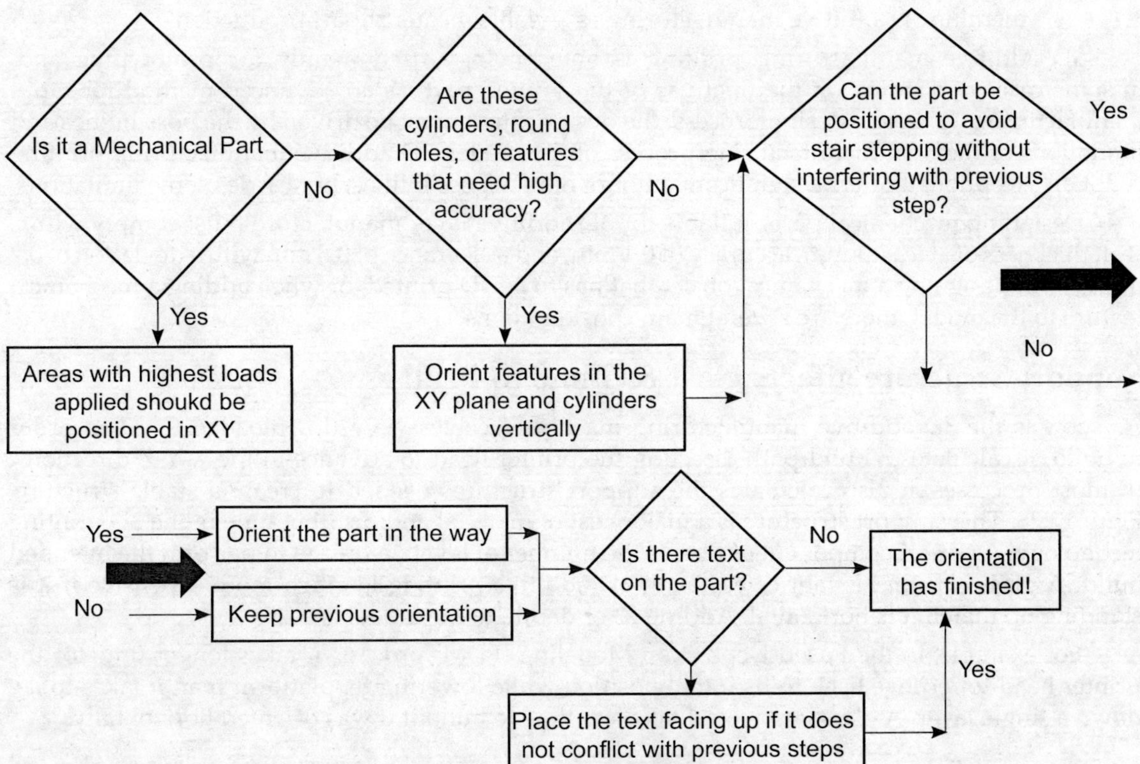

Fig.2.27: Optimizing Mechanical Properties Diagram

CONSIDERATIONS DESIGNING MECHANICAL PARTS

Engineers are faced with mechanical challenges to use 3D additive manufacturing as means of part production. Accuracy and clearance affect the design of parts in 3D additive manufacturing printing. Design engineers must consider the following:

1. Maintain a wall thickness of at least 1 mm.
2. Maintain an accuracy of 0.1 mm.
3. Keep a clearance of 0.6 mm.
4. Allow a deviation of 0.2 mm.

When designing a mechanical part produced through 3D additive manufacturing, the following main constraints must be considered:

1. Clearance is vitally important when designing mechanical parts for 3D printing.
2. Allocate space when parts are required to be cleaned.
3. Design considerations must accommodate the removal of trapped powder in enclosed spaces.
4. When two surfaces are printed in the same orientation that rubs against each other, they will cause substantial friction.
5. A coil spring fabricated by additive manufacturing printing will be much weaker than a traditional spring.
6. Paint changes the mechanical properties of the part.
7. Printed models may be manually corrected, which may defeat the purpose of implementing 3D additive manufacturing as a viable means of part production.

3D additive manufacturing printing is a technology used mainly for prototyping and, in some cases depending on the final use of the printed part, as an advanced method for rapid manufacturing. As in any design process, the design intent must be driven by the possibilities and limitations of the final manufacturing process of the model. 3D additive manufacturing printing is not an exception; it offers a significant amount of new possibilities, but carries some limitations.

Designing mechanical parts suitable for 3D additive layer manufacture must comply with a wall thickness of at least 1 mm, accuracy of 0.1 mm, and a clearance of 0.6 mm with a deviation of 0.2 mm. When designing a mechanical object that has to be 3D printed, or when adding a mechanical feature to the model, there are several main considerations.

Support Structure affects Cost & Time to Print

As soon as the 3D additive manufacturing machine receives via a downloaded CAD an order to build, it calculates a build path directing the printer head to advance in the X-Y-Z directions. In most processes, it also calculates the support structures needed to create a stable structure, Figure 2.28. This support structure is a major cost of material and printing time in the 3D printing technology. A critical component of time is the number of layers required to perform the intended build. A glass cup that is fabricated using 3D additive printing is much more expensive if it is standing up than if it is horizontal. Adding Z, or depth, to the model is costly.

For example, in the Fused Deposition Modeling (FDM) process, it takes longer time for the printer head to retrieve back to its initial position while lowering its platform than it takes to lay down a single layer. A glass cup would take less time to build if it was oriented horizontally.

However, the glass cup may not be the only object that will be printed in that print run. Usually a cluster of items are printed in the same print run. The 3D additive printer is tasked to calculate the most efficient path to print all the items in one run. Even though it would generally be less expensive to print a single glass cup vertically, it is unlikely that all the glass cups will be oriented vertically.

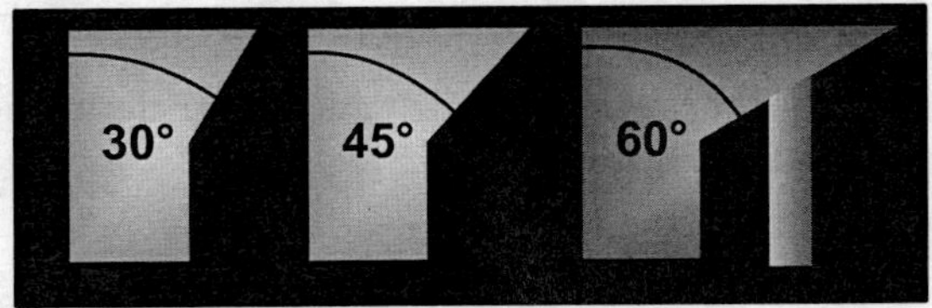

Fig.2.28: Model Orientation Rule Requiring Structural Support for Overhanging Walls

The price is based on the volume/weight of an object model. Therefore, the actual pricing depends on a host of different factors. The greater the volume/weight, the more cost there is to print the part. The total cost is affected by the complexity of the part, the time to clean it, the time to orient and assemble it, the handling, and the costs of preparations.

Clearance

The amount of clearance depends on the functionality of the part. The higher the clearance, the more likely the part fits. However, when developing complex moving mechanical parts, an increase in clearance leads to greater inefficiency in the part. An acceptable balance between clearance and efficiency is crucial to the design. For a gear, a gearbox, or a spring, it is recommended to have as minimal clearance as possible. However, when designing any other assembly, it is recommended to maintain more clearance. The minimum recommended clearance is 0.6 mm. Maintaining less clearance than 0.6 mm may cause the parts to fuse.

The reason clearance is especially important when designing parts and 3D printing a part using another technology such as 'Selective Laser Sintering (SLS)," ass discussed below. Utilizing using SLS technology is enabling the Laser to melt the metallic powder, causing it to fuse. Also, shrinkage will lead to tighter clearance in this type of technology, which may also fuse the parts together.

SELECTIVE LASER SINTERING (SLS)

It was developed in the mid-1980s. SLS technology is the most popular technology system. SLS fuses small plastic, metal, ceramic or glass powder particles into a desired 3Dimensional shape using a high power laser. It uses digital data such in the form of a CAD file or scan data to generate cross sections over which a laser is selectively scanned to fuse the powdered material, Figure 2.29.

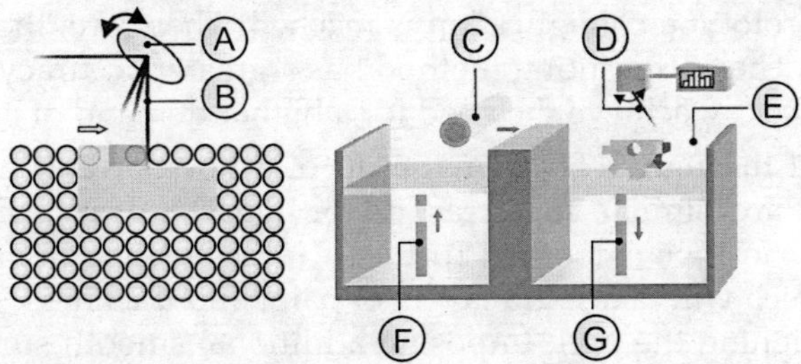

Fig.2.29: Selective Laser Sintering Machine – Courtesy of Ricoh UK Products Limited

A. A:Lens
B. Laser beam
C. Re-coating roller
D. Laser beam
E. Powder bed
F. Powder feed piston
G. Build piston

Once the current layer has been scanned completely, the powdered bed is lowered by a thickness of one layer which allows for deposition of new material on the top following the same repeated procedure.

The following are some advantages of SLS technology over other contemporary technology

1. It can easily make complex geometries because of its compatibility with digital data such as CAD or CAM
2. It has rapidly progressed in limited-run manufacturing to produce end-user parts
3. SLS does not require support structures because of the absence of any sintered material when the part is being constructed, which is not the case with other technologies like Stereolithography and fused Deposition modeling
4. SLS is able to produce parts using a wide range of commercially available materials like polymers, nylon, also glass-filled or with other fillers, s well as polystyrene, metals, steel, titanium, alloy mixtures, composites, and green sand
5. SLS can use different methods to physically melt, such as full melting, partial melting, or liquid-phase Sintering

Using this process, it is possible to achieve a very high density, unlike the contemporary technology. Also, it allows for high productivity in many configurations by packing a large number of parts within the powder bed.

Material Properties and Limitations

The material used in 3D additive manufacturing is also called Selective Laser Sintering or SLS material. Different 3D printing materials and processes have very different material properties, limitations, and strengths. Polyamide12 powder is a white, strong, and flexible material that is selectively laser sintered. The fine powder is applied layer by layer. A laser then hardens the part of each layer that is part of the model. The model is then taken out of the resulting block of powder.

PROCESS LIMITATIONS — SELECTIVE LASER SINTERING (SLS)

SLS is one of the most accurate 3D additive manufacturing printing processes. The process has a layer thickness of 0.1 mm. This is the thickness of each new layer added to each part until the final thickness and the prototype configuration is reached. Therefore, the maximum accuracy on any part will be 0.1 mm. The SLS printing method has a greater accuracy than 0.1 mm. Thus, the expected produced surface can achieve the need for a higher demand of end-user accuracy.

The parts produced through SLS are made up of a powder and have a powdery smooth texture. However, they are similar to terracotta or limestone appearance and may not be completely smooth, nor can it be smoothed further. To reach smoother surface the part may be painted or varnished, which will take extra layers of paint and thickness. The total thickness must be considered when designing the part. Imposing additional smooth surface texture beyond the initial produced surface will add to the final cost of the part.

HOLES AND SELECTIVE LASER SINTERING

Selective Laser Sintering is a very accurate process, but due to the layered construction, there are some specific limitations with holes. If a hole is constructed vertically, it will need too many layers within the diameter of the hole. This may render irregular hole tolerances. Larger holes of 1 mm or more will be more accurate than smaller holes. The wall thickness of a hole is also a factor. Holes for 1 mm thick walls are more accurate than holes for 3 mm thick walls.

Shrinkage Problem

It is evident that all parts printed through the SLS process shrink. The shrinkage depends on the ambient temperature and humidity. The machine manufacturer realized this fact and made provision by making the print larger than it should be to allow it to shrink to the right size.

STEREOLITHOGRAPHY - 3D PRINTING TECHNOLOGIES

There are three separate reliable technologies applying 3D additive manufacturing. They are:

- stereolithography,
- inject-based liquid processing, and
- extrusion-based processing.

The stereolithography technology was the first commercialized "Rapid Prototyping" process, and is the representation of the stereolithography operation. It was invented in 1986; stereolithography concept started the rapid prototyping revolution. It works on the principle of solidifying a photosensitive resin using UV laser light layer-by-layer to develop a 3D object. Stereolithography uses a photo-curable resin that can be classified as an epoxy, vinylether, or acrylate. Acrylics only cure about 75% or 80% since curing stops as soon as the UV light is removed. Epoxies continue to cure even after the laser is not in contact. The device as shown in Figure 1.30 consists of a platform that is moved down as each layer is formed in the tank containing the resin.

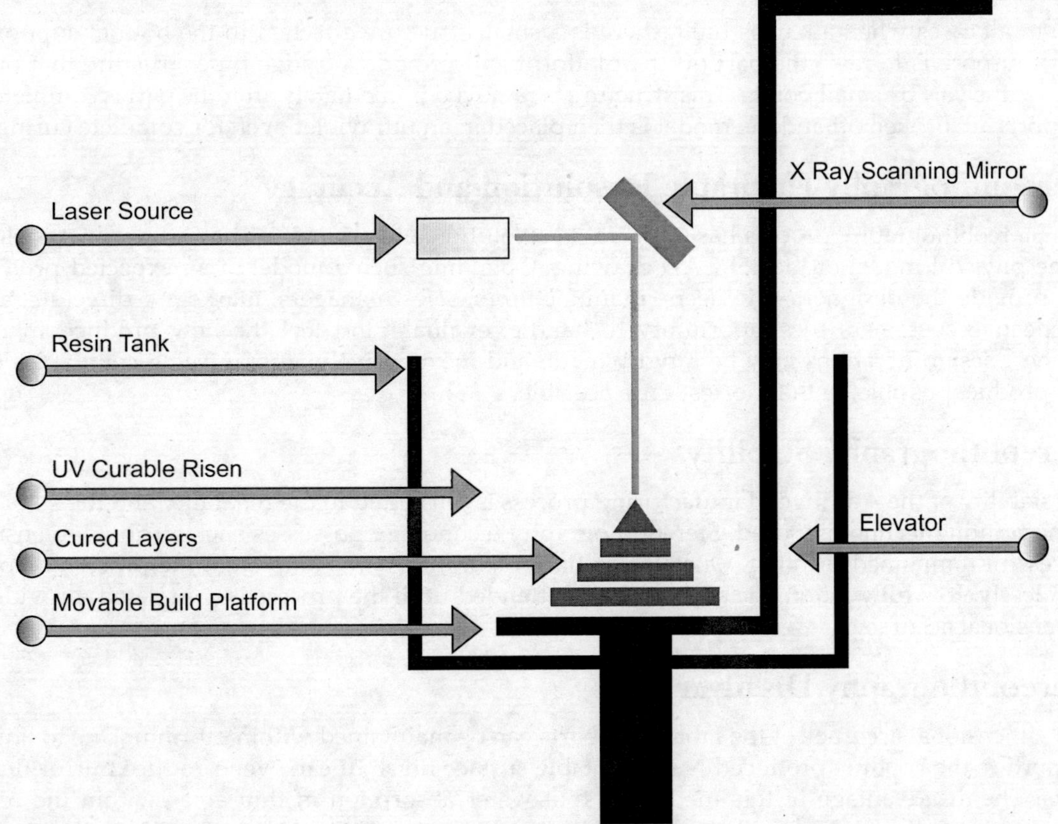

Fig.2.30: Stereolithography Machine

Courtesy: All3D Corporation

The laser light is moved in the X–Y plane by a positioning system. In some cases a support structure has to be created to support the overhanging parts. The stereolithography process converts 3D computer image data into a series of very thin cross-sections, much as if the object were sliced into hundreds or thousands of layers. A vat of photosensitive resin contains a vertically moving platform. The part under construction is supported by the platform that moves downward by a layer thickness. It is typically 0.1 mm for each layer. The device as shown in Figure 2.30 consists of a platform that is moved down as each layer is formed in the tank containing the resin.

A laser beam then traces a single layer onto the surface of a vat of liquid polymer. The ultraviolet light causes the polymer to harden precisely at the point where the light hits the surface. The model is built upon a platform situated just below the surface in a vat of liquid epoxy or acrylate resin. A low-power highly focused UV laser traces the first layer solidifying the model's cross-section while leaving excess areas liquid.

PLATFORM GENERATING LAYERS IN THE Z DIRECTION

The UV laser is controlled by a galvanometer scanner to generate X–Y motion, and thus the table does not need to move in the x and y directions. Next, an elevator incrementally lowers the platform into the liquid polymer as the laser is tracing from the left hand side. A sweeper re-coats the solidified layer with liquid, and the laser traces the second layer atop the first. This process is repeated until the prototype is complete. The solid part is then removed from the vat and rinsed clean of excess liquid.

In all cases when a part is built, there is a small structure attached to the bottom supports. Their purpose is to raise the part off the platform and provide a bridge type structure that only touches the part by small points. This structure is removed immediately after the part is completed. Supports are broken off and the model is then placed in an ultraviolet oven for complete curing.

Stereolithography Favorable Resolution and Accuracy

This Stereolithography process has a favorable resolution and accuracy. The end product is close to the physical model of the 3D CAD drawing. Obtaining such a model of an expected product will provide the designers, engineers, manufacturers, sales managers, marketing directors, and prospective customers the opportunity to handle, evaluate and feel the new product. In this manner, design iterations may be rapidly made and inexpensively, guaranteeing companies the best product possible, in the shortest time possible.

Stereolithography Stability

The stability of the Additive Manufacturing process is attributed to the reliability and the stability of the printing technology used. Stereolithography technology possesses such attribute to ensure successful unmanned printing. Once the initial preparations are completed, the process follows seamlessly. It is fully automatic and can be unattended until the process is completed, providing dimensional accuracy.

Stereolithography Disadvantage

The dimensional accuracy of the fabricated parts can be maintained within ±0.1mm. Due to liquid properties, the product produced has acceptable surface finish. It can even produce mirror finish parts. The disadvantage of this method is that water absorption in thin areas within the resin evaporate over time, which will result in curling and warping the produced part. Also, the system cost is relatively high, and the material available is only photosensitive resin. Accordingly, the parts produced by this technology may not be used for durability and thermal testing.

Usually, the parts in most cases have not been fully cured by the laser inside the vat. When a laser is curing a spot, the energy is forming a cone shape. During processing there are some uncured regions throughout the part. Thus a post-curing process is normally required. The costs of the resin and the laser gun are expensive. Furthermore, the optical-sensor requires periodic fine-tuning in order to maintain its optimal operating condition, which will be also costly. The labor requirements for post-processing are relatively high especially the cleaning process.

THE FUSED DEPOSITION MODELING – FDM

The Fused Deposition Modeling (FDM) process was originally developed by Advanced Ceramics Research in Tucson, Arizona, but the process has been significantly advanced by Stratasys, Inc. of Minneapolis, Minnesota. FDM is a non-laser filament extrusion process that utilizes engineering thermoplastics, which are heated from filament form and extruded material in very fine layers to build each model from the bottom up.

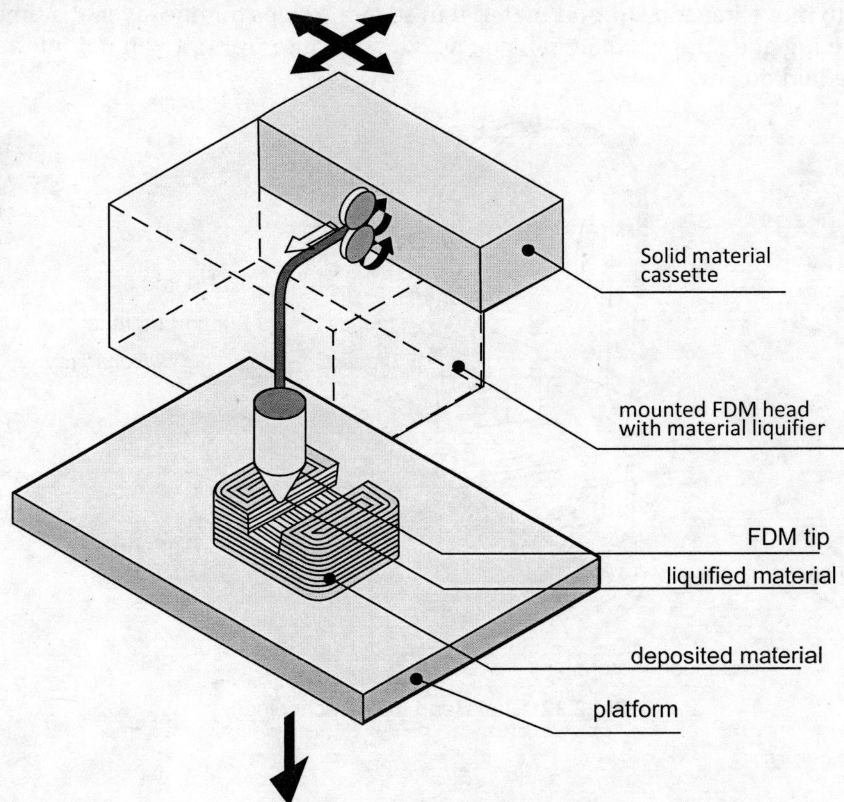

Fig.2.31: Stereolithography Machine

Courtesy: CREAX Corporation

EXTRUDED FILAMENTS OF MOLTEN THERMOPLASTIC

The models can be made from acrylonitrile butadiene styrene (ABS), polycarbonate, polyphenylsulfone (PPSF), and various versions of these materials. Furthermore, the models are tough enough to perform functional tests. The extrusion-based process utilizes filaments of molten thermoplastic that are extruded from a heated tip to build up layers comprising the physical model. Figure 2.31, is a depiction of the head and process in which the material is pulled off the spool, heated just above the melting temperature, and deposited at the desired location. The key steps begin with the filament being fed into the drive which forces the filament into the liquefier. Then the heater block melts the filament and the solid filament is used as a linear piston. Finally, the melted filament is forced out through the tip.

The material used is fed into the head in solid wire form and then liquefied in the head and deposited through a nozzle in liquid form. The extrusion head is able to move in the X–Y plane and is controlled to deposit very thin beads of molten material onto the build platform to form the first layer. The platform is maintained at a lower temperature to ensure the deposited thermoplastic hardens quickly in 0.1second. After the platform lowers, the extrusion head deposits a second layer upon the first. The material then cools and solidifies in place. The speed of the drive wheels can determine the width of the extrusion path that is controlled using the software. The build process lays down both modeling and support material in separate steps for one layer at a time. To switch between modeling and support material, one nozzle will raise up so it will not interfere with the material being laid down.

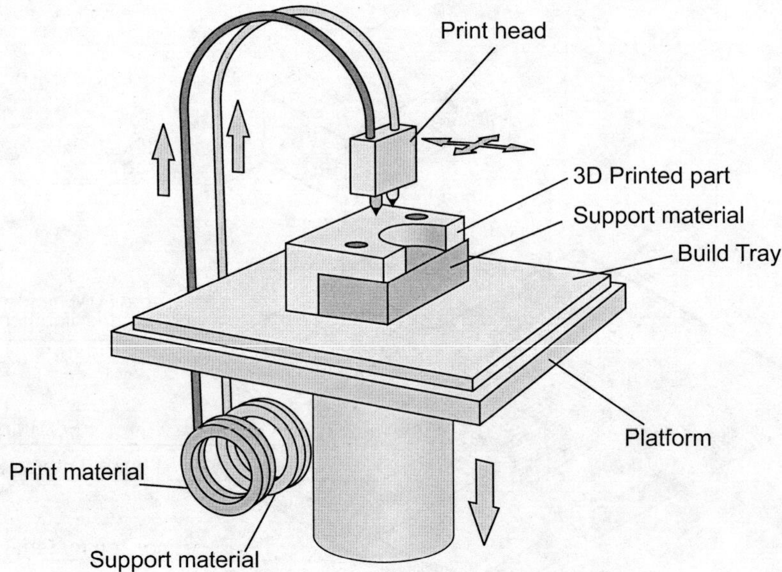

Fig.2.32: Dual Head FDM Technology

Courtesy: dddrop Corporation

Displacement of Z-Axis

The appropriate displacement of Z-axis movement is determined by a setting within the software. The heads are moved in the X–Y plane by a set of high solution linear motors. Figure 2.32, shows a typical extrusion-based process, where plastic filament is used as the material for deposition and is supplied from a coil. This plastic filament passes through an extrusion nozzle that is maintained at a temperature high enough to melt the plastic filament as shown. The nozzles in Figure 2.32, are attached to an X–Y positioning systems or the stage system. As the extrusion nozzle melts the plastic film, the stage moves the nozzle, depositing the material in a layer. Once a layer is complete, indexing is completed in the Z-direction and the process is continued.

FDM Process

In the FDM process, filaments of plastic modeling material and soluble support material are fed from auto-loading carriers in the material bay up to the extrusion head. There, the materials are heated to a semi-liquid state, forced through dual extrusion tips, and precisely deposited onto the modeling base in extremely fine layers, Figure 2.32. The print head moves in X-Y coordinates, and the modeling base moves down the Z-axis as the model and its support material are built from the bottom up, layer by layer. After the build is complete, the support material is removed, and the model is ready to use or finish, if needed.

3D Printing technology is often faster and less expensive than other rapid prototyping methods; it is especially valuable when creating concept and working models early in a design process. Sending a design to an outside service bureau can cost hundreds of dollars or more, whereas a 3D printer using FDM technology can produce the same part in hours for a few dollars.

The small size of 3D printers functions as office equipment that stays within a department. By comparison, large rapid prototyping systems often need to be located centrally and run by a dedicated staff of experts.

Quicker Time to Market

The longer a product stays in the design cycle, the longer it takes to get to market, meaning less profit for the company. Time-to-market considerations were identified as the most critical daily issue facing corporations during 2008-2019. In addition, the process of prototyping presented a time-to-market obstacle.

With increasing pressure to get products into the marketplace quickly, companies are compelled to make quick and accurate decisions during the conceptual stage of design. These decisions can affect almost 80 percent of the product's total cost by establishing material selection, manufacturing techniques, and longevity of the design.

At Graco, Inc., engineers used a "Dimension" 3D printer to experiment with various paint gun and nozzle combinations to create the perfect spray pattern and volume, and to expedite speedy product-to-market. Graco makes paint spraying and texturing equipment for professional use. It designed a spray texture gun based on functional models built with ABS plastic parts from a 3D printer.

FDM Creates Thermoplastic Layers

The Fused Deposition Modeling FDM process was used to reproduce several kinematic models. The process creates a sequence of thermoplastic layers from a filament wound coil that is heated and extruded through a nozzle. The trajectory of the nozzle is derived from the triangle mesh, to scan and fill solid volumes. In order to create functioning mechanisms, a second, water soluble release material is placed in the gaps between the movable parts.

Rapid-prototyping systems, both electric and mechanical, work by incrementally and selectively depositing material from a source onto a substrate. The machines read data describing slices of a computer model and, using one of several methods, lay down successive thin thermoplastic layers of liquid or powdered polymers, ceramics, or metals.

FDM COPY AND ITS ORIGINAL

One of the major differences between the FDM copies and the originals is the clearance between moving parts. The printing process cannot directly leave a zero-width gap. Instead the process either leaves an air gap, if possible, or puts in waste material: Water-soluble material that is later washed away, or for stereo lithography processes, wax-material that is melted or etched away, or un-sintered particles in selective sintering processes. The cle8arance between parts is on the order of 0.4 mm in our case, a rather large gap – an order of magnitude greater than the originals. The large gaps affect the rigidity of the mechanism, add hysteresis effects, and become detrimental when tight surfaces are critical for functionality, such as in pumps. This necessity for large gaps originates from several reasons:

Advantage of the Extrusion Process

An advantage of the extrusion-based process is that it does not waste much material during or after producing the part. Also, it is easy and safe to use. The materials are inexpensive. The system may fit on a desktop to produce non-fragile parts. The extruded based process may use water-soluble support material.

However, it should be noted that the water requires heat, circulation, and sodium hydroxide to dissolve the support structure. The ABS materials used are cost effective. Also, the materials may be easily changed. The binders are relatively inexpensive. Due to nozzle diameter accuracy can be limited compared to the liquid-based processes. Build parts with complicated details is difficult. This method of printing has few limitations such as limited materials, limited size, and unpredictable shrinkage.

EVALUATING FUSED DEPOSITION MODELING (FDM)

In order to validate the 3D Additive Manufacturing system (FMD) several practical and fundamental characteristics of the technology must be carefully studied, which includes design specifications, constraints and other considerations which incorporates also cost analysis, different types of materials used, and dimensions of the parts used to make the 3D Additive Manufacturing Printers.

From the evaluation and analysis of the printer technologies available in the market and their respective advantages and disadvantages, a focus on two technologies which have offer a promising future in development and design of high speed 3D construction. The first technology is the Fused Decomposition Modeling (FDM) concept.

The (FDM) model printer can be used for the creation of 3D objects. It uses computer based models to fabricate 3D objects applying Additive Manufacturing technology. This machine offers self-replicating ability which allows for creation of complex structures and artifacts without the use of industrial infrastructure, which also highlights its economic potential. The other technology is a device which operates like a 2- dimensional printer in the X-Y directions. The additional Z dimension will allow it to move about the vertical axis, which can lead to the creation of 3Dimensional final product. The greatest challenge for the development of this device is the type material which has to be used in building the product.

Layer Resolution – Fussed Deposition Modeling

High-end fused-deposition-modeling (FDM) systems can produce parts with layer resolutions down to 0.005 in., delivering parts that may not be as smoothly as injection-molded parts. However, they have an overall accuracy or tolerance of ±0.0035 in. or ±0.0015 in./in. This falls easily within the average tolerance for an injection molding task, which is typically 0.005 in.

If the additive manufacturing machine's most important requirement is to produce a part with a "class A" surface finish, or sharp edges, then a high-resolution system may be necessary. But smooth surface finishes or sharp edges do not equate to accurate parts.

In general, wall thickness can conservatively be 2 mm; however, it must be 1 mm at the minimum. Wall thickness may appear unimportant. Nonetheless, when designing complex mechanical assemblies such as gearboxes, wall thickness becomes a major determinant for obtaining stable part structure.

Orientation and Order to Build

Orientation is the method in which the model is oriented inside the construction area of the 3D additive manufacturing printer.

DESIGNERS' ACCURACY, REPEATABILITY, AND RESOLUTION

In the realm of additive manufacturing, there are distinctions among accuracy, repeatability, and resolution. Accuracy describes how closely a manufacturing machine's output conforms to a tolerance within a specified dimensional range. Repeatability captures the equipment's capability to produce consistent output, time after time. And resolution refers to the smallest measurement the machine can reproduce.

These concepts are second nature to designers and manufacturing engineers. Nevertheless, there is a widely held misconception about these measurement concepts as they relate to additive manufacturing. Resolution does not, in fact, translate directly to a system's overall accuracy. One simple way to visualize the real-world distinction between the two concepts is to imagine that two measuring sticks of differing length are both marked as 12-in. long, yet the second length is actually 1-in. shorter. The first length is divided into 1/16-in. increments, and its true length is verified at precisely 12 in. Even though the shorter stick reads "12 in.," it is verified to be only 11-in. long. But this shorter measuring length is divided into 1/32-in. increments, which is twice the resolution of the 12-in. measuring length.

The 12-in. ruler with 1/16-in. increments exhibits lower resolution but high ultimate accuracy. The 11-in. ruler with the finer increments does the opposite: It exhibits high resolution but low absolute accuracy.

The same is applicable for additive manufacturing equipment. Some machines make parts like the first ruler with finely resolved features, but lower overall dimensional accuracy. Other machines turn out parts with lower, but acceptable, resolution and excellent accuracy.

When additive manufacturing was in its infancy and used primarily for model making, the distinction between accuracy and resolution did not matter much. It was enough for an early "rapid-prototyping" machine to turn out models that had approximate dimensions at best, as long as the model met the cosmetic goals. Users just needed a model to paint and demonstrate to their marketing department.

Today, much more is required of additive manufacturing systems. They routinely turn out functional prototypes, fixtures, or finished goods that must meet the same stringent accuracy and repeatability standards associated with traditional manufacturing methods such as machining, injection molding, and casting.

Resolution's Role

The confusion between accuracy and resolution is understandable given the incremental way that additive manufacturing machines create parts from CAD models. Some systems build parts from fused layers of a thermoplastic material. Some machines build parts from layers of a photopolymer that have been cured by a light source. Others bind or sinter layers of powdered materials, ranging from starches to metals. Regardless of the build method it is enticing to equate the height of the incremental layer and the width of the smallest feature with system accuracy. The finer the resolution, the more accurate the part is.

Resolution Becomes Critical

Resolution matters greatly. As it relates to accuracy, resolution becomes critical only when part feature size becomes very small. When the application requires micro-scale feature sizes or wall thicknesses, resolution can dictate a system's ability to build the small features accurately. For example, if the feature size is 0.002 in. and the system's best Z resolution is 0.005 in., then layer thickness can constrain part-feature accuracy. This circumstance tends to arise in applications such as jewelry patterns and microfluidics components — applications needing very fine detail that benefit from high-resolution equipment. Today, applications with micro-scale features are a fraction of all additive manufacturing jobs.

Most additive manufacturing machines are used to build parts that measure several inches or more across and have tolerance capabilities no tighter than several thousandths of an inch. At this scale, the size of the material deposits or the "spot size" of the UV light do not control the overall accuracy of the part or the repeatability of the process.

The Necessity for Accuracy

Additive manufacturing obtained its start as a way to produce mostly cosmetic models, but the field has evolved into full-scale manufacturing. Today, additive systems turn out not just models for show and functional prototypes for physical testing, but finished goods, too.

FDM systems, for example, are increasingly employed as a cost-effective way to make manufacturing jigs, fixtures, and other tools. FDM machines also make low-volume production parts that in the past would have been injection molded or machined. Overall dimensional accuracy is paramount for these manufacturing aids and finished goods to function properly.

As additive parts move into applications with more challenging functional requirements, their accuracy tolerances are specified in the same manner as those of traditional manufacturing methods.

However, in manufacturing applications, the accuracy of a single part as it comes out of the system is just one of three critical considerations. The others are the repeatability of that accuracy over many parts and the stability of part dimensions over time.

The Necessity for Repeatability

Repeatability is a vital parameter to the success of additive systems in functional prototyping and direct manufacturing applications, where multiple versions of parts must be made within acceptable dimensional tolerances. It is prudent to know "Just as resolution does not translate into accuracy, so as accuracy does not translate into repeatability. Some systems may have good accuracy, but poor repeatability."

Repeatability considerations may dictate part to part in a single build on a single machine, part to part in multiple builds on a single machine, and part to part in multiple builds on multiple machines. Consistency across the build envelope, from build to build and machine to machine, is critical when manufacturing finished goods. Without process control, dimensional variance will yield unacceptable parts.

To scrutinize the repeatability of the FDM process, two studies analyzed thousands of dimensions over hundreds of parts manufactured on multiple systems. One study showed that the large-format, production- oriented FDM machine had a standard deviation of just 0.0017 in., which means that 99.5% of all dimensions were within ±0.005 in. The other study showed that the multipurpose systems studied — prototyping and production —produced 95.4% of all dimensions within ±0.005 in., for a standard deviation of 0.0027 in. The high repeatability of these FDM systems is paired with long-term dimensional stability.

Long-Term Stability

While the repeatable accuracy of an additive manufacturing system plays a fundamental part in reproducibility, it is not the only parameter. Just as critical is the material stability, which is responsible for part accuracy over time. Environmental conditions, such as heat, moisture, and UV exposure, as well as residual stresses from some additive processes, may cause parts to shrink, expand, or warp.

Some processes use materials, such as photopolymers, that are less dimensionally stable over time than thermoplastics. The materials experience changes in dimensions and mechanical properties, even after the part has been removed from the machine. However, additive parts made from industrial thermoplastics such as ABS or polycarbonate does not exhibit these post-build changes.

The best manner to assess the ultimate accuracy of an additive manufacturing process is to measure parts over time. If parts are needed to maintain their tolerance for years, designers must a series of checks over an appropriate period to verify the material is stable. A series of stress tests may be provided to obtain stability data in a short period of time. These stress tests may include exposing the manufactured parts to an increase in temperature, humidity, conductivity, compression, tension, and other means of fatigue tests, as long as the applied force causes the material to pass its yield point.

Applying FDM manufacturing methods have yielded acceptable repeatability measures when parts were randomly tested in separate humidity environments over a long period of time. The samples were almost unchanged. There was no warping, and the dimensions fell within the range of the original dimensions. The lengths were within ±0.002 in. of the 5.000-in. nominal dimension. Likewise, the 3.000 in. widths were within the same range of –0.003 to +0.004 in.

Printing Multi-Prototyping System

A printing multi prototyping system is capable of dual additive 3D printing using ABS and PLA materials, high-precision milling, extrusion deposition, drop-on-demand metal, including nanoscale printing. It is capable of producing fully functional prototype parts, Figure 2.33.

Fig.2.33: Multiple Print in One 3D Printing Using SLA 3D printer

Courtesy: Formlab Corporation, US

The multi-prototyping printing system is capable of producing prototypes completely in one machine. Not having to move the prototype from one machine to another provides many benefits, including lower upfront cost, faster prototyping, higher accuracy, and lower prototype production costs.

Achieving a Successful Prototype – The 10 Decrees

The following factors must be considered to achieve the desired success in manufacturing 3D printed object.

1. Define clearly if the part is for a prototype or for a real part. This will determine the selected technology to build the part.
2. Establish the main objective of the part: general design concept, form/fit testing, marketing testing, or mechanical resistance testing. Every rapid prototyping technology excels in one of these goals, but may fail in other.
3. Establish external conditions. The part may withstand temperature, contact, abrasion, impact, or chemical attack.
4. Careful study of the material specifications data sheet before printing the part or submitting it to a service bureau.
5. The part required may need to survive in a specific condition. If the part accomplishes its objective, it may be economically efficient to choose an entry-level technology or material instead of using an expensive technology.
6. Unique parts through additive manufacturing. Using additive manufacturing techniques to build parts will enable you to produce unique prototypes, with features impossible to make by conventional methods.
7. Intricate internal channels.
8. Internal moving parts with no assembly required.
9. Single parts made out of several materials, multi-material parts.
10. Multicolor models directly from the additive manufacturing printer.

The final manufacturing technique of the real part must be producible by the model intended to be used in the future. Prototyping a complex part must have the ability to actually create the same part using conventional methods. There are some limitations may arise, nonetheless, it is imperative to have ability to avoid them.

Limitations

1. Support material removal: in some cases, hollow parts will enclose support material with no possible way of removing it after the part is built.
2. Some materials are not suitable for outdoor conditions like UV radiation, heat, or humidity.
3. Resistance to impact: most rapid prototyping materials are not intended for rough manipulation or high impact, as they will likely break.
4. Some technologies do not provide enough surface smoothness for moving mechanisms or interacting parts.
5. If surface finish is important: not all rapid prototyping technologies can provide smooth surfaces.

These recommendations will help to plan ahead, and drive the design according to the final objective of reaching a successful prototype. 3D printing is also known as desktop fabrication or additive manufacturing; it is a prototyping process whereby a real object is created from a 3D design. The digital 3D-model is saved in STL format and then sent to a 3D printer. The 3D printer then prints the design layer by layer and forms a real object.

SLS/FDM WIDELY USED PRINTING TECHNOLOGIES

SLS (selective laser sintering), FDM (fused deposition modeling) & SLA (Stereolithography) are the most widely used technologies for 3D printing. Selective laser sintering (SLS) and fused deposition modeling (FDM), as discussed previously, they use melting or softening material to produce the layers.

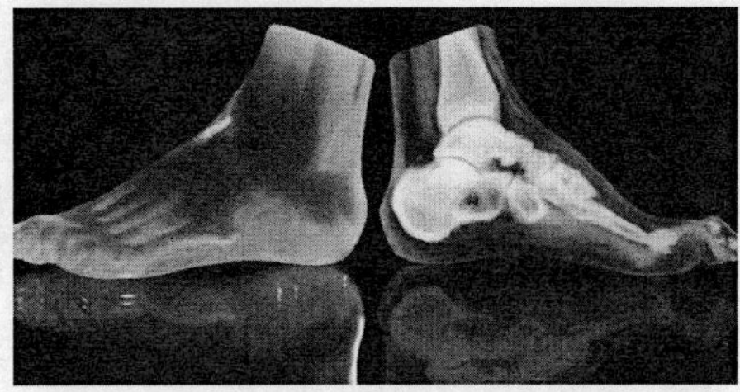

Fig.2.34: 3D Printed Model of a Patient Bone Structure

Courtesy: Tech-Lab-pro

One of the most important applications of 3D printing is in the medical industry. With 3D printing, surgeons can produce mockups of parts of their patient's body which needs to be operated upon, Figure 2.34. 3D printing makes it possible to make a part in hours only. Nowadays, almost everything from aerospace components to toys is built with the help of 3D printers. 3D printing is also used for jewelry and art Figure 2.35, fashion design, art, architecture and interior design, Figure 2.36.

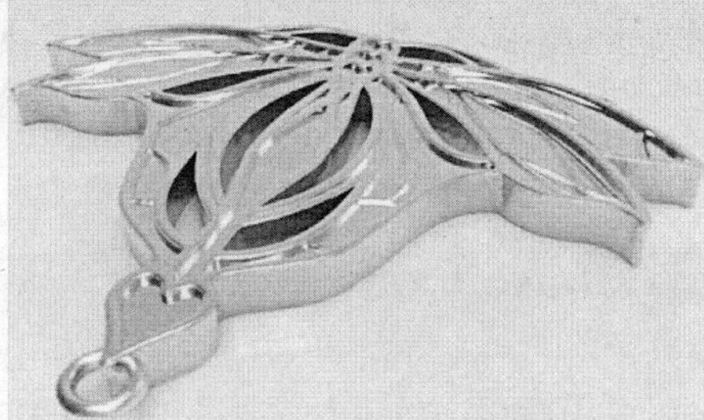

Fig.2.35: 3D Printing Jewelry

Courtesy: ZBrush & Maya

In general, 3D printers are compact and smaller than rapid prototype machines. They are ideal for use in offices. They use less energy and take less space. They are designed for low volume reproduction of real objects made of nylon or other plastics. That also means 3D printers make smaller parts. Rapid prototyping machines have "build" chambers at least 10 inches on a side whereas a 3D printer has less than 8 inches on a side. However, a 3D printer is capable of all the functions of rapid prototyping machine such as verifying and validating design, creating prototype, remote sharing of information, etc.

Many different materials can be used for 3D printing, such as ABS plastic, PLA, polyamide (nylon), glass filled polyamide, stereolithography materials (epoxy resins), silver, titanium, steel, wax, photopolymers, and polycarbonate.

Fig.2.36: 3D Printing Architecture, City of the Future

Courtesy: RTF Rethinking The Future

ENHANCING THE DESIGN PROCESS

2D to 3D CAD Printing

Affordable, quick, and easy-to-use 3D printers are changing the face of product design and development, bringing this additive fabrication technology in-house for many designers and manufacturers. Thanks to simple software and advanced technology, it is now just a matter of hours for a computer-aided design (CAD) drawing to become a three-dimensional model.

Reduce Development Time

In-house 3D printing eliminates shipping delays and reduces administrative slowdowns that sometimes occur with sourcing prototypes from external services. Manufacturers thus have greater flexibility to manage their own production.

Cost Saving of 3D Printers

The acquisition cost of a 3D printer can be as little as $15,000 (USD) versus as much as $500,000 for a rapid prototype machine, making 3D printers a good option for companies of all sizes. Annual operating costs are generally lower, too.

"EOIR Technology" corporation created a camera and mount for gun-sights on Bradley fighting vehicles for the Mississippi National Guard with its "Dimension 3D Printer." Having found that the functional models built with ABS plastic were sufficiently tough, the company manufactured 40 camera mounts using the 3D printer to be used as final production parts. This saved time and reduced the cost of parts.

Manufacturing costs for these components would have exceeded $100,000. For less than $40,000, EOIR acquired a CAD software package, a 3D printer, and ABS modeling materials. In addition to the obvious cost savings, giving a toolmaker a prototype along with a CAD drawing can result in a more accurate part as well as a more accurate price quote, sometimes saving companies up to 8 percent of tooling costs, according to industry experts. Other costs to consider are printer maintenance, which can run between $3,000 and $9,000 per year, and material costs that vary depending on use.

In-House Convenience

3D printing technology can create multiple, ready-to-use models right from a desktop, making it especially convenient for companies needing a high number of models. The printers are versatile, however easy to use and require no special training. Users just load the printer software onto a Windows network and start printing. Most software accepts major 3D file formats, including ".stl," ".wrl," ".ply," and ".sfx files." Also, some accept "CT" and "MRI" diagnostic data, protein molecule modeling database data, and digitized 3D-scan data.

Increased Data Security

Corporate owners, and executives are concerned about security today. While sending confidential "STL" files to an outside vendor is generally safe having a 3D printer in-house removes any worry about risking intellectual property.

Product Testing

3D printers can produce models with fine feature detailing, while strong enough to stand up to rigorous testing. At "MSA Auer," the world's leading manufacturer of high-end safety products and gas measuring systems, models must be durable enough to withstand planned drops onto concrete from a defined height. "MSA" Auer designers cannot risk a fracture of their safety equipment. They have produced acceptable parts with the necessary strength produced with 3D printers.

Testing Form, Fit, and Function

Because leading 3D printers use production-grade thermoplastics, companies can test the form, fit, and function for every design iteration. A true industrial thermoplastic, ABS is widely used in thousands of applications. ABS models will not warp, shrink, or absorb moisture, and they can be drilled, tapped, sanded, and painted.

The ABS material used in Dimension 3D Printers is environmentally stable, so overall shape and part accuracy don not change with ambient conditions over time, unlike resins and powders in other 3D printing technologies. Dimension's ABS has been tested for the following characteristics:

- Tensile Strength: 5300 psi (37 MPa)
- Tensile Elongation: 3.0%
- Flexural Stress: 7600 psi (53 MPa)
- IZOD Impact, notched: 2.0 ft-lb/in (106 J/m)
- Heat Deflection: 204 F (96 C)

Benefiting the Design Process

Today, 3D printers are used by designers and engineers for concept development and product design. They help users and potential customers see and feel an actual part, rather than rely on their imagination to bring a computer image to life. Models are also used as a visual aid to support tooling development.

With 3D printing, companies can experiment with new ideas and numerous design iterations — without extensive time or tooling expense — to determine whether product concepts are worthy of additional resources. The time between the concept and the production becomes significantly less when utilizing 3D additive manufacturing technology.

Time was essential for Mayo Clinic physicians in Rochester, Minnesota, who asked the Computer-Aided Drafting Technology instructors at Rochester Community and Technical College to print 3D models of the "bile duct" and "liver" of two conjoined twins, Figure 2.37. Physicians used these models to help them evaluate the twins' medical condition and determine appropriate surgical procedures for separation. After a team of 18 doctors successfully separated the twins in an 11-hour procedure, physicians used the models again to show the media how the procedure was accomplished.

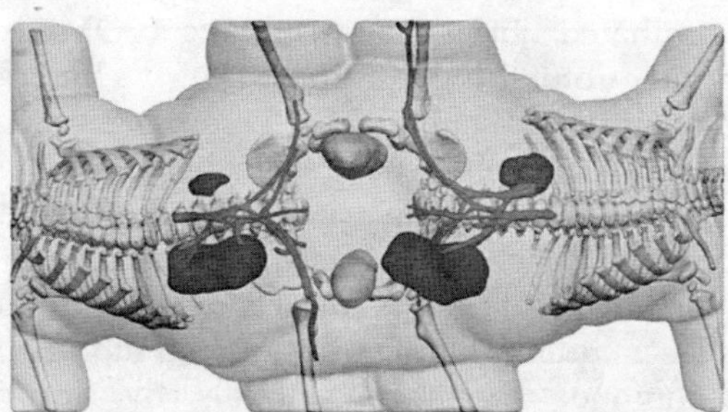

Fig.2.37: Doctor Helped separating Conjoined Twins

Courtesy: 3D Systems, Inc."

IDENTIFICATION OF DESIGN FLAWS

Successful product design requires review and input from many sources. With in-house 3D printers, design teams can review concepts earlier with others who may provide feedback. This real-time collaboration with engineering, marketing and quality assurance can lead to early quality suggestions, enabling designers to make adjustments throughout the design process and follow-up testing.

Early and frequent communication and collaboration can reduce substantial cost. The late changes will cost substantially more than early changes in the design process. The cost of changing a product in the concept stage may cost only $1, but as the design progresses, so does the cost of making changes. By the time a product is in production, the change that would have cost $1 in the concept stage now costs $10,000.

3D printing provides a highly cost-efficient means of producing numerous design iterations and immediate feedback throughout the critical beginning stages of the development process.

The ability to refine form, fit and function quickly can significantly affect production costs and time to market. This can create a distinct competitive advantage for those companies who include 3D printing as an integral part of their design process.

Lower costs will continue to expand the 3D printing market, especially in schools and small to medium sized businesses. The speed, consistency, accuracy and low cost of these printers will help companies reduce time-to-market and maintain a competitive edge.

Types of Prototyping Technologies [1-head]

The major rapid prototyping technologies are classified as follows, EBM and 3DP technologies are discussed in details in the following chapters:

1. Fused deposition modeling (FDM),
2. Selective laser sintering (SLS),
3. Photo Stereolithography (SLA),
4. Laminated object manufacturing (LOM),
5. Electron Beam Melting (EBM),
6. Three Dimensional Inkjet printing (3DP).

LAMINATED OBJECT MANUFACTURING (LOM)

Laminated object manufacturing (LOM) is a rapid prototyping manufacturing technology that allows the building of 3D representations. It was established where plastic or metal is coated with adhesives and glued together on top of each other. Once the solid lump is made, it is cut into the desired shape utilizing a laser cutter. This 3D printing and fabricating technique creates 3D models by adding, depositing and hardening raw material in consecutive horizontal layers to generate a solid object. The Figure 2.38, is detailing a building platform [1]. It is provided with thin construction material from a feed roller [2]. The utilized material is often paper, plastic or metal foil. Subsequently each layer, the building platform is dropped and it is advanced with new layer of material. The building material is usually delivered with heat-activated glue and a heated lamination roller [3] is passed over the building platform to join each new layer.

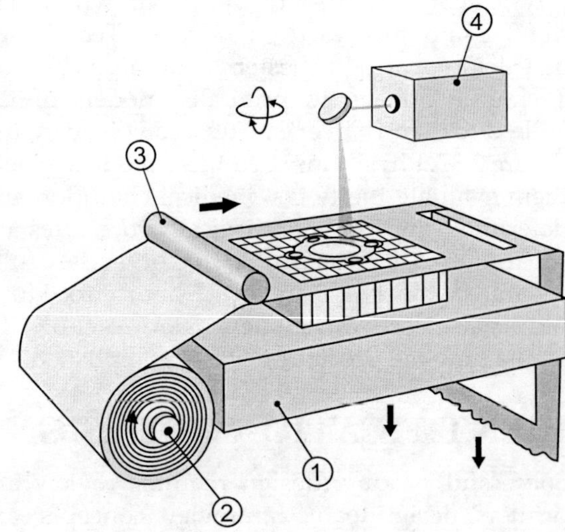

Fig.2.38: Laminated Object Manufacturing LOM

Courtesy: Manufacturing Guide, Sweden

A laser cutter [4] or a knife, trims a cross section of the model as well as a grid design in the excess material to make it easier to eliminate as a post process.

Some machines have an integrated color printer which enables printing of full color models. The following is the methodology of the process:

1. The sheet is glued on to the surface and subjected to a heat roller on which the dimensions of the prototype are traced on using a laser.
2. The unwanted parts of the material are discarded and the completed layer moves down to facilitate the next layer.

New material is put in position and the platform is raised to receive the new layer. This process is repeated as many times as required.

The main advantages offered by this technology are its low cost due to easy availability of raw material and easy construction of large parts as no chemical reactions are involved.

The main disadvantage of this method is that it provides less dimensional accuracy, than some other technologies like Stereolithography.

SOLID-FREEFORM SYSTEMS

The solid-freeform systems are commonplace tools for industrial designers. They are used mostly by automakers to create prototypes of car parts, from engine blocks to side-view mirrors; by appliance manufacturers to model products such as air conditioners and microwave ovens; and by consumer electronics companies to model Blue tooth headsets, cell-phones, and other products. The industrial rapid-prototyping systems use inks, pastes, and suspensions that are combinations of filaments, powders, flakes, precursors, cross-linkers, binders, solvents, dispersant, and surfactants, whose properties—including viscosity, density, melting point, and surface tension—are tailored to particular applications.

PARAMETERS CONSIDERED IN 3D PRINTED PRODUCT DESIGN

Some of the vital parameters are stated below for designers to consider when developing a new 3D printing prototype, which leads to a low volume production (job or batch production):

Product Release

Ensuring the surfaces does not fuse together while printing.

Product Warping

Because the prototype is built up of many layers, each originating as a molten liquid, there is the problem of thermo-elastic warping. Thus maintenance of plane surfaces in the case of joints with tight clearance represents a technical challenge.

Product Friction

The FDM models do not have any lubrication, so a large gap helps minimize friction. This is especially true for processes that have low surface finish.

Removing Product Support Material

Most Rapid prototyping processes require secondary processes for removing support material, either using solvents, melting, blowing or etching. If gaps are too tight, these solvents cannot reach the material to be removed and/or the dissolved material cannot flow out. Consequently, the design is typically modified to allow easy pathways for removal of release materials.

Product Material Strength

Another issue is the strength of the rapid prototyping material itself. For example, brass handles models have an appealing shape but has a fragile ABS plastic. Several aspects affect the strength of the printed mechanism. Various materials, which can be used, affect the durability of the resulting mechanism. A typical ABS product is far weaker than the original. Weak points and stress concentration points may need to be reinforced.

Product Deposition Pattern

Since the deposition processes is layered and, and since the FDM processes are also fibered, the material's mechanical behavior approximates laminated material (e.g. composites) or fibrous material (e.g. hardwood) more than it does a solid material. Thus the original material properties specified by the manufacturer may be somewhat deceptive, and the true properties end up being largely dependent on the orientation of the part and exact deposition pattern. It is therefore desirable to set the mechanism at a state such that it may be printed with fibers along the length of load gradients. New materials and deposition processes may alleviate this need in the future.

Product Compliance

The elasticity of the material may be useful when printing compliant mechanisms such as springs. However, in such components the geometry interacts with the material properties, so that the geometry must be changed to achieve certain kinematic behavior. This is demonstrated in the printing of the spring loaded-ratchet mechanism, in which leaf-springs lengths and width needed to be adjusted to produce the proper functionality.

Currently, rapid prototypes can fashion plastics, ceramics, and certain metals into almost any kind of mechanical structure, including sliding and rotary kinematic joints, links, springs, gears, ratchets, nuts, and bolts, with relative quality that may be good enough for functional testing. Many solid prototypes incorporate a tough, rigid plastic, called acrylonitrile-butadiene-styrene (ABS).

Electronics maker "Logitech Inc.," in Fremont, Calif., used a 3D Additive Manufacturing machine, to make an ABS prototype of its Bluetooth headset and then attached weights to the boom microphone to test the design for strength. Diebold Inc., in North Canton, Ohio, builds automated teller machine prototypes made of ABS or polycarbonate and tests their endurance in rain, sleet, snow, and extreme temperatures. "Grotell Design Inc.," in New York City, makes prototypes of all its watch components, from bracelets to bezels, using thermoplastics, natural and synthetic waxes, and even fatty esters in desktop solid free form fabricators made by "Solidscape Inc.," in Merrimack, N.H. Like countless other technologies, Additive Manufacturing systems are becoming faster, more economical, and achieve better performance.

Printing Rates and Limitations

Today's machines have several obstacles to overcome before they may become fame in industry. Additive Manufacturing has typical printing rates of about 1 cubic inch per hour. It can take as long as a day to produce even a small simple plastic cubic shape through one of slow speed Additive Manufacturing systems. In general, the systems may be somehow difficult to use, requiring a fair amount of specialized knowledge. In addition, the cost may vary from US $15,000 to $750,000. Furthermore, a single free form machine cannot make parts composed of different materials. And even if a machine existed that could handle multiple materials, the commercial software needed to command the tools to apply plastic on one layer and metal on the next does not exist yet.

Also, the range of materials of Additive Manufacturing systems can work with, while expanding, still does not include materials that produce motion in response to electrical impulses, and chemical systems and separators that can be used to make batteries.

Additive Layers with Thin Distribution of Powder

Three-Dimensional Printing functions by building parts in layers. From a computer (CAD) model of the desired part, a slicing algorithm draws detailed information for every layer. Each layer begins with a thin distribution of powder spread over the surface of a powder bed. Using a technology similar to ink-jet printing, a binder material selectively joins particles ejected from the print head where the object is to be formed. The ejected substance joins the powder in a plane forming a single layer. A platform that supports the powder bed is lowered so that the next powder layer can be spread and selectively joined. This layer-by-layer process repeats itself until the part is completed. Following a heat treatment, unbound powder is removed, leaving the fabricated part.

COMBINING POWDERS AND BINDERS

There are 3D Additive Manufacturing Machines combine powders and binders with unprecedented geometric flexibility. The support gained from the powder bed ensues that the overhangs, undercuts and internal volumes can be created, as long as there is a hole for the loose powder to escape, Figure 2.39.

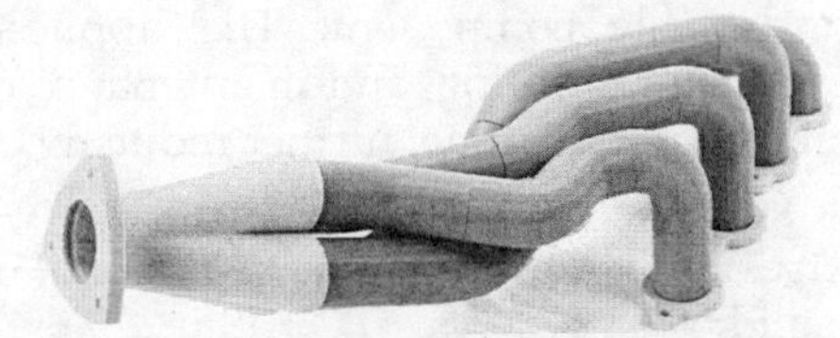

Fig.2.39: 3D Printing Product with Holes to Escaper Product Powder

Courtesy: LulzBot Corporation

3D Printing can form any material that can be obtained as a powder - which is just about any material. Different materials can be dispensed by different print heads. 3D Printing can exercise control over local material composition. Material can be in a liquid carrier, or it can be applied as molten matter. The proper placement of droplets can be used to create surfaces of controlled texture. Also, the proper placement of droplets controls the internal micro-structure of the printed part.

Powder Processing 3D Printing Machines

There are numbers of 3D Additive Manufacturing machines surpasses conventional powder processing because there are no tooling or geometric limitations needed. Because of the great flexibility in handling a wide range of materials and because of the unique ability to locally tailor the material composition, the 3D Additive Manufacturing Printing offers structural components with unique microstructures and capabilities.

3D Additive Manufacturing is readily scaled in production rate through the use of multiple nozzle technology which has been commercially developed for printing images on paper.

Three-Dimensional Printing has led the field of Additive Manufacturing in the creation of functional parts and tooling directly from a CAD model. It was the first technology to achieve the fabrication of ceramic parts, and pioneered the direct fabrication of ceramic molds for casting.

STRUCTURAL CERAMIC PARTS

Most recently, various corporations have pioneered the fabrication of structural ceramic parts using the 3D Additive Manufacturing printing process. The "3DPTM" technology is used extensively in structural 3D printing parts to substantially reduce the time to market for new products, enhance product quality by improving the coupling between design and manufacturing, and lower product cost by reducing development and tooling costs. Furthermore, the flexibility of the process makes totally new technologies and applications possible and has already generated novel solutions to engineering problems. 3D Printing is at the forefront of the coming revolution in manufacturing brought about by Rapid Prototyping.

Parallel Data Processing

The software processes data in parallel with the printing of the part. Unlike the other processes, while the 3D printer deposits the first layer, the software slices and processes the fifth layer.

Although the processing time may seem to be fast, it is often only a fraction of the total time it takes to build the part. It can actually take up to an hour to prepare a job with multiple parts using some additive technologies.

Inject printing technology printers produce very little waste. The unprinted powder surrounds and supports complex parts during printing. Users can reuse all unused support powder. Thus, printed-part volume becomes the basis for all part-creation costs. The use of an off-the-shelf print head allows for inexpensive, quick replacement of the system's primary consumable component. The application of modular design techniques to the printer's electronics, printing, and maintenance components makes the printers efficient to maintain with minimum downtime, further reducing costs.

Unlike a liquid-based process, which changes material from liquid into solid state, an extrusion-based process feeds material in solid wire form and then melts it into a shape and forms a solid.

Assessments: Laminate Object Manufacturing - Fused Deposition Modeling

Products made by additive manufacturing are increasingly replacing the products made by classical procedures of polymer processing. The current market requires the products to be of good mechanical properties, low prices, and less complicated geometry. With the procedures such as Fused Deposition Modeling and Laminated Object Manufacturing it is possible to produce a product of relatively low price and good mechanical properties.

In practice it is impossible to avoid using the products in various atmospheric conditions (e.g. UV radiation and humid environment) either intentionally or unintentionally. The UV light, the Sun being its usual source, and the average humidity of countries worldwide can vary from 20% to 90%, depending on the weather, day and geographic location. It is therefore necessary to determine how UV light, after a longer exposure, affects the final mechanical properties of the products. A comparison of tensile strength and elasticity module of the test specimens in FDM and LOM procedures has been performed.

Procedure - FDM and LOM

The Stratasys cheapest model of the device is Dimensional "uPrint" from the "uPrint" family. The device is based on the principle of FDM procedure and can make parts of "ABSplusTM" material, but only in one color (Ivory). The characteristics of the uPrint device are: build size: 203 x 152 x 152 mm, layer thickness: 0.254 mm, machine size: 635 x 660 x 800 mm, machine mass: 76 kg.

The test specimens made by LOM procedure are made of PVC film. The test bodies in LOM procedures are made on the SD 300 Pro machine, produced by "Solido," is a machine which can produce transparent prototypes of PVC film, has small dimensions, and is practical for use in offices.

The characteristics of the SD 300 Pro machine are:

1. precision: 0.1 mm (in axis),
2. layer thickness: 0.168 mm,
3. working area: 160 × 210 × 135 mm,
4. machine size: 465 × 770 × 420 mm,
5. machine mass: 45 kg.

The printers have similar working volumes and price, which makes them favorable for small companies. They belong to the class of personal printers, which means that in service there is no health-harmful impact on the environment, and the installation is simple. The characteristic of both machines is the making of products of smaller dimensions with recyclable material.

Effects of Extrusion Speed and Temperature on Surface Roughness In FDM built

Surface roughness remains one of the weak functions of FDM technology in general, while the problem is especially pronounced in the domain of 3D printers for personal use. Extrusion speed and extrusion temperature are two controllable parameters which, other than layer thickness, have greatest impact on the surface quality of FDM-built parts. The influence of extrusion speed and temperature on the arithmetic average of the roughness profile (Ra) of FDM specimens is factual. Factorial experiments were used with two replicas and two center points. The results indicate a dominant, statistically significant influence of extrusion speed, as well as the pronounced nonlinearity of effects.

Today, Fused Deposition Modeling (FDM) is one of the most popular additive manufacturing technologies. It is used for rapid prototyping as well as for manufacture of an ever larger number of final products. Its popularity is due to several advantages, such as: simplicity and ease-of-use, choice of materials, easily replaced and non-toxic filament, as well as cost-effective maintenance. However, due to limitations which are primarily related to filament and extruder nozzle diameter, as well as some deficiencies which are intrinsic to the process of filament extrusion, FDM technology yields relatively low surface quality compared to traditional stereolithography (SLA) or PolyJet technology. For this reason, it is interesting, from the technological point of view, to investigate controllable factors which have dominant impact on surface quality, in order to achieve the best possible surface quality before post-processing.

In a study on surface roughness in the domain of FDM technology, Surfaces at various angles of inclination (0-90o) were built while testing their dimensional accuracy and surface roughness parameters. Factorial analysis was used to examine the influence of several technological parameters on "Ra" in "FDM-"built parts.

In a comparative study of five additive manufacturing technologies surface roughness of (SLS, FDM, DLP, 3DP and PolyJet) have investigated as one of parameters, using ABS as material. A benchmark model was proposed, which is composed of 3D modules of various geometries, dimensions, locations and orientations. SLA demonstrated that highest surface quality and accuracy was produced, followed by SLS, LOM and FDM. Also, a benchmark model was proposed for measuring surface quality depending on the angle of inclination, "▯," where "▯" was varied from "0° to 90°." Based on the obtained data, significant difference in surface quality has been observed based on applied technology, layer thickness and material, as well as the angle of inclination of the measured surface.

One common trait observed of the employed industrial-class machines, the price of which is usually on the $300. On the other hand, modern trends in development of additive technologies are oriented towards a newly emerged market of 3D printers for personal, desktop application, which are within the available price range and are often able, to produce quality parts which find application in various walks of life.

One of the most popular 3D printers in this class is the Makerbot Replicator 2, which is based on FDM technology and can work with 0.1, 0.2 and 0.3 layers, allowing user to vary a broad range of parameters.

Additive Manufacturing Market - $8.41 Billion by 2020

According to a new market research report published by "Markets and Markets the 3D printing market" is expected to grow at a CAGR of 26% from 2014 to 2020, and reach $8.41 billion in 2020.

The elements behind the rapid upsurge in the 3D printing market are the innovative and advanced technologies, customized products, governments funding, wide unexploited application market, rapid development of the products at a low cost, and less time-to-market.

In addition, the upcoming, in 2014 expiration of patents with respect to the Selective Laser Sintering (SLS) is also believed to provide a further impetus to the growth of the 3D Printer market.

As of 2013, the Americas hold the largest revenue share followed by Europe- in 3D printer machines, materials, and related services. However, Europe is also expected to surpass Americas in terms of the 3D printing market revenue by 2020. The foremost factors accountable for the expansion of 3D printing market include new and improved 3D printing technologies, wide range of materials (polymers, metals/alloys, sand, ceramics, living tissue), government funding, broad application scope, and increased awareness regarding the benefits of 3D printing over traditional techniques, injection molding and CNC machining. However, APAC is the fastest growing and most promising market for 3D printing due to high industrial growth, technological awareness, supportive government policies, and financial investment by the governments in R&D. The major companies operating in this market are 3D Systems (U.S.), Stratasys (U.S.), Arcam AB (Sweden), Exone (U.S.), and others.

3D Additive Manufacturing Market by Technology

3D Additive Manufacturing has been evolving as a technique to create 3D models and prototypes, in many industries, namely, automotive, aerospace, healthcare, and consumer products- in order to investigate the possibility of completing a project in lesser time and with few resources. However, in the last two decades, 3D Additive Manufacturing has made a radical shift from rapid prototyping to rapid manufacturing, mainly, because of its advantages over traditional manufacturing practices such as injection molding, CNC machining, and vacuum casting. These advantages include innovative designing, high adaptability levels, less time to market, and less tooling requirements. 3D Additive Manufacturing is also found to be quicker and less expensive. The ability of 3D Additive Manufacturing to print almost any geometry with a variety of materials makes this technology a preferred choice, especially, in those markets which are characterized by high individualization, low volume, and high value; such as aerospace and healthcare. The major driving factors that support the exponential growth of 3D printing include – the development of new and improved technologies, financial support from governments, large application areas, rapid product development at a low cost, and ease in development of custom products.

The global 3D Additive Manufacturing is segmented on the basis of technology, material, application, and geography. The technology segment comprises of

1. Stereo lithography (SLA),
2. Laser Sintering,
3. Electron Beam Melting (EBM),
4. Fused Disposition Modeling (FDM),
5. Laminated Object manufacturing (LOM),
6. Three Dimensional Inkjet printing (3DP), and other proprietary technologies.

The global 3D Additive Manufacturing market by material comprises polymers, metals/ alloys, and others ceramics, sand, and paper. The application segment includes aerospace industry, automotive consumer products, healthcare, government and defense, industrial/ business machines, education & research, and others, arts, architecture. Health care and aerospace are the two fastest growing application areas for 3D printing. The stringent requirements such as light weight and accurate and precise design requirements for airplanes parts are the major driving factors behind the growth of 3D Additive Manufacturing in the aerospace industry, over traditional manufacturing methods. However, the increasing medical procedures' volumes due to the growing population, rising income levels, low labor costs, and increasing health awareness are responsible for the tremendous growth of 3D printing in healthcare sector, as well.

The market is also categorized into four major geographic regions, namely

1. Americas,
2. Europe,
3. Asia-Pacific, and
4. Rest of the World.

Developing economies such as China and India provide substantial growth opportunities for the 3D printing manufacturing technologies, majorly, due to the rise in lifestyle and general income levels. For instance, China is aggressively taking initiatives such as huge investment and government funding in R&D, to promote 3D Additive Manufacturing as a manufacturing technique.

THE DEMAND FOR 3D ADDITIVE MANUFACTURING

The categories of 3D Additive Manufacturing market are based on various applications, technology, materials, and geography; it also covers the revenue foretold from 2014 to 2020. It describes the demand for 3D Additive Manufacturing in various regions. The applications describe mapping the 3D Additive Manufacturing market with respect to the growth potential.

On the basis of the applications

3D Additive Manufacturing is used for various applications such as in the automotive, aerospace, consumer, healthcare, government & defense, industrial, and education and research sectors, Figure 2.24

On the basis of the Technologies

The 3D Additive Manufacturing technology is segregated into Stereo lithography (SLA), Laser Sintering, Electron Beam Melting (EBM), Fused Disposition Modeling (FDM), Laminated Object manufacturing (LOM), Three Dimensional Inkjet printing (3DP), and the likes.

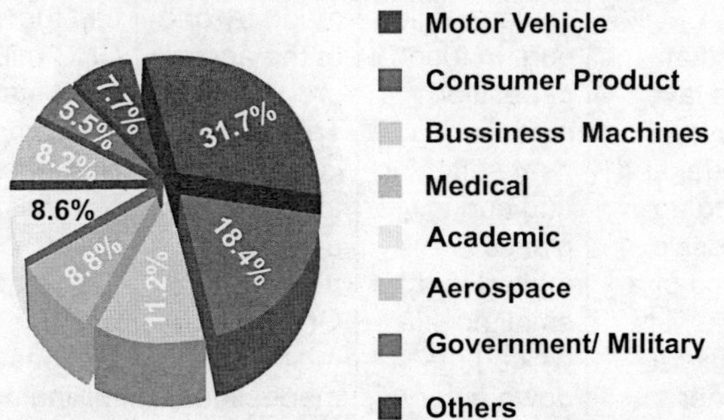

Fig.2.40: 3D Additive Manufacturing Distribution by Application

On the basis of the Materials

The materials used in 3D Additive Manufacturing are:

Polymers,

1. metals & alloys,
2. powder, and others.

The 'metals' based 3D printing holds the future prospect of the market.

On the basis of the geographical regions

Geographical analysis covers:

1. Americas,
2. Europe,
3. Asia-Pacific, and
4. Rest of the World

The USA currently, leads the 3D Additive Manufacturing market. Additive Manufacturing is a technology which has developed in the last decade. Through the use of modern solid modeling CAD packages, combined with laser systems and new materials, one can now generate solid parts directly from the computer model without human interaction. In addition, these parts are generated in a matter of hours, and are accurate to a few thousandths of an inch. Compare this to handmade prototypes which can take weeks or months to create and tend to contain errors. Table 2.2 provides a detailed comparison of 3D advanced and traditional technologies.

Table 2.2: Comparison between 3D Printing Technology and ddvanced Traditional Technology

STEREOLIT HOGRAPHY - SLA	LAMINATE OBJECT MANUFACTU RING - LOM	FUSED DEPOSITION MODELING - FDM	SILICONE RUBBER MOLDING/ URETHANE CASTING	CNC-MILLING	INVESTMENT CASTING
Once a pass is completed, another layer of the liquid is added to the existing part, and the process repeats until the full part height is achieved. SLA parts are extremely accurate, and tend to have excellent surface finishes. A variety of SLA materials are available for different purposes, including waxes, plastics, and flexible elastomers.	Laminate object manufacturing builds a part by taking individual sheets of paper that have a layer of glue on one side, and building up successive sections of a part. LOM is very useful for larger parts, as the process builds these much faster than the other processes.	Fused deposition modeling is a process that most closely resembles a miniature glue gun. In fused deposition modeling, a heat softening and curing plastic is melted in a small nozzle which puts down a very fine bead wherever the solid part is supposed to be.	This is a two stage process: First, a master prototype is made by one of the above processes. This master is then used to make a silicone rubber mold. Once the molds are made, liquid urethane rubber is placed inside the mold and allowed to cure.	When a single metal part is needed, and it is too complex to fabricate by hand, we can turn to CNC milling. CNC stands for Computer Numerical Control. In this process, the computer controls the motions of a milling machine, guiding it in the fashion of the old saying "cut away everything that doesn't look like the	If very complex steel parts are needed, or many steel parts are needed, in- vestment casting may be the technique of choice. Investment casting begins with a wax pattern piece, which can be made by the stereolithography process. This pattern is coated with a very fine plaster-like material called investment. Layers of this

Most SLA materials do tend to be somewhat brittle, which means they break easily, as we have proven on many occasions. Stereolithography is a method that employs an ultraviolet laser to cure a thin layer of liquid plastic into a solid. The process operates by taking a thin layer of the light-sensitive liquid plastic and passing the laser beam over the points where the part is solid.	LOM is not as effective for very small parts as the flexural stiffness of the paper is usually too low for small parts. LOM parts are accurate, and very easy to sand and paint. LOM parts, if untreated will tend to absorb humidity and can crack because of it. LOM parts also have different strengths in different directions due to the paper layers.	FDM parts have a rougher surface finish than an SLA part, but typically are stronger and more durable.	Once cured, you have a urethane copy of your master prototype. The process allows the creation of a larger number of prototypes (up to about a dozen, possibly more) rapidly and accurately. Additionally, urethane castings are typically more durable than any of the master prototypes, and can be made faster and for a much lower cost.	part." While the theory is simple, the actual application is more complex. The process yields a single part, but this part can,theoretically, be made out of anything, and if correctly made, can perform in an identical fashion as the real part. CNC parts, however, tend to take longer than the other processes, and are usually more expensive. CNC parts become more cost effective as they grow larger, and if a wider variety of materials can be used (aluminum parts typically cost less than steel).	investment are built up until a thick layer is formed. The wax is then melted out in an oven, leaving behind a cavity in the investment that is a perfect copy of the master part. This cavity can then be filled with any molten material, brass, aluminum, steel, stainless steel, even exotic alloys such as inconel. Once the material cools, the investment is broken away and a finished part is left. Investment casting can be used to make an unlimited number of parts, all of which are very accurate to the design. Investment cast parts need little to no finishing work in most applications before use. However, investment cast parts are typically more expensive than other processes, and are limited to parts that are normally no more than 12" in any direction.

REFERENCES

1. Perez, C. L., "Analysis of the surface roughness and dimensional accuracy capability of fused deposition modelling processes", International Journal of Production Research, Vol. 40, No. 12, pp. 2865-2881, 2002
2. Galantucci, L.M., Lavecchia, F., Percoco, G., "Experimental study aiming to enhance the surface finish of fused deposition modeled parts", CIRP Annals - Manufacturing Technology, 49(2), pp. 189-192, 2009
3. Ulbrich et al., "Comparison of five rapid prototype techniques (SLS/FDM/DLP/3DP/Polyjet)", Innovative Development of Virtual and Physical Prototyping - Bartolo et al. (Eds.), Taylor & Francis Group, London, ISBN 978-0-415-68418-7, pp. 573-580, 2012
4. Mahesh, M. et al., "Benchmarking for comparative evaluation of RP systems and Processes," Rapid Prototyping Journal, 30(2), pp.123-135, 2004
5. Upcraft, S., Fletcher R., "The rapid prototyping technologies", Rapid Prototyping Journal, 23(4), pp.318-330, 2003
6. Vasudevarao et al.,"Sensitivity Of RP Surface Finish To Process Parameter Variation", 11th Solid Freeform Symposium, Austin, TX, Proceedings, pp. 251-258, 2000
7. Soloman, Sabrie "Sensors Handbook, 2nd Edition," McGraw-Hill Publishing Co., NY – ISBN 978-0-07-160570-0
8. Soloman, Sabrie "Sensors and Control Systems in Manufacturing 2nd Edition," McGraw-Hill Publishing Co., NY – ISBN 0-07-059626-3
9. Soloman, Sabrie "Affordable Automation," McGraw-Hill Publishing Co., NY – ISBN 0-07-059633-6
10. Soloman, Sabrie " Sensors Handbook, 1st Edition," McGraw-Hill Publishing Co., NY – ISBN 0-07-059630-1
11. Soloman, Sabrie "Sensors and Control Systems in Manufacturing 1st Edition," McGraw-Hill Publishing Co., NY – ISBN 978-07-10572-4

3 Fundamentals of Additive Manufacturing

This chapter provides the details of vital processes and their corresponding techniques in optimizing each process to obtain successful Additive Manufacturing output products at an affordable cost, desired speed, and production rate. Also, this chapter provides the foundation of 3D printing in Microelectronics.

COREPROCESSES - THEROLEOFPHOTO-CURINGTECHNIQUE

Optimizing the Photo-curing Technique - VAT Polymerization – SLA/FLP

What is VAT Photopolymerization?

It is a liquid made of polymer substance used in making objects to be constructed layer by layer, and then use ultraviolet (UV) light to the resin, during which a platform dedicated to move the object downwards once a layer is cured.

The most common forms of Vat Polymerization are SLA (Stereolithography) and DLP (Direct Light Processing).

A simplification of 3D printing categories may lead to better understanding the types of available 3D printing techniques to enable the designer and the user to tailor the planned product prototyping. Selecting the most appropriate 3D printing technology will undoubtedly lead to better prototype, which is as close as possible to the intended final product, not only in appearance but also in function, durability and cost. It may also ensure the project time of delivery is at hand at the most appropriate time. Management and decision makers will ensure successful plans for schedule implementations when the physical prototype is touched and seen and held for engineering and marketing evaluations.

SIMPLIFICATION OF 3D PRINTING TECHNOLOGY

The contents of this table is the theme of this chapter, which is specifically tailored to present the types of most applicable 3D printing technology and their corresponding implementations in various industries, including medical and dental industries.

Table 3.1 – Recommended Utilization of 3D Printing Technology

3D Printing Technology	Application	Sample Available System	Status
Material Extrusion Technology	Fused Filament Fabrication – Figure 3.1	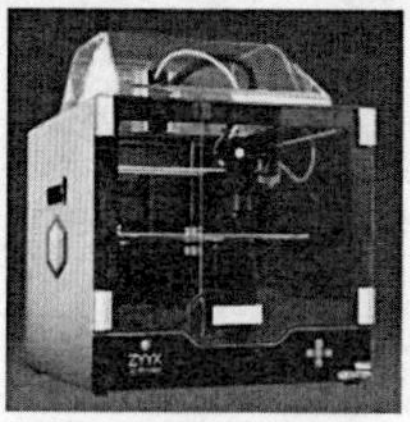 **Fig.3.1:** Fused Filament Fabrication FFF 3D Printer *Courtesy: ZYYX™ 3D Printer*	Will be discussed
Vat Polymerization	Stereolithography SLA Digital Light Processing DLP, Figure 3.2	 **Fig.3.2:** Stereolithography – Digital Light Processing SLA/ DLP Printer *Courtesy: All3DP*	Will be discussed
Powder Bed Fusion – Polymers	Selective Laser Sintering, Figure 3.3	 **Fig.3.3:** Selective Laser Sintering SLS Printer *Courtesy: ProMakker*	Will be discussed
Material Jetting	Material Jetting DOD, Figure 3.4	 **Fig.3.4:** Material Jetting Printer *Courtesy: Courtesy Mimaki*	Will be discussed

Binder Jetting	Bonder Jetting, Figure 3.5	Fig.3.5: Binder Jetting Printer *Courtesy: GE*	Will be discussed
Powder Bed Fusion – Metals	DMLS- SLM- EBM, Figure 3.6	Fig.3.6: Electron Beam Melting Printer *Courtesy: GE*	Will be discussed

VAT POLYMERIZATION

Vat Photopolymerisation can also be known as (in alphabetical order):

1. Continuous Liquid Interface Production or CLIP
2. Scan, Spin and Selectively Photo-cure Technology or 3SP
3. Solid Ground Curing or SGC
4. Stereolithography or SL
5. Stereolithography Apparatus or SLA® (3D Systems Corporation)
6. Two-Photon Polymerization or 2PP

These different nomenclatures represent once type of technology. They are interchangeable.

Digital Light Processing (DLP)

Vat polymerization uses Digital Light Processing DLP. It consists of a set of chipsets arrangements based on the technology that was originally developed in 1987, Figure 3.7. The original DLP-based projector was used in a variety of display applications from 3D printing, to medical applications.

DLP technology is used in in additive manufacturing - 3D printing as a light source to cure resins into solid 3D objects.

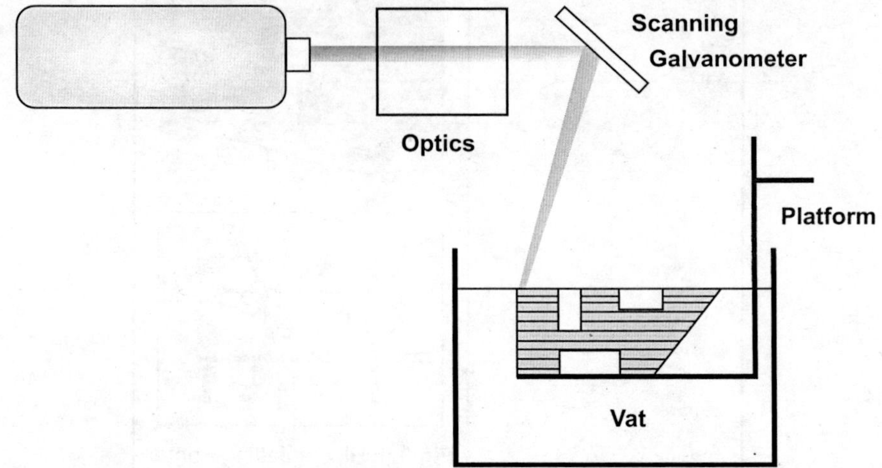

Fig.3.7: Vat Polymerization – 3D Printing Technology

Courtesy: Canda Makes

Producing parts through Direct Light Processing (DLP) is identical to parts produced through Stereolithography (SLA) 3D printing technique. However, the major difference is that in the DLP is utilizing a digital light projector screen to project a flash light on a single image of each layer all of the object at once. The projector in this event is a digital screen, while the image of each layer is composed of square pixels, resulting in a layer formed from small rectangular bricks called voxels.

DLP can achieve faster print times compared to SLA. In the SLA, an entire layer is exposed all at once. Light is projected onto the resin using light emitting diode (LED) screens or a UV light source. This UV light source is being directed to the object build surface by a Digital Micro-mirror Device (DMD). A DMD is an array of micro-mirrors, generating the light pattern projected on the surface of the 3D printed object.

SLA VS. DLP

The fundamental difference between SLA and DLP is the light source difference in each technology used to cure the resin.

SLA printers use a point laser compared to the voxel approach that DLP printers use. The resolution of DLP is 1024 x 780, while the resolution of the SLA printer, which uses a laser, is 130 - 150 micron spot size. The disadvantage of using laser point in SLA printer is that it takes longer time to trace the cross section of the object compared to the DLP printers, which are capable of exposing the cross section in a single flash. This property makes the DLP technique faster than the SLA technique when printing an identical object.

Printer Characteristics

Vat Polymerization printer parameters are unlike Fused Filament Fabrication (FFF) printers. The printer parameters are fixed and cannot be changed. Typically, the operator inputs are part of orientation, as well as support location, layer height, and material. The operator inputs are specified at the slicing stage. However, most printers' adjustments are adjusted automatically, based on the type of material used.

Layer height and light source resolution govern the surface finish and accuracy of the part. Most Vat Polymerization printers produce parts with a layer height varies between 25 - 100 microns. However, for high resolution and finely detailed prints, the DLP projector can be replaced with narrow beam projector lenses. This allows the beam to print smaller layers at a faster rate, and a higher level of details.

Object Orientation

The Vat Polymerization printers have the ability to produce objects in two different orientations. One orientation is placed bottom-up, while the second possible orientation is placed top-down as illustrated in Figure 3.8. Both SLA and DLP printers can perform their tasks in either way of object orientation. The design configuration is depending on the manufacturer. However, bottom-up printers have their light source positioned below a resin vat with a transparent bottom. Initially, the object build-platform is positioned to allow only a single layer thickness between at the base of the vat and platform. The light source, which may be laser, UV, or LED screen cures and solidifies the thin layer of resin. Also, a special coating is added to stop the resin from sticking to the base of the vat. Once the first layer is cured and firmly stuck to the build platform, the printer performs a separation step to separate the cured first layer from the base of the vat. The base then moves one thickness layer up Figure 3.9).

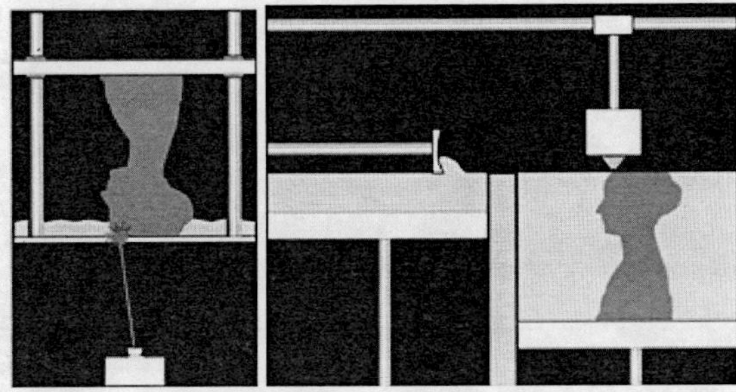

Fig.3.8: Printing Top -Down or Down-Top Orientation

Courtesy: Freely Published Media

Depending on the machine, this separation stage can involve peeling, sliding, rotating or shaking the vat. After the separation stage, a new uncured layer of resin fills the gap. Some bottom-up machines utilize wiper to spread a layer of resin across the base of the vat to ensure uniform coverage, mix the resin and remove any debris of the cured resin spots. The process is then repeated with the build platform moving up one layer thickness and separating the newly cured layer from the base of the vat until the object is completed.

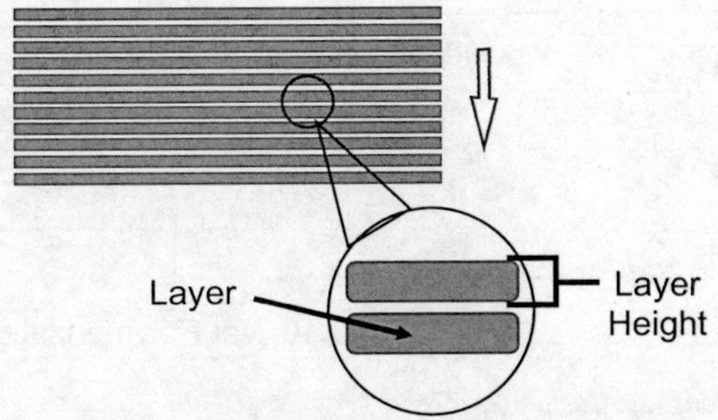

Fig.3.9: Build Platform Drops Down Layer by Layer of Equal Height

The success of a bottom-up print is dependent on reducing the forces on the newly printed layers during the separation stage. The separation stage creates areas of high stress along a potentially razor thin edge, which can lead to part failure and warping when the part sticks to the bottom of the vat instead if the build plate. Bottom-up printers generally have a non-stick coating. This is applied to the base of the vat to assist with the separation stage however; this must to be replaced regularly to ensure it performs adequately. The wiper also assists in oxygenating resin helping to improve non-stick performance.

Top-down printers position the light source above the build platform. The build platform begins at the very top of the resin vat with a thin layer of resin coating. The light source cures the thin layer of the resin. Once the first layer is cured, the build platform moves down a single layer thickness, resin re-coats the previously cured layer and the process is repeated. As the build progresses, the build platform continues to lower into the resin vat.

Once the build is completed the object will be completely submerged in resin. The object is then raised out of the resin and removed from the build platform. Like bottom-up machines, the first layer is often the most critical in a build. Accordingly, it is vital that this first layer must be successfully adhered to the build platform. For top-down technology printers, it is important that a uniform layer of resin coats the build surface after each downward movement of the build platform. To achieve this, top-down machines must ensure that the resins have adequate viscosity.

Vat Polymerization Wiper Device

Some Vat Polymerization printers machines make use of a wiper device to spread the layer of uncured resin over the surface. Also, the platforms must also move slowly into the resin to ensure no air bubbles are developed, which have a detrimental effect on print quality. Build platforms are typically perforated to reduce the disruptive forces on the platform, Figure 3.10.

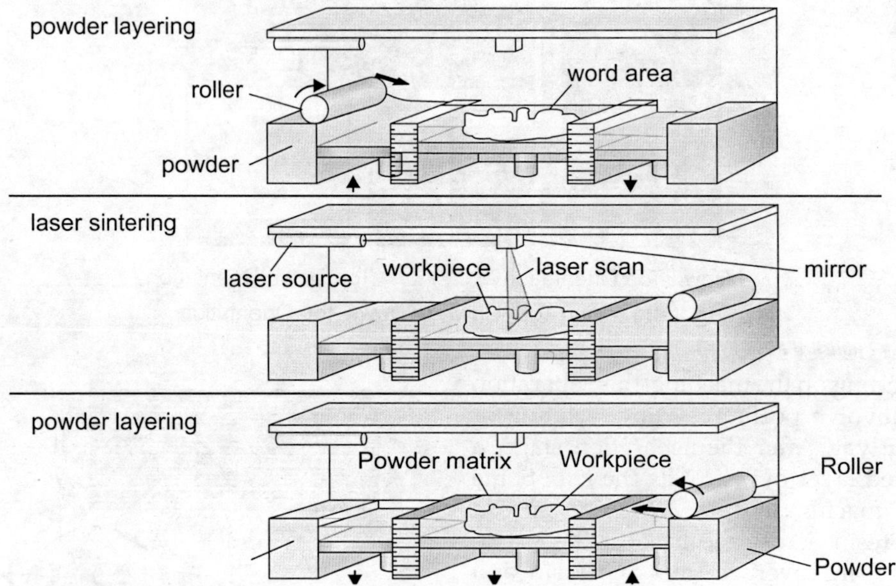

Fig.3.10: Vat Polymerization Process –Wiper Mechanism (Roller)

Courtesy: Custom Part Net

Support Structures

Vat Polymerization objects require support structures similar to most 3D printing technologies. The location and amount of support depends largely on the type of 3D printer being used. For top-down printers, support requirements are similar to FFF with overhanging features and bridges needing material to hold them up, and allow them to accurately be printed.

The support of the Bottom-up 3D printers is more intricate than other types of 3D printers. A large horizontal surface of the build layers can cause excessive forces when the print is separated from the base of the vat during the removal stage, Figure 3.11. If the during the removal stage is inhibited to easily separate the build platform from the vat, the print may then fail. Accordingly, the objects are printed at an angle, thus the support structure is reduced, which causes a primary concern. For either method of printing, the support structures are always printed in the main build area in the vat, and must be removed manually after printing.

Fig.3.11: Failed Structural Supports

Dimensional Accuracy and Warpage

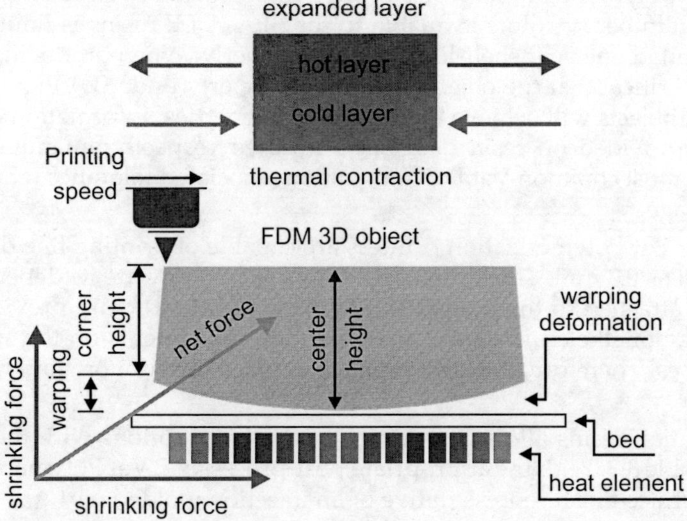

Fig.3.12: Warping Deformation - Method of Calculating the Deformity

Courtesy: International Journal of Mechanical & Mechatronics Engineering

Dimensional accuracy is the most important issue corresponding to the accuracy of 3D object printing. However, objects produced though Vat Polymerization may warp if the platform is large, combined with printing a large flat surface, Figure 3.12. The light focused on each layer of the object will cause each layer to shrink during solidification, Figure 3.13. Stress between the two layers arises as one layer shrinks on top of a previously solidified layer. This will cause slight deformity. Therefore, adequate support is vitally important to anchor at-risk sections, and may limit or diminish the deformity. Also, object orientation and limiting the size of the flat layers play important factors. The separation of the bottom-up printers may cause dimensional discrepancies during the removal stage. Additionally, the forces during the separation stage may cause the soft print to curve, caused by the accumulated load of each layer. Resins that possess malleable properties are at a greater risk of warping. This will result in poor object accuracy for high-end applications.

Fig.3.13: 3D Printed Part Suffered Warping Deformation

Table 3.2: Vat Polymerization Accuracy Requirements

Parameter	Description
Dimensional Tolerance	±0.5 (Lower Limit: ±0.15)
Shrinkage Warpage	Large Flat Surface, Long Unsupported Span Likely to Shrink or Warp
Support Requirements	Essential to Achieve Accurate Part

Limited Shelf-Life

Vat Polymerization technologies use thermoset photopolymers materials unlike FFF, materials to produce objects. The polymer materials are supplied in the form of viscous liquids or resin easily cured by laser. The number of colors available to the SLA/DLP resins is limited. Photopolymer resins also have a limited one year shelf-life, if stored properly. When producing objects using Vat Polymerization, it is critical that the objects are cured properly under UV light after 3D printing. This will ensure that objects will achieve their optimal properties. Information on the optimal UV exposure times are provided on resin datasheets by their respective manufacturers. Table 3.2, presents some of the most common Vat Polymerization associated tolerance requirements to avoid Shrinkage and warpage.

Post-processing Vat Polymerization printers are capable of printing fine detailed prints with feature sizes as small as 0.3 mm. The limitations of this technology are its dependency on support structures, which is attached to the model. While the support structure provides stability to the object being printed, nonetheless, it leaves marks and possible dents on the surface, which may create uneven surfaces. Therefore, it is best practice to place the supports on the least visible part of the model.

Vat Polymerization resins allow for a range of finishing options with the most common of these described in Table 3.3. With the appropriate post-processing, Vat Polymerization objects may provide a smooth surface finish representative of an injection molded part.

Table 3.3: Common Post Processing Options for Vat Polymerization

Condition	Post Process	Description
Compulsory	Support removal	Vat Polymerization support material is always printed in the same material as the build. Support is printed as a series of tower structure~ that narrow at the tip to aid in removal. Support is generally broken off by hand or cut using tools. This typically results in small nibs on the surface of the print.
Surface finish	Sanding/ wet sanding	Because of the smooth surface that Vat Polymerization technologies produce, sanding is often only required at the areas where support was attached to the print. 600 grit sandpaper is recommended. Wet sanding is recommended for areas where build lines are present or when it is desirable to have an injection mold-like smooth finish. For wet sanding, a grit of 800 - 1000 is recommended. Wet sanding offers the best surface preparation for painting.
Aesthetic	Mineral oil finish	Mineral oil helps to assist in hiding any white/light spots on the print creating an appealing finish. It is usually applied after the sanding process (not wet sanding) and results in a glossy finish. This finish is well suited for mechanical parts reducing friction and lubricating the surface. A clean cloth is used when applying the oil.

	Spray painting	Spray painting helps to conceal layer lines often eliminating the need to sand the unsupported side of the model. The coating also protects the model from yellowing by limiting UV exposure. Color paints or clear acrylics are typically used. It is not recommended that flexible resin parts are painted.
Aesthetic	Polished	Polishing of Vat Polymerization parts is possible but requires parts with a simple geometry (ideally large flat surfaces like watch faces or clear enclosure cases). It is a very labor intensive process requiring the surface to be sanded using increasing grit levels of sandpaper (concluding with 2000 grit). The surface is then polished with a plastic polishing compound. If completed correctly, it is possible to produce a fully transparent appearance on parts printed with clear resin.

The main benefits of Vat Polymerization are the smooth surface finish, the high accuracy, and the fine details this technology is capable to produce objects.

The smooth surface finish makes SLA one of the best suited 3D printing technologies for replicating or producing injection molded-like prototypes. Also, the SLA technology is often adapted to 3D print visual models, where a smooth surface finish is desirable, such as figurines, enclosures, hand held consumer products etc.

Fabricating Detailed Parts

Vat Polymerization is also one of the most dimensionally accurate techniques of 3D printing, due to its ability to produce objects of high details, where object accurate tolerances and intricate features are needed, such as producing jewelry and dental industries. The downside of the SLA 3D printing is the photopolymer properties used. The SLA Photopolymers usually produce brittle objects that do not have the impact strength or durability if compared to injection molded objects. Therefore, their use for producing functional objects is limited. Additionally, SLA/DLP typically printed objects have limited life. They experience a loss of mechanical properties over time and degrade in the presence of sunlight. Accordingly, coatings are applied to extend their life. These material limitations are the main factor that Vat Polymerization technologies have not been widely used for functional applications.

The main difference between industrial and desk top Vat Polymerization 3D printers is that the industrial build environment machines are used a regulated environment for greater control over resin behavior during printing. Industrial machines have a smaller laser spot size of higher DMD resolution, enabling the machines to produce more accurate parts, and utilize a large range of engineering materials. Vat Polymerization technology strength is their scalability, while most desktop Vat Polymerization printers are similar in size. The industrial machines vary in size significantly. The SLA machines in particular, have very few limitations when scaled up to several large build sizes, irrespective of their slow build time. SLA machines represent some of the largest 3D printers in use today. Most large industrial Vat Polymerization machines are top down configurations. The bottom up separation forces increase substantially for large builds. The Industrial machines offer a greater range of engineering application for specific materials that are often proprietary to a specific printer. They will often have slight variations in properties allowing engineers to select the exact material that is suitable for an application

Medical and Dental Applications

Desktop printers can produce parts to an accuracy of 100 - 250 microns, while industrial machines are capable of 10 - 30 microns. Higher accuracies in general, lead to a better surface finish. The level of accuracy and surface finish of desktop machines can produce objects suitable for most applications. Industrial Vat Polymerization machines are best suited for larger objects and full scale prototypes, where a very high level of object accuracy is required in particularly for medical or dental applications, as well as in mid-level production runs as in the case of jewelry, Figure 3.14 and Figure 3.15.

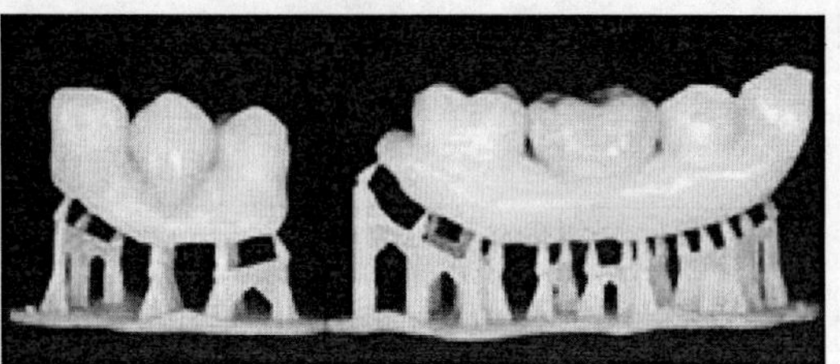

Fig.3.14: 3D Printed Teeth
Courtesy: Formlabs

Smooth Surface Finish

Vat Polymerization for 3D parts printing are most suitable for visual applications and prototyping where a smooth surface finish and high accuracy is desirable. Some of the most common applications of Vat Polymerization are presented below.

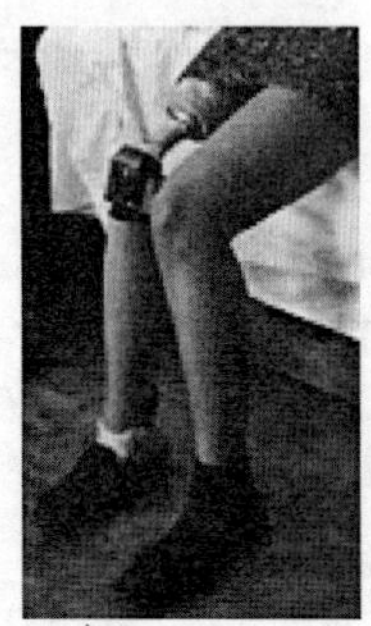

Fig.3.15: 3D printed prosthetic
Courtesy: All3DP

Injection Mold and Prototypes

ThesmoothsurfacesproducedbyVatPolymerization often adopted for producing injection molded prototypes. Designers are able to quickly 3D print a design to display it without requiring to invest expensive tooling. Jewelry's investment-casting often utilize Vat Polymerization technologies, Figure 3.16. The accuracy and intricate details of jeweler parts make it ideal using Vat Technologies to create smooth surfaces for the this industry.

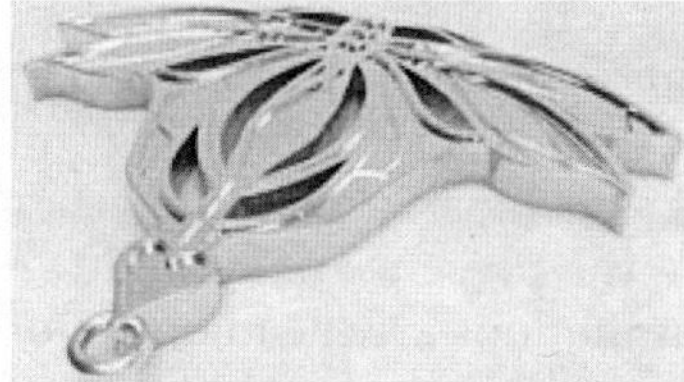

Fig.3.16: 3D Printed Jewelry

Orthodontics Models

The dental industry has adopted Vat Polymerization for a range of applications, Figure 3.17. Vat Polymerization is used to 3D print dental models, surgical guides, dental appliances, crowns and bridges. Vat Polymerization becomes a positive disruptive technology to produce parts with available hygienic materials with to a high level of accuracy and detail, specifically dental resin.

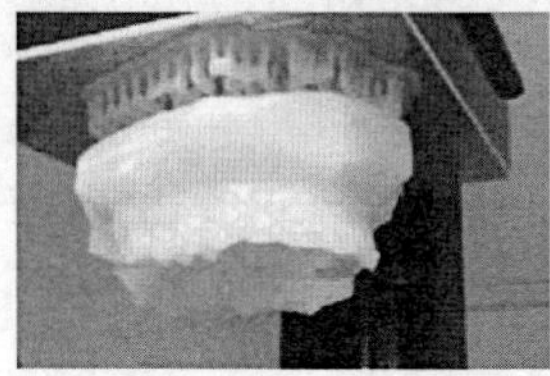

Fig.3.17: Vat Polymerization SLA Printing

Hearing aids

Vat Polymerization has been successfully used in custom built Hearing Aids as a notable success implementation of the technology. There ae in excess of 20,000,000 patients wearing hearing aids produced with Vat Polymerization technologies. The ability to 3D print organic smooth surfaces required for a hearing aid at a substantial low cost than traditional techniques, Figure 3.18. Vat polymerization technology is currently 3D printing 97% of all hearing aids produced.

Fig.3.18: 3D Printed Hearing Aids

New Developments

Although Vat Polymerization has been in the market for a longer time as it is the oldest 3D printing technology, nonetheless it has seen little innovation in the last decade. Though, continuous printing is one of the most celebrated innovations in the field of 3D printing.

CONTINUOUS LIGHT PROCESSING - CLP

Carbon Continuous Direct Light Processing (CCDLP) produces parts in a similar way to the DLP. However, CCDLP relies on the continuous motion of the build plate in the upward Z direction. One company that commercially utilizes this 3D printing technology is "Carbon." The company's Digital Light Synthesis™ technology, which is similar to CCDLP, was enabled by a process called "Continuous Liquid Interface Production" (CLIP). Carbon's M-Series printers use a process-specific photopolymer in conjunction with an oxygen-permeable window to create a "dead zone" of uncured resin at the bottom of the vat. This development resulted in preventing the bottom of the print sticking to the vat. Also, this development eliminated the need for the separation required in the separation process necessary for most bottom-up printers. Additionally, this development allows faster build times as the printer is not required to stop and separate the part from the build plate after producing each layer. This advantage enabled the parts to become isotropic. There are no individual layers produced during printing. The Carbon printers are exclusively employing bottom-up machines.

POLYMER POWDER BED FUSION TECHNOLOGY

Selective Laser Sintering (SLS) technology uses powder bed fusion through a thermal source to induce fusion between powder particles, at a specific location of the build area, Figure 3.19, to produce a solid part. Most powder bed fusion techniques employ mechanisms for applying powder as the part is being printed. This results in the final component being encased in powder. Referring to the simplification chart of 3D Printing Technologies stated below:

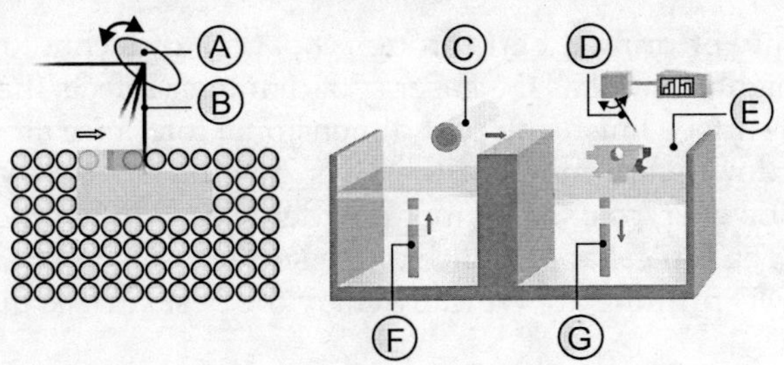

Fig.3.19: SLS 3D Printer

Courtesy: FARSOON

Technology	Recommended Use	Status
Material Extrusion Technology	Fused Filament Fabrication	Discussed
Vat Polymerization	Stereolithography SLA Digital Light Processing DLP	Discussed
Powder Bed Fusion – Polymers	Selective Laser Sintering SLS	In progress
Material Jetting	Material Jetting DOD	To Follow
Binder Jetting	Bonder Jetting	To Follow

THE SELECTIVE LASER SINTERING (SLS)

The Selective Laser Sintering (SLS) process involved in heating a polymer powder, which is placed in a bin to a temperature slightly below the melting point of the polymer, Figure 3.19. A typical layer of 0.1mm thick is deposited onto the build platform though a recoating blade. A CO2 Laser beam is applied to scan through the surface to sinter and solidify the scanned cross section of the surface. The Laser is then focused to the precise location through a pair of galvanometers. As the entire cross section is scanned the build platform advances downward one layer in thickness height, Figure 3.20.

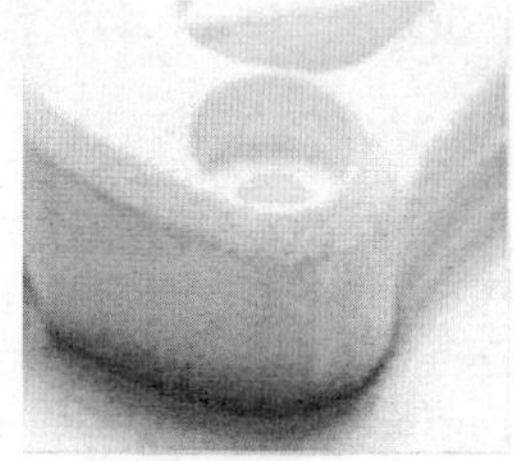

Fig.3.20: Visible Surface Texture Grains Produced by SLS Technique

The recoating blade deposits a new layer of powder on the top of the recently scanned layer. Once again, the Laser begins to sinter the successive cross section of the part onto the previously solidified cross sections. The process is repeated until the part is fully printed.

One of the major advantages of the Selective Laser Sintering is that the un-sintered powder remains in place acting as a structural support for the part being printed.

Accordingly, a solidified 3D printed part is produced in the middle of a bin filled with polymer powder. This type of technology employs multiple parts to be simultaneously printed. Additionally, the considerable volume of the polymer powder can be reused. Therefore, optimizing the built volume to accommodate as many parts as possible to orient and fill the bin volume plays an important part in optimizing the process of printing batch of parts simultaneously.

Once the sintering process solidifies the batch of cooled parts in the bin, the bin is unpacked. The un-sintered powder is collected to be reused. The solidified parts are usually cleaned by blast of air.

SELECTIVE SINTERING LASER PRINTER

The quality of printed part through this type of technology is dependent on several parameters. Layer height, as well as the Laser spot diameter affects the production printing quality of the part. The part surface finish and part dimensional tolerance are directly impacted by the Laser spot size and the downward move of the layer height. Finer polymer powder will induce better surface finish. However, coarse polymer powder is easy to handle, while it produces surface finish of less quality. The typical appearance of the surface finish of SLS is lusterless, while the parts produced through 3D printing are typically enjoy the best surface quality if the part if facing downward.

Minimum Operator Interface

Printer manufacturers usually set up the printer settings to operate the machine with minimal operator interference except in the process of loading bin powder, unloading and cleaning parts, as well as monitoring the print progress. The machine may be automatically adjusted depending on selecting the type of build materials. The operator in this case selects the build material, and then the machine is consequently adjusted its parameters according to the manually selected materials. The SLS machines are autonomous during the heating and cooling processes.

Bin Volume Packing

SLS produces parts in a batch production environment. The efficiency of part production and the number of parts produced is impacted directly by the optimal arrangements of parts and their corresponding orientation in the bin volume. The time to print is directly proportional to the height of the bin, considering that the Layer height is 0.1mm, and the bin height is 400mm.

Accordingly, cycle parts are 400/0.1 = 4000 Cycle regardless the size of the parts, and number of parts in the bin.

Printers are equipped with optimizing routines to ensued part volumes and their corresponding orientations are at maximum dense capacity in the given bin volume. It is beneficial to fill the bin at a maximum capacity before each printing cycle begins. The bin configuration allows for 5mm clearance between the maximum height of parts in the bin, and the top edge of the bin. Typically, the bin size is 300 mm x300 mm x300 mm. There are other bin sizes that can reach up to 550 mm x 500 mm x 750 mm.

Layer Adhesion

Adhesion between layers is vitally important in 3D printing and as much as important when using SLS technology. For instance SLS creates parts layer by layer, the adhesion between layers must be robust, and cohesive to ensure part manufacturing integrity. Initial heating of the build powder followed by Laser sintering exposure causes the powder molecules to fuse together in multiple directions, resulting in homogeneous parts. The properties of the parts produced by SLS PA12 machine are stated as follows:

Tensile Strength	48 MPa in XY direction	42 MPa in Z direction
Tensile Modulus	1650 MPa XY direction	1650 MPa in Z direction
Strain at Break	12% XY direction	4% in Z direction

This is consistent with most powder suppliers.

When adding glass and carbon, as in glass filled polycarbonate to SLS polymer for parts, the materials strength can be reduced as much 40% in the build direction. Therefore, the user must be well equipped to select materials appropriate to SLS materials for a specific application to avoid inconsistent part strength produced in the same batch.

Dimensional Accuracy

As in several 3D printing technologies SLS parts may experience shrinkage and warping during part printing production. SLS sinters each layer, which is fused with the layer below as it cools. The cooling process, which causes the newly printed layer to shrink, causing in return pulling up the underlying layer -curling the part upward to clash with the recoated layer during the powder spreading stage. Accordingly, it is prudent to orient large flat parts at an angle, or even vertically to reduce curling consequence of curling effects.

Limiting Shrinkage or Warpage

To limit the likelihood of parts warping or shrinking during manufacturing using the SLS printers, a heated build chambers is incorporated to raise the temperature of the powder just below the sintering temperature. This enables monitoring the temperature gradients, in particularly in large SLS parts, where the bottom of the part may have cooled down, while printed top layers may still remain heated at an elevated temperature.

The cooling step in the SLS technology represents a vital process in printing applications. Additionally, further cooling the part in the polymer bin assists in reducing the warping effect. Usually, parts are left in the powder bin to cool slowly, sometimes up to 50% of the total build time before removing.

POWDER BED FUSION MATERIALS

Powder Bed Fusion exhibit more stable behavior during the sintering phase when using materials with a low thermal conductivity. Laser sintering technology uses thermoplastic polymer known as polyamide (PA). This thermoplastic polymer is used exclusively in the SLS applications to produce parts. Polyamide parts have excellent long-term stability and good chemical resistance, such as the common commercial polyamide known as "nylon."

The price of the SLS Standard PA12 "nylon "powder may vary from $50-$60/Kg depending on the material type. While SLS powders are usually white color, however; it can be obtained in grey or black. Also, produced parts can be dyed in different colors. All powder based methods of manufacturing-care must be highly considered in handling any form of loose powder. Small grain size of SLS powder can be a health hazard to operators. . The airborne and particles can cause respiratory problems. Correct safety gear must be worn.

In order to obtain a greater SLS parts mechanical properties, as well as heat/chemical resistance with a better appearance, nylon can be mixed with different materials, such as aluminum, glass, carbon and graphite to form a composite powder.

The SLS process has no significant loss of mechanical properties when 50% of un-sintered powder per print is unused allowing it to be recycled.

Post Processing

Table 3.4 provides the common practices of post processing activities to ensure SLS parts are printed to a high level of dimensional accuracy, have good strength, and function as stated in the engineering specifications. The surface finish of the SLS parts is contributed to the characteristic of the Powder Bed Fusion process.

Table 3.4: Common Post Processing Options for SLS

Type	Post Processes	Description
Compulsory	Loose powder removal	Parts are removed from the build chamber and all powder is removed from the part with compressed a1r. The surface is also cleaned via plastic bead blasting to remove any un-sintered powder sticking to the surface. This finish is inherently rough, similar to medium grit sandpaper (satin-like matte finish that is slightly grainy). This is the best surface finish for paint1ng or lacquering.
Surface "fInIsh		For a smoother surface texture, parts can be polished in media tumblers or vibro machines. A tumbler contains small ceramic chips that vibrate against the Object gradually eroding the outer surface down to a polished finish. This process does have a small effect on part dimensions and results in rounding sharp edges. It is not recommended to tumbled parts with fine details and intricate features. It is not recommended to tumbled that parts with fine details and intricate features.
Aesthetic	Dyeing	The fastest most cost effective method to color SLS prints is via a dye process. The porosity of SLS parts makes them ideal for dyeing. The part is immersed in a hot dye color bath with a large range of colors available. Using a color bath ensures full coverage of all internal and external surfaces. Typically, the dye penetrates an approximate depth of 0.5 mm. Eventfully; the surface may continue to wear until the original surface is exposed.

	Painting and lacquering	SLS parts can be spray painted and/or coated with a lacquer (varnish or clear coat). Via lacquering it is possible to obtain various surface finishes, similar to high gloss surface or metallic shine. Lacquer coatings can also improve wear resistance, and reduces the likelihood of the paint drying time or surface hardness, water tightness and limit marks and smudges on the surface of the part. Due to the porous nature of SLS it is recommended that 4- 5 very thin coats are applied to achieve a final finish instead one thick coat. These results in faster drying time and reduce the Likelihood of the paint of lacquer running.
Functional	Water Tightness	A correctly sintered SLS part will have some inherent water tightness. Coatings can be applied to further enhance the tightness feature. Silicones and vinyl-acrylates have been shown to provide the best results. Polyurethane (PU) is not recommended for water proofing SLS parts. If water resistance is required a clip coating method is recommended.
	Metal Plating	SLS parts can be electroplated. Stainless steel, copper, nickel (or a combination of both), gold and chrome can be deposited and chrome can be deposited on the surface of parts to increase strength or electrical conductivity. Parts are cleaned and a conductive layer of material is applied to the surface. The parts then go through traditional metal coating procedures. The plastic can be retained as structural support or burnt out to create thin- walled parts 25 to 125 microns thick.

BENEFITS AND LIMITATIONS

SLS is best suited technique for producing high strength and functional parts with complex geometries. The isotropic nature of parts and high their level of accuracy enables the SLS technology adopted for the production of end use parts. Although the produced part through SLS technology are not of higher quality as parts produced through Vat Polymerization or Material Jetting. The SLS process requires no support structure of parts. Therefore, it gave this technology additional advantage over the other 3D printing technology. The support structure does not need to be removed after printing providing a better overall surface finish.

Nevertheless, the significant drawback in SLS printing technology is the cost of its industrial machines. The cost may exceed $250,000, and may require a high level of skilled operators. Also, the demand of material handling protocol, as well as the lead times to obtain a batch product is much longer than other 3D printing technologies.

SLS lead time is considerable compared to traditional subtractive technology such as CNC machines. The necessary heating and cooling processes prior, during, and post printing, on a full batch of 300 mm x 300 mm x 300 mm bin takes an average of 22 hours in addition to another 12 hours of cooling time before parts can be handled for post processing.

At present SLS machine efficiency has increased by allowing for the removal of powder bins to be heated and cooled while parts are extracted from the machine. SLS parts also have a rough surface finish unless post processed.

Common Applications

The versatility of the SLS technology enthused end users and manufacturers to use the technology for a large range of applications, as stated below.

Functional parts

(a) SLS printing can produce acceptable functional parts with acceptable tensile strength
(b) SLS produces parts to withstand normal loads when placed in service.
(c) SLS can produce parts with complex geometries
(d) SLS uses standard well-known materials - like PA12 "(Nylon)."

Low run part production

(a) SLS is cost effective,
(b) low production run
(c) SLS provides parts engineering feedback, design modifications, and part performance.
(d) SLS prints multiple parts in a full powder bin.
(e) SLS machines can manufacture variety of sizes considering the economies of scale.

Complex ducting (hollow sections)

(a) SLS can create parts with hollow sections, unlike several other 3D printing technologies
(b) SLS is ideally suited for the low run production of complex ducting and piping.
(c) SLS is capable of printing parts for various applications.

NEW DEVELOPMENTS

Recently, HP recently embarked on developing 3D printing space using their current hardware. HP spent a decade of research and development to manufacture a new HP 3D-SLS printing technology, which is different from SLS, however; its applications are similar. HP is the first corporation to bring SLS to the desk top practice. Also, "FormLabs" corporation is developed a similar SLS printer, "Fuse 1" SLS printer.

Multi-Jet Fusion - HP

The HP Multi Jet Fusion (MJF) printer works in a similar method to other Powder Bed Fusion technologies in additional to an extra step added to the process. The addition is a "detailing agent." It is a layer of build powder, which is first applied to a work area. A fusing agent is then selectively applied, where the particles are to be fused together, similar to how "Binder Jetting" works.

The fusing agent improves the heat energy absorption from the heat source. At the same time, a localized detailing agent is applied. The detailing agent reduces fusing at the part boundary. This produces sharp smooth edges. The work area is then subjected to heat fusing energy to fuse and solidify the powder particles. The heat source scans the build in a linear fashion instead of as a single point. The process is then repeated layer by layer until a complete part is formed.

The Multi Jet Fusion printer is 25% faster due to the substantial reduction in cooling time after printing, and the method the heat energy is applied compared to traditional SLS printers.

Undoubtedly, the future generations of the Multi Jet Fusion printers will offer the ability to produce full color prints based on the color of the binder that is jetted onto the powder.

Fig.3.21: HP Multi Jet Fusion Color Printer

HP Multi Jet Fusion is positioned as a formidable competitor for low- to medium volume production contributed t the improvements made in the speed of the printer, and the ability to create functional polymer parts at adequate mechanical properties, Figure 3.21.

DESKTOP SLS

In an attempt to introduce low cost competitive desktop SLS printers a range of printers have recently been developed. Accordingly, Sharebot SnowWhite, Sintratec Sl, Sinterit Lisa, and more recently the "FormLabs" Fuse 1 were introduced.

The use of a fiber laser has lowered the cost of the printer as well as reducing the required power used in machines.

MATERIAL JETTING – DROP ON DEMAND (DOD)

Material Jetting is one of the additive manufacturing - 3D printing technology developed to use droplets of photopolymers to be cured and solicited when exposed to UV light. The 3D printed part is built up one layer at a time. The nature of the "Material Jetting" process allows for different materials to be printed in the same part. This technology often uses support structures printed also during the build process.

Technology	**Recommended Use**	**Status**
Material Extrusion Technology	Fused Filament Fabrication	Discussed
Vat Polymerization	Stereolithography SLA Digital Light Processing DLP	Discussed
Powder Bed Fusion – Polymers	Selective Laser Sintering SLS	Discussed
Material Jetting	Material Jetting— Drop-On-Demand (DOD)	In progress
Binder Jetting	Binder Jetting	To follow
Powder Bed Fusion (Metal)	DMLS/SLM, EBM	To follow

Material Jetting Technologies

Material Jetting printer sputters a blast of substantial amount of micro droplets of polymers successively following a specific path to build a uniform a cross sectional layer of part. The jetted photopolymer droplets in the layer are solidified once exposed to UV light, Figure 3.22. Multiple layers of multiple cross sections are built upon one another to create a solid part. After a layer has been deposited and cured the build platform drops down one layer thickness and the process is repeated to build up a 3D part.

Material Jetting operations deposit build material in a linear rapid manner, Figure 3.22. Accordingly, Material Jetting printers have the capabilities to print multiple parts in a single line without inhibiting the build speed. Once the space allocated of the parts are correctly placed and optimized within each build line, the Material Jetting system can competently 3D print parts at a much quicker rate than other 3D printing technologies. Material Jetting technology deposits material in a line.

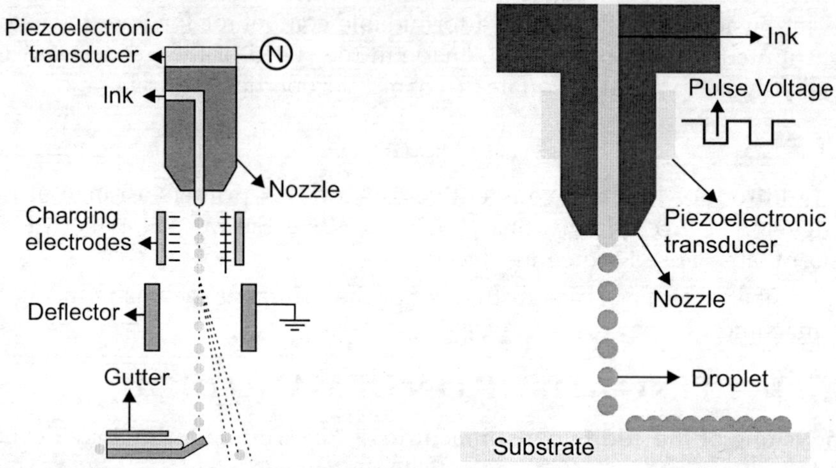

Fig.3.22: Material Jetting 3D Printer – Drop-On-Demand

A support structure is required in "Material Jetting" printers in particularly, when printing simultaneously during the build from a dissolvable material that is removed during post processing. Multi-material printing with multiple different colors can be uniquely used in "Material Jetting" 3D printing technology.

DROP ON DEMAND (DOD)

"Drop-On-Demand" (DOD) printers have two unique print jets. One print jet is to deposit the build photopolymers materials. The other print jet is to deposit dissolvable photopolymers support material, Figure 3.22. Similar to other additive manufacturing techniques, DOD printers follow a set path and jet material in a single moving point to generate the cross sectional area of a component layer by layer.

Also, DOD printers employ a fly-cutter that skims the build area after each layer is produced to ensure a perfectly flat surface before printing the next layer. DOD technology is used to produce patterns of casting and mold making applications.

PRINTER CHARACTERISTICS AND PARAMETERS

Material Jetting printers jet out build, as well as support material structure to create parts. The size of the Jet droplet is directly related to the print head jet diameter, together with the layer height influence the surface finish and the minimum size of a part. The "Material Jetting" system is one of the most accurate forms of 3D printing, which is capable of producing parts with layer heights as low as 16 microns. Consequently, it results in very smooth surfaces. The maintenance of the print head is important to avoid clogging when the small jet diameters are blocked. The "DOD" machines have special systems located precisely in place to clean the jets and electronically notify operators when cleaning is needed.

The build material must be kept in a liquid state to successfully dispose its jetted droplets. Most "Material Jetting" machines heat up the resin to an optimal temperature of 30 - 60°C to ensure that the droplets viscosity of the photopolymer is consistent during printing.

The "Material Jetting" machines automatically adjust machine parameters based on the selected printing material, similar to the SLA and SLS machines.

SUPPORT STRUCTURES

Material Jetting possesses a unique quality, which enables parts to be printed in two different materials. One material is used for the main build, while the other material is a dissolvable material and used to print the structural support. The structural support which is made of dissolvable material can be easily removed with light agitation. The support structure if removed carefully during the post processed it may not show left over marks on the finished part.

The structural support in this type of technology, rather than the lattice or tower styles adopted by Fused Filament Fabrication and SLA, involves a large amount of material used, which causes increase in build time and cost. Also, printing this type of structural support requires that part orientation must be flexible than other 3D printing technologies. Additionally, part orientation must be considered with the purpose of minimizing support material dependence.

Dull vs. Glossy

"Material Jetting" has the ability to print parts in either dull or glossy settings, Figure 3.23. The dull setting will add a thin coating of support across the entire part surface, regardless of orientation or structural requirements. The glossy setting will only use support material where required, such as overhangs, and drafts.

Fig.3.23: Object Partly Printed in Glossy and Partly Dull Texture

While the advantages of printing in a glossy setting produce smooth and shinier surface finish on the areas of no structural support, and a reduction in material usage for the build, there are also few disadvantages, which include:

A non-uniform finish on parts, is exhibited and a few rounding of sharp edges, and corners on top surfaces, the additional material usage, require additional cleaning time, and a softer surface. This softer surface often leads to weakness of a small or a thin wall.

However, the advantages of printing in a dull setting are maintaining part accuracy as a whole, as well as maintain a uniform surface finish.

DIMENSIONAL ACCURACY

"Material Jetting" is considered the most accurate form of 3D printing as there is no heating involved in the printing process, except during the initial heating of the resin to maintain printing consistencies preventing part warping, and avoid shrinkage.

"Material Jetting" begins to lose accuracy as part size increases due to photopolymers shrinking as they cure. This effect becomes more exaggerated with larger parts. Most dimensional inaccuracy issues are related to having a design of thin wall that are printed below printer specifications. "Material Jetting" prints support structure as a solid structure from a soft secondary material that is dissolved and removed after printing. The solid nature of the support structure results in surfaces printed to a high level of accuracy.

Care must be taken when handling parts produced through "Material Jetting" as they can warp and become dimensionally unstable as a result of exposure to ambient heat, humidity, or sunlight.

MATERIALS

"Material Jetting" technology uses thermoset photopolymer resins to produce 3D printed parts. The resin is jetted in micro-droplets on the build platform and then cured by a UV light. "Material

Jetting" technology requires materials with a low viscosity to successfully be jetted in droplet form. This requires that most resins are heated up from 30 - 60°C, depending on the printer and material, while printing. As stated previously, "Material Jetting" requires two different resins when printing, utilizing different color cartridges to produce colored 3D prints. The resin proprietary cost of cartridges used in "Material Jetting" machines vary between $300 to $1000/kg.

POST PROCESSING

Once "Material Jetting" support structure material is removed it produces most desirable natural surface finish with respect to all 3D printing technologies. A range of common post processing techniques for "Material Jetting" parts are shown in Table 3.5.

Table 3.5: Common post processing options for Material Jetting

Type	Post Process	Description
Compulsory	support removal	Material Jetting uses a water soluble support material. Initially the part Is soaked In cold or room temperature water to soften the support material before removing by hand. For smaller, difficult to reach areas a water jet stream is used. Fragile parts should be cleaned. Using a low pressure fan stream. Parts are then soaked In a sodium hydroxide solution to remove any remaining support material from the surface, followed by a water rinse.
Surface	Sanding	Material Jetting parts generally only require sanding if parts are going to be coated. Parts are sanded with 320-grit wet sandpaper until a smooth, paint-ready surface is achieved.
finish	dyeing	Dyeing of Material Jetting parts Is generally applied to clear and translucent parts to change their appearance without losing Transparency. Parts are Submerged in a Water-dye mix for coloring.
Aesthetic	Painting	After sanding, a fast drying primer (lacquer based primers or paints are recommended) is applied to the surface before painting. The primer will provide a superior surface for the paint to adhere to. Paint should then be applied in a number of thin coats.
Functional	Clear coat	A clear coat can be applied to the surface of Material Jetting parts to Improve wear resistance. As with palnt1ng, sanding Is recommended before application. A lacquer based clear coat should be used and applied in several thin coats. When used in conjunction with paint this combination of coatings greatly helps to Increase the life of Material Jetting parts.
	Metal coating	A metal coating can be applied to the surface of Material Jetting to provide a decorative finish or a hard surface for wear. After sanding, a conductive coating is applied to the surface of the part. The part then goes through standard metal coating procedures.

Benefits and Limitations

"Material Jetting" has three main benefits stated as follows:

1. "Material Jetting" creates a near homogeneous part when the layers are cured throughout the printing process.

2. Parts produced by "Material Jetting" have a quality surface finish, compared to injection molded parts.
3. "Material Jetting" is provides dimensional stability and accuracy compared to most of 3D printing technologies.

   Accordingly, "Material Jetting" is often used for realistic prototypes, which is closely representing end-parts.

Material Jetting uses photopolymers to produce parts. Unfortunately, the produced parts have poor mechanical properties, poor compression, and poor tensile strength. . "Material Jetted" parts are not as strong as other 3D printing with polymer materials. The brittle nature of the acrylic based resin can be a problem for functional testing. Also, low heat deflection temperature is a negative factor for most real world testing or functional applications. Additionally, for rubber-like materials, the lack of elongation represents additional negative factor when testing for a rubber elongation test.

"Material Jetting" is one of the most expensive techniques used in 3D printing compared to the other 3D printing technologies, contributed mainly to the high cost of material. The need for printing a structural support in "Material Jetting" machines as a solid mass, adds a large amount of waste that further adds to the already high material cost.

COMMON APPLICATIONS

The smooth surface and the high accuracy combined with the diverse range of available materials, resulted in using "Material Jetting" technology to create very realistic prototypes to appear like the real part. Some of the most common applications of "Material Jetting" are presented below.

Full color visual prototypes

"Material Jetting" has the ability to print high detail, full color models that accurately represent a final part. This allows designers and prototype developers to obtain a unique insight into the final look of the part.

Medical models

"Material Jetting" technology is also used to rapidly produce medical models. Patient specific data can be collected and stored, and then printed to provide physicians and orthopedic surgeons rare perspective on patient's anatomy. Medical models play an important role in training and preparing physicians for medical procedures used for visual, discussions, opinions, or educational purposes.

Injection mold-like prototypes

The smooth surface and high level of detail offered by "Material Jetting" often consider the technology as a useful tool to verify injection molded designs. Parts can be quickly printed to enable the designers the chance to check form, clearance, fit, assembly before investing in elaborate and expensive tooling.

Low-run Injection Molds

Simulated ABS is a material often used for the production of low run injection molds as illustrated in Figures 3.24. The high temperature resistance material used in "Material Jetting" machines provided developers the ability to accurately produce complex geometries. Also, material property enabled the technology to become popular in the injection molding industry.

Fig.3.24: Low Run Injection Molds Using Material Jetting Technology

NANO PARTICLE JETTING - XJET

The most notable development in the "Material Jetting" application has been accomplished by the innovation of an Israeli company "XJet," who focused on metal printing through "Material Jetting" at "nano" scale. Nano Particle Jetting "NPJ", by "XJet," utilizes a liquid containing metal nanoparticles loaded into the printer as a cartridge. Extreme thin layer of "nano" particles jetted onto the build tray, similar to "Material Jetting." High temperatures inside the build envelope cause the particles to bind enabling the jetting liquid to evaporate leaving behind metal parts. These 3D prints are then sintered in a furnace to create a fully dense printed part.

Producing parts by this method offers three main advantages:

1. The ease of removing the support material provides a high degree of design freedom
2. The "nano" particle size and thin layer thickness provide intricate features that can be are easily produced
3. The printer cartridge system used to allow a safe handling of materials, unlike handling powder-based systems

BINDER JETTING

Table 3.5: Presentation Status of the 3D Printing Technology in this book

Technology	**Recommended Use**	**Status**
Material Extrusion Technology	Fused Filament Fabrication	Discussed Chapter 1
Vat Polymerization	Stereolithography SLA Digital Light Processing DLP	Discussed Chapter 2
Powder Bed Fusion – Polymers	Selective Laser Sintering SLS	Discussed Chapter 3
Material Jetting	Material Jetting— Drop-On-Demand (DOD)	Discussed Chapter 3
Binder Jetting	Binder Jetting	In progress
Powder Bed Fusion (Metal)	DMLS/SLM, EBM	In Progress

"Binder Jetting" is the technology of depositing successively a binding agent onto a powder bed to form a part, one layer at a time. The layers bind to one another to form a solid part. "Binder Jetting" is a versatile 3D printing technology, which is used in a wide range of applications. "Binder Jetting" can be separated into two categories as follows:

1. sand printing
2. metal printing.

"BINDER JETTING" TECHNOLOGIES

"Binder Jetting" system prints in a similar fashion as the SLS technology, however, requiring an initial layer of powder on the build platform. "Binder Jetting" uses a laser to sinter powder. It moves a print head over the powder surface depositing binder droplets of a dimeter of 80 microns. The droplets bind the powder particles together to produce each layer of the part. Once a layer has been printed, the powder bed is lowered and a new layer of powder is spread over the recently printed layer. This process is repeated until a solid part is generated, Figure 3.25.

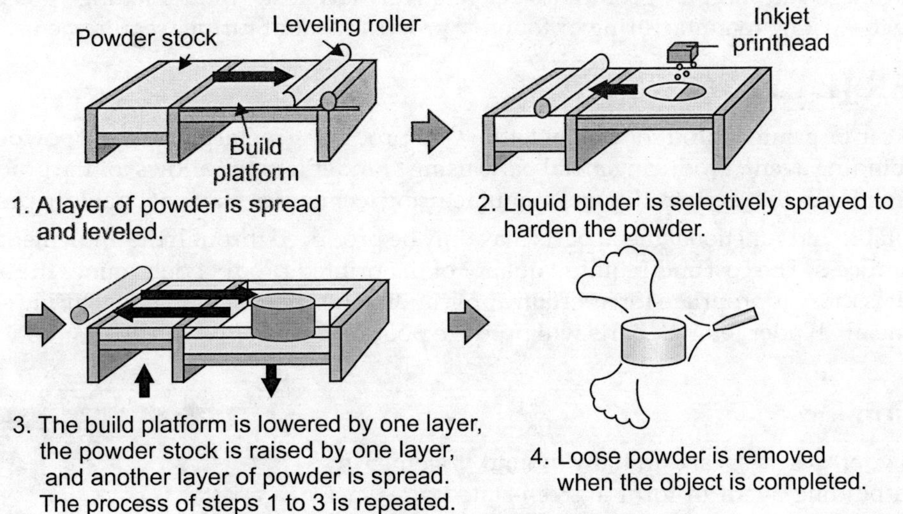

Fig.3.25: Binder Jetting Printing Technique

Courtesy: TCaST® Corporation

The part is then left in the powder to cure and gain strength. The part is then removed from the powder bed, while the unbound powder is removed by compressed air.

SAND "BINDER JETTING"

"Sand Binder Jetting" is a low cost method for producing parts from sand extracted from sandstone or gypsum. The two most common methods of sand printing are described below.

Full Color Models

Models of full color presentation can be printed using a plaster- based powder with a binder liquid as a binding agent. The print head jets the binding agent, while a secondary print head jets color to create colored models, Figure 3.26. Once the printed parts are fully cured they are removed from the loose unbonded powder, and then cleaned. Mechanical properties may be enhanced parts may be exposed to an infiltrating material. There is a large range of infiltrating materials available. Each infiltrating material provides different mechanical property. Additional coatings can be added to improve color appearance.

Fig.3.26: Binder Jetting System Printed Object Using Color Printing Sandstone

Sand Casting Cores and Molds

"Binder Jetting" may also be used for the production of sand cast molds and cores. The molds and cores are generally printed with sand for special applications. After printing, the cores and molds are removed from the build area and cleaned to remove any loose sand. The molds are then ready for casting. After casting the mold is broken apart and the final metal component is removed.

The major advantage of producing sand casting "cores and molds" through "Binder Jetting" is producing large and complex geometries at a relatively low cost. "Binder Jetting" process is able to integrate to existing manufacturing or foundry process without further requirements.

"BINDER JETTING" - METAL

Metal parts can be printed through "Binder Jetting" for producing metal parts. Metal powder is using a polymer binding agent. Producing metal parts using "Binder Jetting" allows for the production of complex geometries that even traditional manufacturing techniques may not be able to produce.

Acceptable and functional metal parts may only be produced through the implementation of a secondary process. The cost and required quality of the printed product determines the secondary process that is most appropriate for a certain application. The absence of manufacturing secondary steps, the metal "Binder Jetting" parts will produce poor mechanical properties.

Infiltration

"Metal powder particles are initially bound together by means of a binding agent to form a green-state part. Once the parts are fully cured, they are removed from the loose powder and placed in a furnace, where the binder is burnt out leaving voids throughout the part at 60%. A bronze material is then used to infiltrate the voids through a capillary action, (Figure 3.27, resulting in parts with high density – 95% and acceptable tensile strength. "Binder Jetting" metal parts have lower mechanical properties than metal parts created through Powder Bed Fusion.

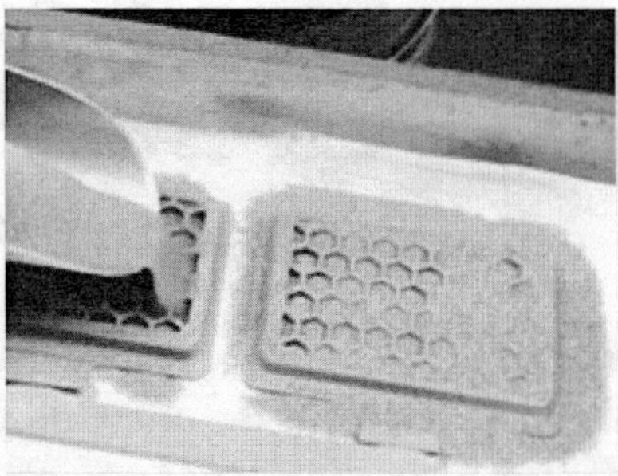

Fig.3.27: Stainless Steel part is covered with sand. Bronze Particles are poured to infiltrate the sand
Courtesy: ExOnc

Sintering

Metal parts may be printed without infiltration. After printing is completed, green-state parts are cured in an oven, enabling the parts to be handled. Parts are then sintered in a furnace to a higher density level, which is greater than 97. Non-uniform shrinkage property is a negative issue during sintering. This property must be considered at the initial design stage.

PRINTER CHARACTERISTICS

Printer parameters

"Binder Jetting" is an effective characteristic combinations of SLS, and "Material Jetting" technologies. It is developed to use powdered material, and a print head together to jet a binder agent to create solid parts. The quality of the surface finish of the parts depends upon the specified layer height, the jetted droplet size, and the powder size. "Binder Jetting" does not require support structures to be printed as parts are surrounded by powder during the printing process. This feature reduces post processing times, and the amount of material consumed per print.

Parts strength

In the absence of material infiltration and sintering the "Binder Jetting" prints parts at limited tensile strength. In spite of implementing material infiltration parts, it still exhibits limited tensile strength and acceptable material elongation. The material breaking point is inferior to the breaking point of part printed by "Powder Bed Fusion. "Binder Jetting" prints are only used as functional parts when secondary processes like infiltration or sintering are implemented with the exception of sand casting.

"Powder Bin Packing"

"Binder Jetting" makes use of a powder bin to print parts without heat. Thus, the process of removing parts avoided the complications associated with differential cooling, which often leads to warping or distortion. This desirable feature allows for multiple parts to easily be printed during the printing process. Design and printing of metallic parts becomes feasible for the low to mid-volume manufacturing of parts prior to applying infiltration or sintering as secondary processes. Large "Binder Jetting" machines have some of the largest build volumes of all 3D printing technologies, up to 1800 × 1000 × 700 mm. This dimensions often used for sand casting of mold production.

DIMENSIONAL ACCURACY

Colored parts caused by sandstone are produced with a layer height of 100 microns. Cores and molds are printed with layer heights ranging from 240 to 380 microns. Several printers are able to print layer heights as low as 50 microns to produce parts of smooth surface if required. printing time does not increase as the heat process is not required during printing the parts. The produced prints exhibit dimensional stability and attractive colors.

When a secondary process is implemented either through infiltration or through sintering process the resulting thermal shrinkage varies between 0.8 – 2%.

POWDER MATERIALS

Printed parts by "Binder Jetting" technology are printed through powder and a binding agent. There are varieties of powder suitable to each print application. The unused powder in the bin can be recycled.

Post Processing

Post-processing of parts made through "Binder Jetting" is limited. Implementation of some infiltration materials and the sintering process may not be adequately applied. Parts always remain in the powder bed after printing for a period of time to allow the binder agent to fully cure. Both sand and metal "Binder Jetting" processes require excess powder to be removed from the part, Figure 3.28. Metallic parts, which make use of secondary processes, like infiltration or sintering, can be post processed using traditional metal techniques.

Fig.3.28: Loose Powder Removed From "Binder Jetting" Part

Courtesy: ExOnc

Most sand casting cores and molds require the removal of excess powder to be ready for casting. Some binders require thermal post processing to improve the tensile strength. The colored models are either dipped or coated with an infiltrating material to improve mechanical properties. The "Binder Jetting" technology directly provides colored printing without the need for additional colorization. However, printed parts can be coated in a clear lacquer to improve wear resistance and improve surface finish.

BENEFITS AND LIMITATIONS

As a result of the absence of the heating process within the "Binder Jetting" operation parts do not suffer the consequence of residual heat stress attributed to rapid heating and cooling. Consequently, the operating cost is low, even with printing large parts.

The binder agents used to bind the sand or metal powder is inexpensive. The material used in making sand cores and molds is an inexpensive silica sand. Also, the powders used in metal printing significantly increase the cost of "Binder Jetting." However Binder Jetting cost is significantly lower than the cost of "Powder Bed Fusion." Also, "Binder Jetting" allows much larger parts such as printing die casts.

The main limitation of Binder Jetting is the mechanical properties of the parts. The printed components removed immediately from the print bed are fragile. A secondary process must be achieved to ensure acceptable and functional painted part, with acceptable surface finish.

COMMON APPLICATIONS

Some of the most common applications of "Binder Jetting" are presented below:

Full Color Models

(a) Although full-color parts printed though "Binder Jetting" are not functional.

Nonetheless, the ability to print in full color propagates many practical applications.

(b) The full color allows for realistic prototypes to be produced, showcasing the appearance of a final part before investing in production.

(c) Also, parts can be used to showcase areas of stress gradients allowing designers to gain a unique perspective into the performance of a part.

Fig.3.29: Intricate Part Printed by "Binder Jetting" Technique

Courtesy: ExOne

Sand Casting

(a) The production of large sand casting patterns is one of the most common uses for "Binder Jetting.".

(b) The low cost, and speed of the process make it an excellent solution for elaborate pattern designs that would be impossible to produce using traditional techniques, Figure 3.29.

Fig.3.30: Oil and Gas Stator 3D Printed by Produced "Binder Jetting"

Courtesy: ExOne

Functional Metal Parts

(a) The secondary processes that are used in conjunction with "Binder Jetting," such as sintering or infiltration, allow for the production of functional metal parts, Figure 3.30.

(b) The large range of metals available and the ability to create complex shapes, make "Binder Jetting" a viable solution for designs that would be very expensive and difficult to produce.

POWDER BED FUSION (METALS) DMLS/SLM, EBM

Table 3.6: Presentation Status of the 3D Printing Technology in this book

Technology	Recommended Use	Status
Material Extrusion Technology	Fused Filament Fabrication	Discussed
Vat Polymerization	Stereolithography SLA Digital Light Processing DLP	Discussed
Powder Bed Fusion – Polymers	Selective Laser Sintering SLS	Discussed
Material Jetting	Material Jetting— Drop-On-Demand (DOD)	Discussed3
Binder Jetting	Binder Jetting	Discussed
Powder Bed Fusion (Metal)	DMLS/SLM, EBM	In Progress

Solid parts can be 3D printed by "Metal Powder Bed Fusion" technology. Thermal source is used to induce fusion among the powder metal particles. "Metal Powder Bed Fusion" is printing part one layer at a time. Additional powder is introduced through a unique mechanism in most "Powder Bed Fusion" technologies. Powder is adding while part is being constructed to produce a final part encased in the metallic powder. The heat source causing fusion is using either lasers or electron beams as the main source of different energy.

POWDER BED FUSION TECHNOLOGIES DMLS/SLM

"Direct Metal Laser Sintering" (DMLS) technology and "Selective Laser Melting" (SLM) technology have been developed to print 3D parts through a similar technology of "SLS." The major difference is that the "DMLS" and "SLM" technologies are used solely in the production of metallic parts.

"DMLS" technology does not melt the metallic powder, but instead it heats it to a point that can fuse the molecules together. SLM uses the laser to melt the metal powder forming a homogenous part. This results in a part that has a single melting temperature. The main difference between DMLS and SLM is that the DLMS produces parts from metal alloys, while the SLM from single element metal, such as titanium.

DMLS and SLM processes require structural support, in order to limit and minimize the possibility of distortion that may occur, despite the physical support the surrounding powder provides. DMLS/SLM parts are at risk of warping due to the residual stresses produced during printing through the required high processing temperatures. However, 3D printed parts are heat treated after printing, while parts are still attached to the build plate, to relieve the residual stresses.

EBM

In contrast to "Powder Bed Fusion technologies," "Electron Beam Melting" EBM uses a high energy electronic beam instead of laser photons to induce fusion between the particles of metallic powder. A focused electron beam scans across a thin layer of powder, causing localized melting and solidification over a specific cross sectional area. These layers are built up to create a solid part.

"EBM" has a superior build speed, contributed to its higher energy density. Also, EBM speed is favored over SLM and DMLS technologies. EBM enjoys the smallest size features in part configurations, powder particle size, layer thickness, and better surface texture. Parts printed by EBM technology are produced in vacuum using conductive materials.

PRINTER CHARACTERISTICS

Printer Parameters

The quality of surface finish and accuracy of the printed parts produced through DMLS/SLM technologies relies heavily on laser spot size, powder geometry and layer height. Also, the manufacturing printing operation is highly dependent on skilled operators. The industrial metal additive manufacturing machines often require strict operating procedures, and continuous calibration. Also, they require careful material handling, and post processing. The machines require periodic maintenance.

Support Structures

DMLS/SLM printers require support structure. The generation of high temperatures requires support structures to connect unsupported geometry to the build platform. The same support structure acts as a heat sink during the energy generation. The support structure is therefore becomes an essential factor to consider when designing parts for metal 3D printing.

Surface Quality

Some designs require one side of the part to have a smooth surface finish, often is referred to as the presentation side. If a high quality is imperative, a post processing secondary operation is required. There are some steps that can be taken when selecting part orientation to improve the surface quality of the print. Usually, 3D printing upward facing surfaces of a part will have a sharper edges and a better surface quality than downward facing surfaces.

A visible "stepped" effect may occur on an angled surface depending on the layer thickness. In order to avoid the effect of "stepped" surface, the angle of any surface on the printed part may not be greater than 20° relative to the horizontal. The surface finish off the printer is approximately 8.75 Ra μm, depending on the material, layer thickness and orientation of surface.

DIMENSIONAL ACCURACY

DMLS/SLM uses a laser to selectively sinter or melt metal powder to produce metal parts. Metal printing produces parts one layer at a time in a controlled, heated environment on industrialized machines. This layer-by-layer construction combined with the operating high temperatures creates extreme thermal gradients, which produces high thermal stresses. Although metal printed parts, face thermal stresses, the printed parts are still produced with high dimensional accuracy. Nonetheless, parts are built up on a solid metal plate are required to be removed by parting them off from the plate.

MATERIALS

The metallic powder particles are either partially or completely melted. Materials with ability to be welded are good materials to be used in 3D printed parts within "DMLS" and "SLM" technologies. The appropriate materials used in these technologies are aluminum, stainless steel, titanium, cobalt chromium and Inconel. Additionally, more precious materials can be used in SLM technology such as gold, platinum, palladium and silver for the production of jewelry. The thermal conductivity of each material plays a role in the ability to fuse it to form a solid part. Designs with intricate features, and a high level of details in a tight tolerances may be 3D printed with materials of low thermal conductivity such as stainless steel, or titanium. The high cost of metal powders is one of the main drawbacks of DMLS/SLM technologies. The cost of stainless steel "316L" can exceed $450/kg.

BENEFITS AND LIMITATIONS

The strength of the DMLS/SLM process lies in the manufacturing of complex, modified parts where a high level of customization is required, or where geometries are needed that traditional manufacturing techniques are unable to produce. These involve topology optimization for weight reduction, in particularly for aerospace and automotive applications. Also, similar weight optimization must be considered for organic geometries as in medical and dental applications. Also, 3D parted parts are also made from well-established materials, whose behavior is well understood.

The main limitations surrounding DMLS/SLM are the necessary cost of the materials, and the cost of the printers. Additionally, the cost of the build size represents further limitations. Accordingly, Traditional manufacturing techniques may be the most cost effective solution for some applications, such as generic washers, fasteners, and very large parts.

The build size is additional restriction. The largest metal 3D printers have a small build volume when compared to conventional manufacturing build sizes, Table 3.6. Also, metal parts require a significant understanding of designing for 3D printing.

Table 3.6: Comparison between Powder Bed Fusion and Binder Jetting for Printing Custom Metal Parts

	Powder Bed Fusion	Binder Jetting
Tolerance	± 0.1mm	0.2mm
Maximum Part Size	250mm x 150mm x 150mm	400mm x 400mm x 750mm
Strength/Weakness	• Highly Accurate • Great Mechanical Properties • High Cost • Limited Part Size	• Low Cost • Very Large Parts • Mechanical Properties not as good • Limited Accuracy

All applications designed for conventional manufacturing may not be converted to a 3D printing solution. If a part was originally designated and designed for conventional manufacturing, then it is more likely not a viable candidate for 3D printing, Table 3.6. When designing for DMLS/SLM, it is often useful to consider how 3D printing may be integrated with traditional manufacturing to work synergistically, for example, the critical and complex portion of a design may be apportioned for 3D printed, while another simpler portion of the design may be apportioned for CNC machining and then assembly.

METAL POWDER BED FUSION VS. BINDER JETTING

When deci8ding between "Metal Powder Bed Fusion," and "Binder Jetting" the tradeoff is the cost on one side and the dimensional accuracy, as well as and the mechanical properties. "Metal Binder Jetting" parts can be up to 10 times less expensive than "Powder Bed Fusion," however the dimensional tolerances, as well as the mechanical properties are inferior.

Also, if the size of a design exceeds the build size of the Powder Bed Fusion," and 3D printing is the technology of choice, "Binder Jetting" is the price competitive option for producing metal parts.

COMMON APPLICATIONS

DMLS/SLM is best suited for 3D printing applications of low volume, highly complex parts, in particularly for applications and geometries where traditional manufacturing is unable to produce the parts.

Dental Applications

DMLS/SLM has become a popular option in the dental industry with the direct metal printing of crowns and bridges, Figure 3.31. 3D printing has the capability to produce metallic parts resulted in a substantial reduction of the lead time. DML/SLM has the ability to produce a considerable number of custom parts in a single print setting to further accelerate the production process. Additionally, the design freedom of DMLS/SLM technologies overcame their high cost their corresponding materials and systems. These technologies have reaped the benefit of manufacturing unique and custom designed individual parts instead of high volume repeatable component manufacturing.

Fig.3.31: Metal Crowns and Bridges in Dental Applications

Medical Applications

The medical industry has embraced metal 3D printing. The custom design of patients' need can be precisely printed according to patient's anatomy to include unique surface characteristics, as porosity, to stimulate bone growth, Figure 3.32. DMLS/SLM technologies provide a range of common medical metals suitable for sterilization. However, the high cost of DMLS/SLM printing technologies sanctioned exclusively their used for low volume custom parts.

Fig.3.32: Hip Implant with Porous Outer Surface in Medical Application

Aerospace and Automotive Applications

According to the vital need for weight reduction of parts for the aerospace and automotive industries, the design parameters are criterial to adopt there fundamental requirement. DMLS/SLM is an ideal solution in those industries. DMLS/SLM technologies have few design constraints allowing geometries to be produced that would have been historically very expensive and most likely impossible to manufacture. This impossibility has enabled the 3D printing of complex parts with a high strength to be made with high performance metals, such as titanium or aluminum alloys.

REFERENCES

1. R. Wicker, E. MacDonald, "Multi-material, multi-technology stereolithography", Virtual and Physical Prototyping, Vol. 7. Issue 3, pp. 181-194.
2. C. Gutierrez, R. Salas, G. Hernandez, D. Muse, R. Olivas, E. MacDonald, M.D. Irwin, R.B. Wicker, K. Church, M. Newton and B. Zufelt, "CubeSat Fabrication through Additive Manufacturing and Micro-Dispensing," Proceedings from the IMAPS Symposium, Nov. 2011.

3. Castillo S., Muse D., Medina F., MacDonald E. and Wicker R. (2009), "Electronics integration in conformal substrates fabricated with additive layered manufacturing", Proceedings of the 20th Annual Solid Freeform Fabrication Symposium, University of Texas at Austin, Austin, TX, pp. 730-737.
4. De Nava E., Muse D., Castillo S., Alawneh M., Navarrete M., Lopes A.J., MacDonald E. and Wicker R.B. (2008), "3D off-axis component placement and routing with solid freeform fabrication", Proceedings from the 19th Annual Solid Freeform Fabrication Symposium, University of Texas at Austin, August 3-5, pp. 362-369.
5. Soloman, Sabrie "Sensors Handbook, 2nd Edition," McGraw-Hill Publishing Co., NY – ISBN 978-0-07-160570-0
6. Soloman, Sabrie "Sensors and Control Systems in Manufacturing 2nd Edition," McGraw-Hill Publishing Co., NY – ISBN 0-07-059626-3
7. Soloman, Sabrie "Affordable Automation," McGraw-Hill Publishing Co., NY – ISBN 0-07-059633-6
8. Soloman, Sabrie " Sensors Handbook, 1st Edition, McGraw-Hill Publishing Co., NY – ISBN 0-07-059630-1
9. Soloman, Sabrie "Sensors and Control Systems in Manufacturing 1st Edition," McGraw-Hill Publishing Co., NY – ISBN 978-07-10572-4

4 Design for 3D Printing

DESIGN FOR 3D PRINTING

The information provided are specifically tailored to assist in the decision making process for selecting the most appropriate printing method utilizing additive manufacturing technology to fabricate an object through 3D printing technique for a desired object for a specific application.

TECHNOLOGY SUMMARY TABLE

Table 4.1, is a reference that can be used to make an informed comparison between many 3D printing technologies commonly used in the market today. While the provided table is simplified, it contains a high level comprehensive categorized data as discussed below. The technologies are categorized by their ability to produce parts that have either functional or visual purpose.

Table 4.1: Property Comparison between 3D Printing Technologies

	Material Jetting	Binder Jetting	DMLS / SLM
Material group	Photopolymer resin	Sand or metal powder	Metal powder
Common materials	Standard Cast able Transparent High temperature	Stainless/Bronze Full color sand Silica (sand casting)	Aluminum Stainless steel Titanium
Support material required	Dissolvable	No	Yes
Strength/ weakness	Best surface finish Full color and multi-material Available Brittle, not suitable for mechanical parts Higher cost than SLA/DLP for visual purposes	Low-cost Large build volumes Functional metal parts Mechanical Properties not as good as metal powder bed fusion	Strongest, functional parts Complex geometries Small build sizes Highest price point of all technologies
Dimensional accuracy	± 0.1 mm	±0.2 mm (metal) or ±0.3 mm (sand)	±0.1 mm

Common applications	Full color product prototypes Injection mold-like prototypes Low run injection molds Medical models	Functional metal parts Full color models Sand casting	Functional metal parts (Aerospace and automotive) Medical Dental
	FFF	SLA/DLP	SLS
Material group	Thermoplastic Filament Medical Models	Photopolymer resin	Thermoplastic powder
Common materials	PLA ASS PEI TPU	Standard Cartable Transparent High temperature	Nylon 6 Nylon 11 Nylon 12
Dimensional accuracy	±O.5 (lower limit ±O.5 mm)	±O.5 (lower limit: ±O.15 mm)	±O.3 (with a lower limit of ±O.3 mm)
Common applications	Electrical housings/ enclosures Form and fit testing Jigs and fixtures Investment casting patterns	Injection mold-like polymer prototypes Jewelry (investment casting) Dental applications Hearing aids	Functional polymer parts Complex ducting (hollow designs) Low run part production
Support material required	Yes, dissolvable available	Yes	No
Strength/ weakness	Low-cost Functional parts (non- commercial) - Limited dimensional accuracy for small parts Print layers likely visible on surface	Smooth surface finish Fine feature details - Brittle, not suitable for mechanical parts	Functional parts, good mechanical properties Complex geometries - Longer lead times - Higher cost than FFF for functional applications

GUIDELINE SELECTING PRINTING TECHNOLOGY

In order to select the most appropriate 3D printing system for the intended applications, a guideline is stated below to consider when making an unbiased decision:

1. 3D printing parts made of different types of polymers, it is more beneficial to conduct the comparison between "Fused Filament Fabrication FFF" and Thermoplastics, which is often referred to "Selective Laser Sintering –SLS"
2. The SLS technology has the superior characteristics at a higher cost of material. However, the higher cost is a justified price for highly intricate and complex functional parts. SLS is the preferred technology for 3D printing in particularly due to its ability to produce intricate and complex parts.

3. As for the non-functional aesthetics parts, it is highly recommended to compare "Stereolithography SLA," and "Digital Light Processing DLP" to "Material Jetting - Thermoset Photopolymers," where Material Jetting has the greater control of dimensional accuracy and surface finish. Material jetting is often associated with a higher price.
4. As for metal parts it is recommended to compare "Binder Jetting" vs Direct Light Modeling (DLM) / Selective Laser Modeling (SLM). These 3D printing technologies vary greatly in characteristics and cost. "Binder Jetting" prints metal parts at a price of 10 times less than "Powder Bed Fusion." However, the dimensional tolerance, and the mechanical properties are not as readily acceptable as DLP/SLM technologies. Nevertheless, "Binder Jetting" technology is the only choice if the part configuration is larger than the allowable build volume of the DLP/SLM technologies.

The recommendations stated above employ several 3D printing technologies; however, the recommendations are not absolute. They are recommended in normal 3D printing application considering the cost and the expected quality of the surface finish. There are circumstances on which general recommendations may not be applicable, e.g., Utilizing "Fused Filament Fabrication FFF" can be used for low cost functional parts in place of "Stereolithography SLA," and Light Digital Processing DLP.

DESIGN CRITERIA

Once the technology is selected to yield optimum quality, price, aesthetics, and volume, part design becomes the vital parameter to ensure successful project implementation.

The design of the parts for a chosen 3D printing technology has also guidelines that must be followed to ensure return on investment, and a timely delivery to comply with customer or corporate expectations.

Designing for 3D printing is a delightful exercise to reap the benefit of a new part rapidly created and demonstrated to fulfill certain demands and requirements varies from launching a new product, reduce the cost of part evaluating, test new engineering system, test a new part configuration, or evaluate new material.

Common Errors in Part Design

The most common error in part design for 3D printing technology is the ill-fated part inability to be printed according to the followings:

- inaccurate configurations,
- inappropriate material,
- lack of supporting part during printing,
- inability to extract the part from the printing system, and
- the 3D printer is inadequate.

There are several more causes that part design failed to be printed. Nonetheless, most failures are contributed to 3 categories;

I. Part failure,
II. 3D printing machine failure,
III. Material failure.

While design freedom is one of the strengths in additive manufacturing - 3D printing, designing for 3D printing requires specific design fundamentals that must be observed. The design recommendations, together with clear understanding to the processes involved in each technology of the 3D printing will enable producing parts according to well-developed specifications.

GENERAL DESIGN CONSIDERATIONS FOR 3D PRINTING

Undercut

As it is well known in subtractive manufacturing in CNC milling machines, it is a fundamental requirement to allow for a recessed undercut, and an affective consideration to the cutting tool path.

Draft Angle

In a similar manner, molded parts require draft angles and ejector points to retract molded parts. Equally, there are a number of fundamental elements that are relevant to 3D printing.

Design Factors

The most important factors in 3D printing success are stated as follows:

1. The layer height,
2. Part Shrinkage
3. Part Warping,
4. Part Support
5. Fillets

Layer Height

Every 3D printing technology relies on parts being additively manufactured one layer at a time. While there are a number of factors that contribute to the quality and surface finish of a part, the height of each layer often referred to as Z-axis. The resolution of the Z axis is smallest increment of distance that the system can move up or down, Figure 4.1.

Fig.4.1: 3D Printing Resolution

Since surface texture is directly proportional to the resolution of the forming device, the smallest layer in 3D printing represents the ability of the dispensing nozzle to dispense the lowest possible layer thickness. The smallest thickness is therefore, the 3D printer resolution. The smallest layer thickness is the system resolution. When a part is printed with a smaller layer height it will produce the smoother surface the system is able to produce. Accordingly, fine-printing details are the consequence of a better accuracy. Nonetheless, the print time will be increased, Figure 4.2.

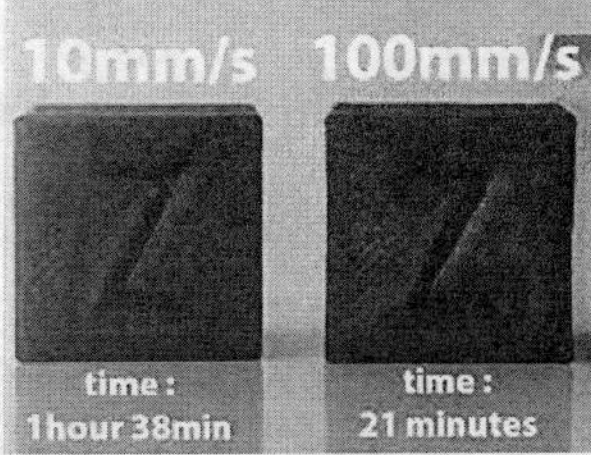

Fig.4.2: Time increases with the increase of Layer Fine Resolution

It is known thicker layer heights will produce parts at a faster rate with layers visible lines on the surface of the printed part, contributed to the poor printing resolution.

The relationship between the build time and the layer height is proportionally linear.

The print that produces a part with 50 µm layer height will take twice as long to print as a part with 100 µm layer.

Most 3D printers have a default layer height with only special cases requiring parameters outside of these defaults (e.g. very high detail models or very fast build times). Common layer heights are presented in Table 4.2. These serve as a useful reference for a measure of "surface smoothness" for each technology, with Material Jetting being the smoothest and FFF generally having the roughest finish.

Table 4.2: Common Layer Height Per Technology

Technology	Common Layer Thickness µm
FFF	50-400
SLA/DLP	25-100
SLS	100
Material Jetting	10-30
Binder Jetting	100
DMLS/SLM	50

Layer Default

Table 4.2, provides common layer heights as default applied in most 3D printing technologies. The default layer height produces adequate printed surface. If aesthetics, or tight fit to interlocking with mating part is an important factor in part design, the layer height may be reduced to produce a smoother surface with greater tolerances. However, the total build time as well the cost will rise. Rapid and quick prototype, often, where appearance is not a critical parameter, a thicker layer height may allow the parts to be produced at a faster rate at a lower cost. However, layer lines will appear visible unless additional post-process is adopted.

Shrinkage and Warping

The Shrinkage and warpage of 3D printed parts are expected, in particularly when heat is associated with the printing process. The temperature and curing time are the vital causes of shrinkage and warping.

Shrinkage and warping are directly caused by the residual stresses generated by the thermal energy required as a fundamental parameter to achieve 3D printing of part. The part differential cooling of its thermal energy produces residual stresses form within a printed part.

Technologies use elevated temperatures to produce parts and are most vulnerable to heat induced warping and shrinkage. Part design must accommodate for the shrinkage and warpage phenomena occurred due to the residual thermal stresses produced during printing. Parts produced through the following technologies that suffer most consequence are listed below:

1. "Material Extrusion – in Fast Filament Fusion, FFF"
2. "Powder Bed Fusion in Selected Laser Sintering, SLS"
3. "Direct Metal Laser Sintering, DMLS," and
4. "Selective Laser Melting – SLM"

When a certain section of the object is being printed using heat energy, it is also subjected to cooling. The cooling rate of one section usually is not consistent with the cooling rate of another section. This differential cooling energy creates differential thermal stress. It results in one area is contracted more than another area. Naturally, cooling creates contact, while contract creates shrinkage. Also, differential shrinkage causes warpage. The contraction also pulls surrounding areas creating internal stresses. High stress causes the part to warp. In some instances, extreme warpage causes crack.

Curing

The Photopolymers used in Stereolithography-SLA, Digital Light Processing-SLA, and Material Jetting require elevated temperatures during printing, in addition to the initial heating of the resin to the recommended ideal working temperature. These types of technologies must be solidified by a curing process. The produced parts are consequently subjected to curing stresses.

As each layer is exposed to the curing light source, it shrinks during solidification. When one layer shrinks on top of a previously solidified layer, there would be stress between the two layers. The result is a pulling stress on the surrounding solid layer initiating the curing of part.

Avoiding Shrinkage and Warping

There are several ways to avoid or reduce the possibility of warping or shrinkage, which is adversely affecting part integrity. Most differential cooling stress are associated with varying of part sections. Some sections may have been designed with thicker walls, while other sections within the same part are designed with thinner walls. When a thick section is directly connected to a thin section it results in faster cooling, happens in the thinner section than in the larger section. This causes the section to warp with respect to each other. Accordingly, designers must strive to design consistent wall thickness of the same part.

Temperature Controlled Environments

Most industrial printers are equipped with temperature controlled build environments to printing parts. The temperature controlled environments greatly reduce the prospect of warping or shrinkage.

Flat Surfaces

All 3D printing technologies are disposed to warpage when a large flat surface within the print is required to be printed. Large flat surface should be avoided in all the 3D printing technologies, when possible. The designer may consider dividing the part into multiple components with smaller flat surfaces of the same part to be content with assembling smaller parts to ensure printing integrity.

Anchor Support

Part support must be located in a strategic location to maintain part stability and consistency during the 3D printing operation. Parts that are at risk of deformation, shrinkage, or warpage are in need of stable means to be anchored to the build plate. Considerations to find the optimum support configurations and the location of the anchor are very important, in particularly, the amount of material used to print the support, to ensure part integrity. Anchor support plays an important role in stereolithography –SLA, digital light processing, and Direct Metal Laser Sintering - DMLS, and Selective Laser Melting – SLM.

Support Structures

Fig.4.3: SLA Printed Part With Support Structure

Depending on the specific 3D printing technology and the complexity of the 3D model, which is printed additively, layer by layer, the 3D print may require support structures as each layer must have a firm platform to build upon. Most technologies essentially, require support to ensure that the model is capable to be printed successfully and accurately, Figure 4.3.

The support structures affects the final result and greatly influence the deciding factors in selecting the appropriate technology to employ. Most support structures have some adverse effect on the surface finish quality. They require post-processing to remove, which that may result in imperfections and surface degradations. Fast Filament Fusion – FFF and Material Jetting offer dissolvable support, which allow simplicity in the removal of the support structure. However, printing powder and sand based technologies do not require support material, except in printing metal.

Support is often placed below the required model to print, resulting in upward facing surfaces. This type of orientation provides a superior surface finish. Nonetheless, when the model aesthetics or part "fit and form" characteristics are imperative, it is preferred to orient the model with the most critical surfaces facing upward. This recommendation is not applied to Selected Laser Sintering - SLS printing.

Fillets - Radii

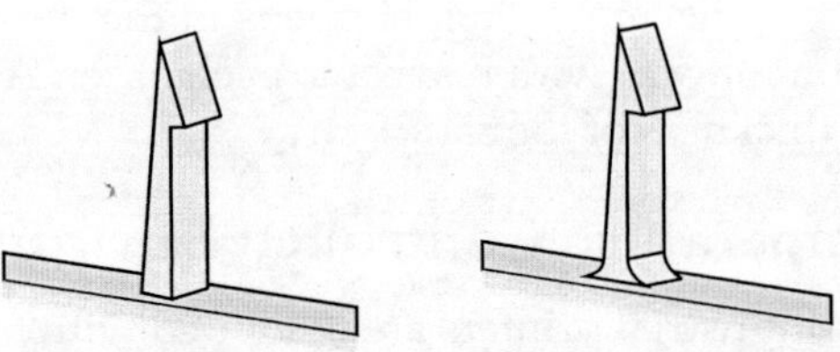

Fig.4.4: Recommended Fillets and Chamfers

The presences of fillets/radii are common feature in every mechanical engineering design, included in the design of 3D printing parts. Fillets are a fundamental characteristic in design aspect to assist reducing the stress concentrations at corners and edges enabling easier parts printing. Fillets can also assist in the removal of parts from the build plate produced through Fast Filament Fusion - FFF.

Printing technologies inherently, produce a "natural fillet" on all edges and corners. Fast Filament Fusion – FFF will produce an outer radius equal to that of the nozzle radius. Selective Laser Sintering - SLS printed parts will typically have a radius of approximately 0.4 mm, which is equal to the laser spot size on all sharp edges. It is highly recommended to incorporate fillets/radii should in the design of 3D printed parts wherever possible. A minimum of 2 mm radius is recommended in most 3D printing technologies. However, a 45° chamfer is also recommended to incorporate at the outer edges of the build plate, instead of fillet/radii as it does not require any structural support material, Figure 4.4.

Description of 3D Printed Features

Table 4.3 offers a description of each of these. It should be noted that not all the features presented in Table 4.3, apply to all 3D printing technologies (e.g. FFF parts do not require escape holes).

Table 4.3: 3D Design Features and Associated Description

Feature	
Feature 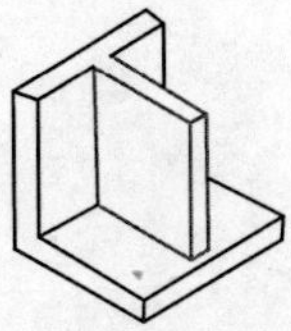 **Supported Wall**	Require a minimum thickness of a wall that is connected to other structures on at least two sides, to ensure little chance to warp.
Feature 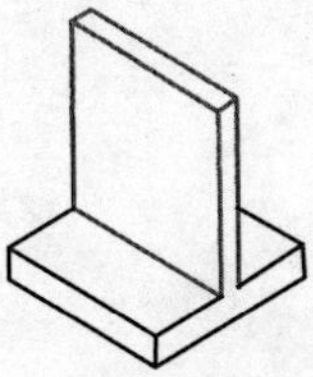 **Unsupported Wall**	Wall of maximum thickness that are connected only side, will have a high degree of warpage or detachment
Feature 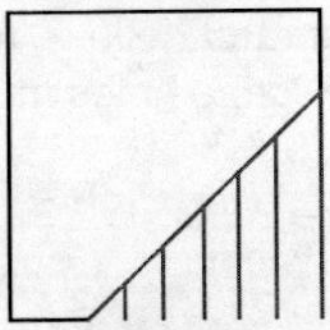 **Overhang**	The minimum angle of a wall relative to the horizontal, must have a supporting structure
Feature 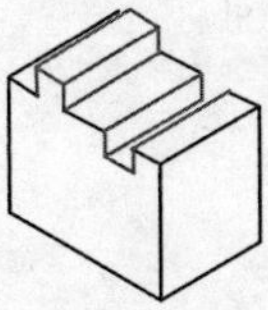 **Embossed and Engraved Details**	The minimum feature depth/height that may include texts, imprinted or recessed into the model. The details are at risk of fusing with the rest of the model while printing if they are too small
Feature 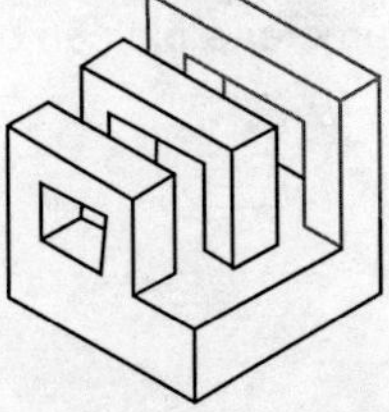 **Horizontal Bridges**	The maximum Bridge length between two points on a model that can be successfully printed without the need for supporting material

Feature 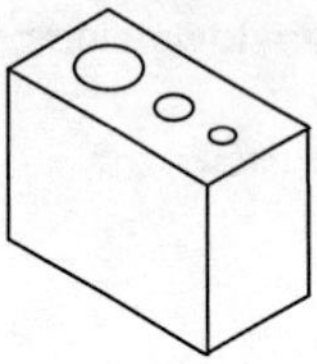 **Holes**	The minimum diameter that a selected technology can successfully print
Feature 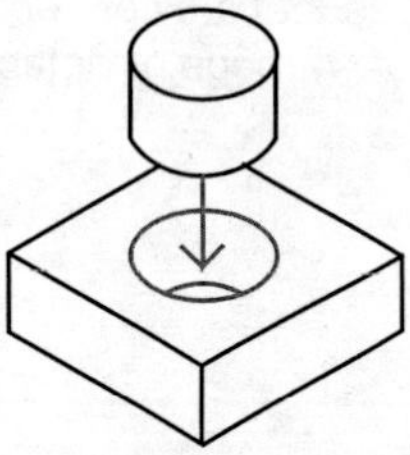 **Holes**	The recommended clearance is based on the required Standard. If no standard is specified, the connection is assumed an interlocking standard.
Feature 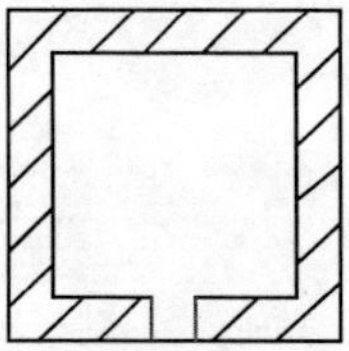 **Holes**	The minimum size of escape hole must allow for the removal of the build material, and to save weight, and often cost. Parts can be printed hollow. To remove build material after printing, an escape holes must be included.
Feature 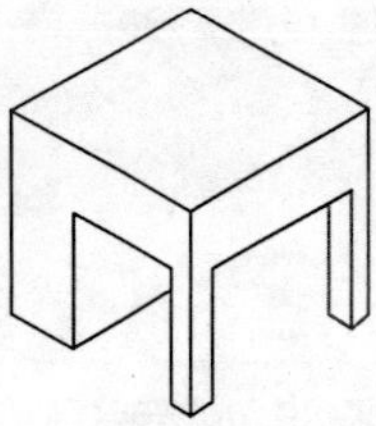 **Holes**	The minimum thickness of any feature is required to ensure it will print successfully
Feature 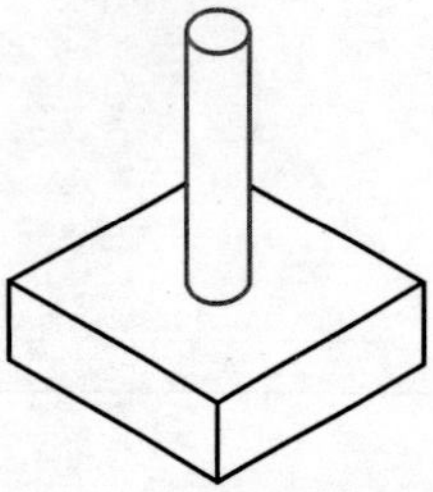 **Holes**	The minimum reliable diameter a pin can be printed

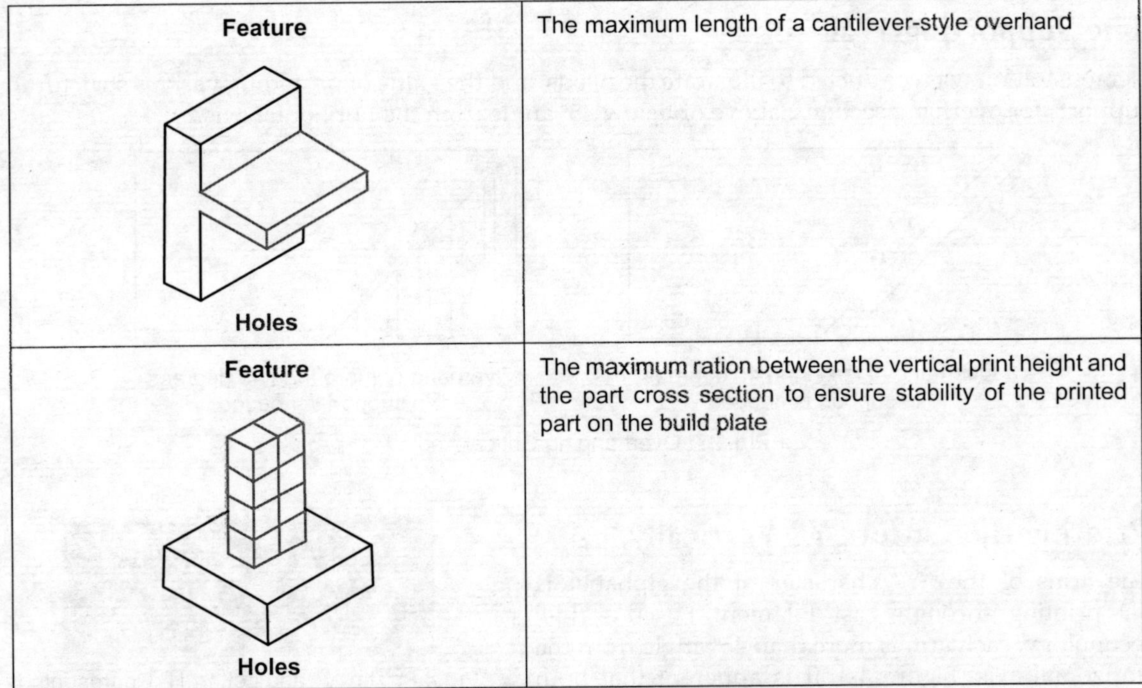

Feature Holes	The maximum length of a cantilever-style overhand
Feature Holes	The maximum ration between the vertical print height and the part cross section to ensure stability of the printed part on the build plate

Designing for Fast Filament Fusion

Fast Filament Fusion - FFF technology is considered the simplest technology to apply. It is, however, limited by the number of design limitations imposed as rules that must be carefully deliberated. Most of these limitations are the consequence of the anisotropic behavior of Fast Filament Fusion - FFF as parts are being printed, as well as its needs for support material.

Fig.4.5: Failed Horizontal Overhangs

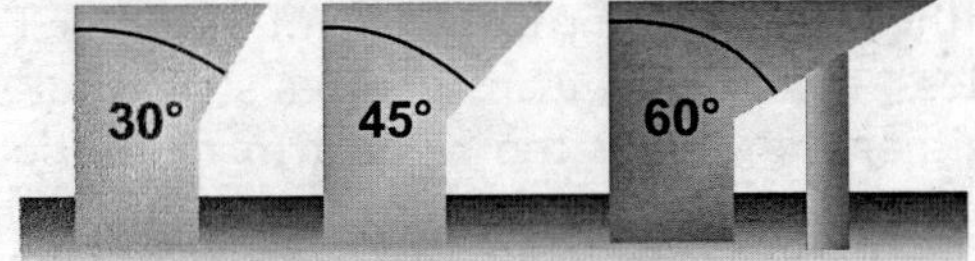

Fig.4.6: A Supported Overhang

Support Structures and Part Orientation

Overhang section is a common feature on Fast Filament Fusion - FFF in 3D printing technology. Overhangs occur when the printed layer of material is partially supported by the layer below. Also, angled and curved walls are examples of overhangs. When a feature is printed with an overhang of 45° or less, relative to the horizontal, it may drop and requires support material beneath to support it in place, Figure 4.5. A wall of critical angle must be supported, as illustrated in Figure 4.6

When an overhand is below 45° the support structure allows overhanging to be printed accurately, Figure 4.6. Once the part containing the overhang section of less than 45o angle, the structural support must be removed by a secondary process, Figure 4.7. This may have an adverse effect on the surface of the part. However, using a dissolvable support can alleviate this problem. Also, the location and amount of support required material is dependent upon part orientation.

The Supporting Wall

Demonstration was conducted to illustrate the needs, and the value of providing various structural supports for overhang sections, above or below 45° angle from the horizontal axis.

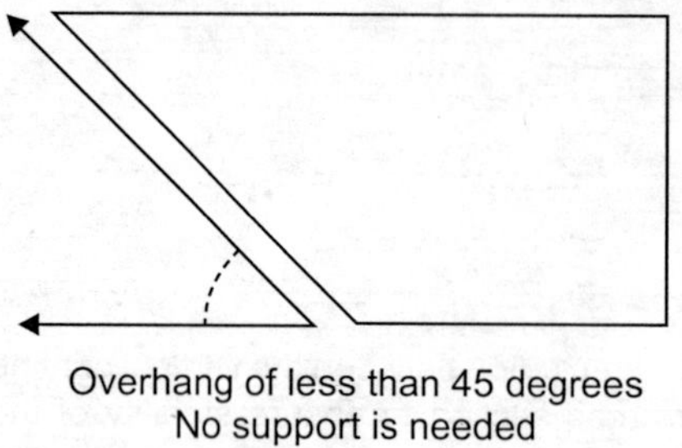

Overhang of less than 45 degrees
No support is needed

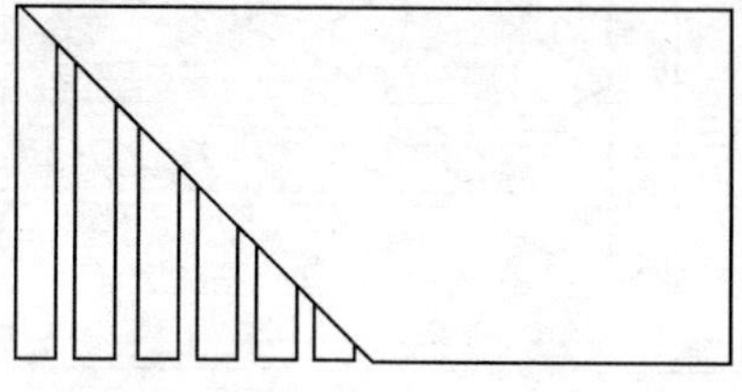

Overhang of more than 45 degrees
No support is needed

Fig.4.7: Overhanging Critical Angle

Printing the Latter "Y" Vertically

The arms of the "Y" character of the alphabet is 3D printing through Fast Filament Fused – FFF technology, each arm is more than 45° angle from the horizontal axis, Figure 4.8. It is apparent that both arms do not require structural support, Figure 4.9.

Fig.4.8: The "T" and Letter "H" Represented Failure for the lack of Support Structure

If the center horizontal bridge of the letter H is less than 10mm, it does not require structural support. If the center horizontal bridge of the letter H is more than 10mm, a structural support is required to prevent drooping.

Fig.4.9: Letters YHT Overhang Remedy

Printing the Letter "T" Vertically

If either the left or the right arm of the letter T is 3 mm or less, none of the arm requires structural support. If either the left or the right arm of the letter T is more than 3 mm, both the arms require structural support. The right and left cantilevers of the letter T have nothing beneath either to support the overhang, Figure 4.8.

Often structural support my not be required throughout entire length of the curved arch, as illustrated in Figure 4.10. This type of selective support is applicable to curved surfaces, as the angle changes over the length of the surface. The cost of material is reduced and the time to print becomes shorter, when selectively placing the support structure where it is needed. The arch illustrated in Figure 4.9, requires a limited amount of support structures placed in the needed location, where the angle drops below 45° to allow it to be printed accurately. The angle of the arch is the tangent of any point on it. The angle is measured from the horizontal X-axis.

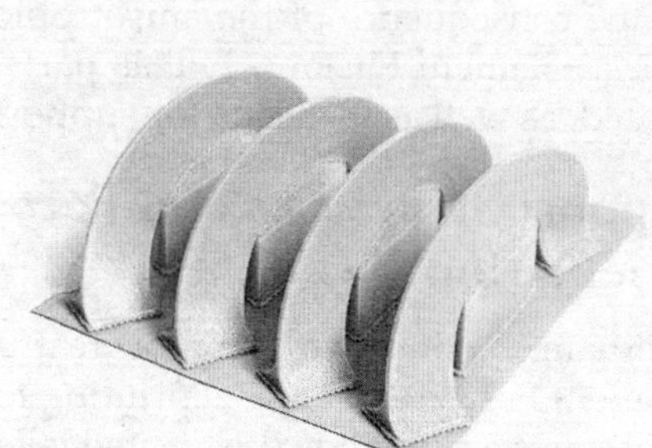

Fig.4.10: Part of the Arch Requires Structural Support

Courtesy: Dinsmore

Types of Fast Filament Fusion - FFF Support

There are three types of Fast Filament Fusion - FFF support structures, listed as follows:

1. Dissolvable support Structure
2. Zigzag Shaped Support Structure, Figure 4.11
3. Tree-like support structure, Figure 4.12.

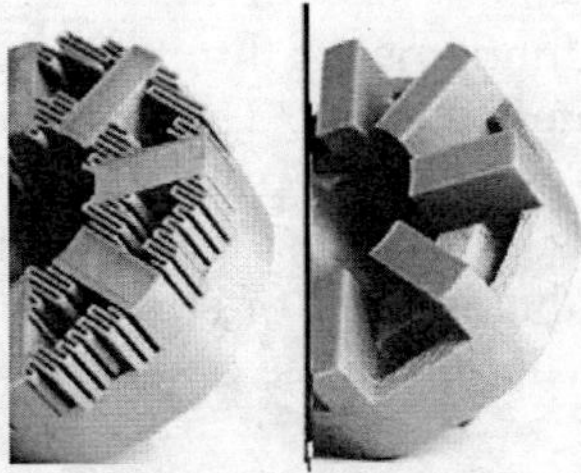

Fig.4.11: Zigzag (Accordion) Structural Supports Before and After Removal

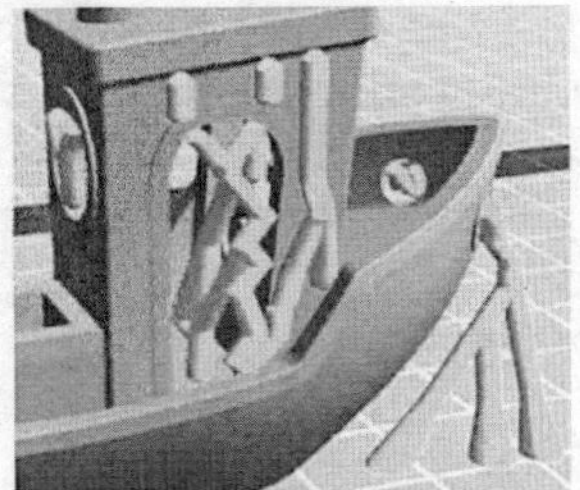

Fig.4.12: Tree-Like Structural Supports

Fast Filament Fusion - FFF technology can be equipped with dual printing heads. It is recommended to dedicate one print head to print the water or chemical dissolvable structural support, while the other head conducts its major task in part printing. This approach will result in a better and smoother surface finish, Figure 4.10. Dissolvable support will increase the cost of a build, due to the added cost of the support material.

Adequate layer adhesion between the primary printing material and the dissolvable support material must be matched to ensure relative seamless surface transition. The most common Fast Filament Fusion employs PLA as the main build material with support material PVA, which is dissolvable in warm water; and ABS as build material with support material of HIPS, which dissolvable in Limonene and isopropyl-alcohol.

Accordion Structural Support

When using single head printing the most common support material is the "Accordion" shaped material, Figure 4.10. Although the use of Fast Filament Fusion - FFF technology with single head is used due to its simple geometries, but the "Accordion" support structural support material requires more material than "Tree" structural support material type, Figure 4.11. Therefore, the cost of printing increases.

Tree-Like Structural Support

While, the Tree-like structural support is less popular than the other alternative supports it is nonetheless, preferred by some 3D printers. The advantage of tree support is that it has less contact with the print, which can result in a better surface finish. The disadvantage to tree support is that it offers less stability, often making it unsuitable for complex prints that are support-dependent. Dissolvable support will always give the superior surface finish, but does increase the build cost and time. The experienced printer operator is best positioned to make a judgement on the type of support that is best suited for printing a specific design.

Slicer programs will generate support for a print based on the way a part is orientated, vertically, horizontally or on a sloped angle. However, understanding that the top surfaces and the surfaces in contact with the build plate will have the best surface finish can define how a part is orientated.

The fundamental limitations of some structural supports are their inability to dissolve in appropriate solutions. The effect of this limitation determines the type surface finish expected of a build. Also, post processing can overcome this issue by imposing additional clean process. A designer must consider prioritizing the expected accuracy a printed part when limiting or avoiding the amount of structural support required for the build. Reducing structural support material will reduce the cost of the build, but sacrificing the print quality.

Bridging

One exception relating to Fast Filament Fusion – FFF requires structural supports to create support bridges to accumulate layers till the build is completed. Bridging occurs when the printer is required to print on a span supported only at the ends of the span. Once the printer head passed the one point of support there may be nothing to accumulate the deposited material on. Therefore, it is required to "bridge" the span, Figure 4.13, illustrates several bridges printed through Fast Filament Fusion with increasing spans. Additionally, the type of materials used have viable effect on the length of the bridge. Bridges that are less than 10 mm in length do not require structural support materials.

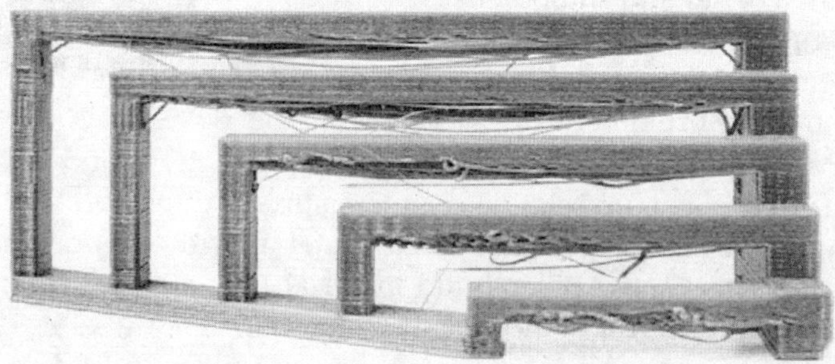

Fig.4.13: Increasing from 20mm to 60mm Print Span Affects the Print Quality

Anisotropy

The major limitation of the Fast Filament Fusion - FFF printing is the anisotropic nature of the produced printed parts. The "Anisotropic" materials demonstrate its varying mechanical properties to expand in various directions during material deposition process.

The adhesion between the layers defines the strength of the Fast Filament Fusion - FFF part. Timber is a good example of an anisotropic material. When chopping wood, it is much easier to split wood in the direction of the grain, rather than chopping perpendicular to it. FFF parts behave in a similar manner.

The consistency of adhesion of layers can be well maintained if the printer is well calibrated prior to the printing operation. The attendant operator is responsible of the printer performance and settings as described in its operating manual. The Fast Filament Fusion – FFF printing system deposits layers, which are pressed down layer by layer upon one another to create the appropriate mechanical adhesion. However, the lack of continuous material paths due to a malfunction in the print head and the lack of adequate force between layers, contribute to the weakness of the Fast Filament Fusion – FFF parts. Since the layers are printed as rounded rectangles, between each layer there are small ebbs. These ebbs create stress weaknesses, where a crack begins as the part is placed under a load. This behavior can greatly affect the performance of a part as presented in the datasheet shown in Table 4.4.

Table 4.4: Mechanical Properties at Different Print Orientation of Fuse Filament Fabrication – 3D Printing Technology

	Printing Horizontally		Printing Vertically	
Infill	50%	100%	50%	100%
Tensile Strength (MPa)	4.4 ± 0.6	6.5 ±1.8	17±0.8	29.3±0.8
Force at Break (MPa)	2.7±1.8	7.8±1.3	2.3 ±0.8	26.4±1.8
Elongation at Max. Force %	0.5±0.1	2.3±0.1	2.3±0.9	2.4±0.1
Elongation at Break	0.5±0.2	0.7±0.1	4.8±0.9	3.7±0.9
Relative Tensile Strength MPa/g	0.7±0.1	0.8±0.2	2.5±0.1	3.0±0.1
Elastic Modulus (MPa)	1031±53	1358±139	1072±38	2030±45

The test data shows that the horizontally printed part has a tensile strength nearly 5 times greater than the vertically printed part when printed with 100% infill. When using Fast Filament Fusion - FFF to produce functional parts, the print orientation must be carefully selected to ensure the anisotropic behavior of the part does not adversely affect the performance. When the orientation of the build direction is parallel to the subjected load the printed part will stand few times greater load. Also, the designer must consider the information presented in datasheets, and find optimum ways to apply to the primary filament to create the desired 3D printed part.

Infill

Fast Filament Fusion - FFF printers are printing with a low density infill. Also, it creates a low density core. The infill percentage is directly imparting the time to print. Part printed faster leads to cost effective, as long as the part strength is within the engineering design specifications. The slicer's default program of most Fast Filament Fusion - FFF prints parts with a 20% infill, which is adequate for major 3D printing applications. The design specifications of a printed part enable the engineering designer to specify the optimal infill percentage. When part "Form" is the main requirements, the part may be printed with low infill, 10%, consequently, this preference saves cost and time to print. Nevertheless, printing a support bracket-carrying a load may require a full infill of 100%, which is fully dense.

The standard infill is printed as a rectangular shape. The four most common infill shapes are shown in Table .4.5.

Table 4.5: Standard Infill Geometries of Fused Filament Fusion

Infill Geometry	Description
	Standard Infill pattern for most FFF prints. Has strength in all directions and is relatively fast to print. Requires the printer to do the least amount of bridging across the infill pattern.
	Used when strength is needed in the direction of the walls. Triangular infill takes longer to print.
	A good choice for designs that needs to be soft, to twist, or to compress particularly when printed with a soft rubbery material or softer nylon.
	Popular infill is very strong providing strength in all directions.

The application of the part and the amount of subjected load applied to the part determines the Infill parameters. While 20% is the default infill percentage, increasing this value can have a significant impact on part strength. It is recommended to fasten or fit the printed part into a minimum of 50% infill. The increase of cost and time is therefore, justified as increasing the strength is critical parameter, while a lightweight structural print is still desirable. The honeycomb or triangular infills are the best configuration to maintain high strength at low cost.

There are several slicing software routines provides the ability to change the infill percentage throughout a print. A higher infill density can to be printed at the desired locations where higher strengths are required. Areas adjacent screw points, and areas subjected to load. Can receive higher density infill, while lower density infill can then be quickly printed areas where strength or stiffness is not critical.

Holes

The Fast Filament Fusion – FFF often prints the vertical axis holes smaller than the intended design diameter, caused by the reasons listed below:

1. As the nozzle head prints the perimeter of a vertical cylindrical hole, the nozzle compresses the deposits on the newly printed layer down onto the existing build layers to help improve adhesion.
2. The compressing force from the nozzle deforms the extruded round layer shape from a circle into a wider and flatter shape, Figure 4.14.

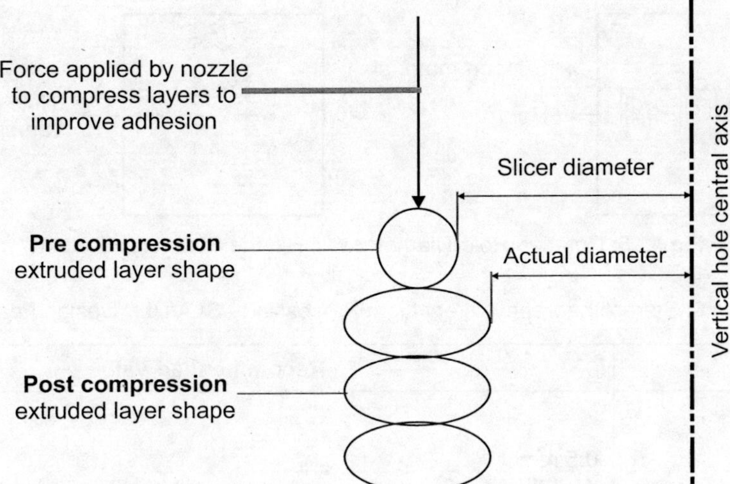

Fig.4.14: Slicer Dimeter is Slightly Different than Actual Diameter

3. This increases the area of contact with the previously printed layer to improve adhesion. Also, it increases the width of the extruded segment.
4. As a result, the cylindrical hole printed diameter is decreased This particular phenomenon greatly affects smaller diameters than large diameters. This is occurred due to the ratio between the nozzle diameter and a small hole-diameter is larger than the ratio between the same nozzle diameter and a lager hole-diameter.

The under-sized hole-diameter depends on the following parameters:

1. The printer,
2. The slicing software,
3. The size of the hole
4. The printing material used.

The reduction in diameter of vertical axis hole will often create unacceptable hole-tolerance. The demand for tighter tolerance will demand for drilling to accurate dimensions through applying a secondary machining process to the printed hole. The Fast Filament Fusion FFF printing may encounter further limitations during horizontal-axis hole printing, depending on the hole-dimeter. Structural support material are required if the hole is large enough. Lack of structural support material to reinforce the upper curved wall of the horizontal hole will cause it to collapse. Alternatively, it may encounter poor tolerance and poor surface-finish.

It is often difficult to remove the structural support of a printed hole. The following recommendations are summarized below to achieve successful hole-printing, and the removal of its corresponding structural support:

1. The need for structural support my not be required by rotating the build direction by 90 degrees, Figure 4.15
2. Print blind-hole first as top priority; it is difficult to process - Optimize the axis direction of the blind hole
3. Hole's diameter less than 3 mm, does not require structural support
4. Prioritize the rest of the holes in order of their importance.

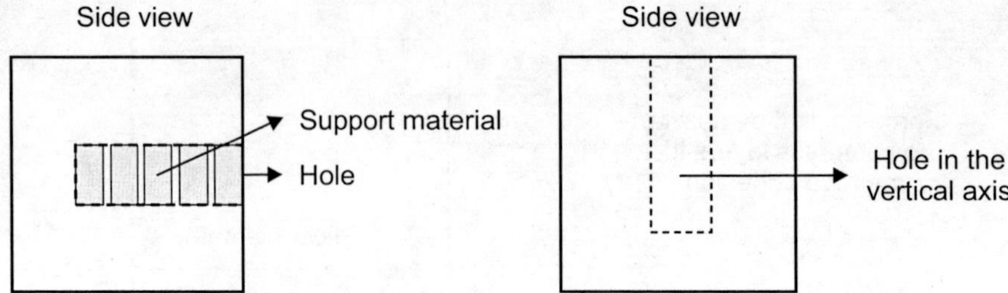

Fig.4.15: Rotating Hole-Diameter Will Rid the Structural Support

Table 4.6: Stereolithography/Direct Light Processing - SLA/DLP Design Features

Feature	Recommended Value
	0.5 mm Wall thickness is dependent on the wall length with longer walls requiring thicker sections. Supported walls should never be thinner than 0.5 mm.
	1. 0 mm To avoid warping or detachment from the model during printing, unsupported walls should be no thinner than 1.0 mm.
	1. 0 mm Any unsupported overhangs must be kept less than 1.0 mm in length and at least 19o from level.
	Embossed x 0.1 mm, engraved; 0.4 mm Embossed details should be a minimum of 0.1 mm above the surface while, engraved details should be a minimum of 0.4 mm below. Embossed details will typically print clearer than engraved details at lower sizes
	4mm To allow resin to effectively drain from inside hollow sections, escape holes should be a minimum of 4 mm in diameter. As many escape holes as possible should be included in a design with holes in the highest and lowest point of a build being the optimal solution.

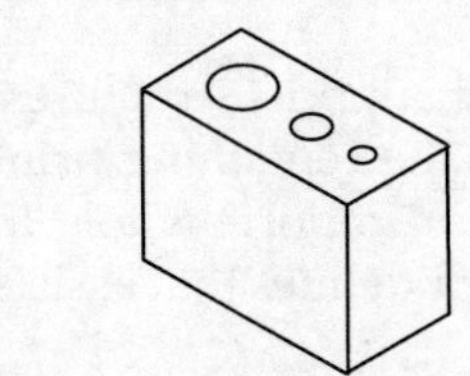	0.5 mm Holes with a diameter less than 0.5 mm in the x-, y-, and z-axis may close off during printing.

Designing for SLA/DLP

The Stereolithography (SLA) and Digital Light Processing (DLP), which often refers to as "Vat Polymerization techniques (SLA/DLP)" are using a light source to cure a photopolymer resin. SLA/DLP is best suited for parts that require a high level of accuracy and a better surface finish. The support structure of the part is fundamental to this type of SLA/DLP technology for successful priming. The design recommendations emphasize on the support location of part to ensure adequate surface finish.

Support Structures and Part Orientation

The bottom up, and top down SLA/DLP printers require support structures. The support structure secures the model to the build platform, and assists in preventing possible warping by reinforcing overhangs and other complex features. The SLA/DLP slicing programs is equipped with the capabilities to allow for automatic configuration of support based on the design of part.

The SLA/DLP technologies can only print support structure in the main build material. Other 3D printing technologies are able to print the support structure in a secondary dissolvable material. The support structure in this event is removed from the final part. Therefore, the structural support material is configured to be as simple as possible to remove easily, and to save the cost of additional material. The support structure takes a shape of tree-like structures, which narrows to a small area, Figure 4.16.

Fig.4.16: Support Structure is Still Attached to the Print

Printing the support structures has an after print structural support lines when using SLA/DLP technologies. These marks may have adverse effect on surface texture. The marks are often sanded to generate a smooth surface. The structural support is affected by part orientation. If aesthetics is a vital criterion, the designer ought to avoid presenting the parting line where it is facing the direct view. This is particularly relevant for visual prototype parts. The majority of traditional injection molding technology exhibits parting-off line in molds. The parting off lines can be seen even in dental applications, and hearings aids, where a smooth surface is essential.

It is therefore, imperative that the designer considers tools-accessibility to reach the support structures for removal. Complex details in inaccessible areas will make removal of support material difficult and increase the likelihood of damage to the printed model. The bottom up and top down printer configurations of the support structure requires innovative approaches.

TOP-DOWN SUPPORT STRUCTURES

Top Down Approach

The top down printers have a few design restrictions. The parts are permitted to be orientated in any direction with a flat alignment. The selection of the alignment is often optimized by the least amount of support, and the lowest number of layers, reducing print cost and time.

Bottom up Approach

Parts produced by bottom up printers require more complex orientation and support structures. When the print is separated from the bottom of the vat, there is a risk that the remaining print is stuck to the build plate resulting in part failure. Part orientation plays an important role in ensuring that the printed part is not stuck to the printing plate. There several guidelines that assists in orienting the bottom up printer:

1. Orient the part's centerline parallel to the longest axis of the front of the printer
2. Orient the part in an optimum way to minimize layer cross sectional area to reduce the forces during the peeling stage.
3. Orient the enclosed cavities away from the face of the resin reservoir
4. Orient small and intricate part to build layer upon previous layers to reduce dependence on support material to prevent damage during support removal

Several types of SLA / DLP slicing programs capable of accomplishing an automatic orientation of parts and generate support based on part geometry. However, if a part has a surface where contact with the structural support is undesirable, the designer may select to orient the part in another desirable position.

The default part orientation is the suitable orientation for bottom up printing methodology, Figure 4.17:

1. Align the part so that the longest axis is parallel to the X-axis
2. Rotate the part 60° around the Y-axis
3. Rotate the part 30° around the Z-axis
4. Generate support material

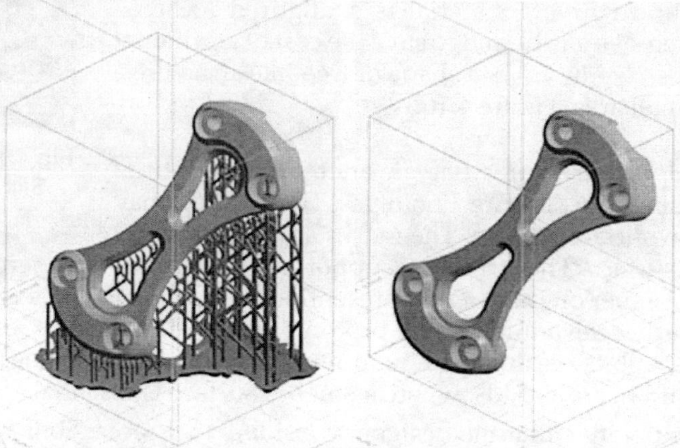

Fig.4.17: Optimal Orientation of Bottom Up SLA/DLP Part Printing - Rotate 30° around the Z-Axis (Bottom Right) – and Then Generate Support (Bottom Left)

Hollow Sections

Printing a hollow part is best suited SLA / DLP technologies. SLA / DLP technologies are favorable among designers when developing Hollow designs as they decrease material use and lower print costs. When designing printing hollow sections using SLA, hollow sections do not require internal structural support as it is difficult to remove.

The use of hollow sections may result in trapping resin and air as illustrated in Figure 4.18. An escape hole must be considered during the design stage of the hollow part. The minimum escape hole must not be less than 4mm in diameter to allow the resin to easily drain out during the alcohol washing stage during post processing. If the diameter of the hole is not suffient enough to let the gas escape a portion of the print resin may not fully cured.

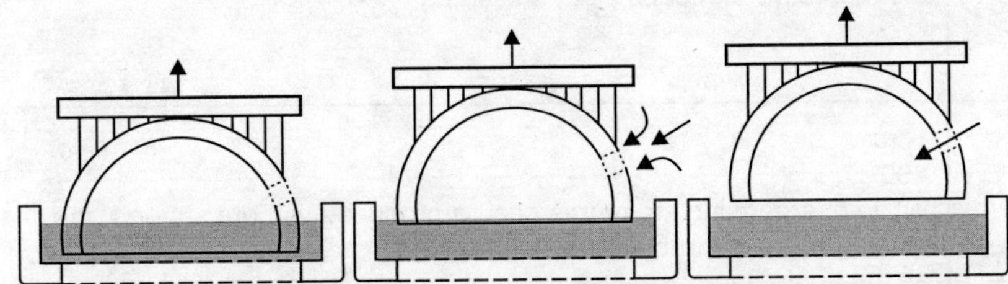

Fig.4.18: Incorporating Escape Hole into Hemispherical Print Avoids Failure in Thin Walls

The optimum location of the escape hole is as vital as the diameter of the hole. Accordingly, the escape hole must be located opposite to each other if design configuration permits. A pair of holes is recommended, nonetheless, some designs can only accommodate a single escaping hole. Holes can be located in a hidden position when the part is in its natural orientation. Holes may be located at the optimum lowest portion of the design, as long as the resin drainage is not obstructed.

SLA-DLP Design Table

Table 4.7: Stereolithography/Digital Light Processing SLA/DLP Recommended Standard Values

Feature	Recommended value
	0.7 mm - 2.0 mm Wall thickness varies by material. For standard PA12, walls should be 0.7 mm. For composite materials like glass, graphite or carbon-filled powder the minimum recommended thickness is 2 mm.
	1 mm height/depth Details should be least 1 mm below or above the surface to esnsure they are visible. Engraved details are generally more visible for SLS printing than embossed details.
	1. 5 mm All holes should be larger than 1.5 mm diameter to avoid over sintering.

	0.1 mm - 0.3 mm Clearance is dependent on the type of connection. For connection parts a clearance of 0.1 mm (0.05 mm each side) is recommended. For moving parts (shafts or hinges) 0.3 mm (0.15 mm each side) should be used.
	>30° While overhanging structures can be produced without support, the likelihood of the inner corners of structures fusing together increases as the angle decreases. Because of this, overhangs should be kept above 30°.
	50 mm long x 1 mm thin Unsupported edges are easily printed by SLS However, if they are too long and thin they are at a high risk of breaking during powder- removal. Unsupported edges should be no longer than 50 mm and no thinner than 1 mm.
	1x 10 mm or 2 x 5 mm To save weight (and sometimes costs) SLS parts are printed hollow. To remove un-sintered powder after production escape holes must be included. Escape holes must be a minimum of 5 mm diameter.
	0.8 mm Features (pins, protruding sections, fins etc.) should be a minimum of 0.8 mm in size to ensure they are able to be printed

Designing for Selective Laser Sintering - SLS

Polymer powder bed fusion applications are well used in Selective Laser Sintering Technology, where laser is used also to sinter the powder and solidify it to produce parts. SLS printing has a design freedom advantage to allow multiple parts to be printed simultaneously to produce parts from functional Polymers as Nylon.

SLS part design does not require structural support. The powder surrounding a part acts as structural support during printing. The unneeded structural support is an advantage of the SLS technology. The vital design criterial within SLS technology is reducing or eliminating the likelihood of warping or distortion.

Shrinkage and Warping

An overall dimensions increase in the SLS designs by 3 - 3.5% during the printing stage. Overall dimensions shrink during a cooling stage. A conversion data can be set to accommodate the shrinkage effects during the cooling of the polymers. The shrinkage is predictable and does not affect the design of a part, as it routinely considered into account by the printer software.

Warping, print distortion, and poor surface finish are the main concerns in print failure in SLS technology of large flat surfaces. The designs that have thick and dense areas connected to thin structures are particularly at risk of warping. Several ways have been devised to alleviate the warping and distortion of SLS parts. Much of these ways relate to similar practices often implemented by the injection molding industry.

Part Orientation

Orientated parts in the powder bin are left to the aptitude of the operator. Horizontal printing of a long thin component in the powder bin creates a temperature gradient across the part which can lead to warping through differential cooling. The distortion is contributed to a long starting and finish distance of the laser path. The parts are often oriented in a position that will allow heat to dissipate at the fastest rate. The size of the build volume also plays a role in oriented parts. The designer investigates the maximum build dimensions of the powder bin to ensure a design fits within the build parameters, and that the part can be orientated in the desired direction during printing.

Additionally, part orientation influences the roundness of a hole when using SLS technology. Contrary to the warping of large flat surfaces, holes that are orientated with their axis in the vertical direction are less likely to experience distortion. Vertical axis holes have improved diameter concentricity as the entire cross section of the hole is produced per layer. Areas of consistent circular profile cool at the same rate.

Horizontal axis holes have the circumference of the circle produced one layer at a time, resulting in differential cooling rates. The large horizontal axis holes can become oval. The layer stepping will become visible on the top half of the hole walls. Holes that may require a high level of accuracy, a second machining operation is advisable after printing as a practical solution.

Reducing Part Mass

The manner to limit the likelihood of warping and distortion is to reduce the mass of a design. Thickness reduction of wall, and part cut outs enables to dissipate the thermal energy at a faster rate lowering the potential for warping and distortion.

Recommended design limits are presented in Table 4.8. Reducing the mass of a part will reduce the volume, and consequently reducing the cost of a print.

Table 4.8: Minimum Slot Size Relative Wall Thickness

Slot Width mm	Wall Thickness mm				
0.5	2	4	6	8	10
0.8	√	X	X	X	X
1.0	√	√	√	X	X
1.5	√	√	√	√	√
2.0	√	√	√	√	√

Table 4.9: Minimum Hole Size Relative Wall Thickness

Hole Size mm	Wall Thickness mm				
1	2	4	6	8	10
2	√	X	X	X	X
3	√	√	√	√	√
4	√	√	√	√	√
5	√	√	√	√	√

Over-Sintering

Over-Sintering occurs when trapped and radiant heat fuses un-sintered powder around a part, resulting in a loss of detailed configurations causing the part to be disfigured. Over-Sintering is associated with the presence of small design features as holes and slots, It is possible to determine the minimum printable slot and hole sizes relative to the wall thicknesses presented in Tables 4.8 and 4.9.

The best way to reduce the likelihood of over-sintering is reducing wall thickness. Also, thinner walls dissipate heat at a faster rate, minimizing the probability of fusing unwanted, and surrounding powder.

Powder Removal

A design constraints associated with SLS is the need to incorporate an escape-hole in hollow sections. Once a print is completed using SLA technology, the build material must be removed from inside the hollow section. The SLS powder is removed through compressed air. Having minimum single escape-hole of 10mm diameter or two holes of 5mm diameters in the printed part is highly desirable.

Hollow Sections

SLS does not depend on support material to produce parts. Accordingly, SLS technology is considered the best 3D printing technologies to produce hollow sections. Other technologies may require internal structural support, which is difficult to remove. Hollow sections allow for weight reduction and, causing lower cost of a print. If the addition of escape hole has a detrimental effect on the appearance or performance of a part, the part may be plugged or filled as a secondary process after printing.

It is possible to produce parts with hollow sections that do not include escape holes, by printing parts with tightly packed and un-sintered powder inside the print. This enables the design to maintain a mass equivalent to a solid sintered part, but produced in a much shorter time. When the un-sintered powder technique is used, the produced parts are much weaker than a fully sintered part. SLS parts are printed with solid, fully dense walls of 1-3 mm thickness. When wall thickness is increased or taking a shape of a honeycomb, part strength is improved substantially. A framed structure can be incorporated in the hollow section of the design, in conjunction with the un-sintered powder.

Blind Holes

When printing parts, using SLS technology it is highly recommended to avoid designing blind holes, which does not extend the whole way through a part. The blind holes make powder removal difficult. Creating a through hole in the blind hole removes the powder. However, if creating a through hole within the blind hole is prohibited due to an adverse consequence to the part function it is recommended that a small hole of a minimum of 2mm diameter is included at the base of part, Figure 4.19.

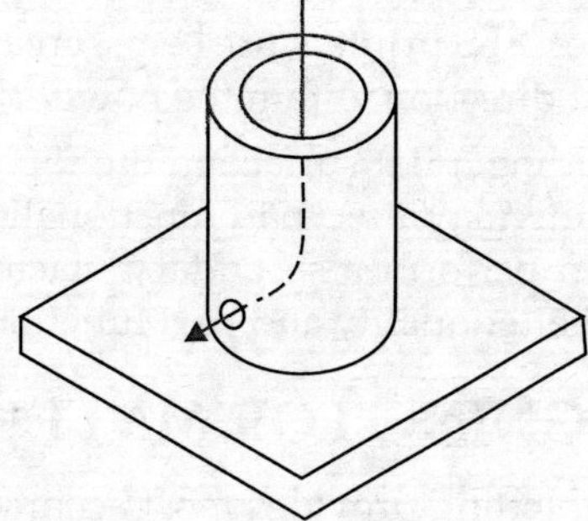

Fig.4.19: 2 mm Hole at the Base of the Hollow Shaft to Assist Powder to Escape

Moving Parts

SLS is one of the few 3D printing technologies able to integrate moving parts into a single build. The surrounding powder makes it possible that the moving parts can be integrated in a single build. The clearance between the moving parts and the bearing surfaces varies with part geometry. As parts become larger they retain more heat, resulting in a higher chance of over-sintering causing the moving parts to fuse together. A shaft less than 10mm in diameter, the minimum clearance of 0.3 mm, or 0.15mm on each side, is recommended between all moving surfaces. For parts greater than 10mm in diameter, the clearance should be increased to 0.5 mm, 0.25mm on each side. Larger clearances are recommended to further reduction in the likelihood of over-sintering.

If possible, it's recommended that dedicated bearing surfaces are added to a design. Also, incorporate and a minimum gap of 4 mm, 2 mm on each side, to maintain stability between moving surfaces. The powder removal in this event is reduced the likelihood of over-sintering. In order to accomplish further powder removal, an escape hole between the bearings surfaces is incorporated.

Comparison with Injection Molding

SLS parts are successfully used as prototypes for determining the form, fit and function of designs that may later be produced in mass production through injection molding. The main differences between designing parts for SLS compared to injection molding are:

1. SLS parts do not need to be removed from a die. SLS is able to easily produce undercuts, negative, draft, and interior features. The ability to produce negative draft is often implemented for securing of gaskets and O-rings.
2. Sharp edges and corners cannot be produced by SLS. The SLS process produces parts that have a radius of ±0.4 mm at all edges and corners. Radii less than 0.4 mm on a design will be printed as 0.4 mm. This is due to the round profile of the laser and the laser diameter.
3. The natural radius produced by SLS offers some stress relief. Areas of concerns that may have high stress a radius of more than 2 mm must be added.

Stepping Effect

Although SLS does not depend upon support to print parts, there are still limitations relating to how parts are built. Layer steps will become visible for angles less than 45°. Angles less than 30° will result in making the stepping appearance visible on the surface, Figure 4.20. This is particularly prevalent on the top surface of horizontal

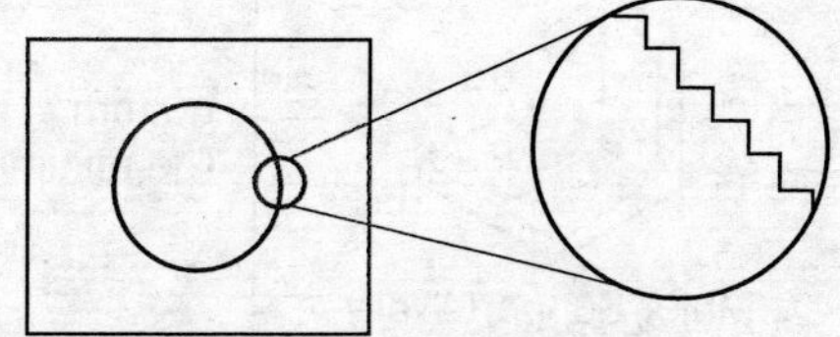

Fig.4.20: Layer steps will become visible for angles less than 45

axis holes. SLS printers use steps, creating stepping effect difficult to avoid. Re-orienting parts in a different direction within the powder bin may avoid the visible steps to be become visible.

Adding a flat surface to the top of a hole will assist in maintaining a smooth surface finish, though this may affect part functionality. If a smooth surface is required, printing the hole undersize, and then perform a secondary machining process after printing is often the best alternative to ensure better quality and part function.

DESIGNING FOR MATERIAL JETTING

Material Jetting provides most accurate 3D printing technology, producing high detail parts with high surface quality, and smooth surface. This technology of Material Jetting can print in full color, of multiple colored materials. Material Jetting is often used to produce high quality visual prototypes.

The absence of heat during the Material Jetting process, as well as the use of dissolvable structural support material allows for a high level of design freedom. The technology has a few rules to design specifications with respect to overall minimum dimensions of part configurations. Most design recommendations for Material Jetting allow adequate space to remove the structural support material.

Support Structures - Part Orientation

As with most 3D printing technologies, Material Jetting requires the use of support material to accurately print parts. As in Fast Filament Fusion technology, the Material Jetting technology requires structural support using a secondary dissolvable wax-like material, which can be removed after the print is completed.

Material Jetting technology requires manual removal of structural support material. This process causes a limitation on the design of parts. Additionally, fully enclosed cavities that will be completely filled with support structural material cannot be removed. Also, long and narrow cavities, or small holes are difficult to clean. Accordingly, holes or channels must be greater than 0.5 mm width. However, channels with a depth-to-width ratio of 2:1 are especially difficult to clean.

Escape holes do not have any significant effect in the removal of structural support material. Therefore, escape holes are not required. Also, the structural support material is printed as a solid print compared to SLA or SLS, where the material being removed is a liquid or powder.

The need for manual removal of structural support using pressurized water can cause damage to intricate and fine detailed models. The minimum size of Material Jetting part is illustrated in Table 4.10.

Table 4.10: Material Jetting Design Features

Feature	Recommended Value
Major Support Walls	1.0 mm The minimum wall thickness for supported walls is 1 mm.

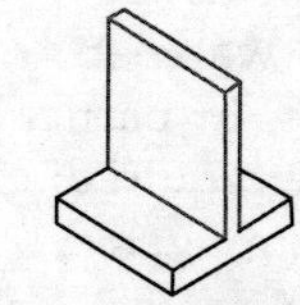 All other Walls	1.0 mm For all other walls (including fins or ribs) the minimum wall thickness should also be no less than 1.0 mm. Care should be taken when removing support material from around these features.
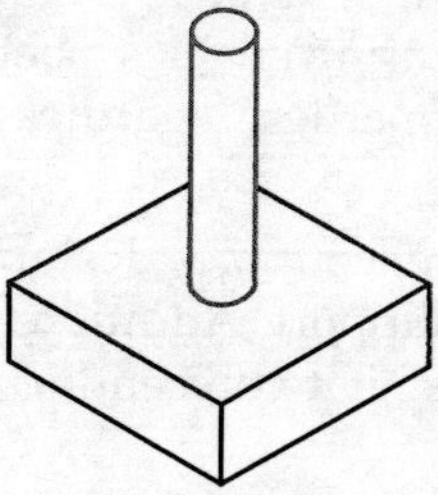 Pin Diameter	0.5 mm A minimum pin diameter of 0.5 mm is recommended. If pins are required to be functional they should be at least 2.0 mm in diameter or an off the shelf pin should be inserted into a drilled hole.
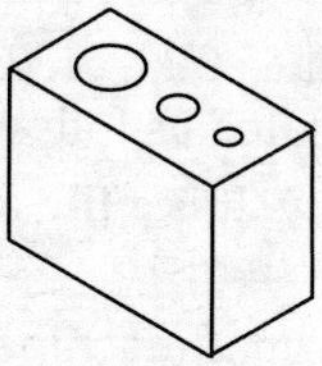 Hole Size	0.5 mm For a hole to be successfully printed the minimum diameter should be no smaller than 0.5 mm. Holes, whenever possible, should be oriented vertically, to maximize the circularity of the feature.
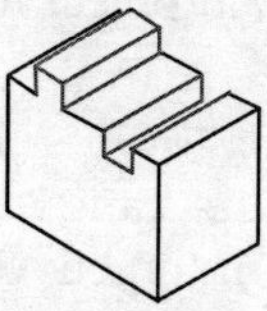 Embossed and Engraved Details	0.5 mm height/depth To ensure details are visible, embossing and engraving should be at least 0.5 mm below or above the surface.
 Feature Size	0.5 'mm Material Jetting is capable of producing part details as low as 0.5 mm. Smaller features are at a greater risk of breaking during post processing.
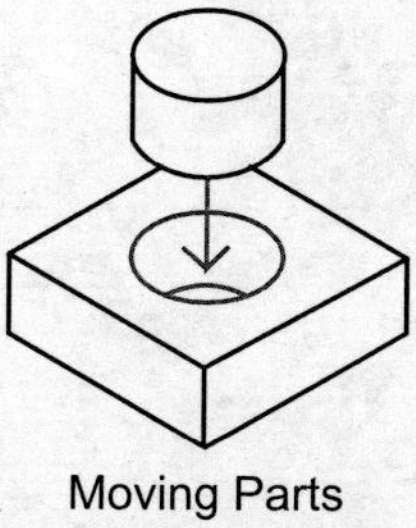 Moving Parts	0.2 mm Assembled parts, hinges and joints should have 0.2 mm clearance around all sides. This clearance must also be accessible to allow for cleaning/removal of the support material that will build in the gap.

The guidelines presented in Table 4.10 would result in printing models strong enough to withstand post processing. Not all printed parts require cleaning with a waterjet system. Occasionally, structural support may be removed manually using small tools, in particularly, intricate small parts with fine details. If stronger and more resilient models are required, glossy finish creates better surface finish, and higher strength. Glossy modes are often stronger.

The removal of support material does not impact the quality of surface finish. The system actually cures 3 layers deep during the print process. As layers are cured, they blend As soon as a layer is deposited, it is 60% cured. When the second layer is dropped, this first layer 95% cured. When the third layer is deposited, the first is 100 cured. This progression of curing enables a better layer bonding, reduces porosity, and results in parts with homogeneous properties, regardless of orientation.

The high cost of "Material Jetting's" structural-support-material often encourages designers to select the optimum model orientation to limit the use of the structural support. Additionally, designers often reconfigure the orientation and the printing procedure to facilitate presenting the best surface finish facing upward.

Full Color Printing

One advantage of Material Jetting is the ability to produce multi-color prints that accurately represent end products. Full color models can be exported as ***three file types***, stated as follows:

STL - An abbreviation of "stereolithography," as described previously. It is **a file** format native to the stereolithography CAD software created by 3D Systems),

OBJ - An object file is a file containing object code, as it relocates format machine code that is usually not directly executable. There are various formats for object files, and the same object code can be packaged in different object files. An object file may also work like a shared library

VRML - Virtual Reality Modeling Language, pronounced virtual or by its initials, originally—before 1995—known as the Virtual Reality Markup Language – It is a standard file format for representing 3-dimensional (3D) interactive vector graphics, designed particularly with the World Wide Web in mind.

STL File - STL is best suited for designs that include discrete colored sections, while OBJ or VRML formats should be used for designs with opaque colors or textures blend.

STL files are assigned color on a per body, or shell, basis with the designer selecting the color of each body within an assembly. Colors are not blended from one section to another.

OBJ – VRML Files

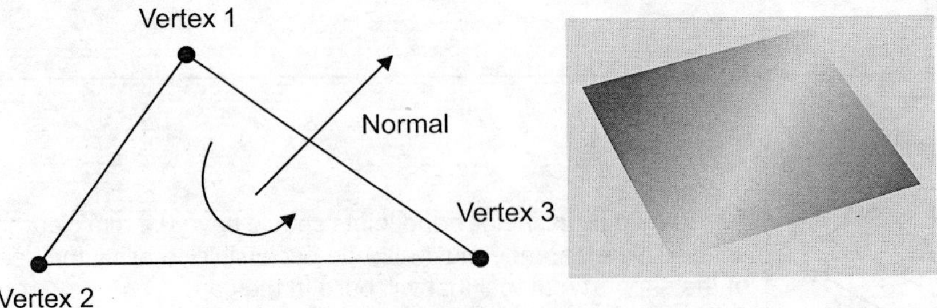

Fig.4.21: Vertex – Mesh Triangle

OBJ and VRML files allow information on colors, textures and materials to be encoded in the file. Both OBJ and VRML files designate color to each face, per mesh triangle, Figure 4.21, or by vertex, the points where mesh triangles meet, allowing color to blend smoothly. This results in highly realistic models. The quality of the color is directly related to the export resolution of the model with higher resolutions resulting in subtler blending when colors change.

OBJ files can be accompanied by a MTL, Material Library, file, which references the materials and colors used in a design. When exporting a model to VRML, the actual file extension will be WRL.

Colors do not bleed into each other, when a minimum shell thickness of 2 mm is used. It is recommended to avoid having support on any colored visual surfaces. Orienting these surfaces upward is the best solution to resolve this issue. Achieving vibrant color can also be accomplished by printing a part in matte, ensuring a uniform finish across the print, and then finishing to a smooth surface, and finally clear coating. Obtaining this type of finish requires careful application and pre-experiments of the finishing process. The result is the most even, and aesthetically pleasing result.

Multi-Material Integration

Material Jetting is the only technology that is capable of printing multi- material parts at once, usually more than two filaments. The functionality of multi- material printing can be utilized in three ways.

Mixed Tray

Mixed tray produces separate parts from different materials on the same build platform. e.g., a flexible rubber part can be printed on the same build platform as a rigid part. This removes the need to change materials between prints, improving efficiency, Figure 4.22.

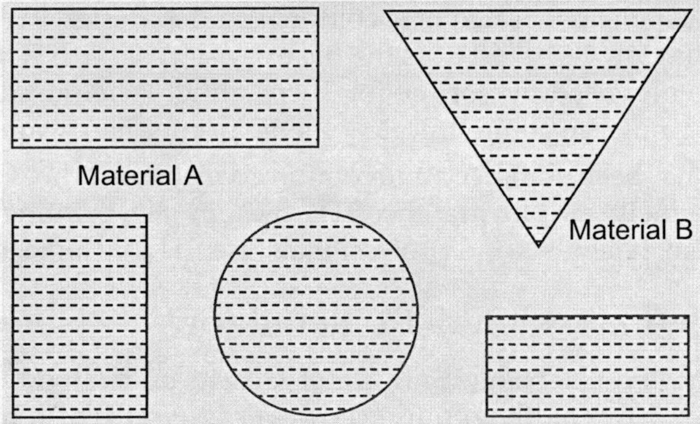

Fig.4.22: Printer Prints Separate Part from Different Material

Digital Materials

Digital materials are the result of combining two or three resins in specific concentrations and microstructures to create a composite material with hybrid characteristics. Combining flexible and rigid materials at specific ratios, made it possible to produce parts with exact properties with custom shore hardness for a specific application.

Mixed Parts

Mixed parts can be made of multiple materials. As an example, Figure 4.23, shows a handheld device with rigid sections (white), and flexible rubber over molding (black,) printed as a single part. Mixed part printing removes the need for assembly of components, increasing production speed. When used in conjunction with full color printing this allows producing realistic designs.

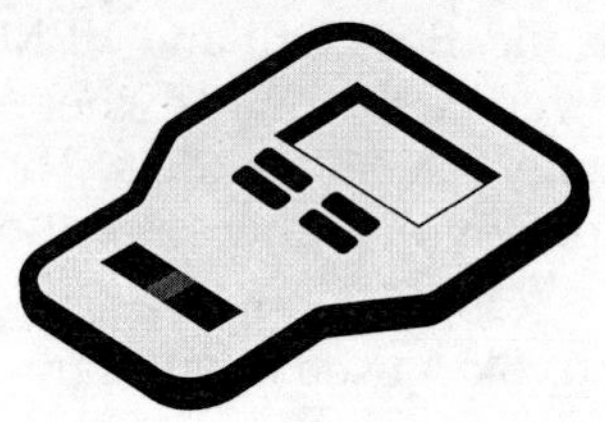

Fig.4.23: Handheld Devices - Rigid and Flexible Sections Printed as a Single Part

Multi-Material Design

Mixed tray and digital materials do not require any extra steps for a designer, other than the designation of the material on a part basis. Mixed part prints require a design to be separated into discrete bodies or shells assigning each body a different material.

Each part within the CAD requires a different material to be modeled separately. The components may be placed together in an assembly as illustrated in Figure 4.22. This aligns all parts to a common origin, defining where each part is located in the design when it is exported. The assembly is then exported using the option that saves each component of the assembly as a separate STL file. These STL files are then imported into the printer software as an assembly and the print material is assigned to each individual part. When creating an assembly using this type of technology it is important to work with an assembly of parts with zero clearance, ensuring that no gaps are present during the printing process. A bill of materials is supplied with the design, clearly annotating the material.

CAD - STL FILES

STL file is directly related to storing digital 3D model information. The STL format describes the surface geometry of a three-dimensional object without any reference to color, texture or other common model characteristics.

These STL files are usually generated by a computer-aided design (CAD) program, as an end product of the 3D modeling process. ".STL" is the file extension of the STL file format. The STL file format is the most frequently used file format for 3D printing. When used in combination with a 3D Slicer, it permits a computer to communicate with 3D printer hardware.

The STL file format has been embraced and reinforced by many other CAD software packages. Rapid prototyping is widely employing the STL files in 3D printing, and computer-aided manufacturing (CM). Also, hobbyists and professionals use STL file alike.

What Does the File Extension .STL Stand for?

The true meaning of the file extension .STL is not clearly understood. It's assumed to be an abbreviation of the word "STereoLithography," however often it is also referred to as "Standard Triangle Language" or "Standard Tessellation Language".

How Does the STL file format Store a 3D Model?

The main drive of the STL file format is to encode the surface geometry of a 3D object. It encodes this information using a simple concept called "tessellation".

Fig.4.24: Tessellation – Similar to Placing Tiles on a Surface without Leaving a Gap

Tessellation

The Tessellation process is tiling a surface with one or more geometric shapes, having no overlaps or gaps. It is a process similar to placing tiles on floor, where tiles are not overlapped, and there is no gap between tiles, Figure 4.24. This is a good illustration of tessellation.

The tiled wall and floor are simple real life examples of tessellation. Tessellation can involve simple geometric shapes or very complicated, as well as imaginative shapes.

The STL file format: Exploiting Tessellation to Encode Surface Geometry

Once the first stereolithographic 3D printer was invented, group of entrepreneurs attempted to find out a way to transfer 3D systems information about 3D CAD models to the 3D printer. They recognized that they could use tessellations of the 3D model's surface to encode this information.

The main concept to Tessellate the is the 2D outer surface of 3D models using very small triangles, which is called "facets," and store information about the facets in a file.

Example: A simple 3D cube can be covered by 12 triangles, illustrated in the image below, Figure 4.25. There are two triangles per face. There are six faces in the cube; it adds up to 12 triangles. Accordingly, to represent the entire cube in a 3D model of a sphere, then it can be covered by many small triangles, also shown in the same image. Illustrated below a complicated 3D Spherical shape which has been tessellated with triangles, Figure 4.25.

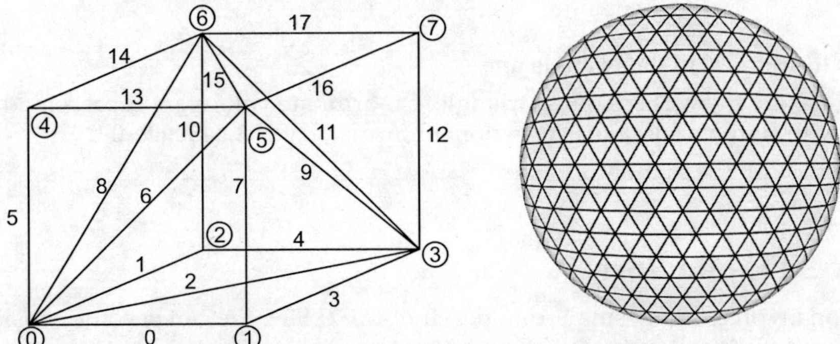

Fig.4.25: Tessellations of Cube and Sphere

3D Systems was realized that if the information could be stored regarding these very small triangles in a file, then this file could completely describe the surface of an arbitrary 3D model. This formed the elementary idea behind the STL file format!

How Does STL File Store Information about Facets?

The STL file format provides two different ways of storing information about the triangular facets that tile the object surface. These are stated as follows:

a. the ASCII encoding and the
b. binary encoding.

In both formats, the following information of each triangle is stored:

1. The coordinates of the vertices.
2. The components of the unit normal vector to the triangle.

The normal vector should point outwards with respect to the 3D model, Figure 4.26.

An STL file stores the co-ordinates of the vertices and the components of the unit normal vector to the facets.

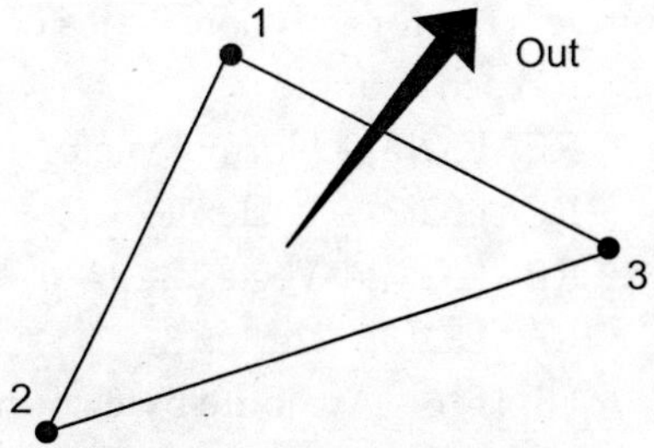

Fig.4.26: A Vertex is Represent by a Triangle (Facet) Coordinate and a Vector Directing Out

4.1 The ASCII STL file format

The ASCII STL file starts with the mandatory line:

solid *<name>*

where: ***<name>*** is the name of the 3D model. Name can be left blank, but there must be a space after the word ***solid*** in that case.

The file continues with information about the covering triangles. Information about the vertices and the normal vector is represented as follows:

facet normal nx ny nz
  outer loop
    vertex $v1_x$ $v1_y$ $v1_z$
    vertex $v2_x$ $v2_y$ $v2_z$
    vertex $v3_x$ $v3_y$ $v3_z$
  endloop
endfacet

Here, ***n*** is the normal to the triangle and

v1, v2 and ***v3*** are the vertices of the triangle. Co-ordinate values are represented as a floating point number with sign-mantissa-e-sign-exponent format, e.g., "**3.245000e-002**".

The file ends with the mandatory line:

endsolid *<name>*

The Binary STL File Format

If the tessellation involves many small triangles, the **ASCII STL file** can become considerable. This is the reason a more compact binary version exists.

The **binary STL file** starts with a **80 character header**. This is generally ignored by most STL file readers. After the header, the total number of triangles is indicated using a **4 byte** unsigned integer.

UINT8[80] – Header

UINT32 – Number of triangles

The information about the triangles follows subsequently. The file simply ends after the last triangle.

Each triangle is represented by twelve 32-bit floating point numbers. Just like the ASCII STL file, 3 numbers are for the 3D Cartesian co-ordinates of the normal to the triangle. The remaining 9 numbers are for the coordinates of the vertices (three each). Here's how this looks like:

foreach triangle
REAL32[3] – Normal vector
REAL32[3] – Vertex 1
REAL32[3] – Vertex 2
REAL32[3] – Vertex 3
UINT16 – Attribute byte count
end

After each triangle, there is a 2 byte sequence called the "**attribute byte count**". In most cases, this is set to zero and acts a spacer between two triangles. But some software also uses these 2 bytes to encode additional information about the triangle. These bytes will be used to store color information.

SPECIAL RULES FOR THE STL FORMAT

The STL specification has some special rules for tessellation and for storing information.

The Vertex Rule

"The vertex rule states that each triangle must share two vertices with its neighboring triangles."

This rule is to be respected when tessellating the surface of the 3D object.

Example: An example of a valid and invalid tessellation is provided according to this rule. The figure on the left violates this rule, and is an invalid tessellation, while the figure on the right is conformant and a valid tessellation, Figure 4.27.

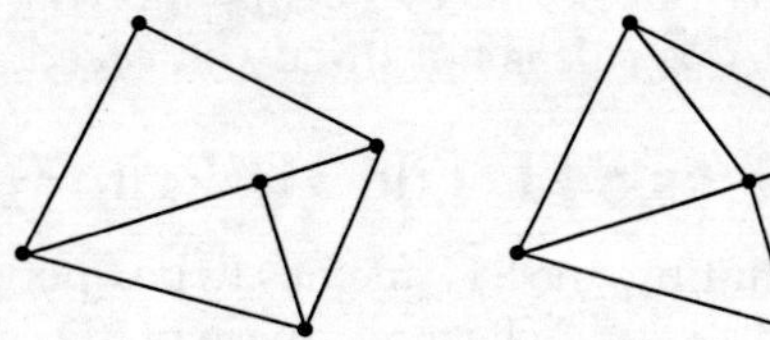

Fig.4.27: Vertex rule for STL files: The figure on the left is an invalid tessellation, while the figure on the right is acceptable.

The Orientation Rule

"The orientation rule states that the orientation of the facet (i.e. which way is "in" the 3D object, and which way is "out") must be specified in two ways."

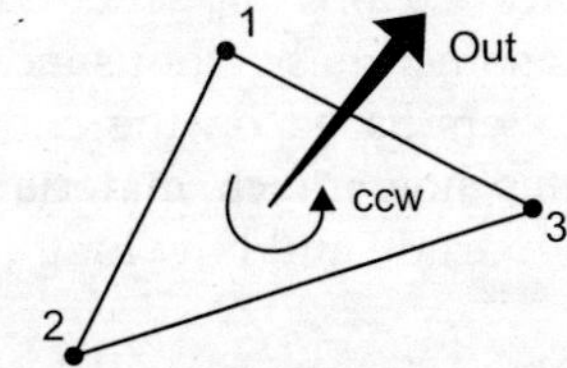

Fig.4.28: the orientation of the facet

1. The direction of the normal should point outwards.
2. The vertices are listed in counterclockwise order when looking at the object from the outside (right-hand rule).

The orientation of each facet is specified in two ways:

1. by the direction of the normal vector and
2. by the ordering of the vertices

This redundancy exists for a reason. It helps ensure consistency of the data and spot corrupt data. Software can, calculate the orientation from the normal and subsequently from the vertices, and verify whether they match, Figure 28. If it doesn't, then it can declare the **STL file** to be **corrupt**!

The all Positive Octant Rule

"The all positive octant rule states that the coordinates of the triangle vertices must all be positive."

This implies that the 3D object lives in the all-positive octant of the 3D Cartesian coordinate system (and hence the name).

The rationale behind this rule is to save space. If the 3D object was allowed to live anywhere in the coordinate space, it would have to deal with negative co-ordinates. To store negative co-ordinates, one needs to use signed floating point numbers. Signed floating point numbers require one additional bit to store the sign (±). By ensuring that all coordinates are positive, this rule makes sure that we are able to use unsigned numbers for the coordinates and save a bit for every coordinate stored value Figure 4.29.

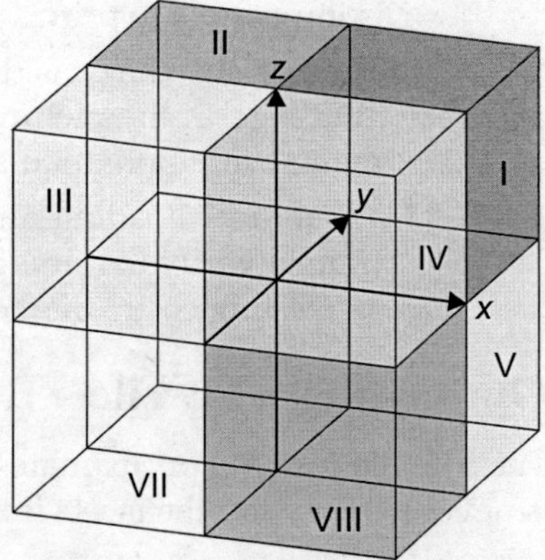

Fig.4.29: Octant I is the all positive octant

The Triangle Sorting Rule

"The triangle sorting rule recommends that the triangles appear in ascending z-value order."

This helps Slicers slice the 3D models faster. However, this rule is not strictly enforced.

How Does STL File 3D Print?

For 3D printing, the STL file has to be opened in a dedicated slicer. A slicer is a piece of 3D printing software that converts digital 3D models into printing 2D instructions for the 3D printer to create a 3D object, Figure 4.30.

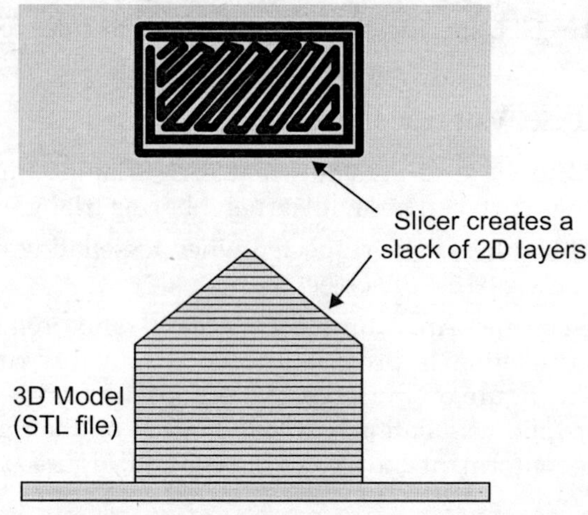

Fig.4.30: The Slicer Creates a Stack of Sliced Layers

The slicer chops up the STL file into hundreds, sometimes thousands, of flat horizontal layers based on the chosen settings, and calculates how much material the printer will need to extrude and how long it will take to do it.

All of this information is then bundled up into a GCode file, the native language of the 3D printer. Slicer settings do have an impact on the quality of the print. Accordingly, it is important to have the right software and settings to get the best quality print possible.

Once the GCode has been uploaded to the 3D printer, the next stage is for those separate two-dimensional layers to be reassembled as a three-dimensional object on the print-bed. This is achieved by depositing a succession of thin layers of plastics, metals, or composite materials, and building up the model one layer at a time.

Printable STL File

1. Not all STL file is 3D printable. Only a 3D design that is specifically made for 3D printing is 3D printable. The STL file is just the container for the data, not a guarantee that something is printable.
2. 3D models suitable for 3D printing need to have a minimum wall thickness and a "watertight" surface geometry to be 3D printable. Even if is visible on a computer screen, It is impossible to print something with a wall thickness of zero.
3. There is also the consideration of overhanging elements on the model. Look at the ALL-3DP logo in the picture above; if the model is printed upright, then overhanging elements with more than a 45° angle will require supports.
4. When downloading an STL file that you have not been created, it is worth taking the time to verify that it is indeed 3D printable. This will save time and frustration, as well as wasting costly filament material.

Optimizing STL File - Best 3D Printing Performance

The STL file format approximates the surface of a CAD model with triangles. The approximation is never perfect, and the facets introduce coarseness to the model.

Fig.4.31: Better Approximated by Tessellations Using Finer Tringles

The perfect spherical surface on the left is approximated by tessellations, Figure 4.30. The figure on the right uses big triangles, resulting in a coarse model. The figure on the center uses smaller triangles and achieves a smoother approximation, Figure 4.31.

The 3D printer will print the object with the same coarseness as specified by the STL file. Obviously, by making the triangles smaller and smaller, the approximation can be made better and better, resulting in good quality prints. However, as decreasing the size of the triangle, the number of the triangles needed to cover the surface also increases. This leads to colossal STL file, which 3D printers cannot handle. Also, it is a cumbersome to share or upload huge similar files.

It is therefore very important to find the right balance between file size and print quality. It does not logical to reduce the size of the triangles to almost zero, because at some point the eye is not able to distinguish between the print qualities.

Most CAD software offers two settings when exporting STL files. These settings control the size of the facets (triangles) and hence print quality and file size.

Chord Height - Tolerance

Most CAD software will let you choose a parameter called chord height or tolerance. The chord height is the maximum distance from the surface of the original design and the STL mesh. If you choose the right tolerance, your prints will look smooth and not pixelated. It's quite obvious that the smaller the chord height, the more accurately the facets represent the actual surface of the model.

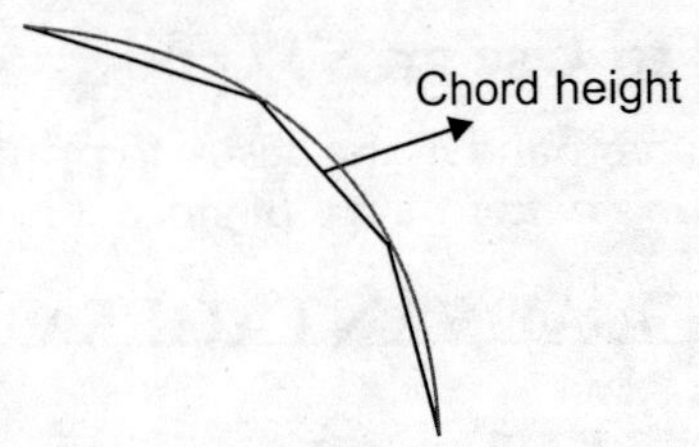

Fig.4.32: The Chord Height is the Height between the STL Mesh and the Actual Surface

It is recommended to set the tolerance between 0.01 millimeters to 0.001 millimeters. This usually results in good quality prints. There is no point in reducing this any further, as 3D printers cannot print with that level of detail, Figure 4.32.

Angular deviation or angular tolerance

Angular tolerance limits the angle between the normal of adjacent triangles. The default angle is usually set at 15° . Decreasing the tolerance (which can range to 0 to 1°) improves print resolution, Figure 4.33.

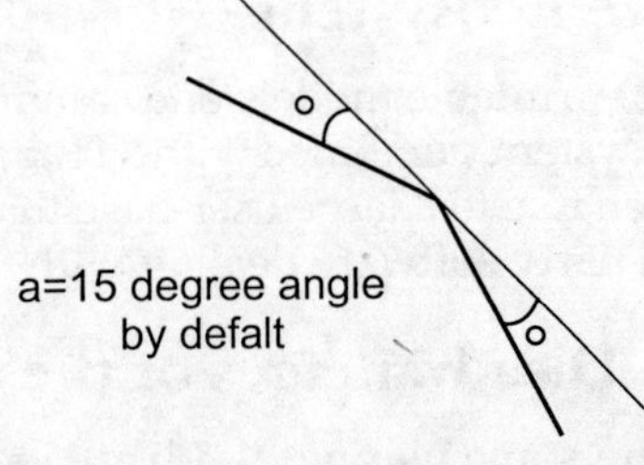

Fig.4.33: Angular Tolerance is the Angle between the Normal of Adjacent Triangles

The recommended setting for this parameter is 0.

Binary or ASCII?

Finally, there is a choice of exporting the STL file in binary or ASCII format. The binary format is always recommended for 3D printing since it results in smaller file sizes. However, if the desire to manually inspect the STL file for debugging, then ASCII is preferable because it is easier to read.

Are there any Alternatives to the STL File Format?

The STL file format is not the only format used in 3D printing. There are over 30 file formats for 3D printing. Most important is the OBJ file format, which can store color and texture profiles. Another option is the Polygon file format (PLY), which was originally used for storing 3D scanned objects.

More recently, there have been efforts to launch a new file type by the 3MF Consortium, which is proposing a new 3D printing file format called 3MF. They claim it will streamline and improve the 3D printing process.

To implement it, Microsoft has partnered up innovative companies like Autodesk, HP, and Shapeways to make their vision a reality. It is far too early to determine if this innovation will become widely adopted.

ADVANTAGES AND DISADVANTAGES OF STL FILE FORMAT OVER OTHER FILE FORMATS

When Not to Use an STL File

The STL file format cannot store additional information such as color, material etc., of the facets or triangles. STL only stores information about the vertices and the normal vector. This means that the desire to use multiple colors or multiple materials for the prints, then the STL file format is not the right choice. The OBJ format is a popular format enjoying good support which has a way to specify color, material etc. Therefore, this is the right choice for this task.

When to Use an STL File

On the other hand, if the desire to print with a single color or material, which is most often the case, then STL, is better than OBJ since it is simpler, leading to smaller file sizes and faster processing.

OTHER ADVANTAGES OF THE STL FILE FORMAT

Universal File

Another advantage of the STL file format is that it is universal and supported by nearly all 3D printers. This cannot be said for the OBJ format, even though it enjoys reasonable adoption and support as well. The VRML, AMF and 3MF formats are not widely supported at this point of time.

Mature Ecosystem

Most 3D printable models one can find on the internet are in the STL file format. The existence of this ecosystem, combined with STL-based software investments made by 3D printer manufacturers, has given rise to a large user-base that is heavily invested in the format. This means there is much of third party software dealing with STL files, which is not the case with the other file formats.

Some Disadvantages of the STL File Format

There are some glaring disadvantages to using STL as well. As the fidelity of printing processes embraces micron-scale resolution, the number of triangles required to describe smooth curved surfaces can result in massive file sizes. It's also impossible to include meta-data, such as authorship and copyright information, in an STL file.

STL/OBJ

If the 3D printing needs are simple, then perhaps there is no reason to move away from the STL file format. However, for more advanced prints using multiple material and color, it is perhaps advisable to try the OBJ or other available formats.

Color in STL File Format

The STL file format cannot handle multi-color models. The reason the STL file format lacks color information is simple. When rapid prototyping evolved in the 1980s, no one thought of color printing. Nowadays, 3D printing materials and processes have evolved rapidly. Some allow printing in full-color. However it is not completely just to state that STL cannot handle colors. It turns out that there are non-standard versions of the STL format that are indeed capable of carrying color information.

Example:

The "V" is CAM and "Solidview" software packages use the "attribute byte count" at the end of every triangle to store a 15-bit RGB color using the following system:

- bits 0 to 4 for blue (0 to 31),
- bits 5 to 9 for green (0 to 31),
- bits 10 to 14 for red (0 to 31),
- bit 15 is 1 if the color is valid, or 0 if the color is not valid, as with normal STL files.

The "Materialize Magics" software, on the other hand, uses the 80-byte header in the binary format to represent the overall color of the 3D object. The color is specified by including the ASCII string "COLOR=" followed by four bytes representing red, green, blue and alpha channel, transparency, in the range 0 – 255. This base color can also be overridden at each facet using the "attribute byte count" bytes.

STL file resources

Some software and resources that can be used for downloading, viewing, editing and repairing STL files.

Downloading STL files

There are many repositories, marketplaces and search engines on the web containing literally millions of free STL files. The user can refer to our regularly updated list: —

- 2019 Best Sites for Free STL Files & 3D Printer Models
- or you can choose one of these models to get started
- 50 Free-Things to 3D Print in August 2019

Opening and viewing an STL file

Opening an STL file is not too complicated. There are several free STL file viewers for this purpose, which can be either used online or as a desktop application, Figure 4.34.

Editing and converting an STL file

Fig.4.34: Opening and Viewing STL File

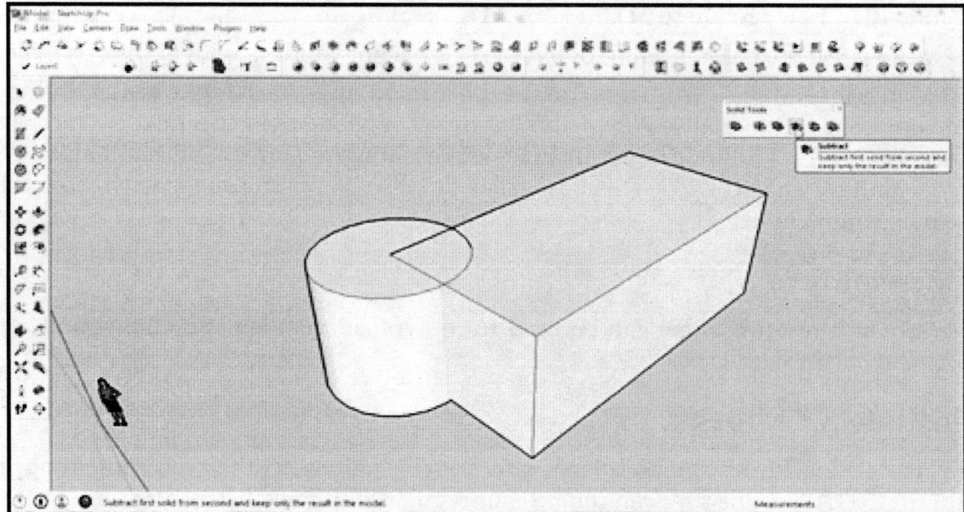

Fig.4.35: Editing and Converting STL File

Yes, it is entirely possible to edit an STL file and convert the STL file to another file format. Because the format is open, there is nothing to prevent from changing the contents of a file. Actually, the process of editing is quite easy.

Repairing an STL file

Adjacent triangles must share two vertices and the right hand rule applied on the vertices should result in the same orientation as the normal vector. If these conditions are violated in an STL file, then it is broken or corrupt.

There are several programs which can help with repairing a broken STL file. For example, Netfabb Basic is a great tool for repairing the most common STL file problems, Figure 4.35.

Conclusion

The STL file format encodes the layout of 3D models. The STL files are optimized for the best 3D printing quality. The STL file format compares with the other popular 3D printing file format, OBJ and when to use each of these formats.

REFERENCES

1. Misael Navarrete, Amit Lopes, Jacqueline Acuna, Raul Estrada, Eric MacDonald, Jeremy Palmer, Ryan Wicker, Integrated layered manufacturing of a novel wireless motion sensor system with GPS, Symposium on Solid Freeform Fabrication (SFF)(18th) held in Austin, TX on 6-8 August 2007.
2. Malone, Evan, and Hod Lipson. "Freeform Fabrication of Ionomeric Polymer-metal Composite
3. Actuators." Rapid Prototyping Journal 12, no. 5 (October 1, 2006): 244–253.
4. Soloman, Sabrie "Sensors Handbook, 2nd Edition, McGraw-Hill Publishing Co., NY – ISBN 978-0-07-160570-0
5. Soloman, Sabrie "Sensors and Control Systems in Manufacturing 2nd Edition," McGraw-Hill Publishing Co., NY – ISBN 0-07-059626-3
6. Soloman, Sabrie "Affordable Automation," McGraw-Hill Publishing Co., NY – ISBN 0-07-059633-6
7. Soloman, Sabrie "Sensors Handbook, 1st Edition," McGraw-Hill Publishing Co., NY – ISBN 0-07-059630-1
8. Soloman, Sabrie "Sensors and Control Systems in Manufacturing 1st Edition," McGraw-Hill Publishing Co., NY – ISBN 978-07-10572-4

5 3D Printing Designing Through Binder Jetting

3D PRINTING DESIGNING THROUGH BINDER JETTING

The "Binder Jetting" 3D printing technique is based mainly on the utilization of a binding agent and sand, such as gypsum or metal powder to form parts. The process allows for the production of full color constructs as well as functional metal parts through a secondary infiltration or sintering process. Also, Binder Jetting is used for the production of cores and molds for sand casting. As "Binder Jetting" does not solely rely on support for the construction of parts, therefore, there is a high level of design freedom when using this technique.

The design considerations of most "Binder Jetting" are related to its low tensile strength of the mechanical properties its produced parts. In particularly, small parts of intricate details are at a high risk of failure and hair cracks prior to the implanting instructions to secondary manufacturing processes.

BINDER

Binder Jetting uses varieties of binder agents applied to sand or metal powder able to alter parts mechanical properties. Several binders require a baking process to fully cure, while other types of binders are water-based to ensure easy burnout before infiltration or sintering. The operator's expertise determines the factor to identify the most appropriate binder for a specific application. Common types of binders include the following binders:

Furan Binder

It is a binder that does not require heat to fully cure, which enables prints to be used for sand casting instantly with no changes to the casting process.

Phenolic Binder

It is a binder that is best suited for high temperature sand casting. The high heat strength also enables thin walls, thin tubes, or pipes to be printed. Parts printed using this process is usually cured using microwave technology.

Silicate Binder

It is a binder based on environmentally friendly, resulting in low gas emissions during the casting process. Also, parts are usually cured using microwave technology. They are best suited for the production of sand molds and cores.

Aqueous-Based Binder

It is an aqueous-based binder, which acts as an adhesive for bonding layers of powdered metal together. This technique is used to facilitate high temperature processing for the production of metal parts.

Green State

Upon the completion of 3D printing task, Binder Jetting parts are in a fragile "green state. "Figure 5.1, where the parts are fragile and delicate. The part consists of sand or powder glued together. Although the "Binder Jetting" process offers a unique design freedom, the part is unable to be handled in the green state. The part may be rendered useless if secondary processes could not be applied.

Fig.5.1: Green State Hemispheres with Hatched Structure Inside. Binder Jetting is Capable of Printing Intricate and fine Detailed Structure

Post processing are typically involving in removing unbound powder with brushes, and air as long as the produced parts are sufficiently strong to withstand the powder removal process. This is an important design consideration. It is important that a designer adheres to the rules presented below.

DESIGN FEATURES

Feature	Recommended Value
Fig.5.2: Wall Thickness	2.0 mm The minimum wall thickness for parts produced via Binder Jetting is 2.0 mm. This allows the part to be removed from the powder and cleaned without being damaged, Figure 5.2.
Fig.5.3: Unsupported Wall	3. 0 mm Unsupported walls (including fins or ribs) are at a greater risk of being damaged during handling and should not be thinner than 3.0 mm, Figure 5.3.
Fig.5.4: Embossed and Engraved Details	0.5 mm height/depth To ensure details are visible, embossed and engraved details should be at least 0.5 mm below or above the surface, Figure 5.4.

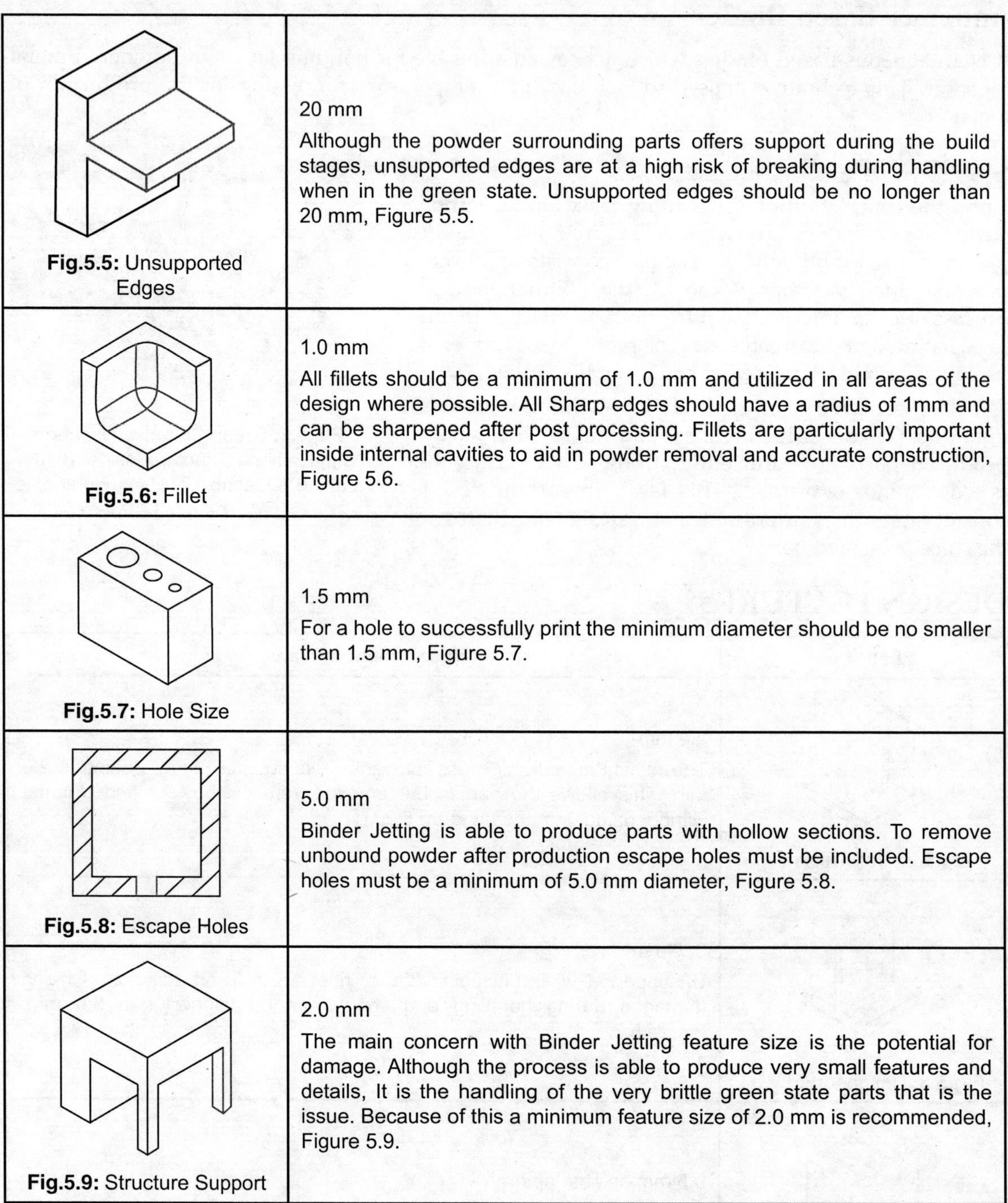

Fig.5.5: Unsupported Edges	20 mm Although the powder surrounding parts offers support during the build stages, unsupported edges are at a high risk of breaking during handling when in the green state. Unsupported edges should be no longer than 20 mm, Figure 5.5.
Fig.5.6: Fillet	1.0 mm All fillets should be a minimum of 1.0 mm and utilized in all areas of the design where possible. All Sharp edges should have a radius of 1mm and can be sharpened after post processing. Fillets are particularly important inside internal cavities to aid in powder removal and accurate construction, Figure 5.6.
Fig.5.7: Hole Size	1.5 mm For a hole to successfully print the minimum diameter should be no smaller than 1.5 mm, Figure 5.7.
Fig.5.8: Escape Holes	5.0 mm Binder Jetting is able to produce parts with hollow sections. To remove unbound powder after production escape holes must be included. Escape holes must be a minimum of 5.0 mm diameter, Figure 5.8.
Fig.5.9: Structure Support	2.0 mm The main concern with Binder Jetting feature size is the potential for damage. Although the process is able to produce very small features and details, It is the handling of the very brittle green state parts that is the issue. Because of this a minimum feature size of 2.0 mm is recommended, Figure 5.9.

Designing extra structures is imperative for fragile and detailed parts to improve stiffness and mechanical support. Obviously, adding structural support to the part will increase the cost, as well as the printing time. However, thinner sections of a design may require increase in size, as illustrated in Figure 5.10.

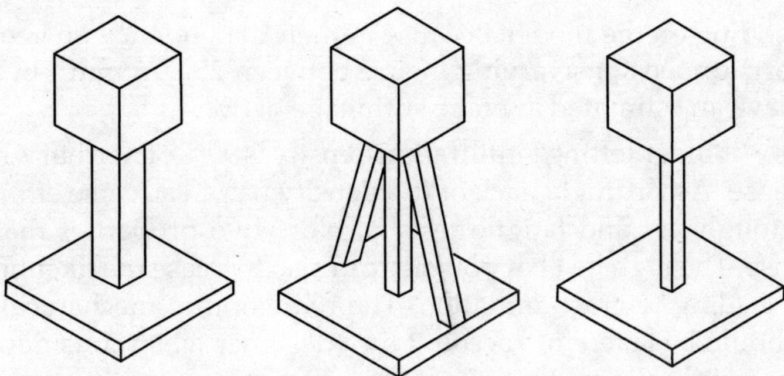

Fig.5.10: Right is a Thin Post – Likely to be damaged during Secondary Process – Middle is More Rigid – Right is Ideal to withstand further operation

The removal of powder is particularly important when the design includes hollow sections. Escape holes of a minimum of 5 mm must be included in the design to allow for the removal of loose sand or metal powder.

Sharp edges and corners in particularly, are at high risk of chipping or hair cracking during powder removal, handling, or heating. A deign consideration must add fillet of at least 1 mm to eliminate the stress concentration of sharp corner to minimize failures, and avoid chipping or damage from handling during post processing. However, sanding may also compensate for sharp edges. Sanding will add the necessary filet needed to avoid stress concentrations

Full Color Print

"Binder Jetting" produces full color prints by jetting ink and binding agent onto sand or gypsum powder simultaneously. After printing, the *green state* parts are cleaned of any excess powder. The parts are then coated with cyanoacrylate super glue sealant to improve part strength and enhance vivid colors. Also, a second epoxy layer may be added to further improve strength and color appearance. Nonetheless, even with these extra steps, full color "Binder Jetting" parts are still brittle and cannot be recommended as functional parts.

Color may be applied to "CAD" models by two methods:

1. ***Apply to each face*** - Applying color to a model on a "per-face" basis is rapid and easy to implement. The result will produce print in fewer details. Most CAD software routines allow color to be assigned to each face of the design, and the file can then be exported as a "VRML - Virtual Reality Modeling Language.
2. ***Using a texture map*** approach to apply color is extensive and more complicated. It must be considered as an alternative if the software is simple. Texture files are written in "PNG or TGA format." Once all textures have been assigned the model is exported as a "VRML" or "X3D" file.

Infiltration and Sintering

"Binder Jetting" produces functional metal parts through secondary processes namely, infiltration and sintering process. After removing the excess powder, *green state* parts are placed in a furnace to enforce the binder agent. Parts are then infiltrated with bronze or sintered to near full density.

Both infiltration and sintering are not recommended for designs where a high level of accuracy is required, as the processes typically results in non-uniform shrinkage of parts. This effect will be relatively considerable on long flat surfaces. Accuracy and tolerance are hard to predict. Tolerance can vary greatly depending on the model, because of their dependency on geometry. The average shrinkage is 2% of parts dimensions varying in size between 25 - 75 mm - in any direction. Parts greater than 75 mm have an estimated average shrinkage of 3%.

Parts produced by "Binder Jetting" infiltration density is 90% of normal while sintering is 97%, which are not fully dense. Accordingly, lack of full density has an adverse effect on the mechanical properties. Fracture, toughness and fatigue resistance are two properties that often suffer when parts are not fully dense. The high level of porosity can also increase the likelihood of delamination between print layers leading to crack initiation. The reduction in mechanical properties and the increase in porosity should be taken into account by a designer when considering "Binder Jetting" metal parts. However, applications where mechanical performance is critical, DMLS and SLM offer a superior solution.

Stilts

A unique aspect to the infiltration process is the need for stilts to allow bronze to infiltrate the part. Expert operator may decide the best stilt location on a design. Stilts require a smooth surface to be placed in an optimum location, with a minimum surface area of 15 mm x 10 mm. The designer may place the stilt at an optimum location. Stilts are removed after infiltration, leaving behind a scarred area on the produced part.

DESIGNING FOR SELECTIVE METAL LASER SINTERING – SELECTIVE LASER MELTING DMLS/SLM

While the level of part complexity that metal printing is able to produce and/or exceeds that of traditional manufacturing techniques, there are certain design rules that must be followed.

The isotropic printed metal parts can be produced from common engineering metals, such as stainless steel. Their prints provide strong and functional parts. However, the condition under which metal printing is cost effective depends largely on the geometry of a design. Many of the design constraints are associated with metal printing methods to limit the likelihood of warping or deformation, as well as incorporating of support structures.

Supports Part Orientation

Metal printing utilizes support in 3 different ways, stated as follows:

1. The support offers a platform for the next layer to be built upon. Sections that are not attached to the build plate, support structures are first printed, and then the solid sections are built on top of the structures.
2. The Support structure anchors the part to the build plate, which increases stiffness, and hold thinner features in place. This is critical due to the high temperature gradients that metal parts experience, and the resulting residual stresses that can cause warping or deformation.
3. The lattice geometry of support acts as an excellent heat sink, drawing heat away from the printed part and allowing it to cool at a more controlled rate.

Support for metal parts is characteristically printed in a lattice structure similar to Fast Fused Filament FFF to save on material costs and build time. An optimum quality of surface finish can be achieved by adding structural support when an overhang features at angles less than 45° from the horizontal.

Normally, part orientation may determine the type and the configuration of the structural support in an effort to reduce the amount of support material used. The high likelihood of warping of metal printed parts, the requirement for support to successfully complete a print takes precedence over part orientation, as illustrated in Figure 5.11. As with most 3D printing technologies, the upward facing surface will have the best surface finish.

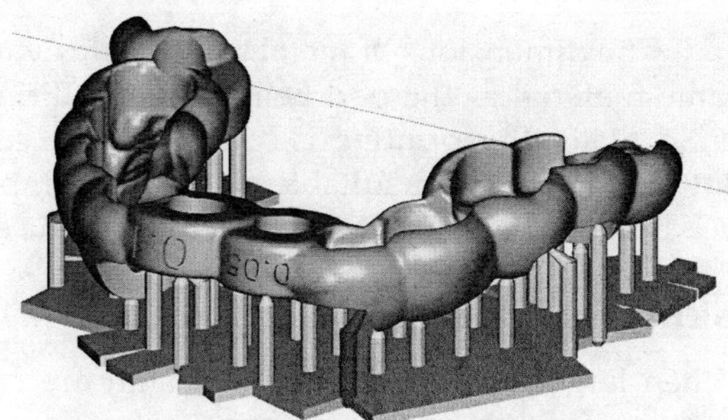

Fig.5.11: Dental Parts Attached to Build Plate and Structure Support

Courtesy: Sprintray, USA

Although support for metal parts is critical to ensure parts are printed within specified and accurate tolerance, it is much more difficult to remove metal than removing polymer in 3D printing. The lattice type of the structural support may make it easier to remove; nonetheless, second processes using cutting tools are required. Also, metal support has a more detrimental effect on the surface of the part. The produced surface is in need of a second process, such as part sanding to achieve the expected surface finish quality as well as the rest of the print. However, the more structural support included in part design, the more accurately the part is printed, but with a higher cost of material, and increase in post processing time.

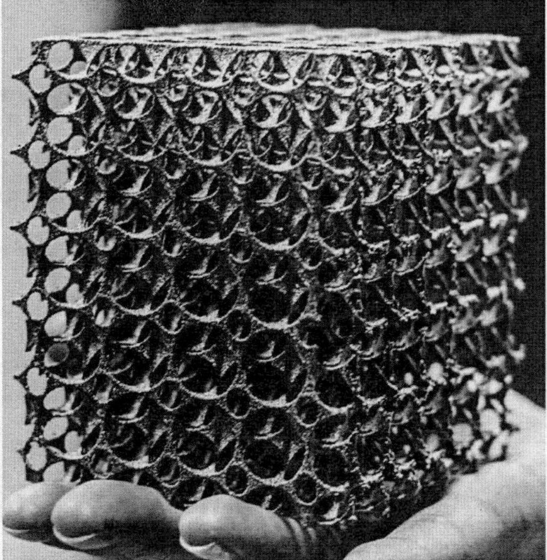

Fig.5.12: Powder Removed from Metallic Part

Hollow Sections

Large hollow sections are not suited for metal printing using polymer based powder, as in Fast Fusion Filament – FFF, due to the lack of adequate structural support to successful build. Stereolithography – SLA is used to construct hollow sections. The design of the structural support may not be required, instead the use of escape holes are needed to allow for the removal of loose powder, Figure 5.12. The minimum diameter of the escape holes should 5 mm.

Build plate

When printing metal printed parts the build plate is attached to the build part during printing. The build plate serves as a foundation for the construction of the part and acting as a heat sink. The build plate gives the required rigid support to create stable printed parts, Figure 5.13.

Fig.5.13: Build Plate Rigid Support

The dimensions of metal build plates vary from 20 to 40 mm thick. It is constructed from the same material as the part being printed. Printed parts are bonded and must be cut off from the build plate after printing is complete. The cut off process may use milling machine, wire/EDM cutting, or any other suitable means. The impact of this extra step on the surface finish of the part needs to be considered when orientating the design on the build plate.

Skin and Cores

When large thick sections of a design are required, metal printing may not be the best manufacturing solution. A large solid metal part is required the necessary design for metal printing utilizes "skins and cores" illustrated in Figure 5.14. Printing parts in this manner reduces the printing time is significantly reduced, and maintains part stability without sacrificing surface quality.

Fig.5.14: Cross Section of Solid Metal Part Representing Skin and Core

Skins and cores are exposed to different laser intensities and times to print, which results in different densities. The skin of the part is scanned to create a fully dense outer layer for maximum strength and hardness, while the core scanned at a greater speed resulting in greater porosity. Skin thicknesses are typically 1 - 3 mm.

Designer should allow for a possible secondary machining operation to create a threaded hole into the most appropriate location. If a threaded hole is added to the design after printing, it may lie in the core section of the print. There may be less dense material to cut for a thread. It is advisable to 3D print a hole. Printing holes will ensure that the hole--walls lie in the skins. It is recommended that hole-design is undersized by an amount 0.6 mm, which is 0.3 mm each side. This ensures accurate and stable hole-print, and then drilled after printing.

COMMON DESIGN PARAMETERS

Channels

The ability to produce internal channels within parts is one of the advantages of metal printing. The addition of channels allows a part to cool uniformly and helps to reduce the weight. It is recommended that channels configurations must not be more than 8 mm in diameter. Larger diameter than 8mm, the channel geometries begin to deform, causing print failure to the top surface of the channel.

If a required channel is above 8mm in diameter, it is recommended that the geometries presented in Figure 5.15 are used. These geo- metrics allow the channel diameter to be maximized while also maintaining a uniform surface finish.

Fig.5.15: Channel Configurations Change as the Diameter Increase beyond 8mm

Fillets

Fillets are a critical part of any design. Fillets are particularly important for metal printing due to residual stresses generated during printing by the high temperatures gradients subjected on parts during printing. Also, fillets allow for features to be built up gradually, with a smooth transition from thick to thin regions helping to distribute heat gradually throughout the build. Fillets are particularly important in hollow cavities. It is highly recommended to make fillets as large as possible. Also, cavities are difficult to print and remove its powder.

Build size

Designers who are familiar with traditional subtractive technology in metal manufacturing are often surprised by the limitations of the build volume for most metal printers. The average build size is 200 x 200 x 200 mm, Figure 5.16. Designers must consider these size limitations when deigning metal printing parts. Printing larger metal parts may be best suited to use "Binder Jetting" technique.

Fig.5.16: SLM 3D Printing Machine Size - 280 x 280 x 365 mm

Also, the size of a printed part plays an important role 3D printing cost and time to produce. The size and the perpendicular height to the build direction cause the rise in cost. Utilizing metal printing cost effective, consider only the relevant sections of a design to be printed with extra volume being avoided wherever possible.

METAL PRINTING DESIGN TABLE

Design Rules Summary Table

Summary of the design rules corresponding to the 3D printing technology are presented in the Table 5.1, below. The chart can be used as a reference to determine which technology is best suited a particular design. The values presented in this chart are general recommendations and may vary based on specific printer type or the material used.

Table 5.1: Summary of 3D Panting Design Rules

	Supported Walls	**Unsupported Walls**	**Support**	**Embossed & Engraved Details**	**Horizontal Bridges'**
FFF Fused Filament Fabrication	0.8	0.8	45°	0.6mm wide – 2mm high	10mm
Stereolith-ography & Direct Light Processing	0.5mm	1.0	Support Always Required	0.4mm wide & high	X
Selective Laser Sintering	0.7mm	X	X	1.0mm wide & high	X
Material Jetting	1.0mm	1.0mm	Support always Required	0.5mm wide high	X
Binder Jetting	2.0mm	3.0mm	X	0.5mm wide & high	X
Metal Printing	0.4mm	0.5mm	Support Always Required	0.1mm wide & high	2.0mm

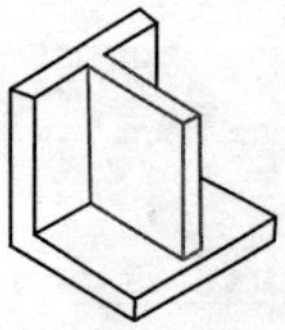

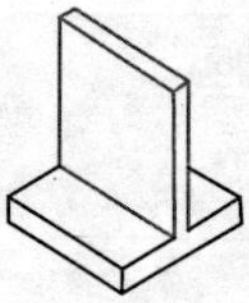

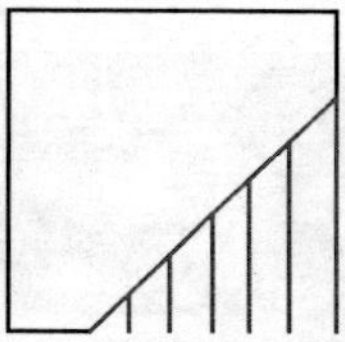

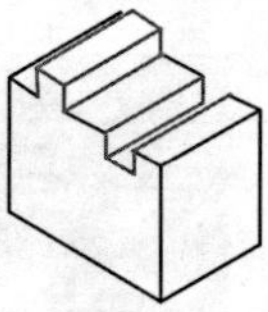

	Hole	Connecting Moving Part	Escape Hole	Minimum Feature	Pin Diameter
FFF Fused Filament Fabrication	ϕ0.2mm	0.5mm	X	2.0mm	3.0mm
Stereolithography & Direct Light Processing	ϕ 0.5mm	0.5mm	4.0mm	0.2mm	0.5mm
Selective Laser Sintering	ϕ 1.5mm	0.3mm for moving parts & 0.1mm for connection	5.0mm	0.8mm	0.8mm
Material Jetting	ϕ 0.5mm	0.2mm	X	0.5mm	0.5mm
Binder Jetting	ϕ 1.5mm	X	5.0mm	2.0mm	2.0mm
Metal Printing	ϕ 1.5mm	X	5.0mm	0.6mm	1.0mm

The design freedom of 3D printing have provided shorter lead times, and substantial cost reduction, which allowed the technology to disrupt many aspects of traditional manufacturing.

TOOLS FOR 3D PRINT DESIGNS

The design rules for each technology provides a guide line to 3D Print design parts, and select the most optimum technology to accomplish the 3D printing efficiently and most economically in the desired time. Also, "Computer Aided Design - CAD" has become is the most common method of producing the designed part through 3D printing.

Three basic steps are highly recommended to produce printed parts, summarized as follows:

1. CAD modeling techniques.
2. 3D model production techniques
3. Topology of part optimization and reverse engineering.

CAD Design

3D modeling or CAD allows engineers and designers to generate the drawings needed for digital manufacturing, producing realistic models of parts and assemblies. These models can be used to digitally "form, fit, and function," or for running complex simulations. A wide range of parameters can be simulated, such as strength or temperature resistance, before any physical model has been created, enabling a much faster and less expensive workflow. There are three main methods of CAD modeling presented as follows:

A. Solid modeling,
B. Surface modeling and
C. Sculpting.

Solid Modeling

Solid modeling creates 3D models to appear exactly as actual parts, with a workflow that is similar to the processes used in traditional manufacturing. The model begins with a solid block of material, where sections are added or removed to produce a final shape. A range of operations including extrusions, cuts, and rotations can be used to produce a design. Solid modeling is usually parametric, which can be changed, and saved at every stage of the modeling processes. It can be edited at any time during the design phase, allowing the model to be perpetually updated.

Assembly modeling is also an important part of solid modeling, allowing the design components to be assembled together, forming complex models, Figure 5.17. Rotational features can be applied to assemblies, allowing the analysis and evaluation of dynamic performance.

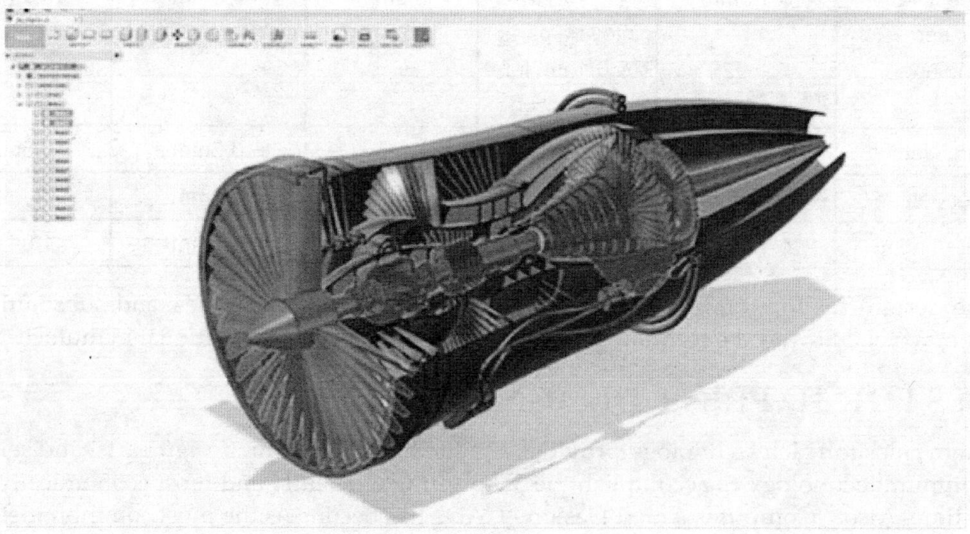

Fig.5.17: Solid Modeling Used in Complex Design – *Autodesk Fusion 30*

Surface Modeling

Surface modeling is used for designing varieties of shapes, as plane-wings, a vehicle body or a machine covering. This type of surface forming is often refers to as free form geometry. It is simple and convenient to create free-form geometries using this type of CAD software, when compared to solid modeling, Figure 5.18. Solid modeling correlates all operations on movements in 3 dimensions; therefore, the creations of complex organic curves can be difficult. Surface modeling typically places a series of points (or poles) over a surface and allows them to be manipulated to form a desired shape. Although this does offer more design freedom, the lack of constraints can lead to problems with accuracy or manufacturability.

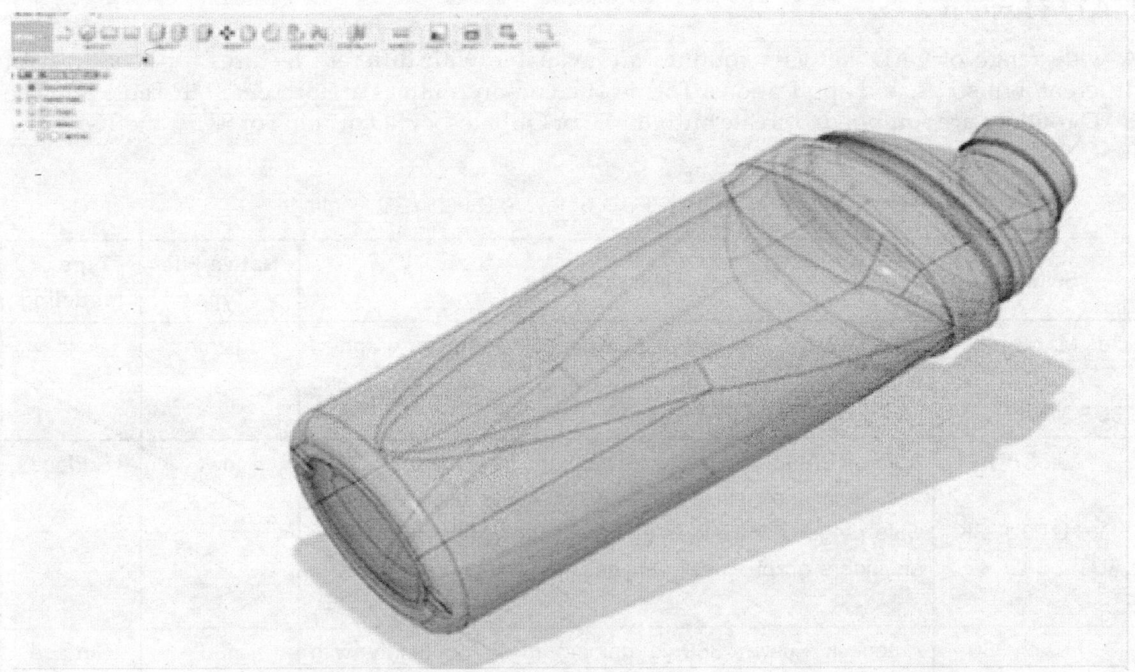

Fig.5.18: Free Form Surface Modeling Container– *Autodesk Fusion 360*

Surface modeling contains only surfaces of a part with no solid interior, as a shell. Once a design encompasses enough surfaces to become fully enclosed, it can be "filled" to be used for 3D printing. This type of surface modeling is not parametric. Therefore, surface modeling can exceptionally be difficult to incorporate changes in the design, when iterations are required.

Sculpting

Sculpting is known as organic modeling, which is used for creating free form surfaces with intricate details. This type of sculpturing modeling includes characters, jewelry or organic shapes found in nature such as trees or rock formations.

Most sculpting software was developed considering the classical approach in sculpting. Digital sculptors may start creating the object from a simulated ball of clay, and then use a pressure sensitive drawing monitor to manipulate the object. The design process becomes complete when a digital brush representing a classical sculpting tools, as a rake or thumbs to move, add or remove material, Figure 5.19.

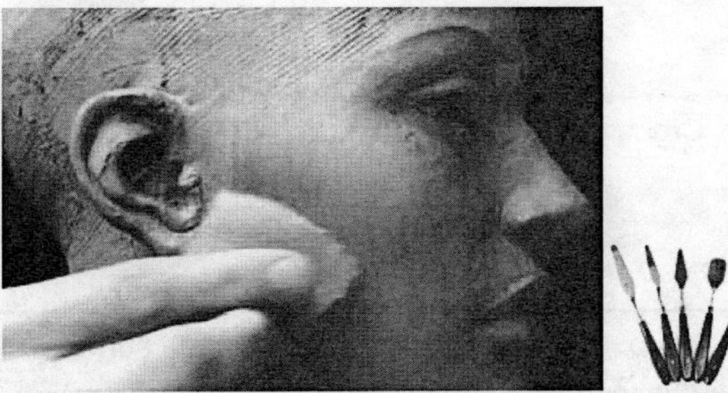

Fig.5.19: Block of Clay - Sculptor with Traditional Working Tool

Using sculpturing tools, artists may create sculptures consist of a sufficient number of polygons counted in tens of millions, allowing intricate details displayed in a design, while old traditional sculpturing technique used simple geometric shapes in of significantly less polygons counted in thousands to produce a model.

CAD Design

A wide range of CAD software routines are available with different features for a number of different industries, and applications. The most common routines are presented in Table 5.2. The CAD routines are equipped to create output STL or OBJ files for 3D printing or STEP and IGES files for CNC manufacturing.

Table 5.2: Industry CAD Software Used in 3D Printing

Software	Description	Native File Type	Type of Modeling
3ds Max 3 AUTODESK **3DS MAX**	Autodesk.3ds Max is a professional 3D computer graphics program for making 3D animations, models, games and images.	.3ds.max	Solid & Surface
AutoCAD A AUTODESK AUTO CAD	Autodesk AutoCAD, software package for 20 and 3D CAD, has been used since 1982. AutoCAD is used across a wide range of industries by architects, project managers, engineers, graphic designers and many other professionals.	.dwt .dwg	Surface
Fusion 360 F AUTODESK FUSION 360	Autodesk Fusion 360 is gaining more popularity with engineers and designers. It is similar to "Solidworks" with the addition of integrated manufacturing sculpting tools. It is also available for free for students, enthusiasts, hobbyists, and startups.	.f3d	Surface
Inventor **I** AUTODESK INVENTOR	Inventor is one of the most popular programs available, offering professional 3D mechanical design, drawing tools and product simulation tools.	.ipt .iam .idw	Surface
Onshape Onshape	"Onshape" is a full internet based CAD software package. It makes extensive use of cloud computing, with compute-intensive processing and rendering performed on cloud based servers.	Cloud only	Surface
Creoe ptc	PTC Creo is a suite of design software with a focus on product design for discrete manufacturers. The suite consists of apps, each delivering a distinct set of capabilities within product development.	.prt .asm	Surface
Rhino Rhinoceros	Multi-use, free-form surface modeler for engineering, architecture and jewelry design.	.3dm	Surface & Solid
Sketchup SketchUp	Entry level software that is easy to use, but with basic features. Mainly used for applications such as architectural models & interior design.	.skp	Solid

Solidworks 7- **DS SOLID WORKS**	Industry standard engineering software used for part and assembly modeling. Includes simulat10n features as well as drawing and assembly tools.	.sidprt .sidamsiddrw	Surface & Solid
Solid Edge Solid Edge.	Solid Edge provides solid modeling, assembly modeling and 20 orthographic view functionality for mechanical designers. Solid Edge also integrates with several product lifecylce management programs.	.prt .asm	Surface
ZBrush ZBRusH	ZBrush is a digital sculpting tool that combines 3D/2.5D modeling,texturingandpainting.Themaindifferencebetween ZBrush and more traditional modeling packages is that it is more skin to sculpting.	.0bj	Sculpting

TOPOLOGY OPTIMIZATION

Topology optimization is a method used to optimize the geometry of a part. This deals with minimizing a part's mass, while maintaining its structural integrity. The process involves analyzing the loads that are applied to the part during operation to determine where part of the mass may be removed. The optimization is often used as a guideline to generate the means to create a part based on a bulk design to improve the performance of an existing design.

Topology Optimization Parameters

Before applying topology optimization to a design, the following information is required:

1. ***An existing design: Topology*** optimization can only remove mass where it has been modeled by the user. A predefined "workspace" of initial bulk design is required, Figure 5.20.

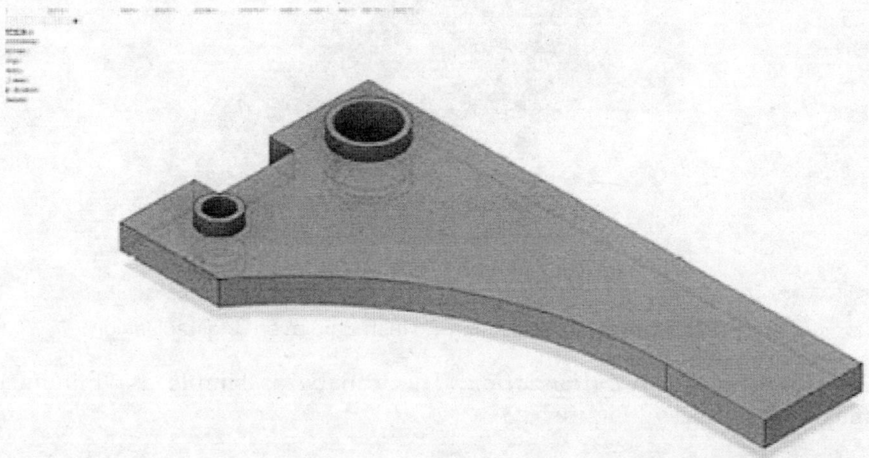

Fig.5.20: Topology Optimization – Initial Design Configuration

2. ***Loads and constraints:*** A part is required to be mechanically loaded before it can be optimized. The direction, magnitude and position of the loads acting on the part have to be well defined. Also, the part material must be well defined, Figure 5.21.

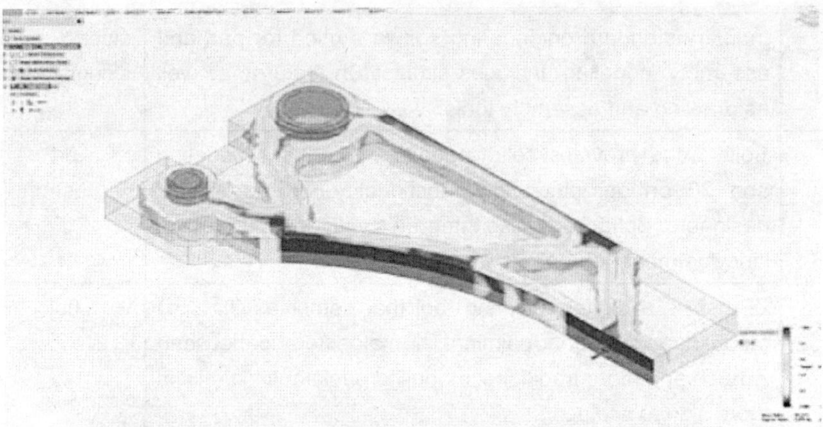

Fig.5.21: Topology Optimization – Load Assessment

3. ***A constraint for the optimization:*** The optimization has to be constrained within a set of limits by maintaining a specific part stiffness and strength, while reducing part mass.
4. ***Manufacturability:*** Consider how the part is produced and the relating manufacturing constraints, such as undercuts using 3D printing. Some software routines able to apply constraints to ensure the design is optimized and can be made by a specific production method. Figure 5.22.

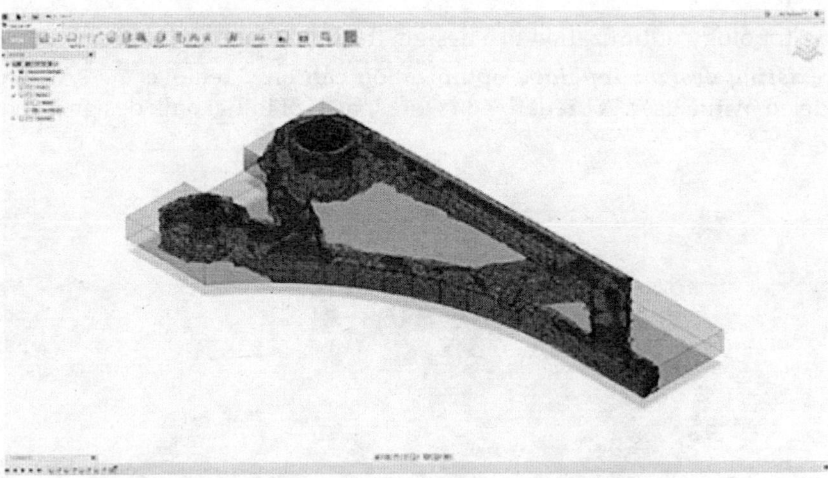

Fig.5.22: Topology Optimization – Mesh Laid over Original Design

The objective of the optimization: This can be as simple as "minimize mass" or "maximize stiffness," Figure 5.23.

Topology optimization is effective when used in 3D printing. The organic geometries that topology optimization produces are ideally suited 3D printing. Topology optimization geometries are difficult to produce when using traditional manufacturing techniques like CNC machining.

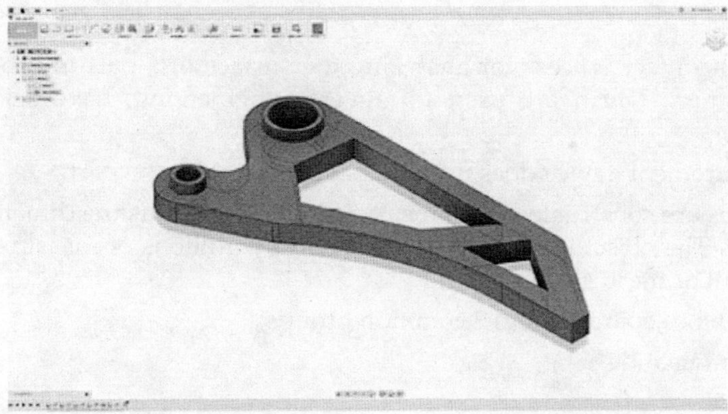

Fig.5.23: Topology Optimization – 50% Weight Reduction

Benefits and Limitations

Topology optimization is best suited for industries where parts are highly loaded, and required to be lightweight, in particularly, in automotive and aerospace industries. Utilizing topology optimization early in the design process can greatly assist guide the design towards the best solution.

Topology optimization can be used to refine a design. Once an original bulky draft design was configured to withstand loads, an optimization process begins to reconfigure the shape, allowing a more detailed design to be reached. This process can be reiterated few times if needed to finalize the required design.

The topology optimization is that the software used to perform the simulations does not provide a "one-click" solution. Basic knowledge of "Finite Element Analysis (FEA)" may be required during the initial set up stage. The loads and constraints must be applied correctly. A clear understanding of how a load may be applied to the part that may have a significant impact on the integrity of the final printed part.

Reverse Engineering

Reverse engineering is the process of studying existing parts or products to gain insight into how they are designed and manufactured. It involves complete disassembly and documentation of all parts and assemblies, followed by computer digitization to recreate the parts as 3D files.

Reverse engineering, while it is considered illegal for many products, many corporations and foreign institutions and foreign governments resort to reverse engineering techniques as a form or intellectual piracy infringing upon intellectual properties. The writer is emphatically discouraging such practices. However, there are numerous products that are no longer having the same intellectual proprietary privileges. Furthermore, their patents and trademarks may have been expired, which place them as an "open source" for copying and perhaps modifications, to further advance the technology.

Some of the most common applications of reverse engineering include:

a. The generation of 3D files that represent complex and organic surfaces,

b. Verifying parts to check dimensional compliance and the measurement of parts that are no longer in production.

Reverse engineering can be separated into two main categories:

a. 3D scanning

b. Physical measuring.

3D Scanning

3D scanning is a noncontact process for analyzing the surface of a part to produce a 3D model. 3D scanning techniques, which are used for reverse engineering, have several fundamental commonalities:

1. The measurement device does not come in contact with the part.
2. Digital files are constructed from measured data as well as the dimensions of the physical object. The object is digitally represented by millions of measurements, in form of either point or mesh elements.

3D scanning can be separated into 2 common groups:

a. laser scanning and
b. CT scanning.

Laser Scanning

Laser scanning surveys the surface of an object and captures data represented as a collection of points, which are then used to generate a 3D surface. This enables parts to be precisely measured to generate and digitize their exact 3D models for reproduction. The vast number of the collected data points through the noncontact characteristics of laser scanning; this laser scanning method is the best method suitable for free-form of non-uniform surfaces.

Fig.5.24: A Handheld 3D Scanner Generating 3D Digital Image of the Motor Cycle

Laser scanners can either be handheld, Figure 5.24 or fixed, requiring the part that is being scanned to be manipulated. Scanners can also be mounted on robotic arms for accurate surface tracking and high repeatability

CT Scanning

Industrial "Computed Tomography - T" scanning uses X rays to create an accurate representation of a component. CT scanners work by placing an object on a turntable between an x-ray tube and a detector. The detector captures multiple x-ray images of an object as it rotates 360 degrees, acquiring the outer dimensions, internal geometry and density within the object's walls. The series of 2D images are then run through a reconstruction algorithm that creates a 3D volumetric model.

CT scanners are relative occupy large area. It is an expensive industrial machine. This is another type of reverse engineering system. The CT scanning technology is capable of inspecting a part both internally and externally, in a non-destructive manner. It creates highly detailed 3D models with complex geometry, Figure 5.25. Parts can be scanned to reveal any imperfections or voids, which could lead to failure once the part is in use. CT scanning is also used for dimensional accuracy verification in manufacturing. A part is scanned and the

Fig.5.25: Computer Tomography Scanning BMW Vehicle

Courtesy: BMW

corresponding 3D model produced from the scan data is then laid-up over the original 3D model of the design. Automated software then detects any variations in dimensions and determines whether they are within an acceptable range.

Physical Measuring

Physical measuring is the process of measuring specific points on a component relative to a datum point to produce a 3D model of its configurations. Physical measurement techniques vary from 3D scanning methods, the main characteristics are as follows:

a. Require direct contact with the object being measured
b. Can be more accurate than 3D scanning techniques

Physical measuring can be separated into 2 common groups:

1. CMM
2. Manual measurement.

CMM

A Coordinate Measuring Machine (CMM) uses a probe to physically contact certain features of a part digitally registering each touch point. These points are then verified against a 3D model. CMM is used to verify dimensions of parts rather than to digitizing them, but it is also possible to generate a point cloud, which can be converted into a 3D file. This method is best suited for simple parts where a high level of accuracy is important. CMM's are often used in batch manufacturing, where a sample of parts are selected from a batch and measured to verify compliance.

Manual Measuring

Manual measuring is a much simpler and accessible method of reverse engineering in which a 3D model is created by manually measuring features of a part. Each measurement is manually recorded and used to produce the 3D file in CAD. It is much slower than 3D scanning, Table 5.3.

The following table summarizes each of the methods of reverse engineering:

Table 5.3: Reverse Engineering Techniques

	Time	Accuracy	Cost	Best Suites for
Laser Scanning	Fast	Medium	Medium	Free Form Surfaces of medium details and non-uniformity
CT Scanning	Fast	High	Very High	Internal and external inspection
CMM	Slow	High	High	Precision engineering part
Manual Measurement	Slow	Medium	Low	Simple geometries

APPLICATIONS OF FAST FUSED FILAMENT

The ability to produce functional parts from adequate polymers Fast Fused Filament has been adopted for a range of applications. Jig and fixture are becoming a common products to be produced though 3D printing of FFF technology.

Jigs and fixtures

Jigs and fixtures are workpieces used to aid in positioning the parts in a precise location to perform further operation such as the assembly of parts. Traditionally, jigs and fixtures are CNC machined to a high tolerance allow a part to be accurately held and located a desired position. The level of customization required for jigs and fixtures usually results in long production lead times and high costs. Jigs and fixtures geometries are often unique and somewhat cumbersome to machine. Although 3D printing was originally developed to print rapid prototyping parts of low volume production, the improvements in the quality of printed parts, enhanced with the strength of the available range of supplied engineering materials has enabled the Fast Fusion Filament – FFF to print manufacturing of functional jigs and fixtures. Using FFF a design may be printed, tested and placed on the assembled line in a short time to meet the production demands of another known products. Operator's feedback is valuable to be incorporated in the design iteration process leading into consecutive final design for reliable tools.

A prominent automotive manufacturer produces 100,000 cars/annum. The jigs and fixtures aiding the assembly were 3D printed for daily use on the assembly line. The jigs and fixtures along with the 3D printing machines were successfully validated. Multiple desktop FFF printing systems are still in operation since 2014. All the required supporting assembly tools, including jigs and fixtures are totally produced in-house since the installation of the multiple FFF 3D printing machines.

The dependency on external manufacturers was gradually eliminated. Consequently, jigs and future tooling cost was reduced substantially. The transition to 3D printing saved the automotive manufacturer in excess of **990%** of the tool development and manufacturing costs. Additionally, the time to prepare the assembly tools were substantially reduced, Figure 5.26. In 2016, the facility saved an estimated **$205,000** on jigs and fixtures, where it was expected to reach in excess of **$330,000** in 2017, in addition to substantial reduction in time. The 3D printed tools are ergonomic, which yield greater operator engagement, due to their valuable feedback to improve tool functions incorporated into design iterations.

Competitive Low-Volume Production with FFF

A well-managed corporation in Denver, Colorado, USA has successfully used additive manufacturing 3D printing technology for rapid-prototyping of low volume production runs of plastic parts. The corporation was given a project to custom design and manufactures a cap that turns and locks into place over a USB insert. The cap would protect a USB flash drive outdoor. Originally the cap was manufactured through injection molding. While injection molding is unequaled producing high volumes, 3D printing can be a competitive solution for low production runs. The initial costs relating to tooling of injection molding can be exorbitant. As the required volume is only 200 parts per year, the 3D printing alternative was identified to be a cost effective solution.

The primary focus was to ensure the cap fitted correctly. The industrial Fast Fused Filament FFF printers produce parts in an enclosed and controlled environment, resulting in highly accurate parts with high repeatability. Using the 3D printing additive manufacturing machines the corporation was able to produce the flash drive cap to a dimensional tolerance of "0.005" – "0.127" mm of the critical dimensions. Additionally, the build size of the industrial FFF printer allowed for a large number of caps to be printed in a single run, leading to further cost reduction. FFF machine in this event has provided the ability to produce functional parts from reliable engineering polymers. Initially, component was manufactured using Nylon 12 polymer. According on the annual volume, cost and overall performance of the part, it was suggested to switch material to acrylonitrile styrene acrylate - ASA polymer. ASA has the best surface finishes for FFF thermoplastics, high accuracy, low shrink rate and UV stable, which makes it ideal for outdoor applications. The manufacturing of part using FFF technology met all the engineering specifications at a lower cost and better quality part printed part using Nylon 12, Figure 5.26.

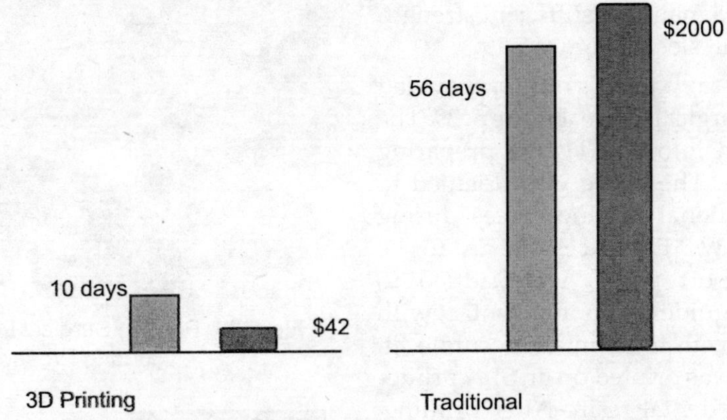

Fig.5.26: Substantial Reductions in Cost and Lead Time

APPLICATIONS OF STEREOLITHOGRAPHY (SLA) AND DIGITAL LIGHT PROCESSING (DLP) 3D PRINTING SLA/DLP

The smooth surface finish, and the high accuracy that stereolithography (**SLA**) and digital light processing (**DLP**) 3D printing offer, combined with the sizable range of materials available, have enabled adopting SLA, DLP, and SLA technologies to the medical and dental idustry. Also, the affordability of SLA/DLP desktop machines has made the technology highly accessible. The global adoption of 3D SLA/DLP printed technologies spread their use to surgical, dental, and hearing application.

Accurate, Custom Surgical Dental Guides

Modern dentistry relies heavily on the ability to produce small, smooth, complex components that fit well inside the patient's mouth. Every set of teeth is unique to the individual patient; thereby every dental appliance has to be custom-made. The ability to meet these constraints, 3D printing is employed in dental applications. The dental and medical industry are accounting for over 18% of all 3D printing revenue annually.

3D printers are capable of producing accurate parts that meet the pressing level of customization in the dental industry. Also, parts are printed with a very smooth surface to improve patient comfort, and reduce the amount of post processing that is required. Additionally, many SLA/DLP resins have been specifically chemically engineered to withstand sterilization processes, offering a high level of biocompatibility.

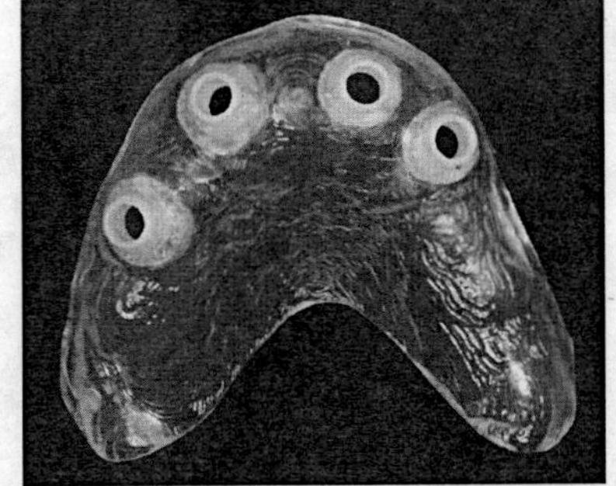

Fig.5.27: 3D Printed Custom-Made Dental Guide – Precision Guide during Dental Procedure

The manufacturing of surgical instrumentation guides have been widely adopted the use of 3D printing technology in the dental industry with traditional guides having high manufacturing cost and long lead times, Figure 5.27. A desktop SLA printer was utilized to assists with the placement of an implant for a missing tooth. A surgical drill guide was required to accurately position a drill to the correct location during the dental procedure, Figure 5.28.

Figure 5.28, the needed guide was printed to fit comfortably inside the unique shape of the patient's mouth, and maintain sufficient strength to house a metal drill sleeve, Figure 5.29.

The 3D scan of the patient's mouth provided a 3D model of the surgical guide, Figure 5. 29. The design is converted into an STL file preparing for the 3D printing. The guide was oriented to minimize cross-sectional peeling forces during printing. This allows the excess resin to be drained. Few support points were added to surfaces where the guide was not in contact with the surface of the teeth to maintain accurate fit. The surgical guide was printed on an SLA printer using well defined dental resin. After printing, *the guide* was washed in isopropyl alcohol, dried, UV cured, and then the supports were removed. The stainless steel metal drill sleeve was inserted into the printed guide hole to complete the *guide* fabrication. The guide was sterilized through an autoclave process to prepare it for the procedure, Figure 5.30.

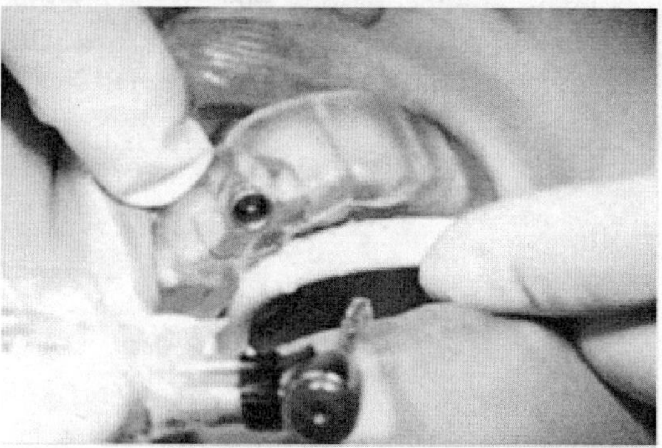

Fig.5.28: Printing Surgical Dental Guide Used during Procedure

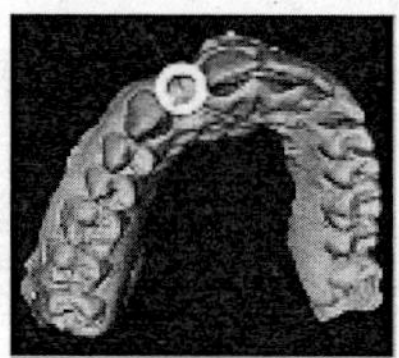

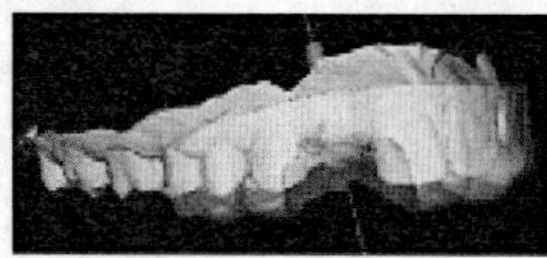

Fig.5.29: Guiding Metal Sleeve Insert

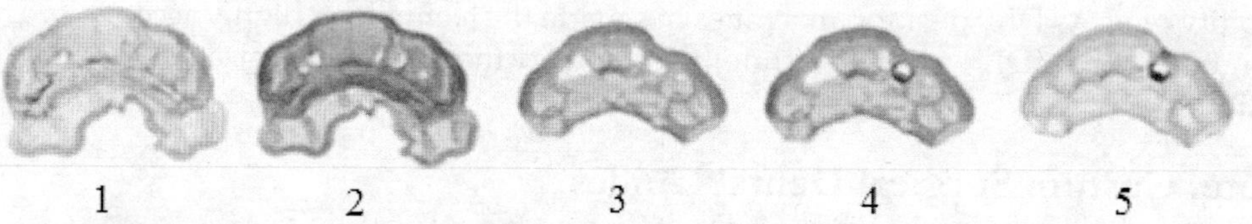

Fig.5.30: "Formlab" 3D Printing

Cutesy: Formlab

1. Printing
2. Post UV Cured
3. Support Removed and Sanded
4. Metal Sleeve Added
5. Denture Sterilized

The final guide was produced in house at a significantly lower cost when compared to traditional methods. Also, devising a guide has significantly decreased procedure time, and cost, eliminating flap advancement, drill angle determination and tissue re-approximation. The use of custom guide turned a traditionally 75 minute procedure into only 20 minute.

The improvement in quality and cost of desktop 3D printers has lowered the barrier to entry, allowing smooth, accurate parts to be in-house easily printed, while improving accessibility to 3D printing for smaller dental laboratories.

Hearing Aids - 3D printing's Case Study

Today, over 15 million people are wearing 3D printed hearing aids. 97% of all hearing aids globally are produced using 3D printing. Not only has 3D printing technology significantly reduced the cost of custom hearing aids, but also produced accurate, smooth, complex surfaces. Additionally, it has reduced ill fitted customer returns from 40% to 5%, leading to greater customer comfort.

The traditional means to manufacture a hearing aid consisted of a several manufacturing steps involved extensive manual labor and a significant amount of lead time. Traditional manufacturing of hearing aid starts with making an impression mold which is taken directly from the ear canal of a patient using a flexible material. The formed impression creates a negative mold, which is made of rigid silicon material. The shell enables the actual hearing to be casted using certain type of acrylic. The casted acrylic is left to be curd. The final acrylic hearing aid had holes drilled for the placement of the electronics. Excess acrylic debris was cut off before sanding the surface with fine sandpaper, and then polishing the surface to reach acceptable smooth surface finish for patient comfort. A skilled worker spent about one day on each individual hearing aid.

Thanks to the 3D printing the hearing aid production has been rapidly accelerated with substantial cost reduction. A 3D scanner is used to scan the patient's ear to produce an accurate three-dimensional image. The generated 3D model created from the 3D scanner data enable the designer to make modifications to the configurations of the product. In addition the designer creates the necessary cavities and attachments for electronics, which can be easily integrated in the design. The "CAD" software converts the scan onto a file that the 3D printer can interpret and duplicate. A batch production allows multiple hearing aids to be printed in a single build within 2 hours, Figure 5.21. The 3D printer allows the 3D file of the original impression to be saved should the hearing aid need to be replaced.

Fig.5.31: 3D Printed Hearing Aids Attached to the Build Plate

The traditional production of hearing aid is considered an art form instead of material science, manufacturing technique, and controlled tolerances. 3D printing technology has transformed a manual, labor-intensive industry into an automated, high accurate and economical industry. It has reduced the lead time from weeks to few days, or even hours, while producing a superior product.

APPLICATIONS OF SLS

The Selective Laser Sintering - SLS technology has the ability to produce parts with relatively high tensile strength from materials of known polymers. Also, SLS has the capability to produce parts at low to mid volume production. This type of technology is used for a range of functional products with tight tolerances and reliable applications. SLS technology has been used in many custom design applications. One of the custom designs is the design and build of a camera, and the design of a functional bicycle. This is to replicate more expensive models and functional bike accessories. Both products capitalize on the production capabilities of SLS printers.

Custom Panoramic Camera

The need for a camera capable to capture images with much larger negatives than the current standard format allowed entrepreneurs to turn to 3D printing technology. Also, the lack of budget to purchase such a unique camera to achieve this goal has inspired developed to undertake the development of this unique camera. Moreover, a design agency in London, England turned to 3D printing to study the possibility of creating a custom solution.

Traditional manufacturing techniques such as CNC and injection molding were considered. The 3D printing technology was selected due to its design freedom. The 3D printing technology can competently allow complex design geometries that would be exceedingly difficult with traditional machining. Additionally, 3D printing can easily incorporate complex geometries in the design, at low cost in spite of the low volume. CNC machining would have been expensive if it was feasible to machine complex geometries. The injection molding alternative would require exorbitant initial investment of tooling.

The camera corporation selected SLS as the most suitable technology, planned to make it from a functional polymer, Figure 5.32. The SLS technology can easily print complex organic shapes. Also, it can competently produce the required strength of polymer. These are few of the factors that governed the decision to select SLS technology. SLS technology produces parts with a dimensional accuracy of ±0.3mm, with a lower limit of ±0.2 mm, making it suitable for applications where components need to be tightly assembled together.

Fig.5.32: SLS Printed Specially Designed Panoramic Camera

After 3D printing the camera, it was subjected to various forms of post processing, including sanding, coating with special primer and a coat of sealant.

The camera is built to shoot 6 x 14 negatives on 120 mm film. Its modular of eight part design enabled it to integrate different lens housings to facilitate different shooting settings.

The camera corporation hopes to make this unique camera accessible to more people by turning it into an affordable product, making this kind of photography available for both professional and amateur alike.

Bicycle Accessory

A bike chain reattachment tool is required once a bike chain comes off during a commute, often resulting in late arrival and a messy and oily situation. The needs inspired an entrepreneur design and manufacturing a solution that enables high quality parts at an acceptable strength to be produced. Also, the part and the market must to be tested without a large initial investment. At first, a desktop FFF printer was used to produce low-cost 3D printed prototypes of the product. This made it possible to test multiple design iterations of the product rapidly on a variety of bikes and gear configurations.

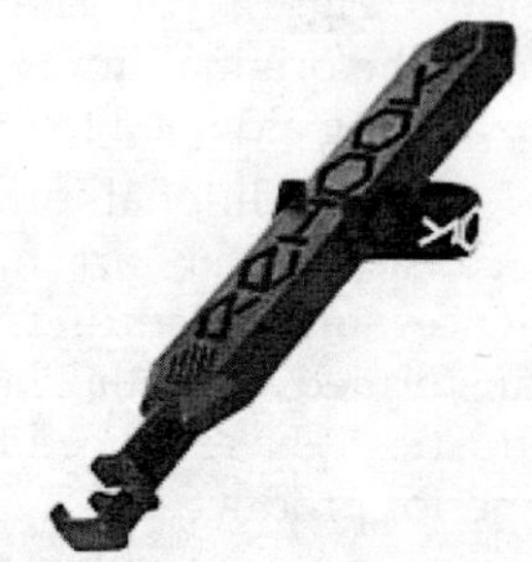

Fig.5.33: 3D Printed Bike Chain Attachment

SLS was selected as the manufacturing solution for the initial production of "Rehooks," as the technology offered a range of strong, functional materials and was capable of mid volume production, Figure 5.33.

The 3D Printer printed 50 units to be tested for any structural weakness. The testing highlighted a certain weakness in the design. Had the design been produced using injection molding, a design change would have been complicated and costly process to rectify. The use of 3D printing allowed a simple alteration to the design at no cost. Once the alteration was verified, the production material was replaced from carbon-reinforced nylon to graphite reinforced nylon. This has allowed the part to be produced at an even lower weight per part.

As 400 units per month were required, the SLS printing system has the capability to meet such demand for the initial testing. Also, the SLS system was able to meet the ongoing design alterations and development.

Development of the tool, testing the market and launching the new product were achieved within just 10 weeks on a budget of under $5,200. Traditional manufacturing approach would not have been possible to achieve the goals for less than 10 times that cost. 12 months after the initial development of the product, many thousand units have been sold and worldwide, as well as distribution agreements were finalized.

APPLICATIONS OF MATERIAL JETTING

Material Jetting technology has the ability to produce parts from different materials, with the smooth surface texture. There are two different utilization for Material Jetting:

1. The production of functional organs for a neonatal training manikin
2. 3D printed model of a cyborg.

Training the Next Generation of Doctors

The medical industry has been one of the pioneering drivers behind the adoption of 3D printing. There is to-date more than 100,000 acetabular, hip cup implants produced through 3D printing technology. There are 50,000 actual implants implanted into various patients.

Scientists, engineers and medical researchers are in constant discovery to find new ways to utilize 3D printing in medical applications. Various medical schools at different universities have established task forces aiming to increase surgical and procedural success rates for neonatal patients. 3D printing has been utilized to create an optimized training program using artificial "lifelike newborn models" with functional organs, capable of intelligent sensor feedback.

Interaction with anatomical models is a critical part of the training and preparation for surgeons and nurses. It is difficult to realistically reproduce accurate haptic feedback in the neonatal field using current practice lifeless manikins that lack the complexity and the feel of a newborn patient. Several research-works aimed to develop manikins that have all their major internal organs "functioning," while being equipped with sensors to monitor key measurements such as pressure, stress and the impact during trial procedures, such as CPR or intubation.

The lifeless manikin has been manufactured using 3D Bioprinting Technologies states as follows:

1. The ribcage/spine produced using SLS, "Functional" internal organs that are housed within the ribcage/spine, made through the use of Material Jetting.

A soft flexible material was required to simulate the internal organ behavior. Therefore, "Material Jetting" was utilized to print the molds. *VeroWhitePlus polyme*, which is a rigid opaque polymer, was used as the outer mold and *TangoBlack Polymer*, which is a flexible polymer, as the inner cores of the model, Figure.5.34. However, the main parts would be traditionally made though injection molding using silicon. The inner cores of the mold was made also of the same flexible material in multiple components to ease their removal, and prevent damaging the silicon parts. "Material Jetting" was also chosen due to their intricate complex configurations of neonatal organs.

Fig.5.34: 3D Printed Heart in a Manikin with Working Molded Valve

The 3D printed heart required highly detailed working valves within the mold configuration "Material Jetting" was best possible technology to achieve the required components high accuracy. When the ribcage and organs were combined, cameras and sensors were installed throughout the manikin. The fluid ran through the cavities, providing feedback on every part of the model, the instant the system was subjected to various simulated trials, Figure 5.35.

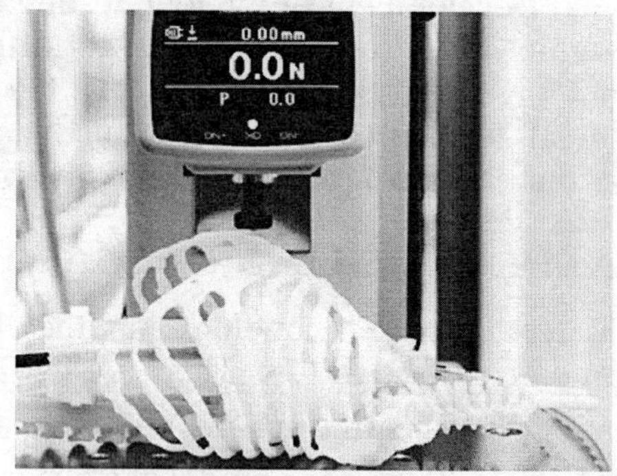

Fig.5.35: 3D Printed Manikin with Interactive Sensors

The developing and advancing in the neonatal can aid medical research in a broader scope. Potentially scientists, medical doctors, and engineers may create realistic patient models of other body parts to strengthen medical training for emergency procedures and pregnancies.

Cyborg - Material Jetting

Hollywood utilized "Material Jetting," the Ultraborg Stiffneck figure was converted to a 3D high detailed physical model depicted from the CAD model.

Fig.5.36: Material Jetting Used to Create "Ultraborg Stiffneck Figure"

3D printing model using "Material Jetting" technology was used to create the highly complex details. The appearance of the model was vitally important to maintain the expected surface texture. All the details were captured from the original design to produce a high-quality finished product, Figure 5.36. A large number of methods for manufacturing the model were investigated. The decision to use Material Jetting was based on two driving factors; namely the resolution and speed. "Material Jetting" facilitated a fast turnaround, while still retaining a high degree of accuracy and details of the final product. "Material Jetting" can produce smooth surface finish and significantly reduce the post processing time, Figure 5.26. The structural supporting materials were removed upon completing the 3D printing process, and then the product surfaces were sanded. Multiple coats of special epoxy paint were added. Additionally, several small detail parts were given a chrome effect using glossy black paint and graphite powder. The final result was an excellent model with a reasonable cost. It would have cost much more if produced by traditionally molding technology, and would have consumed considerable time.

APPLICATIONS OF BINDER JETTING

Binder Jetting is one of the most versatile 3D printing technology with the ability to produce full color models, functional metal parts, sand casting molds and cores. Binder Jetting case study is presented using 3D printed sand casting molds to cast a metal part.

Sand Casting

Traditional technology correlates many metal parts to sand casting as it is the only way to manufacture. However, this traditional sand casting methodology is one of the least accurate methods of manufacturing when compared to CNC or die casting.

The 3D printing of sand casting molds and cores provides a superior approach as manufacturing alternative to sand casting approach. "Binder Jetting" technology does not depend on support structures, sand molds and cores. The part can be produced with a high degree of design freedom through "Binder Jetting." The "Binder Jetting" system can print large components, as well as printing complex molds, Figure 5.37.

Fig.5.37: Vertical Pump Impeller Cast from Bronze – by Binder Jetting

Courtesy: ExOne

3D printing was utilized to produce a sand casting mold for the vertical pump impeller. The impeller was cracked and corroded. It was required to be replaced. The pump was over twenty years old and no longer in production. Using blue-print drawings, a 3D model of the impeller was produced to create the needed mold. The dimension of the impeller pump is 1.27m in diameter and weighed approximately 900 kg. The modular design for the mold was 3D printed preparing for casting.

The final mold was produced using a combination of silica sand and "Furan" binder. The "Silica" sand is one of the most common varieties of sand available worldwide. It is derived from quartz crystals. Also, it is used in a wide range of applications, including the creation of molds and cores for industrial castings. The 3D printing molds used common silica-sand, similar to the silica-sand used in traditional casting processes at the foundries. Additionally, when used with furan binder, it is immediately ready for casting. The components of the mold were printed in few days.

Fig.5.38: ExOne Exerial Fabricating Sand Casting Molds – The Biggest 3D Printer - 2200 x 1200 x 600mm

APPLICATIONS OF - DIRECT METAL LASER SINTERING - 3D SELECTIVE LASER MELTING - DMLS/SLM

Two case studies that have utilized metal printing are presented: a satellite antenna that had parts consolidated from 100 to 1 - and a 3D printed racing car bracket that utilized topology optimization to produce a design. The largest 3D Printer know today is ExOne-Exerial Fabricating San Casting Molds – The dimensions of the Printer are - 2200 x 1200 x 600mm, Figure 5.38.

Consolidate Satellite Antenna

The design freedom of the 3D printing technology caused the elimination of tedious manufacturing processes such as, tool paths, undercuts, and assembly access, allowing full assemblies to be produced as in a single 3D printed operation. Consequently, reducing part count has reduced the maintenance and service requirements.

Micro-antenna product is required to be produced for high performance aerospace and defense applications. A complete redesign of a high-bandwidth is required, incorporating the directional tracking antenna array, which is known as "aKa-band 4 x 4 Monopulse Array." The was design verified, and the 3D print was performed. The components were printed in a single print using the SLM technology "Concept Laser."

Manufacturing antenna systems made through conventional methods, require a complex, multistage process that can take an average of eight months of development and additional six more months to build.

The use of 3D printing, allows the use of a variety of aluminum metals for antenna products. Aluminum has relative surface conductivity, low weight, corrosion resistance, and good strength under shock and vibration.

Fig.5.39: Satellite Antenna – Printed by DMLS Technology

The 3D printed of metal parts had the same mechanical properties as a solid piece of wrought material, Figure 5.39. 3D printed parts can have the same "coefficient of thermal expansion (CTE)" as produced in traditional technology.

Table 5.4: Improved Performance of Satellite Antenna

Part Count	**From 100 Discrete Pieces to 1 – Integrated Assembly**
Lead Time	Reduced from 11 months to 2 month
Weight	Savings is over 95%
Production Cost	Reduced 20-25%
Non-recurring Cost	Reduced by 75%

A profit analysis was conducted make a manufacturing comparison to produce microwave antennae through either 3D printing or through traditional technology. The results of the study are presented in Table 5.4.

Optimization - Racing Car Suspension Bracket

When 3D printing is combined with topology optimization, will result in producing high performance part. Topology optimization was used to generate a design for a highly loaded suspension bracket for an electric race-car. The bracket functions as the main connection piece between the wheels and the chassis of the car. The part was loaded with 400Kg force.

Topology optimization cycle began by constructing a rough initial design. The design incorporated all connection points to the other parts on the car. The geometric constraints were made as large and heavy as possible to allow a maximum optimization freedom. This type of model was used to fine-tune the inputs of the optimization algorithm, which were loads, and constraints. A number of possible design solutions were produced by running the topology optimization simulation through several assumptions. It was determined that producing the part as a single component was not feasible. The design must be split into assembly to allow maintenance and ease of part replacement. Re-running the simulation with the assembly design, a viable simple solution was reached to implement and manufacture.

A "CAD" design was made to match the topology optimization results. The performance of the model was then evaluated using certain simulations technique. The simulations provided the possible way to verify that the stresses in the "CAD" model that matched the stresses generated by the topology optimization.

Based on the design of the components, the only possible method of manufacturing was 3D printing. SLM metal printing technology was selected. The material used for the bracket was Grade 5 Titanium. The combination of material and production technique resulted in a lightweight part with a yield strength of 850 MPa, which has less than half the weight and twice the strength of an equivalent part machined from steel, Figure 5.40.

The part was printed as designed. It was used on electric race-car. The part performed well the total racing season after winning twice the first place in 2 out of 3 competitions.

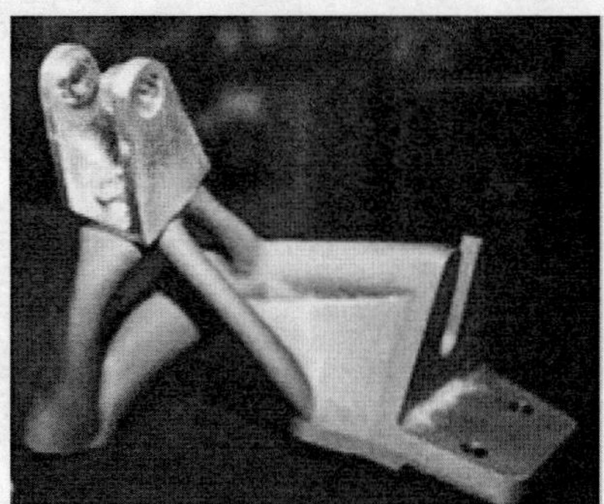

Fig.5.40: Output Geometries of Topology Optimization Process - (Left) Original – (Right) Material Removed due to Optimization

3D PRINTING MATERIALS

After 3 decades of research, innumerable filament reels, and thousands of hours printing, scientists and engineers were able present a collaborative effort to compile the 3D Printing Materials Directory, dealing with hundreds of the most popular materials suitable in 3D printing use today. These collaborative efforts will assist the user in selecting the best material for the next project or advance the quality of the prints with instructions from worldwide experts to rapidly sort the materials based on their features, for a detailed comparison.

Once the user have selected a material, a detailed article of the pros and cons, hardware requirements, best practices, pro-advice, and example projects are provided. Whether the user is a novice to 3D printing or an advanced user looking to experiment with a new material, this directory has all is needed to ensure project success in material choice.

PVA

PVA is frequently known for its capability to be dissolved in water and is often used as a support material for complex prints.

PVA, or Polyvinyl Alcohol, is a soft and biodegradable polymer that is extremely sensitive to moisture. When exposed to water, PVA will essentially dissolve, which makes it a very useful support structure material for 3D printing. When printing extremely intricate forms or ones with partially enclosed voids, PVA supports can be used and easily removed by dissolving in warm water, Figure 5.41.

Fig.5.41: PVA Used in 3D Printing

Typical supports may have been difficult to print or remove in these circumstances. PVA can also be used as a classical material if there is a need to make rapid prototypes.

Advantages

- Great water dissolvable support material
- No special solvents required
- No additional hardware required

Disadvantages

- Moisture sensitive
- Airtight storage containers required
- Greater chances of clogging if the nozzle is left hot when not extruding
- Expensive

Hardware Requirements

Before 3D printing with PVA ensure the 3D printer meets the hardware necessities recorded below to ensure the best print quality.

Bed

Temperature: 45-60 °C

Heated Bed Optional

Enclosure Not Required

Build Surface

PEI

Painter's Tape

Extruder

Temperature: 185-200 °C

No special heat required

Cooling

Part Cooling Fan Required

Best Practices

These instructions will aid to reduce the chances of typical 3D printing issues associated with PVA such as absorbing moisture, warping, and fumes.

Since PVA is highly responsive to water, storing the filament in a moisture-free container is mandatory. PVA that has acquired water from the air will incline to fizz and blow during printing, which will devastate the quality and strength of the completed part. When completed with the PVA reels, ensure to take them away from the printer and store them in air-tight containers along with a desiccant to remove any moisture from the container. If the filament has already absorbed too much moisture, the filament can be dried using a commercial dehydrator, or an oven set to the lowest possible temperature for a few hours. This will draw the water out of the filament without melting it, allowing it to print smoothly and without bubbles.

Use Ozone Shield

When using PVA as a support material for the PLA prints, think through adding an ooze shield with at least two outlines. This will have the benefit of preparing the PVA extruder after every tool change, which will ensure this extruder is prepared to print at the start of the layer. Moreover, the ooze shield will latch any oozing from both nozzles to prevent these plastic deposits from getting into the part. Once the print is finished, the ooze shield can be detached and thrown away. If you find the ooze shield is too weak or tends to separate easily, consider adding an extra ooze shield outline to thicken the defensive shield.

Dense Support Layer

PVA is expensive; it is prudent to minimize the quantity of material that is used for the support structures. Several 3D printers are equipped with a useful feature called "Dense Supports," which enables the use of the PVA material only at the interface layers between the part and the supports. This can effortlessly provide a PVA material savings of 90% or more, while still permitting the supports to easily be dissolved and separated from the part. The dense supports also generate almost solid interface layer, which increases the surface quality on the underside of the print.

Instructions

In multi-extruder arrangements, cooling down the PVA which is usually heated during idle times can prevent oozing and clogging

Set the support upper separation distance to zero layers for a better surface finish. The PVA will be dissolved away, so it may not matter if it bonds to the part while printing. This same method can be utilized for the PVA rafts.

Take Initiative

Since the user understands the basics of printing with PVA, the user is ready to attempt the initial print. The followings are some sample and known filament brands to assist thee user to take bold initiative.

Common Applications

- Removable supports or rafts
- Dissolvable applications
- Decorative parts

Sample Projects

- Hilbert cube
- Spring
- Parrot Pendant

Popular Brands

- Ultimaker PVA
- Matterhackers PVA
- eSun PVA
- HobbyKing PVA

HIPS

Fig.5.42: HIPS Used in 3D Printing

"HIPS" is a lightweight material most commonly used as a dissolvable support structure for ABS models.

Overview

"HIPS," or High Impact Polystyrene, is a dissolvable support material that is normally used with ABS. When being used as a support material, HIPS can be dissolved in d-Limonene, leaving your print free of any markings caused by support removal, Figure 5.42. HIPS has many of the same printing characteristics as ABS, making it a logical dual extrusion partner. Not only is HIPS excellent for supporting the ABS prints, it's also dimensionally steady and somewhat lighter than ABS, making it an excellent choice for parts that would end up becoming worn out or used in applications that can benefit from the lighter weight.

Advantages

- Low cost
- Impact and water resistant
- Lightweight
- Dissolvable by d-Limonene

Disadvantages

- Heated bed required
- Heated chamber recommended
- High printing temperature
- Ventilation required

Hardware Requirements

Before 3D printing with HIPS ensure your 3D Printer meets the hardware requirements listed below to ensure the best print quality.

Bed

Temperature: 100-115 °C

Heated Bed Required

Enclosure Recommended

Build Surface

Glue Stick

Glass Plate

Kapton Tape

PET Sheets

Extruder

Temperature: 230-245 °C

No special hot-end required

Cooling

Part Cooling Fan Not Required

Best Practices

These instructions may aid the user to reduce the chances of collective 3D printing issues related with HIPS such as warping and poor bed adhesion.

"Kapton" tape can be used on top of the heated build platform, generating an ideal surface for HIPS to stick to the surface. Each layer of "Kapton" tape is about 0.1mm thick, so endure accounting for this in the initial layer settings. The user can make this adjustment in mny 3D printing machines by increasing the global Z-axis offset on the G-Code tab of the process settings. "PET" sheets can also be placed over a glass bed to significantly improve the glue characteristics while maintaining the surface finish smooth. These solutions will all aid to alleviate warping from your HIPS prints.

Printing with Elevated Temperature

To keep "HIPS" from shrinking rapidly, the user can attempt to increase the temperature surrounding the 3D printed part. This can be achieved by encompassing the build environment and permitting the heated bed to slowly heat the air to the desired temperature. Some modern printers may include a heated-chamber, which makes this process even simpler. If the user is not able to enclose the printer's build volume, the user may also try attempt using an ooze shield, as this will create a shell that encases the model, enabling the air within that shell to remain at a higher temperature. This shield may also behave as a wind block to keep colder air from rapidly cooling the freshly extruded polymer.

Prevent Layer Separation

Printing with HIPS can necessitate some additional maintenance to ensure that each layer of the print is firmly bonded to the prior layer. If not, the user may find out splitting or cracking between layers. To improve the bonding between layers, think through dropping the layer height or raising the extruder temperature. Both of these modifications will aid to improve the adhesion between layers.

Instructions

Some HIPS filaments include additives that make the filaments easier to print, but these same additives can frequently make the material harder to dissolve and remove if the user plans to utilize HIPS as a support material.

To improve bed adhesion, plan on using ABS slurry. This mixture can be generated on by mixing pieces of ABS filament with acetone and then applying the mixture to the print bed surface.

Take Initiative

Since the user understands the basics of printing with HIP, the user is ready to attempt the initial print. The followings are some sample and known filament brands to assist the user to take bold initiative.

Sample Projects

Dissolvable Material Pivot Test

U-Joint

Popular Brands

- GizmoDorks HIPS
- FormFutura EasyFil HIPS
- Matterhackers HIPS
- eSun HIPS

ABS

ABS is a low-cost material, great for printing tough and durable parts that can withstand high temperatures, Figure 5.43.

Fig.5.43: ABS Used in 3D Printing

Overview

ABS (Acrylonitrile Butadiene Styrene) has an extended history in the 3D printing world. This material was one of the first plastics to be used with business 3D printers. Many years later, ABS is still a widespread material on account of its low cost and good mechanical characteristics and will hold up to excessive usage and wear. "LEGO" building blocks are manufactured from this material for the same specific reason Figure 5.43. ABS also has a higher glass transition temperature. The material can endure much higher temperatures before it begins to distort. This makes ABS an excellent choice for outdoor or high temperature applications. When printing with ABS, ensure to use an open space with good ventilation, as the material inclines to have a slight stench. ABS also inclines to shrink as it cools. Accordingly, controlling the temperature of the build volume and the part inside can have major benefits.

Advantages

- Low Cost
- Good impact and wear resistance
- Less oozing and stringing gives models smoother finish
- Good heat resistance

Disadvantages

- Heavy warping
- Needs heated bed or heated chamber
- Produces a pungent odor while printing
- Parts tend to shrink leading to dimensional inaccuracy

Hardware Requirements

Before 3D printing with ABS ensure your 3D Printer meets the hardware requirements listed below to ensure the best print quality.

Bed

Temperature: 95-110 °C

Heated Bed Required

Enclosure Recommended

Build Surface

Kapton tape

ABS Slurry

Extruder

Temperature: 220-250 °C

No special hot-end required

Cooling

Part Cooling Fan Not Required

Best Practices

These instructions will aid the user to reduce the chances of collective 3D printing issues related with ABS such as warping and fumes.

One of the most typical print quality issues with ABS is warping. As the plastic cools from its extrusion temperature reaching the room temperature, this change in temperature is the bases the plastic shrinks and contracts. This may be particularly troublesome for the first layer, as this change in size can frequently lead to part separation from the bed, corrupting the print.

The user may minimize this harmful effect by using a proper build surface heated to 110º C. The build platform will transfer some of its heat to the first few layers of the object, which will prevent the layers from shrinking, contracting and separating from the bed. It is also common to set the extruder temperature about 10 to 20 degrees higher for the first few layers of the print, which may also help decrease the risk of separation. Many printers give the user complete control over the bed and extruder temperatures; Accordingly, the user may easily set the required values on a per-layer basis utilizing the "Temperature" tab of the process settings.

While these changes may help with the bottom layers of the print, taller parts may have issues as the layers grow further away from the bed. However, as printing larger parts, consider adding an enclosure around the printer to maintain a higher temperature around your print. The enclosure may also avert wind drafts that could rapidly cool the part during printing.

Utilizing Brims and Rafts

When printing bulky parts, or thin delicate parts, the user may find that it is still troublesome having these parts to properly adhere to the bed. In these instances, incorporating a brim or a raft to the print may be an excellent way to anchor these parts to the build platform and avoid warping. A brim will add several rings of plastic around the model on the first few layers, generating additional surface area to secure down the edges of the part. Utilizing a raft will indeed print a whole new polymer structure underneath the print, which may be removed after the print is finished. The rafts in various printers were heavily optimized allowing them to print faster and use more material; however, the user may still find that a brim is faster for larger parts.

Ventilation is Vital

Printing with ABS is recognized to manufacture a strong odor with fumes that could potentially be harmful if inhaled in large quantities. Evade confined spaces and place the printer in a well-ventilated area to avert these issues. Modern 3D printers may incorporate a separate air-filtration system or HEPA filter that can deal with these fumes. If the printer doesn't include these features, consider opening a window or using a flexible air duct readily available at the local hardware store to help route the fumes outside.

Instructions

Bed adhesion can be enhanced by utilizing ABS slurry. You can make this slurry by mixing small portions of ABS filament with acetone and applying the mixture on the bed.

When using a dual extrusion print, PLA can be a good break-away support material as it does not adhere strongly to ABS.

Take Initiative

The user may view some common applications below, select from a typical sample project, or even view popular filament brands if the user desires to stock up on low cost material.

Common Applications

- Cases or Project Enclosures
- Toys or Action Figures
- Automotive hardware

Sample Projects

- Lego Bricks
- Door Catch
- Fidget Spinner

Popular Brands

- Hatchbox ABS
- FormFutura ABS pro, EasyFil ABS, TitanX
- eSun ABS, ABS+
- HobbyKing ABS

PETG

PET and PETG filaments are known for their ease of printability, smooth surface finish, and water resistance.

Overview

"PETG" is a Glycol Modified version of Polyethylene Terephthalate (PET), which is widely used to manufacture water bottles. It is a semi-rigid material with good impact resistance, but it has a slightly softer surface which makes it prone to wear. The material also benefits from excellent thermal properties, permitting the plastic to cool efficiently with almost negligible warpage. There are numerous distinctions of this material in the market including PETG, Figure 5.44, PETE, and PETT. The instructions for this material will apply to all of these PET-based filaments.

Fig.5.44: PETG Used in 3D Printing

Advantages

- Glossy and smooth surface finish
- Adheres well to the bed with negligible warping
- Mostly odorless while printing

Disadvantages

- Poor bridging characteristics
- Can produce thin hairs on the surface from stringing

Hardware Requirements

Before 3D printing with PET / PETG ensure your 3D printer meets the hardware requirements listed below to ensure the best print quality.

Bed

Temperature: 75-90 °C

Heated Bed Recommended

Enclosure Not Required

Build Surface

Glue Stick

Painter's tape

Extruder

Temperature: 230-250 °C

No special hot-end required

Cooling

Part Cooling Fan Required

Best Practices

These instructions will aid the user to reduce the chances of typical 3D printing issues related to PET / PETG such as stringing, oozing, and poor bed adhesion.

Some 3D printers are equipped with a glass bed or blue painter's tape installed on the bed. Although these surfaces might work fine for PETG, it is recommended to utilize a heated build platform for best results. The heated bed may significantly improve the initial layer adhesion, making printing conditions easier for future prints. Many of these heated beds come with a glass surface, enabling the user to print directly on the bed without needing to apply any additional layers of tape or adhesive.

Calibration

One of the few common issues that the user may find out that PETG is stringing. These strings are thin hairs, similar to a spider web, that run between the different surfaces of the 3D print. Averting these strings needs precisely calibrated retraction settings, so make ensure proper setting of the retraction distance and the speed for the best results. Many printers also incorporate many advantageous features that can further reduce stringing. The initial is called Coasting, which works by reducing the pressure in the nozzle before the end of a printing segment. This manner, as moving to the next segment, there is less pressure in the nozzle, accordingly, the user is less likely to observe stringing and oozing during that move. Another exceptional option may be found on the "Advanced" tab of the printer process settings. By allowing the "avoid crossing outline for travel movements" option, the software will automatically adjust the travel movements of the print to stay on top of the interior of the model as much as possible. This allows the strings stay inside of the part where no one may see them, instead of being on the outside of the model.

Preventing Blobs or Zits

When 3D printing at higher temperatures associated with PETG, the user may observe small blobs or zits on the surface of the model. These print-defects generally happen at the beginning or end of each segment, where the extruder has to suddenly start or stop extruding polymer. There are several means to avoid these print defects such as enabling "Extra Restart Distance" or "Coasting" options located in the "Extruder" tab. Various printers also incorporate an option to perform a dynamic retraction, where the filament is retracted while the extruder is still moving. This wholly eliminates blobs that are commonly formed from a stationary retraction.

Instructions

The glossy surface of PETG is particularly useful when utilizing rafts. The part separates easily from the raft and maintains a clean surface finish.

Attempt to restrict the part cooling-fan during the first few layers of the print to avoid warping. This process especially works well for larger prints.

Take Initiative

The user may view some common applications below, select from a typical sample project, or even view popular filament brands if the user desires to stock up on low cost material.

Common Applications

Water proof applications

Snap fit components

Planter Pot

Sample Projects

- Self-watering Planter
- Water Bottle
- Snap Fit Parts

Popular Brands

- ColorFabb PETG
- eSun PETG
- E3D Spoolworks Edge
- Hatchbox PETG
- HobbyKing PETG

Nylon

Nylon is a tough and semi-flexible material that provides high impact and abrasion resistance. It is an ideal choice for printing durable parts.

Overview

"Nylon" (a.k.a. Polyamide) is a popular material in the plastics manufacturing industry, known for its toughness and flexibility. Nylon filaments generally need extruder temperatures near

250 °C, nevertheless, some brands permit printing at temperatures as low as 220 °C due to their chemical composition, Figure 5.45. Several printers do not include a hot-end that can safely reach 250 °C, so these lower-temperature versions can be useful and potentially save the user from needing to upgrade the hot-end. One surmount challenge with Nylon filaments is that they are hygroscopic, which means they readily absorb moisture from their surroundings. Printing Nylon after it has absorbed moisture will lead to many quality issues, thus filament storage becomes very paramount and necessitates special attention.

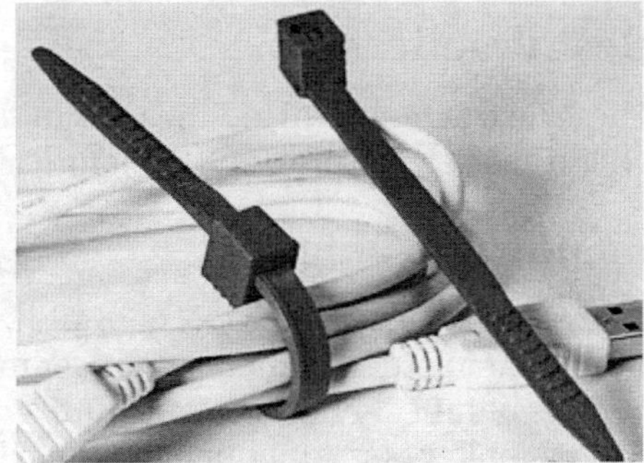
Fig.5.45: Nylon Used in 3D Printing

Advantages

- Tough and partially flexible
- High impact resistance
- No unpleasant odor while printing
- Good abrasion resistance

Disadvantages

- Prone to Warping
- Air-tight storage required to prevent water absorption
- Improperly dried filaments can cause printing defects
- Not suitable for moist and humid environments

Hardware Requirements

Before 3D printing with Nylon ensure your 3D printer meets the hardware requirements listed below to ensure the best print quality.

Bed

Temperature: 70-90 °C
Heated Bed Required
Enclosure Recommended

Build Surface

Glue Stick
PEI

Extruder

Temperature: 225-265 °C
May require All-Metal hotend

Cooling

Part Cooling Fan Not Required

Best Practices

These instructions will aid the user to reduce the chances of typical 3D printing issues related with Nylon such as warping and moisture absorption.

Filament Storage

One of the challenging aspects of utilizing Nylon is the necessity for a particular storage system. Dissimilar to other filaments, the user may not keep the filament reel exposed to the air for extended periods of time. Nylon easily acquires moisture from the environment. Printing the moist filament will result in poor print quality issues like a foggy rough surfaces, or even tiny holes or bubbles on the exterior. These printing matters may also substantially lower the strength and performance of the printed parts. The typical solution for this issue is to remove the Nylon spools from the printer once model is finished, and store the reel in an air-tight container along with some desiccants to remove the moisture from within. Also, there are commercially available storage containers that will keep the filament dry, while allowing it to feed out of a hole in the container.

Prevent Warping

Some high-temperature Nylons are disposed to warping due to the large temperature change between the extruded plastic and the ambient environment. Heated beds can reduce the warping to some extent, but utilizing a printer that has a heated chamber or enclosure would be the ideal solution. Keeping the air around the part at a temperature of about 45 ºC will assist eliminating warping by reducing these temperature variations.

Brims and Rafts

In situations where adding an aftermarket enclosure may not be an ideal option, consider using a brim or a raft to help with first layer adhesion. Adding a brim will add several loops around the bottom layer of the model, generating a larger surface area to secure down the edges of the print. Certain printers have new raft design that can also be useful.

Instructions

If the user does not have air-tight containers for storing the filament, the user may dry the spools just before use by running them through a Food Dehydrator. A few hours in this device will dry the filament significantly.

Nylon makes an excellent filament for cleaning the nozzle. The user may use a technique called "cold pulling" where the use allows the Nylon to bond to debris within the nozzle, and then after it partially cools, you pull the filament (and debris) out of the hot-end.

Take Initiative

The user may view some common applications below, select from a typical sample project, or even view popular filament brands if the user desires to stock up on low cost material.

Common Applications

Plastic Gears

Screws, nuts, bolts

Cable ties

Sample Projects

- Cable ties
- Nut, Bolt, Washer
- Servo Gears

Popular Brands

- Taulman Nylon
- Matterhackers PRO Nylon
- HobbyKing Nylon
- GizmoDorks Nylon

Polypropylene

Polypropylene is excellent for high-cycle, low strength applications due to its fatigue resistance, semi-flexible, and lightweight characteristics.

Overview

Polypropylene is a semi-rigid and lightweight material that is typically used in storage and packaging applications. The semi-crystalline structure of the material allowed the 3D printed parts to heavily warp upon cooling, making it challenging to 3D print. Polypropylene is tough and has a good fatigue resistance making it ideal for low strength applications like living hinges, straps, leashes, etc. A few manufacturers have even created polypropylene blends that have improved toughness making it suitable for practical use. The instructions provided will cover all the different types of polypropylene filaments available for 3D printing, Figure 5.46.

Advantages

- Good impact and fatigue resistance
- Good heat resistance
- Smooth surface finish

Disadvantages

- Heavy warping
- Low strength
- Difficult to adhere to bed and other adhesives
- Expensive

Fig.5.46: Polypropylene Used in 3D Printing

Hardware Requirements

Before 3D printing with Polypropylene ensure your 3D printer meets the hardware requirements listed below to ensure the best print quality.

Bed

Temperature: 85-100 °C

Heated Bed Required

Enclosure Recommended

Build Surface

Packing Tape

Polypropylene Sheet

Extruder

Temperature: 220-250 °C

No special hot-end required

Best Practices

These instructions will aid the user to reduce the chances of typical 3D printing issues related with Polypropylene such as warping and poor bed adhesion.

The Build Platform

Finding a good build surface can be very inspiring when the user requies to print with Polypropylene. The low surface energy of polypropylene makes it exceptionally difficult to get good bed adhesion even with heated beds. Scientists have found that polypropylene adheres well only to itself. Thankfully, packaging tape is a readily available surface that is also Polypropylene based; accordingly, it may be an excellent option. A thick strip of packing tape along with a heated bed can greatly improve the success of the initial layer with this material. Some packing tapes melt around 90 ºC; therefore, it is important to keep the heated bed at a temperature slightly less than the melting point of the tape.

Heated Chamber

Having a good build surface may not be enough to avert printing issues. Since it is known that Polypropylene may warp quite heavily, it helps to have an enclosure of some sort to trap the heat surrounding the print. Utilizing a printer that has a heated chamber will assist eliminating warping. Temperatures of 45-60 °C for a heated chamber will usually work well. Having a standard enclosure may also substantially control warping as used with the heated bed, though the bed will aid heat the air around the part, which is contained by the enclosure. Nevertheless, care must be maintained to control the heated bed temperature to avoid damaging the packaging tape.

Instructions

Although Polypropylene may print well at low temperatures, printing at slightly higher temperatures in the range of 240 ºC may help with adhesion between layers to create a stronger part. Consider using a lower temperature for the first few layers of your print to prevent the part from fusing to the build surface.

Take Initiative with Polypropylene

The user may view some common applications below, select from a typical sample project, or even view popular filament brands if the user desires to stock up on low cost material.

Common Applications

Living hinges

Storage containers

Watch Straps

Sample Projects

- Pebble Watch Strap
- Travel Bottle
- Parts box

Popular Brands

- Ultimaker Polypropylene
- GizmoDorks Polypropylene
- Verbatim Polypropylene

Polycarbonate

Polycarbonate is known for its strength and durability. It has very high heat and impact resistance making it an ideal choice for tough environments, Figure 5.47.

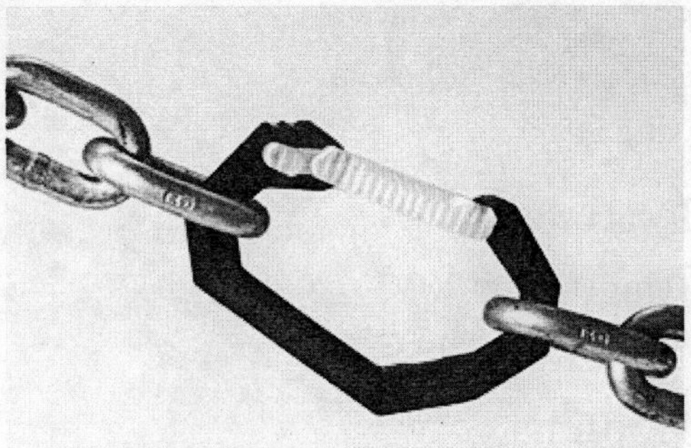

Fig.5.47: Polycarbonate Used in 3D Printing

Overview

Polycarbonate (PC) is a high strength material used for tough environments and engineering applications. It has very high heat deflection, and impact resistance. Polycarbonate also has a high glass transition temperature of 150° C. It will keep its structural integrity up to that temperature, making it appropriate for practice in high-temperature applications. It can also be bent without failure and is often utilized in applications where some minor flexibility is required, Figure 5.47. Most available Polycarbonate filaments encompass additives that permit the filament to be printed at lesser temperatures.

Polycarbonate is exceptionally hygroscopic. It will absorb moisture from the air, which will disturb its printing performance and strength. It should be stored in air-tight, moisture-free containers after opening. It also requires very high temperatures for printing and will exhibit layer separation if printed at too low of a temperature or with excessive cooling. Polycarbonate is often best printed on a machine that has an enclosed build volume and is able of handling high bed and extruder temperatures.

Advantages

- Impact resistant
- High heat resistance
- Naturally transparent
- Bendable without breaking

Disadvantages

- Requires very high print temperatures
- Prone to warping
- High tendency to ooze while printing
- Absorbs moisture from the air which can cause print defects

Hardware Requirements

Before 3D printing with Polycarbonate ensure the 3D printer meets the hardware requirements listed below to ensure the best print quality.

Bed

Temperature: 80-120 °C

Heated Bed Required

Enclosure Required

Build Surface

PEI

Commercial Adhesive

Glue Stick

Extruder

Temperature: 260-310 °C

All-metal hotend required

Cooling

Part Cooling Fan Not Required

Best Practices

These instructions will aid the user to reduce the chances of typical 3D printing issues related with Polycarbonate such as warping, stringing, and oozing.

Build Surface

PEI is the best build surface for Polycarbonate. Standard build surfaces like blue tape and glue stick have adhesives that break down at the high temperatures necessary to print Polycarbonate. Build-take can also be used but tends to permanently adhere to Polycarbonate at temperatures higher than 80°C, and may sometimes warp aggressively enough at this temperature to pull the Build-take off the plate. PEI sheets consistently securre on to Polycarbonate at 110°C, without permanently adhering to the part. Once the bed cools down, the part usually self-releases, or may be easily removed with a spatula.

Stringing and Oozing

Due to the extreme printing temperatures and low cooling fan speeds required to print Polycarbonate, there is a high possibility of stringing and oozing with this material. To avoid this, attempt increasing the retraction distance and retraction speed. some printers include a useful option called Coasting, which will automatically reduce the pressure in the nozzle immediately before the end of a segment to avoid oozing when moving to the next segment. This option can be facilitated on the Extruders tab of the process settings. If the 3D printer can handle it, the user may also attempt increasing the XY travel speed on the "Speeds" tab of the printer process settings. This enables the printer to make faster movements, which has less time for oozing.

Initial Layer Setting

Due to the high warping inclination of Polycarbonate, the first layer is particularly significant when printing with this material. Material scientists concluded that the material seems to stick the best with slow thick extrusions on the first layer. They have recommend using:

- a first layer height of 100-150%,
- a first layer width of 12-150%, and
- a first layer speed of 15-30%.

All of these settings can be entered on the "Layer" tab of printer process settings. Additionally, setting the initial layer to print about 30° hotter than the rest of the part can greatly improve initial layer adhesion.

Smooth Top Layer

High print temperatures cause the extruded filament to droop if it is unsupported. Solid Infill Layers should be printed slowly, at around 50% speed, so that they have the best chance of successfully bridging the spaces between infill lines. At least 4 top layers should be utilized, which will ensure that top layers are as smooth as possible. Infill percentage should generally be kept higher than 25% density to ensure support for these top layers. If a part requires less infill, in order to be lighter or use less material, extra top layers can be added to increase the likelihood of a solid top-most layer.

Calibrate the Bridging Setting

Similar to the point above, bridging segments can also be affected by the high printing temperatures of Polycarbonate. As the polymer is extruded in thin air between either sides of the bridge, it will have a inclination to sag while it cools. Several modern printers added many new bridging features; accordingly, these may be utilized to customize precisely how the bridges are printed for the finest results. Scientists found that using a bridging speed multiplier of 40%, a bridging extrusion multiplier of 120%, and setting the bridging fan speed to come on at 25% offered a good starting point for most materials. The user may wish to ensure the bridging fan speed stays at a low value; else it has an inclination to affect warping and separation due to the rapid temperature change.

Instructions

If parts are overheating, curling, or showing signs of sagging when printing sharp overhangs, consider setting the fan to 0% for the first layer (to ensure good bed adhesion), but then increase the fan speed to 40% or below after about 4 or 5 layers.

If some moisture has been absorbed by the material, attempt lowering the printing temperature to 280°C or below, which may assist prevent artifacts or bubbles in the print from the moisture content.

Take Initiative with Polycarbonate

Now that you know the basics, you are ready to tackle your first print with Polycarbonate. Below are a few suggestions to help you get started.

Common Applications

High-strength parts

Heat resistant prints

Electronics cases

Sample Projects

Carabiner

Platform Jack

Coffee Sleeve

Popular Brands

Polymaker PC Plus, PC Max

Matterhackers Flameproof PC

esun ePC

ASA

ASA is a common alternative to ABS and is excellent for outdoor applications due to its high UV, temperature, and impact resistance.

Overview

ASA, Figure 5.48, also recognized as Acrylic Styrene Acrylonitrile, is a 3D printable polymer with characteristics similar to ABS. It was initially developed as an alternative to ABS that would be more UV resistant by changing the type of rubber that's used in the formulation. ASA is known for high impact resistance, higher temperature resistance, and increased printing difficulty. It's normally utilized in outdoor applications instead of ABS due to its superior resistance to UV and harsh weather conditions. Due to its legacy, ASA still retains many of the same printability drawbacks that are seen with ABS. Warping is still a consistent issue that the user need to consider, as well as the possibly dangerous fumes that the plastic releases during printing, due to the presence of Styrene.

Fig.5.48: ASA Used in 3D Printing

Advantages

- Strong UV resistance
- High impact and wear resistance
- High glass transition temperature

Disadvantages

- Expensive
- Requires higher extruder temperatures
- Requires ventilation due to potentially dangerous fumes

Hardware Requirements

Before 3D printing with ASA user must ensure the 3D printer meets the hardware requirements listed below to ensure the best print quality.

Bed

Temperature: 90-110 °C

Heated Bed Required

No Enclosure Required

Build Surface

"Kapton" tape

PET Sheets

ABS/ASA Slurry

Extruder

Temperature: 220-245 °C

No special hot-end required

Cooling

Part Cooling Fan Not Required

Best Practices

These instructions will aid the user to reduce the chances of typical 3D printing matters related with ASA such as warping, shrinking, and poor bed adhesion.

Build Platform

For ASA to stay anchored to the build surface, Scientist and engineers recommend utilizing a layer of "Kapton" tape on top of the printer's heated bed. Applying this layer of transparent tape will add about 0.1mm of thickness on top of the bed; accordingly, ensure to account for this in the printer setting by increasing the global Z-axis offset on the "G-Code" tab of the process settings. If the printer uses a glass bed, the user may also use PET sheets, which provide similar adhesion but maintain a smooth surface finish for the underside of the part. These modifications will all help to avoid the warping that is commonly experienced with ASA prints.

Prevent Over Heating

Due to the higher temperature that is necessary for printing in ASA, these parts are often disposed to to overheating, which can generate substantial print quality issues. several printers allow the user to define custom temperature values at different locations in the print, therefore, an excellent way to deal with this problem is to print the bottom few layers of ASA with a higher extruder temperature, and then decrease the temperature by 5 degrees a few layers later. The user may repeat this process 1 or 2 more times so that the top of the part may be printed without overheating issues, while the bottom was printed at a higher temperature to help with adhesion.

Fan Slow Speed

When printing with ASA, it can be difficult to overcome the effects of overheating. As stated above, lowering the temperature for higher layers in the print is one option for combating this issue. Another option is to enable the part cooling fan on the 3D printer to run at a very low speed. Values in the range of 10-25% typically work best, adding some additional cooling without inducing excessive warpage or shrinking.

Instructions

Slurry of ABS or ASA and acetone on top of a heated build plate will provide an excellent film onto which the parts can adhere.

ASA can be smoothed using controlled exposure to acetone vapors (a process called "vapor smoothing").

HIPS can be used as a great dissolvable support material for ASA prints.

Take Initiative with ASA

Now that you know the basics, you are ready to tackle your first print with 3D printing with ASA. Get project ideas from common applications and sample projects and check out the popular ASA filament brands below.

Common Applications

Automotive exterior parts

Outdoor electronics housings

Exterior signage

Sample Projects

Birdhouse

Sundial

Sunglasses

Popular Brands

Fillamentum Extrafil

3DXTECH

Matterhackers

Rigid.ink

Flexible

Flexible filaments, commonly referred to as TPE or TPU, are known for their elasticity allowing the material to easily stretch and bend, Figure 5.49.

Fig.5.49: Flexible Materials in 3D Printing

Overview

Flexible filaments are made of Thermoplastic Elastomers (TPE) which are a blend of hard plastic and rubber. As the name suggests, this material is elastic in nature allowing the plastic to be stretched and flexed easily. There are several types of TPE, with Thermoplastic polyurethane (TPU) being the most commonly used among 3D printing filaments. In many cases, these terms are used interchangeably, along with popular brand names such as Ninjaflex. The degree of elasticity in the plastic depends on the type of TPE and the chemical formulation used by the manufacturer. For example, some filaments can be partially flexible like a car tire but others can be elastic and fully flexible like a rubber band. This guide will cover tips to help you with both of these variations of flexible filaments, Figure 5.49.

Advantages

- Flexible and soft
- Excellent vibration dampening
- Long shelf life
- Good impact resistance

Disadvantages

- Difficult to print
- Poor bridging characteristics
- Possibility of blobs and stringing
- May not work well on Bowden extruders

Carbon Fiber Filled

Carbon fiber filaments contain short fibers that are infused into a PLA or ABS base material to help increase strength and stiffness, Figure 5.38.

Overview

Carbon fiber filaments use tiny fibers that are infused into a base material to improve the properties of that material, Figure 5.50. Several popular filaments can be bought with carbon fiber fill including PLA, PETG, Nylon, ABS, and Polycarbonate. These fibers are extremely strong and cause the filament to increase in strength and stiffness. This also means that the 3D printed parts will be much lighter and more dimensionally stable, as the fibers will help prevent shrinking of the part as it cools. Print settings, such as printing temperature, speed, bed adhesion, and extrusion rates will be very similar to the normal settings used for the base material that the fibers were added to (for example, the stock PLA settings would be a good starting point for PLA-based carbon fiber filament). However, due to the added fibers, these specialty materials are more likely to clog and can require special hardware to avoid damaging the printer, Figure 5.50.

Fig.5.50: Carbon Filled Fiber in 3D Printing

Advantages

- Increased strength and stiffness
- Very good dimensional stability
- Lightweight

Disadvantages

- Abrasive and requires hardened steel nozzle
- Increased oozing while printing
- Increased brittleness of filament
- Higher tendency to clog

Metal Filled

Metal filled filaments are made by mixing a fine metal powder into a base material, providing a unique metallic finish and added weight, Figure 5.51.

Overview

Metal filled filaments contain very fine metal powder such as Copper, Bronze, Brass, and Stainless Steel, Figure 5.51. The percentage of metal powder infused in each filament can vary depending on the manufacturer. The presence of this metal powder makes the filament much heavier than standard plastics. This means that the parts printed with metal-filled PLA will weigh significantly more than ones from the standard PLA, despite using the same settings and consuming the same amount of material. Metal filled filaments also tend to be very abrasive as they are extruded through the hot-end. A standard brass nozzle will be too soft and will quickly wear down. Be sure to upgrade to a wear resistant nozzle in order to print this filament effectively. There are other metal-like filaments in the market that may just have metallic coloring added to the filament. These filaments do not contain any actual metal powder, so they do not share many of the same benefits of the true metallic filaments. This article will focus on materials that contain actual metal powders for a realistic metallic weight and feel, Figure 5.51.

Fig.5.51: Metal Filled Fiber in 3D Printing

Adantages

- Metallic finish is aesthetically appealing
- Does not need high-temperature extruder
- Heavier than standard filaments

Disadvantages

- Requires a wear-resistant nozzle
- Printed parts are very brittle
- Very poor bridging and overhangs
- Can cause partial clogs over time
- Expensive

Wood Filled

Wood filaments combine a PLA base material with cork, wood dust, or other derivatives, giving the models a real wooden look and feel, Figure 5.52.

Overview

Wood-based filaments are typically a composite that combines a PLA base material with wood dust, cork, and other powdered wood derivatives, Figure 5.52. Typically, the filament consists of around 30% wood particles, but the exact number may vary depending on the brand. The presence of these particles gives the 3D printed parts the aesthetics of real wood. This filament is also less abrasive compared to other composite filaments such as carbon-fiber filled and metal filled, since wood particles are much softer. There are some wood-like filaments on the market that only contain wood coloring, but no actual wood particles, so these typically have a very different look and feel. This guide will focus on wood infused PLA filaments since these are the most common, but you can use these tips as a starting point for other wood-based filaments as well.

Fig.5.52: Wood Filled Fiber in 3D Printing

Advantages

- Wood-textured finish is aesthetically appealing
- Does not need any expensive wear resistant nozzles
- Aromatic and pleasant smelling

Disadvantages

- Prone to stringing
- Smaller nozzles can end up with partial clogs over time
- May require a larger size nozzle

3D Printing - Powder Materials

The marketplace for 3D printing is self-confident for an explosive progress. Today, a overabundance of 3D printing techniques can form objects from an ever mounting list of materials – photo-polymeric resins, extruded filament, powders of polymers, uncontaminated metals and alloys, etc.

Metal additive developments such as metal *powder bed fusion* and *directed energy deposition* are possibly able of creating high-quality, practical and load bearing parts from a diversity of metallic powder materials. Though, "one-size-fits-all" doesn't spread over well to developed additive manufacturing implementing high value parts and critical applications, it's vital to know the advantages and disadvantages of these procedures and how they apply to a chosen material, or to professional colleague with a skillful talents who does.

Low Cost Metal-Powders

Today, metals are the fastest-growing sector of 3D printing. The development of additive manufacturing is tied to corporate opportunities and, directly, the materials obtainable and their cost. Low-cost metal powders are the important enablers for 3D printing, to understand its potential and renovate manufacturing production.

Accessible metallic powders for additive manufacturing are typically those being utilized in a variety of applications (medical, aerospace, jewelry, automotive, etc...):

- Cobalt-chromium and nickel-based super-alloys,
- Commercially pure titanium and titanium alloys,
- Aluminum alloys
- Precious metals (gold, platinum, palladium, silver).
- Tool and merging steels,
- Stainless steels,
- Copper alloys

Metal powders can differ widely in size, but also in form (spherical to irregular). As a result, treating characteristics in AM metal systems diverge, as well. To guarantee consistency and repeatability of the quality of metal powders and AM developments, the machine builders are working closely with powder suppliers.

Metal AM systems builders (EOS, Concept Laser, ARCAM, etc.) source approved material powders, but many powders are possibly atomized elsewhere and the cost is greater than purchasing directly from a powder manufacturer.

Businesses producing powders for customary powder processes such as *hot isostatic pressing* (HIP), *metal injection molding* (MIM) or *powder sintering* (PS) have a chance to corner some market shares of this thriving sector, if they manage to observe the requirements of *metal additive manufacturing*. Supply chains for 3D printing materials are in flux and, therefore, vulnerable to technological disruption by innovators, Figure 5.53.

Main third party metal powder producers that one may want to investigate when choosing powders are:

- Carpenter, ATI Powder Metals, Erasteel, LPW Technology, Metalysis Technology, AP&C, Sandvik Osprey, TLS, GKN Hoeganaes, HC Starck, Praxair, and Metco

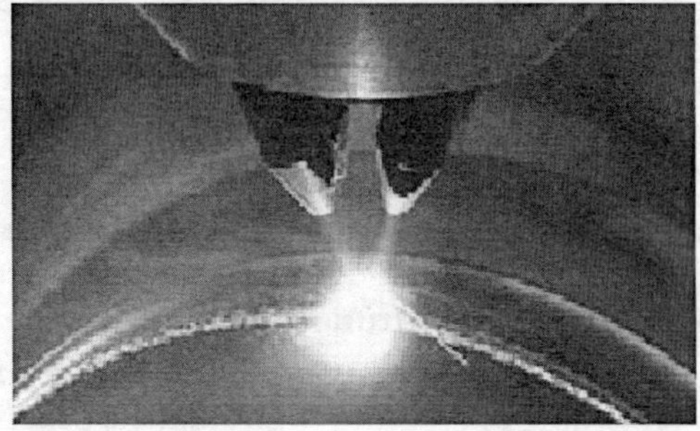

Fig.5.53: Metal Powder Used in 3D Printing

Courtesy: Foundry Planet

Powder Metal Production for Selective Laser Melting (SLM), Electron Beam Melting (EBM), Direct Energy Deposition (DED) and Infiltration

Water atomization is the furthermost typical and cost-effective method to produce metal powders. The high water pressure that impacts more energy into a molten metal flow, joint with a rapid cooling rate, provide increase to powder particles that are hard and irregular in shape. An uneven shape is not as much of desirable for powder particles dedicated to additive manufacturing because it increases the flow time and perhaps decreases the packing density. Though, under certain circumstances, it is likely to yield spherical powders with a particle size supply optimized for additive manufacturing.

Usually, gas atomized powders are favored over water atomization for additive manufacturing and gas atomization has developed the most typical technique to yield metal powders for AM. The feedstock is melted under an air or inactive gas or in a vacuum atmosphere; then, the compartment is back-filled with gas to force molten alloy through a nozzle. High-velocity gas (air, nitrogen, helium or argon) gas impacts into the flowing melt and break it downs into fine droplets.

Interfacial strains naturally spheroidise the surface of molten metal droplets that cool down and fall at the lowest of the atomization tower, where powders are composed. Gas atomization technology delivers competent powders for numerous additive manufacturing procedures, such as *selective laser melting* (SLM), *electron beam melting* (EBM), *direct energy deposition* (DED) and *infiltration*.

Gas Atomization is mostly utilized for Fe, Ni and Co alloys, but is likewise obtainable for Al and Ti alloys. Other variants of this technique available, such as:

1. *Water atomization*: for unreactive materials, yields irregular shaped particles
2. *Plasma atomization*: high-quality and exceptionally spherical powder, limited to alloys that can be shaped into a wire feedstock
3. *Electrode induction melting gas atomization*: suitable for all alloys, but most economic with reactive alloys like Ti. A bar feedstock is rotated and melted by an induction coil before

it flows downwards into a gas stream for atomization. A cheap, clean, and good process for small batches and to produce small diameter powder particles.

4. *Centrifugal atomization*: Good trade-off between *Gas atomization* and *Plasma atomization*, best matched to larger batch sizes of not as much of reactive, low-melting temperature alloys; nonetheless, it can also yield Ni-base super alloy powders

After fabrication, powders can be categorized according to numerous standard techniques used for granular materials:

- Hall flow: Flow rate and obvious density.
- Powder flow and rheological characteristics analysis.
- Angle of repose: Sharpest angle of descent to which powders are heaped without collapsing.
- Tapped density: Bulk density of the powder after consolidation/compression.
- Morphology by scanning electron microscopy.
- Trapped porosity by scanning electron or optical microscopy.
- Laser diffraction: Analysis of the particle size.
- Sieve analysis: Measure the particle size distribution.
- Moisture determination: water instinctively held water on the surface or between the particles of the material.

Chemical composition analysis: amount of metallic, non-metallic impurities (elemental form, or in dissolved form as solid solution or as mixtures).

Finally, powders are wrapped in robust and moisture resistant containers in HDPE. Pure titanium and alloys are packed under argon while other materials are typically packed under normal air atmosphere.

Stable Metal Powder for 3DPinting

Metal powders used in additive manufacturing should have the following characteristic:

1. spherical form to guarantee integrity of flow/coating ability and a high packing density,
2. particle size typically below 50 µm or 150 µm contingent on machine type and surface finish or productivity prerequisite,
3. particle size distribution made to the application and properties,
4. controlled chemical composition and gas content.

The particle size delivery of metallic powder particles influences the density of AM parts. While it's likely to achieve high densities with diverse powder types, the limits of the process must be adjusted consequently. Eventually, the productivity differs. Furthermore, the particle size distribution does not only disturb the density but also the mechanical characteristic and surface quality of the parts.

An important issue in metal ALM is the advanced degradation of metal powders during processing consequently the powder bed being in contact to oxygen and other contaminants. There is large possible demand for cost-effective means of renovating metal powders.

Low Porosity, Fine Microstructure & Uniform Characteristics

Recurrently procuring additively manufactured materials with 100% of the reference density is certainly challenging. Metal additive manufacturing methods can harvest densities in excess of 99%. Some materials are reported with a full density though some others offer a spread of densities.

Density is impacted by the expansion of pores or entrapment of un-melted powders during the layering. Infrequently, *hot isostatic pressing* is utilized to upsurge as-fabricated densities.

From a mechanical stance, porosity (especially "open" porosity) endangers AM parts' splintering toughness and fatigue characteristics. Certainly, under cyclic stress circumstances, porosity or partial de-lamination can start flaws and lead to part failure. For load-bearing uses of the aerospace industry, toughness and fatigue resistance are acute requirements.

When casting metal alloys, the component with the maximum melting point starts to solidify first. As the casting cools from the surface near the center, grains will provide a expressively diverse alloying components' concentration. Concentration will differ throughout the part and grains will formulate in specific orientations. The material characteristics would not be uniform or isotropic.

In metal additive manufacturing procedures minute quantities of material are melted at a time. For alloys, some separation of alloying elements happens but on a much lesser scale. The fast solidification leads to a additional uniform chemical conformation and microstructure all over the part.

Metal AM Enabled Microstructure & Mechanical Behavior

Additively fabricated metal parts can undergo very high cooling rates, generating rise to numerous uncommon effects contingent upon the material, as:

1. Suppression of diffusion-controlled solid-state phase conversions
2. Establishment of supersaturated solutions and non-equilibrium stages
3. Establishment of exceptionally fine, refined microstructures with little elemental separation
4. Establishment of very fine second-phase particles such as inclusions and carbides
5. Occasionally, these influences are necessary but they must be considered on a case-by-case basis.

REFERENCES

1. J.A. Palmer, P. Yang, D.W. Davis, B.D. Chavez, P.L. Gallegos, R.B. Wicker, and F.R. Medina, "Rapid Prototyping of High Density Circuitry," Rapid Prototyping & Manufacturing 2004 Conference Proceedings, Rapid Prototyping Association of the Society of Manufacturing Engineers, May 10-13, 2004, Hyatt Regency Dearborn, Michigan. Also, SME Technical Paper TP04PUB221 (Dearborn, Michigan: Society of Manufacturing Engineers, 2004).
2. F. Medina, A.J. Lopes, A.V. Inamdar, R. Hennessey, J.A. Palmer, B.D. Chavez, D. Davis, P. Gallegos, and R.B. Wicker, "Hybrid Manufacturing: Integrating Direct-Write and Stereolithography", Proceedings of the 2005 Solid Freeform Fabrication, Austin, Texas, 2005.
3. Soloman, Sabrie "Sensors Handbook, 2nd Edition," McGraw-Hill Publishing Co., NY – ISBN 978-0-07-160570-0
4. Soloman, Sabrie "Sensors and Control Systems in Manufacturing 2nd Edition," McGraw-Hill Publishing Co., NY – ISBN 0-07-059626-3
5. Soloman, Sabrie "Affordable Automation," McGraw-Hill Publishing Co., NY – ISBN 0-07-059633-6
6. Soloman, Sabrie "Sensors Handbook, 1st Edition," McGraw-Hill Publishing Co., NY – ISBN 0-07-059630-1
7. Soloman, Sabrie "Sensors and Control Systems in Manufacturing 1st Edition," McGraw-Hill Publishing Co., NY – ISBN 978-07-10572-4

6 Accelerate Design Cycles- Design Verification- Lowering Production Costs

INTRODUCTION

Today's product development landscape is characterized by high demand for diversification, substantial increase in customers' expectations and near constant innovation. To top it all off, product lifetimes are shrinking. The result is driving greater competition among manufacturers with the pressure to do more, better and faster. No Olympian has ever taken home a gold medal without intensive training. In the same respect, no winning product design has ever been the result of beginner's luck. Product development is training for game day and requires cycle after cycle of effort, feedback, and improvement. To stay competitive in an environment that is perpetually shrinking time to market, manufacturers need to meet and break new release schedules in their product development cycles. Rapid prototyping - "3D Printing Technology" offers that opportunity.

WHEN TO PROTOTYPE AND WHY?

The product development process is made up of several iterative loops to arrive at an end product. Each iterative loop provides new knowledge about what works and what doesn't. It can be a time-consuming process, but it is one that cannot be skipped. Prototyping is a key element of product development, and should be brought in at the right stage for optimal impact. When is that, you may ask? As early as possible, for companies with deep and shallow pockets alike, time is the resource in shortest supply.

Rapid prototyping with 3D printing helps companies shorten the time it takes to produce and evaluate physical product models to advance timelines, take advantage of market shifts, and win customers. Just as designers are pushing simulation forward in their process, bringing prototyping into the early stages of product development is a cost-effective way to create a virtual feedback loop, increasing product knowledge and insight with every iteration. An increased frequency of iterations means designers have the time and opportunity to improve designs while still delivering within or before deadline.

3D printing prototypes allows compression of the product design cycle with the potential for far superior products at completion.

SPEED INVENTED PRODUCT TO MARKET

Companies were able to accelerate time to market and deliver greater products Attaining prototypes in a very short time and even in hours, instead of days. Prototypes produced through 3D-printing, enables developer and designers to perform several new iterations processes of a desired design that is prototyped daily. Product numerous evaluation process takes place while a finalized product to enter the market simultaneously.

Corporations needing to assess their success or failure to reduce or improve their product development cycle must observe two key influences:

1. **Lead times**: time from inception to production.
2. **Engineering Effort**: Man-power time from concept development to initial production.

Taking these factors into consideration these two main factors the design can be transformed into a physical product in a very efficient manner as well as affordable through rapid prototyping simulations, Figure 6.1.

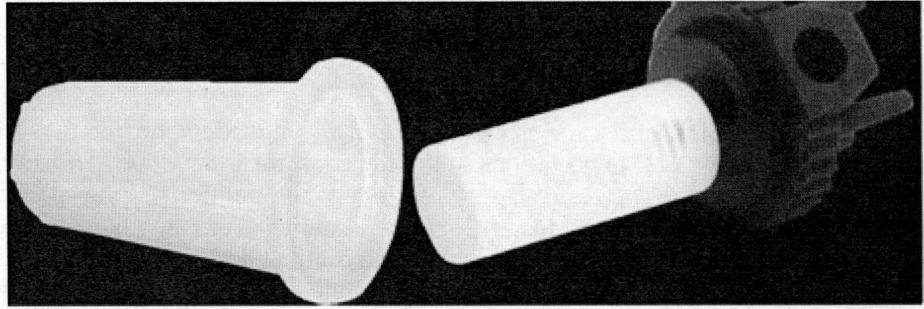

Fig.6.1: 3D printing for rapid prototyping can dramatically improve development cycles by reducing lead times and engineering effort.

AGILE MANUFACTURING

Agile design and manufacturing begins with innovative means to produce product by utilizing physical prototypes made through innovative 3D printing of parts and then face rigorous iterative processes to enforce four fundamental benefits by:

1. Facilitating design modularity
2. Accelerating knowledge generation
3. Advancing communication with process partners
4. Nurturing the values of Product Insight
5. These factors are discussed below.

Facilitating Design Modularity

Logical modules is the logical and effective plan to produce products through rapid prototyping, which is able to promote companies strategy to speed up the design of new products, and advance to produce them. Each design module creates new opportunities and variations of options to explore them simultaneously.

Accelerating Product Insight

Each developed prototype provides new insight impossible to gain without the iteration of prototyping. This insight can be vastly exchanged with the design and development team.

Advancing Communication with Process Partners

Product design and development are often a cooperative process. Forwarding 3D files to be prototyped is a speedy way to clarify design changes and maintains unified thoughts among design and development team members.

Nurturing the values of Product Insight

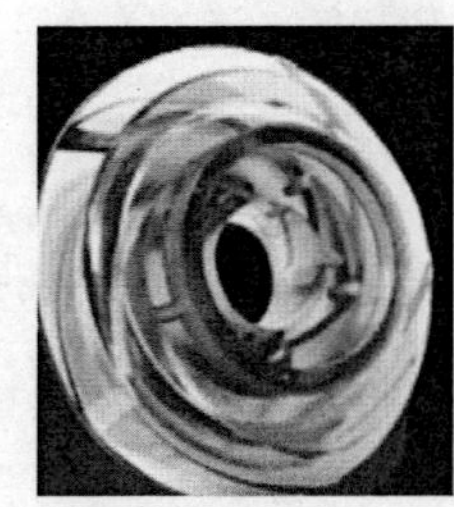

Fig.6.2: Product Iterations until Final Product is Approved

Integrating rapid prototyping as fundamental part of an agile design and manufacturing plans, enables iterative product development process inspires and strengthens the expedition for better product insight, Figure 6.2.

Considerations before Launching a Product

Prototyping is a fundamental process throughout product development and design, the driving purpose they serve is primarily evolves to produce flawless prior to launch. For some products, advanced assessment stages demands new methods to prototyping, while other products permit for greater stability in prototyping techniques, Figure 6.2.

Demands must be raised when choosing a prototyping technology include:

- What is the reason of the prototype?
- Is it for appearance or evaluation?
- What are material characteristics required showing appearance or evaluation to accomplish the desired result?
- How is the time to produce?
- How many units to produce and where do you produce the prototypes?
- What is the total budget?
- Do you need the capabilities in-house, or should you outsource?

Variations of Rapid Prototyping

There are essentially two main types of prototypes:

a. Prototype to look like the end product,

b. Prototype to perform like the end product.

There are much variation in between; however, this simplifies the selections need and to process.

Appearance Type

The appearance type model is a demand for high quality visualization of a product or a design concept. The benefits and aim of appearance type model may vary throughout product innovation and may add worth at any phase.

The appearance initial prototyping models suggest chance to assess and evolve the original designs, while convincing functional appearance models to be implemented at a later phase. It may be used to stimulate consumer feedback, and assist in securing investors.

Prototype Performing As the End Product

Rapid design iterations can transform the design into reality, and a physical proof-of-concept, as well as providing a scaled model to assist in evaluating product development. The design and development teams can rapidly turn CAD files into a realistic parts and assemblies for reviewing the aesthetics, colors, surface texture, and for departmental evaluation. Also, it is to assess the possibility to use for trade shows and sales presentations.

Additive manufacturing technologies and materials enable a range of prototypes, from clear materials that can be tinted and dyed, to elastomeric materials that mimic rubberlike parts, tough gray materials that are ready for painting and finishing as well as full color printing for accurate color evaluation.

Once the prototype to perform like the end product, has been designed, a tailored practical testing procedure is a implemented to affirm that prototype functions as developed. Rapid prototyping gives many benefits to product developers, designers, and production engineers a clear window to assess, the cost and to obtain accurate understanding part or assembly for fitting, assembly process verification, fasteners and joining, fluid and airflow testing and much more.

Additive manufacturing is a powerful tool to enable a robust transparent part to be tested in-situ, and then track and review various functions, such as oil and air flow, as well as assembly checking utilizing both *MultiJet* Printing (MJP) and Stereolithography (SLA) clear materials.

Both technologies of Selective Laser Sintering (SLS) and additive manufacturing materials for enabling production of often used consumer's parts such as living hinges provide actual functioning parts for screwing, pressing and drilling. Hands-on testing of packaging, by both the designers and customer can be consistently tested to the developers and engineers to establish guideline fundamental to meets customer approval.

Using MJP can quickly deliver functional prototypes; equally delivery can be achieved by "Digital Light Printing" (DLP) and SLS 3D printing, with an extensive assortment of materials to meet any prototyping goal.

PACKAGING DESIGN APPROVAL BY 3D PRINTED PROTOTYPES

3D Systems On-Demand assists packaging design corporations to achieve appearance and functionality prototypes with clear SLA 3D printing. Often, food companies are totally aware of how food product must be delivers to consumers to guarantee a high quality experience.

Therefore, when a food company launched its new product a clear 3D printing was employed to successfully prototype the new packaging, which had a number of diverse requirements, such as:

- Easy to extract product
- Functionality
- Re-closure
- Easy of automated fill on the production line
- Satisfy retailers' requirements.

Purpose of Prototyping

Create a functional prototype of clear packaging to test and select the best available option for a new product launch.

Printing

A clear 3D printing is created using SLA 3D printing to print additional copies using cast urethane.

Outcome

- Project cost within budget
- SLA prototypes created in a short time
- 3D Printing using cast urethane process to deliver water-clear parts
- Testing functional and aesthetic prototype

CAD to 3D Printing Technologies for Rapid Prototyping:

3D-printed prototypes can be created directly from CAD in hours compared to the weeks, this process may take using traditional model-making, CNC machining, or tool-based production.

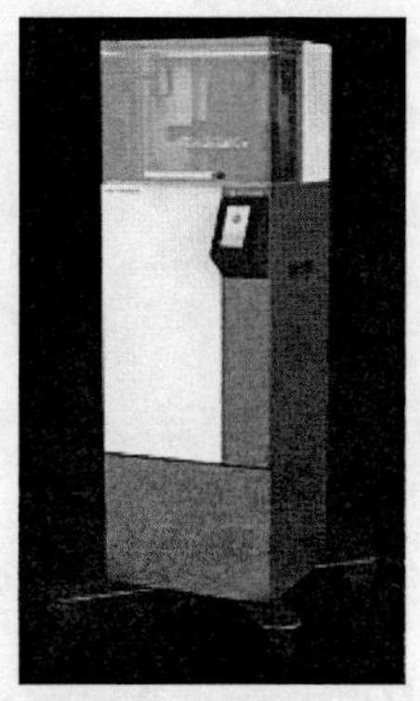

Fig.6.3: "Figure 4 Standalone" 3D printer.

Courtesy: 3D System

Digital Light Printing (DLP)

Affordable prototypes and ease of use, with possible same-day delivery, and even with high quality output can be achieved by industrial 3D printers. The availability of engineering-grade plastic materials are well suited for **Digital Light Printing (DLP)** prototyping. Also, **Digital Light Printing (DLP) can use** elastomeric materials for design verification of rubber-like parts.

1. **Digital Light Printing (DLP) – *"Figure 4"* technology system** conveys exceptional part quality at ultra-fast speed at up to 100 mm/hour. The produced surface finish is excellent. *"Figure 4* Standalone 3D printer" provides simple material changeover to permit versatility in materials and applications within the same printer, Figure 6.3.

Fig.6.4: MultiJet 3D Printing

Courtesy: Konica Minolta

MultiJet Printing (MJP)

MultJiet Printing (MJP) 3D technology provides high speed and easy operation, from file to finished part, for true-to-CAD functional prototypes in a wide range advanced plastic, elastomeric and composite materials, Figure 6.4.

ColorJet Printing (CJP)

3D enable the rapid full vibrant color and inexpensive production of models, convenient for medical models, architectural, communication, medical models. It validates the industrial design phase that need color output for adequate assessment Figure 6.5.

Fig.6.5: ColorJet Printing (CJP)

Courtesy: 3D System

Stereolithography (SLA)

This 3D Technology is suited also for very small printing parts within few mm in size. Also, this technology offers high precision printing, fine part details, good surface finish for a good range of high-quality materials for a variety of applications, Figure 6.6.

Fig.6.6: Stereolithography (SLA)

Coutesy: Formlab

Selective Laser Sintering (SLS) Technology

Selective Laser Sintering (SLS) Technology provides robust nylon materials well suited prototypes of compression test parts, including sports helmets, child car seat designs, and other applications where impact resilience is important, Figure 6.7.

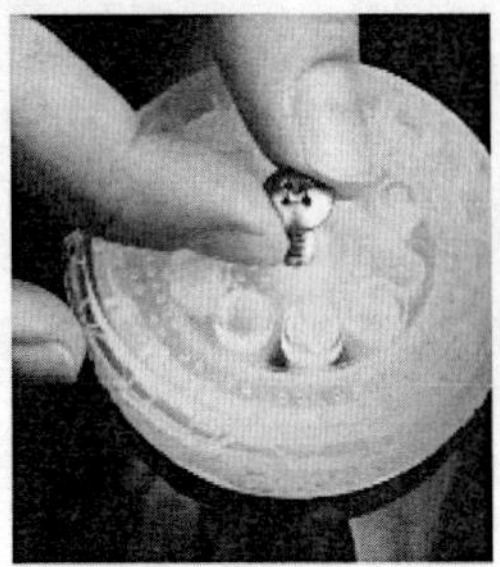

Fig.6.7: Materials for Rapid Prototyping – *SLS 3D Printer – Cutesy Formlabs*

Printing Materials Applications

Many polymer materials were developed for 3D printing application, that can mimic wide array of engineering materials, elastomers, and composites in terms of flexibility, durability, stiffness, toughness, stability, transparency, clarity, look and texture, bio-compatibility, temperature or water resistance, and much more.

Other materials supported by 3D printing can also be used, such as metal and ceramic.

The introduction of new additive solutions and materials means that 3D printed prototypes are being produced even faster than before and are more able to meet increased customer demands for agile product development. Product designers can create prototypes with exceptional accuracy and quality of parts, and greater flexibility to shorten product release schedules.

MultiJet Materials for Functional Prototyping

MultiJet materials are most commonly used to prototype parts that may be used for injection molding, MultiJet printing (MJP) materials can be handled similar to traditional thermoplastics material such as: acrylic, polypropylene, polycarbonate and ABS, Figure 6.8.

All the rigid engineering materials for MJP printers can be press fitted, machined, drilled and tapped, and can deliver robust material properties and functional versatility, with accurate and acceptable surface finishes. These materials are a rigid or semi-rigid thermoset, which will not melt or easily gum-up the nozzle when heated.

The ProJet MJP printers deliver parts in rigid white, black, tan, gray and clear materials, as well as elastomeric materials with highly acceptable elongation under tensile strength.

MJP is an ideal technology for prototyping application if:

1. for need high fidelity, true-to-CAD parts
2. for robust material properties and functional versatility
3. for obtaining good surface texture
4. for repeatable accuracy with fine details and complex geometries, Figure 6.8

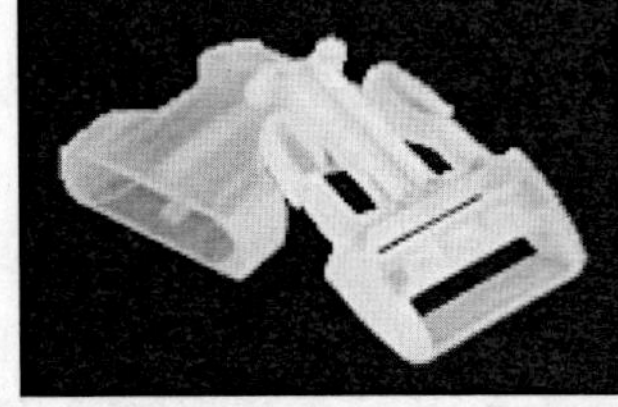

VisiJet Armor M2G-CL
Tough, clear polypropylene-like

VisiJet ProFlex M2G-DUR
Tough, clear ABS-like

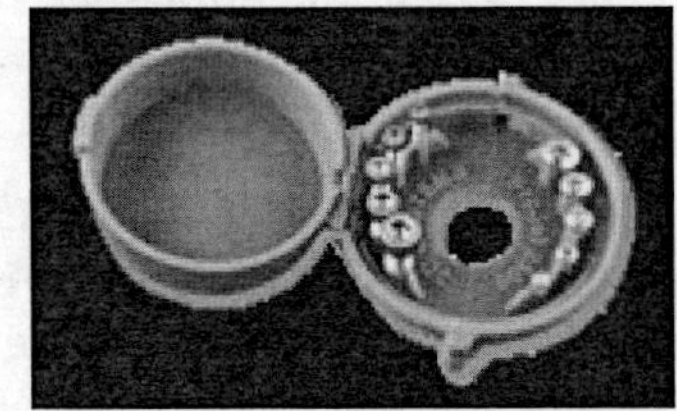

VisiJet M2R-GRY
High contrast, rigid gray

VisiJet M2R-WT
Rigid white

VisiJet M2R-BK
Rigid black

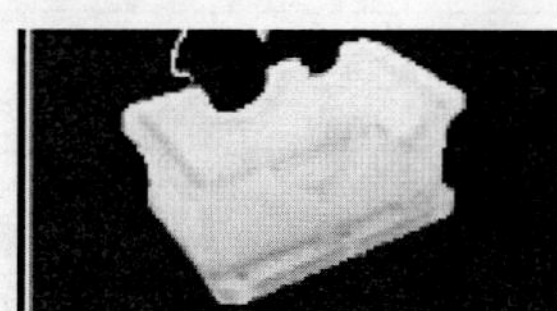
VisiJet M2R-TN
Elastomeric white

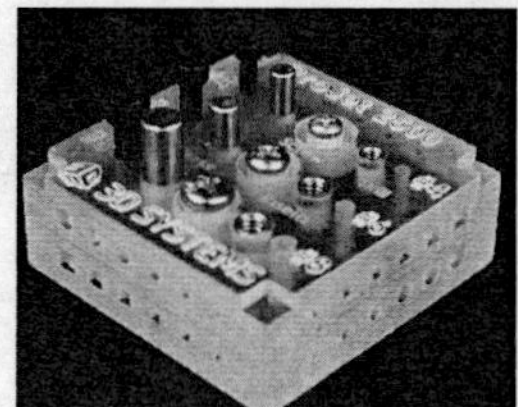

VisiJet M2 ENT
Elastomeric Natural

Fig.6.8: Varieties of Products with Different Materials, Strengths, Colors and Functions-*Courtesy of 3DSystem*

DLP Technology for Rapid Prototyping

This type printing technology is ideal for engineering and jewelry applications. Also, the entry-level industrial printers using this type of technology excels at low-volume, small-part prototyping and direct 3D production across a range of high quality materials.

DLP technology materials provides choices for robust prototype and parts production, elastomeric materials for testing purposes and an ability to cast material for jewelry applications.

Prototypes and Low Volume Production

"*FIGURE 4* –Printer" - Affordable and versatile for low-volume production and fast prototyping, the *Figure 4* printer, which is standalone system provides accurate parts with acceptable quality and accuracy with the ability to produce parts and prototypes within minutes. The "*Figure 4* printer" materials are especially developed and tuned for the production of end-use parts and for fast prototyping applications at print speeds of up to 100mm per hour.

The "FabPro" materials stated below illustrate an expanding range of materials delivering tough and robust plastics, an elastomeric material and have ability to cast material for jewelry applications.

Material	**Sample Product**
FabPro Tough BLK Rigid black plastic	A
FabPro Proto GRY Very fast production for prototypes	B

Figure 4 TOUGH-GRY 10 (High speed, rigid) ______________	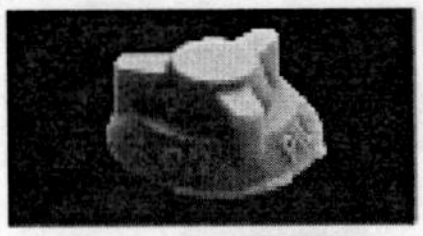 C
Figure 4 Tough Grey 15 Economical rigid gray material	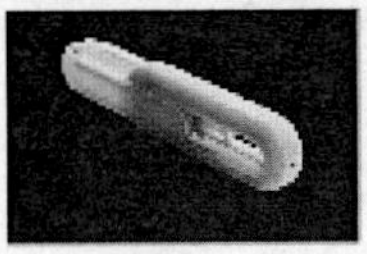 D
FabPro JewelCast GRN Castable material	 E
FabPro Elastic BLK (Elastomeric for rubber-like prototypes)	 F
Figure 4 ELAST-BLK 10 Elastomeric black material for design and test applications	 G
Figure 4 JCAST-GRN 10 Castable green material for Jewelry Applications	 H

DLP Technology - Concept Modeling & Iteration with "FabPro" Printers

In an agile design environment, rapid design iteration demands rapid prototyping for proof of concept, into functional design review. Additive manufacturing can deliver a high speed frequency of prototypes through the FabPro™ 1000 entry-level industrial desktop 3D printer.

DLP Set on Distributed Networks

This Digital Light Printing (**DLP**) 3D printer delivers fast production of prototypes with layers of 30-50 microns for high accuracy and good surface finish. The system can be set on distributed networks. Accordingly, 3D design data can be shared to produce prototypes at any location, enabling fast understanding and iteration of design by engineers and designers:

- New Rubber-like Prototypes with - DLP "FabPro 1000" Printer
- The new FabPro Elastic BLK material is a black elastomeric that delivers compressive properties and shape recovery with a realistic rubber look

With its relative fast printing and curing, DLP - FabPro Elastic BLK materials, Figure 6.9, enable the rapid prototyping of:

- Over molds
- Seals and grommets
- Grips
- Vibration dampening components
- Dust covers
- Push buttons
- Cable stress relief

Fig.6.9: "FabPro" Elastic BLK Materials

Post-Processing Options for Appearance Models

Additive manufacturing materials for rapid prototypes can use many kinds of post-processes and finishes for realistic look. These apply to the SLA, SLS, MJP, and DLP additive technologies.

Fig.6.10: Prototype Painted Lacquer

Panting & Lacquering

Prototypes produced can be painted, lacquered and then finished for realism to show concept car body parts, appliances, medical devices and more, Figure 6.10.

Tinting & Dyeing Clear Glass

Clear materials from both SLA and MJP are very receptive to tinting and dyeing to create realistic prototypes of lenses, headlamps and colored bottles and packaging.

Clear prototypes can be processed to resemble clear glass through sanding and clear coat.

Plating

Many materials are compatible for plating processes used, for example, in interior automotive parts including vents and door handle, consumer goods, appliances as well as power tools.

Same-day Prototyping with – SLA "Figure 4 Standalone" Printer

Simulating injection molded parts in minutes. Rapid SLA "*Figure 4* Standalone" is able to produce injection-molded surface quality parts with shorter lead times. In addition to ease of operation and the ability to produce fast, reliable output at low cost, "Figure 4 Standalone may pay for itself and doubled Rapid Application production capacity for high quality small parts in a few months.

Challenges

Cost-effectively increase production capacity and speed for high quality small parts.

Attempting Solution

Utilizing a Highly Advanced Rapid SLA – "Figure 4™ Standalone," 3D printer with 3D Sprint® software and "Figure 4 TOUGH-GRY 15" material.

Results

- Doubled production capacity on highly detailed small parts since installation
- Return on investment within 4 months
- Parts produced in 30-45 minutes compared to more than 7 hours

Prototyping In-House Versus Outsourcing

Rapid prototyping is a crucial element to bring more agility to the product development workflow. Determining bringing 3D printing in-house versus contracting to a service bureau, or a combining both options.

The benefits of using an on demand manufacturing service include the ability to offload labor while gaining access to a broad scope of material options, possible model sizes and expert finishing techniques.

When ready, an on-demand service provider can also help users transition into full manufacturing as a bridge to production.

Unfortunately, the benefits do not resonate with every user. Anticipate the need to be the same size and material of model just in different design iterations; it may be more cost effective to bring rapid prototyping in-house. Rapid prototyping always include ordering trial models on demand to verify the fit of a specific technology within the workflow.

3D SCANNING TECHNIQUES FOR 3D PRINTING

Quality and reliability substantially matters. In today's hypercompetitive environment, companies cannot afford to lag behind product quality demands. Management cannot lag behind adopting new tools and technologies to ensure ultimate superior product quality.

Determining the true need for 3D scanning for inspection needs, as well as the best manner to implement the technology scanning is the object for 3D printing.

The developer must be totally familiar with the different types of available scanner and their corresponding advantages and disadvantages to reach a well-educated decision of the appropriate scanner with the most economical cost to achieve the expected results qualitatively and quantitatively. The 3D Scanner Landscape provides the following characteristics:

- Must Provide High-level Overview
- Utilizing Accurate Laser Triangulation Scanners Technique
- Structured Light Scanners
- Medium- and Long-range Scanners
- Scanning is not all the requirements
- The need for 3D Scan-native Inspection Software
- The Hybridization of Inspection and Reverse Engineering
- Maximizing the ROI – The DO list
- Maximizing the ROI – The DON'T list
- The 3D Systems Geomatics
- Ease of Integration with the 3D Scanning

Over the past two decades, 3D scanning has become a crucial tool in many manufacturers' measurement and inspection tools. It is a vital metrology method that is reliable for its accuracy, repeatability, speed, and ease of usage. Its noncontact nature and exceptional flexibility make the 3D scanner ideal for measuring a wide variety of parts in a wide range of places. An effective attitude to 3D scanning requires matching the right work piece (model) with the right system and the right software routines to best meet the company's measurement needs. The needs are described as follows:

1. The Real Need - 3D Scanning

3D scanning is becoming an essential component of many companies' quality control strategies. The rapid non-contact measurement technology is the path innovative quality product produced just-in-time.

2. Measuring Complex Parts

By measuring a surface at millions of points, 3D scanning makes it easy to measure even the very complex shapes.

3. Measure Vulnerable Parts

Vulnerable tactile parts deflect or deform when touched. With a 3D scanner, touching part is conducted through photons.

4. Measuring Parts Consumes Considerable Time

3D scanning is exponentially faster at collecting measurements than contact-based techniques.

5. Re-Measure Parts

3D scanning an object is, in fact creating a full and complete digital record, which may be re-measured at any time.

6. Parts Failure

Because 3D scanning measures the entire surface of a part, the part is less likely to face unexpected deviation from nominal.

7. Parts Utilization

With a 3D scanner, provides distinctive view of the parts to make collective wise decisions.

8. Measuring Multiple Locations

Scanners may be relocated at the site of part. Also, software can be easily trails at the site location in the field.

THE 3D SCANNER LANDSCAPE

As there are many types of 3D scanners, each with unique pros and cons. Accordingly, a general guide is compiled and stated below..

There are many different 3D scanners available on the market. The type of 3D scanner that matches the requirements is the major factor to select the scanner to serve the exact needs, Figure 6.11.

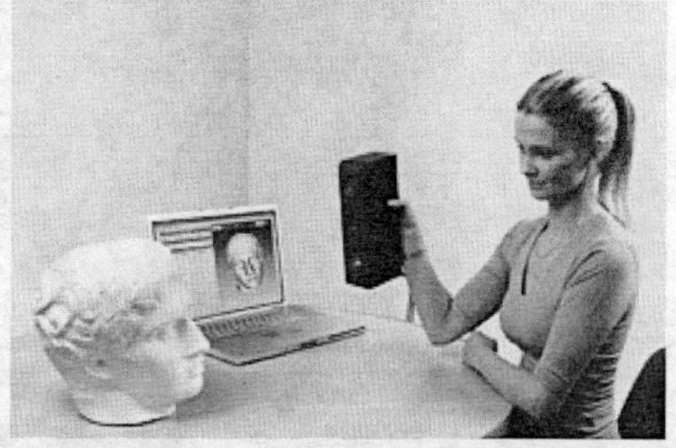

Fig.6.11: Hand Held 3D Digital Scanner – *Courtesy 3der*

Laser Triangulation Scanners

One of the most popular and versatile 3D scan technologies, laser triangulation scanners pass a laser line over the surface of a part, and using straightforward trigonometric concepts, calculate the distance from the sensor to the scanned object's surface, Figure 6.12.

Fig.6.12: Digital 3D Scanner – Structured Light Scanner - Tiangulation – *Courtesy All3D*

Structured Light Scanners

Structured-light 3D scanners project a series of linear patterns onto an object and use sensors to recognize deformations in the patterns that indicate each pixel's distance from the sensor. These systems are often referred to by their light source, with white light or LED blue light being common approaches.

It takes more than a good 3D scanner to obtain reliable measurements, Figure 6.13. Scanning is only the beginning of a long road to creating successful 3D print. The developer needs the right software. Selecting a 3D scanner is important, however, it is undoubtedly useless without software that can effectively work with point cloud data, and let the information flow as needed, Figure 6.14. There are three categories of software that are marketed to users of 3D scanners:

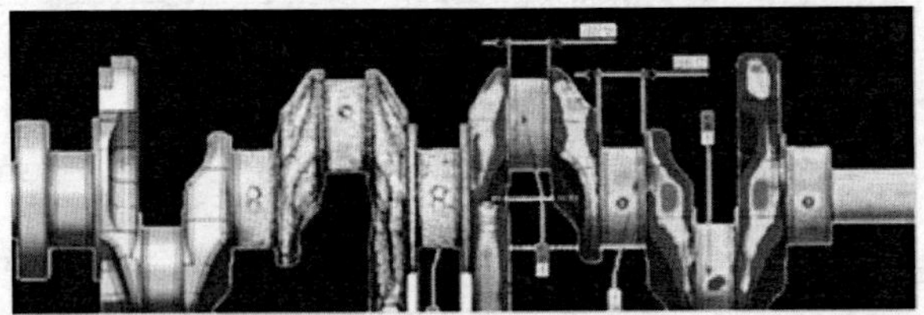

Fig.6.13: Structured 3D Scanner Scanning a Cam-Shaft

Software Abilities

Software which collects abundant amounts of measurement data has become easier and simpler. Modern 3D scanners gather (500,000 - 1 million/s), enabling measuring the geometry of the parts with desired and reliable resolution, leading to consistent reliably of the information to scan through-native software.

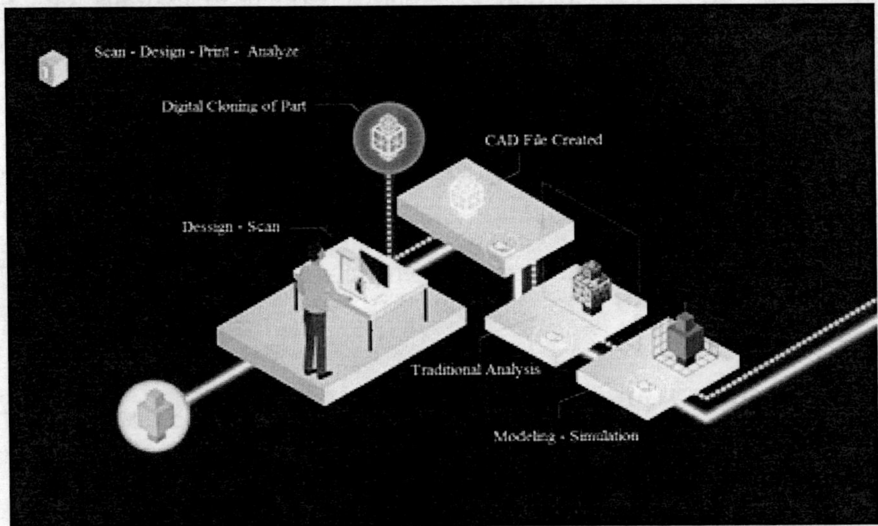

Fig.6.14: (A) – CAD Design – Scan Object

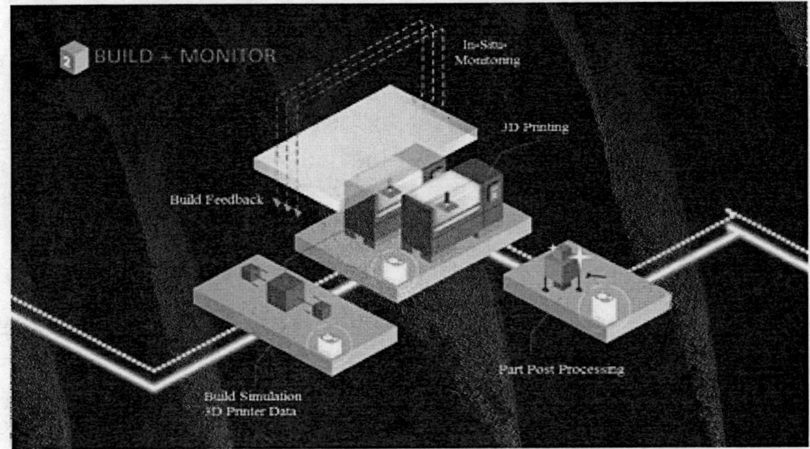

Fig 6.14: (B) – 3D Printing – Build Monitor

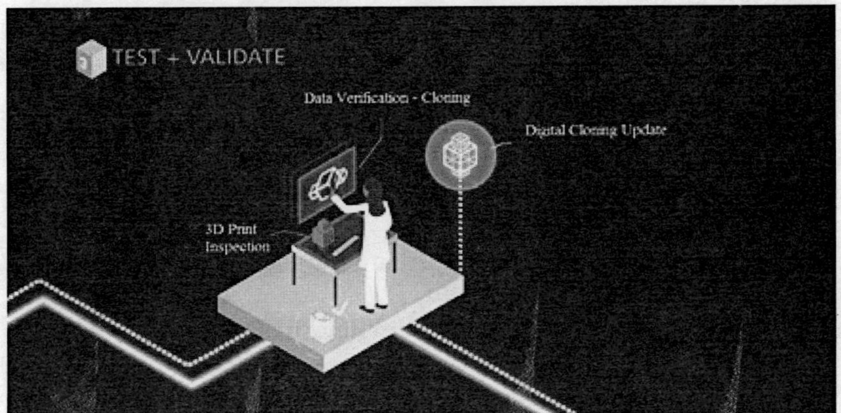

Fig 6.14: (C) – 3D Printing Test and Validating the Build

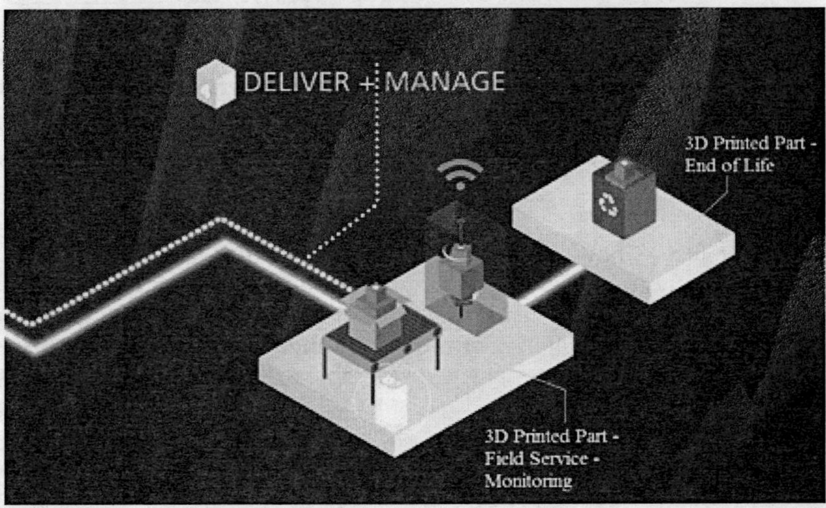

Fig 6.14: (D) – Field Service and Moitoring

Most 3D measurement software was not developed to handle 3D scan data. Accordingly, software built to use data from stationary or portable CMM's or laser trackers is optimized to work with a small number of discrete measurements that are known to be individually accurate. Employing 3D scan data is thought-provoking for various reasons.

Nonetheless, to maximize the full potential of the native scanner, which already exists at the original 3D printer, the user must ensure the resident 3D scanner is consistent, and provide highly reliable measurement, prior to printing or even purchasing the actual printer.

UNDERSTANDING 3D SCANNER FOR 3D PRINTING

Any man, woman, or child can 3D print almost anything. However, having a scanner helps foster even more creativity. The following step by step will greatly enhance deep fundamental to acquire to greatly comprehend the digital scanners for 3D printers.

You desire to gift your tender loving spouse a great anniversary present. She relishes her jewelry wedding box! An impression pops into your head: a 3D model of that favorite possession, eternalized in plastic for her desk at home. You have a desktop 3D printer, so that is totally possible, and it is also attractively inexpensive.

So you take your handheld 3D scanner to her vanity room and use it to scan a digital image that you can now deploy and sooner or later, print straight from your computer. Easy enough, and possible right now if you have the cash for the equipment.

3D scanners are the corner stone to enforcing 3D printing technology into the mainstream. When you can scan anything and turn it into a print -- be it that motorcycle part, or a spare part you need for your toolkit, or a new toy for your innocent child -- the possibilities become endless, and 3D printing makes a much more sense.

Here are pillars to learn about the state of 3D scanning and the progress of the technology. To sum it up: we are presenting some facts herewith..

3D Scanning The Vital Part in 3D Printing - Inhibiting 3D Printing

The fact is that the printer is accessible, while its user-friendly software is one of the aspects of 3D printing that is underdeveloped, and is truly inhibiting and holding the technology back. Part of that fact is the issue of scanning, as it is difficult to figure out how to turn any object into a 3D model that can be deployed, adapted, and then printed.

As the technology advances and someone progresses an easier way to use the software systems for 3D scanning and design, the Printing technology cannot still reach the organic masses. Equally paramount is the expensive cost of a reliable 3D printer. As 3D printers drop in price, hopefully the same matter happens with scanners. Therefore, to catch on with the public, they have to be exceptionally user friendly, and obviously, inexpensive.

Expensive Scanner

That leads to the next fact -- how expensive 3D scanners are. Largely, most 3D scanners are excessively bulky and costly to be used by ordinary consumers. This fact is a contradictory with the culture of 3D printing at present, considering most of the expansions are in desktop 3D printers made for home-based utilization. Majority of the 3D scanners are more than $1,000, which, when added to the cost of the actual 3D printer, is a considerable extra venture.

Handheld Scanners

A portable, handheld 3D scanner is a fact that can scan high resolution images for 3D printing. It can scan 3D art, and game development. Also, it can scan high texture surfaces, the human form, plants and animals, and virtually any object. The hand held scanner cost is about US $1200.

Nonetheless, "3D Systems" corporation also makes a handheld 3D scanner namely "Sense," which can scan objects of any size and shape. Moreover, It is fully integrated with the "Cube" desktop printers, so the entire process is simple. The "Sense" scanner is about US $400.

3D Scanning is Educational Tool

One of the most thought-provoking fact and uses of 3D printing is for education. The Smithsonian has been scanning and archiving its artifacts, and then freeing the designs online for people to download and print. It is a great use for educational purposes -- teachers and students can combine engineering, math, and history collectively for lessons inside and outside of the classroom.

3D Scanning Highlights 3D Printing Usefulness

Majority of people can accustom to open source platforms like "Thingiverse" to download designs to 3D print. Fortunately, It is free, it is easy, and there are countless amount of interesting topics to print for sheer joy. The technology will undoubtedly advance forward, however, 3D printing has to move past the newness stage and become more beneficial in everyday life – individuals old and young alike should consider 3D printed parts, tools, or other objects that are easier and inexpensive to print than buy.

Scanners excavate new opportunities in 3D printing, as a replacement for having to hunt for something wish on an online marketplace -- which in a way, defeats the tenacity of being able to make everything you wish from the blue sky-- you may find whatsoever you wish, or part of what you want, and scan it.

"Matter and Form" Scanner

Often one is not fully certain of what brand to buy, nor is any one forced to buy any particular 3D printer from a specific company. Therefore, "Matter and Form" of 3D scanner may be a good option. It was a crowdfunded product that now runs at $600. It is probably one of the least expensive laser scanners on the market. It utilizes a turntable to scan the model and generate a 3D point cloud, and then it can be downloaded and manipulated on the computer.

Patenting 3D Scanner

As the marketplace remains to be jam-packed with scanners, "MakerBot" company has applied for patients for its "Digitizer 3D scanner," which is now about US$799. This effort enabled the 3D scanner fully integrated with 3D printing enabling more accessibility and simple to use – Two fundamental issues they have achieved; one patent is about the way the "Digitizer" calibrates, and the second way is for the software, which makes it easy for the user to select settings for the scan. "MakerBot" is setting the precedent for the industry.

Smartphone May Soon Function as a Scanner

At present, there is a laser chip that determines the distance and size of an object by measuring the laser light reflected. Since the electronic chip is only a millimeter in size, it may be perfect for fitting in smartphones to 3D scan images with a camera -- which enables 3D printing appear much simpler and more proficient.

3D printing, the theme of 2019 is simple: "*make it useful.*" The last decade has been stimulating for professional and amateur alike of the technology and people in the industry, as it has developed so rapidly it has been hard to keep up.

Difficult with it was, businesses and consumers still had little motive to invest in 3D printers and scanners. Desktop printers and scanners are still novelties, mostly but not all used to make silly toys and gadget accessories. Food printers are captivating, but they are still somewhat out of "Alice and Wonderland," is still a distant dream from being in every kitchen. Bioprinters and 3D Scanners are astonishing accomplishments of technology, but having them in hospitals is still years away.

Although there are still much of 3D Printing hype, but there was also much of evidence to prove that 3D printing and 3D Scanners are a smart investment. The technology is democratizing every industry, allowing people at home -- and at work -- to customize and create their own products. And because of new materials, better software, and more advanced companies, the technology is gradually initially becoming more realistic. Mainly for the following reasons stated below:

Simpler Printer/Scanner Design & Software

Almost every 3D printing supplier enjoy showing off their software platform. Through the end of 2019, the main hurdle 3D printing needed to jump to become mainstream was CAD and Scanner software. The learning curve is just too steep for most people. It is made for engineers. Though, many different companies are making solutions for that. A part of those solutions are scanners that can take any object and turn it into a 3D design.

For instance, "Matter and Form" scanner is a portable scanner that allows designs to be uploaded to any desktop 3D printer model. The "MakerBot" company, is also launching "Cashew 3D," their version of an open source platform for 3D designs. Unlike "Thingiverse," the most popular maker platform, Cashew will use 3D imaging and also be mobile friendly..

More Materials

MakerBot is offering spools of PLA composite materials like wood, stone, and metal. "3D Systems" company is also offering a nylon filament for its desktop 3D printers, and other startups are using carbon fiber and experimenting with metal, ceramic, and more.

The fact that most desktop printers 3D Scanners only worked with plastic filament made it seem like home 3D printing was just a novelty. With the advent of new materials for home printing, the technology actually will become more useful and reach a broader audience beyond makers and hobbyists. With better materials, people can print spare parts, customize things, and create more useful objects.

Small Businesses Manufacturing 3D Scanners

Since its inception 30 years ago, 3D printing has been largely utilized for industrial processes. Those 3D printers are massive, and cost thousands of dollars. The other face of 3D printing -- the face that went avant-garde in 2019 -- is small, compacted desktop 3D printers.

The business motive was absent to utilize the 3D scanners, and mostly businesses that do not have a necessity for very expensive professional printers, but want to print parts for products or do rapid prototyping.

This year will be proclaiming for 3D printing as the technology is beset toward small businesses. Companies like "Ultimaker" are aiming certain models toward a workplace environment, and new materials to print products proposing a better incentive to make an investment in a printer and a scanner.

If businesses are reluctant to purchase their own printer, service agencies will also be an option. "UPS" is now providing on-site 3D printing, and it is expected that tendency to continue to grow in 2020.

Making More Wearable – Better 3D Printers/ Scanners

Wearables and 3D printing and scanners rose at a similar rate in 2019. The main criticism about wearables -- particularly for women -- is that they are hulking and sometimes just total dreadful. And that is where 3D printing may come to the rescue. Home desktop printers and services like "Shapeways" are an ideal way to customize wearables and turn them into trendy pieces of jewelry or accessories. "3D Systems Company," for instance, is actually encouraging this high-fashion, high-tech side of 3D printing by offering partnerships with designers and artists to generate exclusive 3D designs and pieces.

Both wearables and 3D printing/Scanners are becoming even more widespread with the average consumer and more therefore more useful in everyday life.

Classroom - 3D Printing/Scanning

Several companies communicated their strategies to take 3D printing into the classroom. "MakerBot," for instance, is dedicating much of their resources into its "Innovation Centers," which are facilities at universities where there are 30 to 100 "MakerBot" Replicators and 3D Scanners. The company is occupied with universities to educate students on the design procedure and potential with the technology to start to generate courses that utilize 3D printing.

Another instance is the "3Doodler," a 3D printer pen that paints in the air. The company is on its second generation pen, which is much silkier and more instinctive than the first one,, and is aiming a new campaign toward classrooms that want to teach more STEM subjects.

Taking 3D printing/scanning to the classroom in grade-school will endure to be significant as it is a simple way to assimilate STEM subjects into the syllabus while submission another artistic outlet. At the collegiate level, 3D printing/scanning labs offer another admission point into engineering and design, which will be vital in the years to come as the mandate for STEM jobs increases.

3D BIOPRINTER/SCANNER

3D bioprinter to replicate human organs, change the face of healthcare. Researchers, scientists, biologist, and engineers are only strides away from bioprinting tissues and organs to resolve a myriad of injuries and illnesses. If one desires to comprehend how close the medical community is to a quantum leap fast-forward in 3D bioprinting, then one must look at recent work in Stem research.

This team stands at the vanguard of research in 3D bioprinting, as they carefully take steps toward printing a working human heart. As part of this work, the team is also groundbreaking breakthroughs in printing human stem cells -- a change that could remove the intense ethical dilemmas allied with stem cells and potentially takes regenerative medicine to novel heights. The blend of these stem cells and 3D bioprinting is going to help heal or replace injured human organs and tissues, improve surgeries, and eventually give patients far better results in dealing with an extensive variety of illnesses and injuries.

As advanced technology as it is from its surprising background as a military project. It is dreadfully too slow and printing anything with it is a tortuously manual process. The print-head runs on a three-axis robot that does not handle curves very well.

A big part of the research activities is building intricate scripts to tell the system what to print. It is similar to a computer programmer writing in assembly language to give a computer system an exact set of instructions. It is an incredibly laborious process, and it involves extensive patience and breaking new ground in computer communications. Most interesting is extensive knowledge in mathematics, which is essentially a geometry problem.

The outcome is accelerating the work so significantly that what has taken several months will soon take few hours.

This new solution's 3D Printer/Scanner/Manipulator Mechanisms is hardware, namely "Bio-Assembly-Bot" (BAB). It runs as a six-axis robot that is far more precise than traditional robots-3D printer-3D scanner. The real modification, however, is in the software: Tissue Structure Information Modeling (TSIM), which is basically a CAD program for biology. It takes the physical coding out of the procedure and substitutes it with entity bears a resemblance to desktop image editing software. It permits the medical researchers to scan and manipulate 3D models of organs and tissues and then utilize those to reach decisions in diagnosing patients. And then, utilize those same scans to model tissues, and eventually organs to print.

This is a gigantic leap forward in the ability and technology of bioprinting,

Guide to 3D Scanners for 3D Printing

The most collective way to generate models for the 3D printer is to utilize a CAD program to draw them from the point of inception. However, there is another way that enables the user to quickly model real-life objects with extreme accuracy almost instantly – 3D scanning.

3D scanners have been around for many years but, much like 3D printing, it is only recently that they have been affordable to average consumers. The important issue about 3D scanners is that they allow the user to integrate the 3D prints with the surroundings and the objects around more easily and rapidly.

The Use of Digital 3D Scanner

Typically, the main purpose of a 3D scanner is to generate a 3D digital model of the a true real geometry of an object placed in front of it. Combining this ability with 3D printing enables the user to accomplish three main issues:

1. Replicate existing objects
2. Replicate existing objects with changes
3. Generate new objects that perfectly mate to an existing object

Let us illustrate each idea with some examples to shed some light on the issue:

1. If a black "Queen" is lost from the chess set, once could scan the white "Queen" and then print a replacement.
2. If the existing "Knight" appeared fragile in the same chess set, one could scan the existing "Knight" and in a CAD package, like the free "Blender." Polish it to appear as desired, and then 3D print the result.
3. If one desired to create special storage cavities set to all the pieces, one may wish to 3D scan each piece to its precise size and shape. One may transform the scans into 3D design. Consequently, 3D printing can take place.

3D scanners are similar to 3D printers, once the user becomes familiar with the procedures and the outcome numerous possibilities would surface to the point of wondering how one you ever did without a 3D scanner.

Operating the 3D Scanner

When observing the results of a 3D scan, one may be convinced it is like magic. 3D scanning is a combination of advanced technologies and expertise in design. It may appear simple and achievable; however, the novel user needs a little practice to achieve the best results.

A 3D scanner is a device that identifies, collects, analyzes, and displays the shape of solid objects in their digital form. It will capture the structural and surface design of a three-dimensional solid object using laser light and create point clouds from the surface of the object.

The 3D scanning procedure encompasses placing the object on the bed of a digitizer or directing the scanner at the object to be scanned. Specified software guides laser light on the surface while the camera records shape, and distance of the laser line in three dimensions. The resulting data is captured and displays as points on the computer. These points are referred to as "Point clouds." This is the data that can now be used in forming the digital model after scanning and capturing the exact configurations. The process is fairly fast and very accurate.

Point cloud files are formed with the data composed and saved for future references. The file is registered and merged to bring out the presentation of the object. Post-processing may also be required if the scanned geometry is not highly accurate.

Types of 3D Scanners

3D scanners can be either tactile (contact), which represents those that require placing the scanner on the object, or non-contact, scanners that can scan at a distance from the object. Regardless, of the type, they are both available as handled, desktop, or full-body scanners.

Handheld

For a handheld scanner, one must grip on the capturing unit while acquiring data from the object. The intended object to be scanned is placed still, and the scanner is moving around it to capture all particulars. Scanner functions on structured light to calculate the geometries of the subject. Handheld scanners offer several operational modes comprising manual scanning, free scan, and quick hand scan. Hand scanners are more portable, lightweight, and smaller in size equated to desktop 3D scanners.

Notwithstanding being an inventive and fast scanner, its main inadequacy is the need for the object to remain motionless. Any movement can meddle with the data, and therefore it is suitable for scanning motionless objects only. Hand scanners are also not as competent at scanning see-through materials or reflective surfaces, so one is expected to acquire a poor-quality scan.

Desktop Scanners

Desktop 3D scanners are comparable to handled scanners, but they are bigger, weigh more, and can be located on a surface to enable the scanning process. Desktop scanners are the most widespread type of scanner for 3D printing since they are best for scanning smaller objects of the size that one might consider 3D printing.

To use a desktop 3D scanner, one can easily place the object on a designated plate and scan it. The data point clouds are fed into the computer via a USB cable or Wi-Fi. They provide high-quality 3D models for duplicate. The main drawback of Desktop scanners is that they can only scan up to a exact size of object, typically around 60 cm^2.

Full Body

Full body 3D scanning systems are designed to capture the entire body of a human. These types of scanners work by merging structured light and photogrammetry 3D technologies to capture the portrayals. 3D full-body scanners have cameras and light mounted surrounding the intended body to scan. A projector is actually used in this to protect the pattern. The entire body 3D scan captures two photos in series. One is made when the projector is on, and the other when it is off. The first set of photos calculates the geometries while the second one is used for the texture of the body. This is the data that is utilized in generating quality life-like geometries.

In short, full-body systems are dependable, have a fast scanning time, and yield quality textures. Their main constraint, which can easily be controlled, is the succession time between taking photos (150-250 millisecond). In some instances, it can lead to imperfect quality.

PURCHASING 3D SCANNER

Scanning Volume

Buyer must be concerned about the maximum size of an object that can be scanned. One must make sure that the object may fit within the maximum volume of the scanner. It is prudent to have a scanner with a similar scan volume to the 3D printer, so neither will limit the other.

Scanning Speed

Scanning speed is how much time the scanner takes to complete a scan. Usually, it is measured in minutes for the complete scanning volume to be scanned. There is no considerable benefit to speed up the scan speed.

Scanning Precision

The scanning precision is the smallest level of detail that the scanner can measure and replicate. The lower, the better, nonetheless, generally more accuracy dictates a higher price. Also, most 3D printers can only achieve around 0.1mm resolution, so there is no such need to scan anything at a higher resolution than this.

Scanning Weight

The scanning weight placed on desktop scanners, is the maximum weight of the object the canner can support. Many desktop scanners feature a motorized rotating turntable that ensures all 360 degrees of the object have been scanned. The object weight must not exceed the motor torque limitation.

Reviews of 3D Scanners

Summary of some of the top scanners in the market that may be ideal to use, stated as follows:

"Ciclop Lasing" 3D Scanner Kit

This " Open Source DIY - 3D Scanner for 3d printer" is the_lowest Price.

The "Ciclop Lasing" 3D-Scanner Kit is a popular choice among creators in the market as it is a low-price option. The compromise with this 3D scanner that it has to be assembled before scanning, Figure 6.15.

Fig.6.15: Ciclop Lansing 3D Scanner – *Courtesy Ciclop*

Specifications

- It is an open source scanner.
- 200 mm diameter x 8 mm base and a 200 mm diameter non-slip surface

The Ciclop scanner also has several favorable 3D Scanning Properties. They include Scanning Precision of 0.5 mm and a maximum

Supported Weight of 3 KG. The Typical Scanning Time (Configurable) is often 2-8 minutes. The scanner also offers a Scanning Volume Diameter of 250mm by 205mm.

Features

The Ciclop Lasing 3D scanner kit operates by using triangulation technology. It projects a laser beam on to the surface of the scanned object and measures the distance of the laser ray with a 1/8th micro-stepping ability and a scan precision of 0.5mm. This provides exact measurements of point clouds. These specifications make it most appropriate for most detail applications.

It has a 200mm diameter non-slip surface; also it has a large scan area for scanning both small and medium objects. The scanner permits a maximum of 3 kg in weight. Scanning is fast with a typical scanning time between 2-8 minutes. The time is, however, configurable, it will depend on the size of the object and surface area. The "Logitech C270 HD" camera on the scanner captures and records all needed detail, and transmits the data to the computer via the USB.

Advantages

- It has a large number of community users who may offer assistance and tips.
- It is open source with information on mechanical design, software, and electronics readily available.
- It is excellent value for the purchase price

Disadvantages

- It can be a complex for beginners since it requires assembling, and may require some experience to operation effectively.
- Converting the 3D model file requires multiple scanning software to obtain a clear image.

Matter & Form Mfs1V2 3D Scanner V2

The "Matter & Form Mfs1V2" 3D Scanner V2 is a low-cost desktop scanner that continues the trend of "Matter & Form" bringing 3D scanning to the masses, Figure 6.16.

- The scanner has incorporated the "FM Studio" software which is a powerful tool for configuration providing control over the scanning.
- Scan accuracy within ± 0.1mm
- HD CMOS imaging
- 2 Class 1 (eye-safe) lasers
- Maximum Object Size: 25 cm high x 18 cm diameter (9.8□ x 7□)
- Maximum Object Weight: 3 kg
- USB 2.0 high-speed interface
- 100-240V
- Windows 7, 8.1,10 and Mac OS 10.11+
- +Quickscan upgrade software

Fig.6.16: Matter & Form 3D Scanner

Features

The "Matter and Form" Mfs1V2 3D scanner has a good accuracy of up to 0.1mm making it suitable for most 3D printing applications. The hardware side of the scanner is straightforward to use, and features eye safe technology so you can use it at home without any concerns for safety. It is a foldable design, which makes it easy to fold up and store away.

The scanner's software is compatible with most Windows OS including 7, 8.1 and 10 and also Mac OS.

The scanner operates on 100-240V and can hold about 3Kg on its surface scan area. All the scan data is communicated to the PC quickly on a USB 2.0.

The scan point cloud exports include both XYZ and PLY, while the "Mesh Exports" are "OBJ" and "STL" format which is compatible with most "CAD" software.

It features a foldable design that shields both the camera and turntable when it is not in use. This helps to keep the workspace clean and organized. It does not need an additional storage case.

The scanner package includes a +Quickscan upgrade for the MF studio software that provides higher scanning speeds of up to 65 seconds and direct real-time feedback along with scan previews to speed up your workflow.

Advantages

- Easy to use - the powerful software makes scanning easy.
- Faster scan speeds with the upgrade software. Available in a variety of languages.
- Ideal for all people, beginners, and professional alike

Disadvantages

- Several scans required to scan a full 3D object.
- Exporting from point clouds to printable mesh can sometimes be noisy with clean up required.

XYZ PRINTING DA VINCI 1.0 PRO 3 IN 1 3D PRINTER/SCANNER/ENGRAVER

The Da Vinci 1.0 Pro 3in1 is, as the name suggests, three tools in one as it features a 3D printer, a 3D scanner, and an engraver, Figure 6.17.

Specifications

- Wireless
- 3-in-1 3D printer, 3D scanner, and laser engraver
- Build volume; 7.9 x 7.9 x7.9 inch
- Resolution 20-400 microns
- Guided calibration
- Max 120mm/s printing speed
- Material; PLA, ABS, Tough PLA, HIPS, Wood, antibacterial PLA, PETG, compatible with third party material.

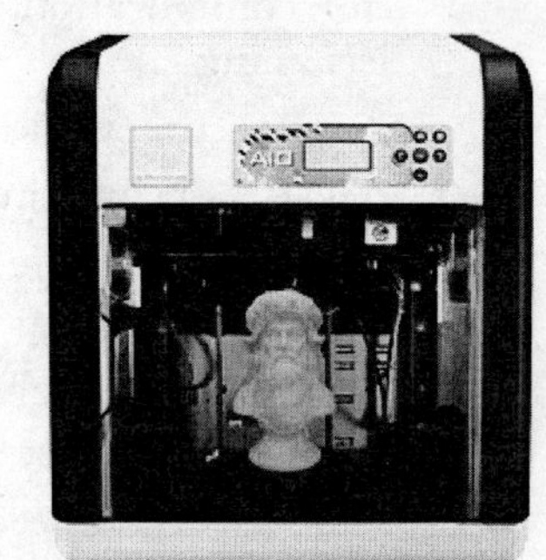

Fig.6.17: XYZPrinting Da Vinci 1.0 Pro 3in1 3D Printer/Scanner/Engraver

Features

XYZ have ensured their device can take care of entire product manufacture from start to finish.

For scanning, you get 20-400 micron of resolution that gives quality output. It works by laser triangulation and is an easy setup.

The scanner rotates in all directions for 3D capture of the entire object in about minutes. Another plus for the device is its compatibility with a variety of industry tools and materials. For Print, you can use all PLA materials – regular, tough, antibacterial, ABS, wood, HIPS, and PETG.

Also, it operates on a WI-FI interface making it fast and reliable. The engrave feature is an optional upgrade with a laser output of 350 MW. It allows you to customize and create your design on wood, cardboard, leather, paper and on the plastic produced by the 3D printer itself.

Additionally, XYZ has an active customer and technical support program for guidance and an online gallery with more than 3000 models for download.

You can configure the 3D printer setting including speed, temperature, and retraction for a perfect outcome.

Advantages

- It's a frustration-free use, with its steps by step easy to follow instructions.
- Online gallery with more than 3000 models for download

Disadvantages

- You will need practice before achieving the best of the 3D printer
- Only operates with Wi-Fi

Xbox One Kinect Sensor

This is one of the most popular hacks to get a cheap entry into the world of 3D scanning, and it works surprisingly well, Figure 6.18.

To turn the Kinect Sensor into a 3D scanner, you simply have to download the Microsoft 3D scan app.

Fig.6.18: Xbox One Kinect Sensor

Specifications

- It is handheld
- Has a 1080HD camera
- Compatible with Windows Mac and Linux
- Video frame rate of 30fps
- Measures 1.8 x 3.1 x5.9 inch and weighs 3.2 pounds

Features

Its high definition camera allows recording of clear captures, but it actually works best with Windows 10. The video frame rate of 30 fps gives it a quality GPU. The scan also offers better calibration of the color mesh by launching the camera app which delivers the best results. It has a non-reflective background which is achieved by maximizing the distance between an object for scan and the background.

Advantages

- Provides high-quality camera for precise, clear images
- An easy setup

Disadvantages

For 3D scanning, you will need to download the software

It's quite heavy than it appears

Handheld Structure Sensor - 3D Scanner for iPad

This scanner can scan autonomously, and it's probably the most sophisticated accessory for mobile devices. It's handheld, portable and the iPad attached structure sensor can scan for 4 long hours, Figure 6.19.

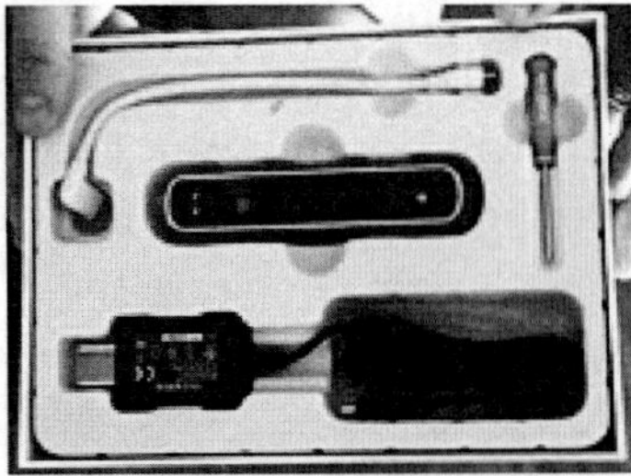

Fig.6.19: Occipetal Handheld Structure Sensor 3D Scanner for iPad

Specifications

- Technology; structured light
- USB cable connected
- Weight of 95g and measures 119x 28x29mm
- Has a 0.5 scan precision from 40cm to 3.5meter range
- Operating systems include iOS, Windows, Android and Linux

Features

The scanning structure sensor can be attached to your mobile device either an iPad or smartphone. You can capture precise, full colored models of people and things in your surroundings. Facilitates indoor mapping, 3D scanning, and mixed reality gaming. It also has a 3D vision that enhances practical tasks like precise measurement of an object or a room. The scanner offers a structured SDK that brings cutting edge computer vision which makes it possible to capture high-quality photos. One can capture any object and turn in into a full-color 3D model.

Advantages

One may experience advanced 3D technology directly from a smartphone handset.

The software is easy to install and use.

Disadvantages

This type cannot work on its own, it must be attached to a device and only works with smartphones.

EINSCAN SE DESKTOP 3D SCANNER

EinScan-SE White Light Desktop 3D Scanner with Tripod

The Einscan SE scanner is designed and created from one of the industry's leading brands in 3D technology. You simply get yourself a powerful tool that is designed for professionals at low costs, Figure 6.20. Its specifications include:

- Has a 0.1mm accuracy
- 8-second scan speed
- A 700mm Cubic max volume
- A fixed auto scan mode and a manual alignment

Fig.6.20: EinScan-SE White Light Desktop 3D Scanner with Tripod

Features

Einscan SE features a free and automatic scan mode running for about 8 seconds per scan and about 2 minutes for a 360-degree overview. It uses automated software that offers automatic calibration; hence no rigid setup is required.

Also, it is easy to start the scan with just one click, and it will auto align and generate the mesh directly. Additionally it has various options of alignment; you can mark points, use a turntable, or manually align.

Einscan SE is a compatible scanner for 3D printing. The auto meshing provides high-quality 3D data for printing. With advanced technology, it can share data with third parties through a data sharing platform access service.

The turntable can manage up to 10 kilos so there is a lot of flexibility with regards to what you can scan at any given time. The scanner also supports Windows 7, 8 and 10 of 64 bits.

Advantages

- An advanced tool with multiple functions but easy to use and apply
- Very flexible in terms of the weight it can hold.

Disadvantages

Operates under particular conditions in terms of software compatibility and computer requirements

Einscan Pro Handheld 3D Scanner

Einscan pro handheld 3D scanner is one of the best 3D scanning choices for both consumer and commercial purposes. It is highly accurate and fast, Figure 6.21.

Specifications

- 4 scan modes including handheld HD and rapid scan, automatic and fixed scan
- Has a scan speed of 90000 points per sec for HD scan and 550,000 points per sec for rapid scan
- An accuracy of 0.1 /0.33, 0.05mm

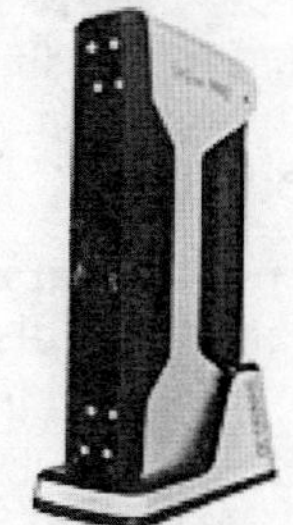

Fig.6.21: Einscan Pro Handheld 3D Scanner

Features

The handheld rapid scan mode commands high speeds. It is therefore recommended for applications requiring high precision, accuracy and fast turn-around time. Also, it operates on both features and markers alignment modes.

Handheld HD scan mode uses markers for alignment and utilizes light scanning which is recommended for large sculptures, automotive parts, etc. The end result is a highly detailed and precise end product. The auto scan mode uses the turntable for small objects for better convenience and easy use.

Furthermore, it is applicable to industrial use, creating models for ornaments, and small artistic pieces. The fixed scan mode enables free scanning. It is recommended for large, complicated parts that are detailed but lightweight, and easy to move around while scanning.

Advantages

- Has various mode options making it usable for a variety of applications.
- Highly accurate and reliable

Disadvantages

- Requires proper learning before use due to cumbersome modes and intense alignments prior operating the scanner to avoid disparity during the scan

Einscan Pro+ with R2 Handheld 3D Scanner

This is arguably one of the most ideal 3D scanners for professionals, and it comes with a full kit of advanced technology, Figure 6.22.

Fig.6.22: Einscan Pro+ with R2 Handheld 3D Scanner

Specifications

- Alignment modes include HD – markers mode; rapid scan- makers, feature, turntable, and manual.
- The Markers, feature Markers, features, and manual modes can be used for both auto and fixed scan.
- 4 scan modes including handheld HD and handheld rapid scan, automatic and fixed scan
- Has a scan speed of 550,000 points per sec for HD scan and 450,000 points per sec for rapid scan
- A single shot accuracy of 0.1 /0.33, 0.05mm for all scan modes consecutively

Features

Handheld rapid scan mode scans quickly with precision and supports markers and feature alignment. This allows it to recognize and align reference points on the model. The high definition mode is ideal for a richly detailed and stationary object. It is very convenient and produces highly accurate precise results.

It is ideal for smaller objects; the auto scan mode is applicable for industrial components while the fixed scan mode is perfect for scanning large objects.

Advantages

- A high-tech multipurpose scanner
- Very accurate and fast

Disadvantages

- Expensive
- Not beginner friendly

Shining 3D Auto Scanner DX-ES Pro

Designed for the dental industry, the Shining 3D auto scanner is cost-effective and lightweight. It comes with a small footprint which makes it ideal for scanning teeth. Blue or white light can be used, Figure 6.23.

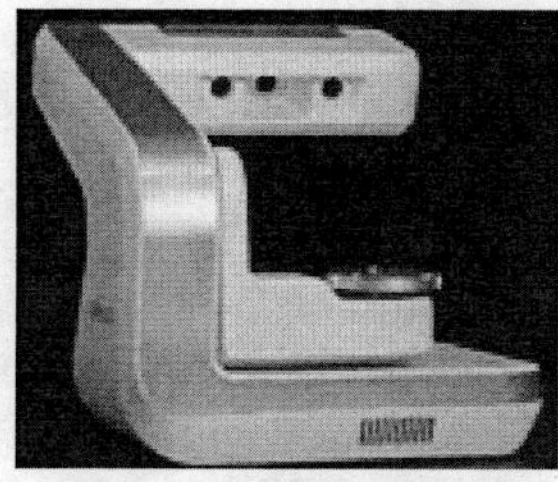

Fig.6.23: Shining 3D Auto Scanner DX-ES Pro

Specifications

- Fully open structure
- Multiple functions
- Clinic mode
- Open data format
- Weighs 11.0 pounds

Features

The open structure features of the shinning DX ES PRO make it compatible with articulators in the market including Kavo and Artex. It comes with an articulator, triple play, texture, and continuous scanning giving you more options for your desired application. The added clinical mode feature makes it an essential aid to dental clinics. It can help and support chair-side restoration solutions. Besides, the open data format makes it compatible with current CAD/CAM software and facilitates STL data exports

Advantages

- High-quality scanner

Disadvantages

- Very Expensive

Einscan SP Desktop 3D Scanner

As an advanced version of the SE, the SP is a faster, more accurate option that offers higher capture volume. It stands out as a capable scanner for a user in need of more precise and quick scans especially a professional creator, Figure 6.24. Its specifications include:

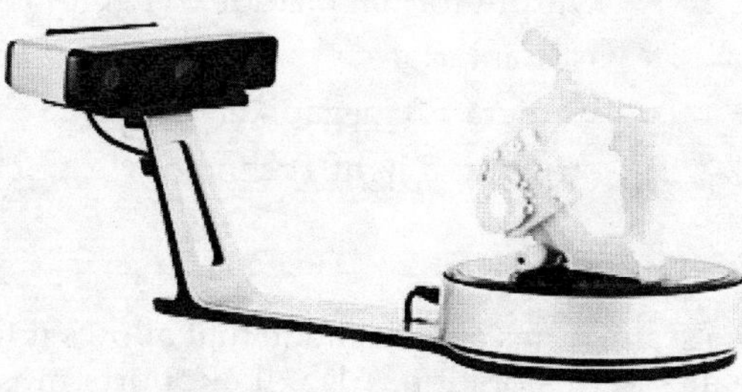

Fig.6.24: Einscan SP Desktop 3D Scanner

- Auto scan / fixed scan modes
- Technology structured light
- Output formats ASC, OBJ, PLY STL
- Sing shot accuracy < 50 microns
- Higher accuracy 0.05mm on a single scan
- Scan time 4 seconds
- Auto scan for 360 degrees is 1 minute
- Alignments; manual, feature, mark points, and turntable mark point
- Compatible with Windows 7, 8, and 10 on a Dual core i5 or higher

Features

The scanner has incredible speed; in fact, it achieves a 360-degree scan in 1 minute and a single scan in 4 seconds. It also offers quality scans with an accuracy of 0.05 and a high resolution of 0.17 mm point distance. This allows the scanner to capture finer details. It also provides different calibrations to alignment for an optimal quality scan. It's quite easy and fast to set up and use too. The inclusive one-click 3D scanning option and the auto-alignment features make it ideal for beginners.

Advantages

- Has a faster turnaround time due to its improved features
- You get to scan faster and produce quality 3D models
- Price friendly

Disadvantages

- Gathers dust easily on the lens, this may cause blurred during capturing an objects.

Einscan Dental 3D Scanner DS-EX

The Einscan dental 3D scanner DS EX, as the name indicates, was developed for the dental market to provide efficient solutions cost-effectively, Figure 6.25. Its specifications include:

Fig.6.25: Einscan Dental 3D Scanner DS-EX

- Accuracy of less than 15 mm
- Scanning time; bite is 13 sec, lower and upper jaws 30 sec, dies 30 sec
- Scanning range is 100 x 100 x 75 mm
- Scanning principle, structured light
- Dimensions 260x 270 x 420mm
- Output format include STL, OBJ
- USB interface
- Camera 1.3 megapixel
- structured light Technology

Features

The scanner has a clinic mode that allows it to be applied directly into dental clinics and facilitates chair side restoration. Also, it supports most articulators available in the market like the Artex, Kavo, Bio-art, and others. The scanner also features a continuous scanning mode using a LAN network. This helps in facilitating the separation of scanning and data processing.

The compact size is more favorable ass it is fully open structure that makes it easy to clean. Its open data format options allow exporting of STL data and are highly compatible with current market software. It comes with various scan options including Texture scan, articulator scan, and Triple tray scanning. The features are applicable to different levels, including dental labs, colleges, clinics, and others.

Advantages

- A beneficial and efficient tool for the dental industry.
- Adopts the latest technology for cost-effective and productive solutions

Disadvantages

- Complex due to various advanced features

Einscan Pro 2X Plus Handheld 3D Scanner

This scanner offers some of the most versatile scan modes, Figure 6.26. They include handheld rapid scan, handheld HD scan, fixed scan with turntable, and fixed scan without a turntable. Other tech specs include:

- Scan accuracy of 0.05mm for HD scan, Rapid scan 0.1 mm, 0.04 for fixed scan
- Scan speed of 20 fps in HD, 30fps in Rapid Scan
- 3D Print Compatible: Yes
- Output Format: OBJ, STL, ASC, PLY, 3MF, P3
- Scanner Weight: 1.13kg
- Supported OS: Windows 7, Windows 8, Windows 10, 64 bits

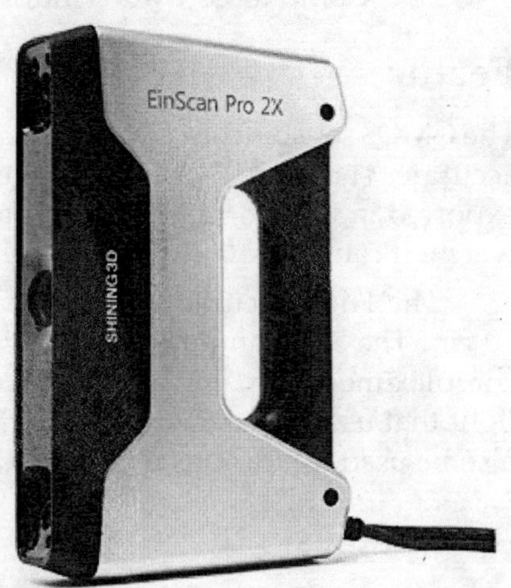

Fig.6.26: Einscan Pro 2X Plus Handheld 3D Scanner

Features

The scanner is equipped with a new operational mode that enhances the HD scan mode. The extended scan range will perfectly capture 3D models more efficiently. The higher scan speed translates to a more elevated, precise, and faster scanning. This will, in the end, assist the user in producing high quality 3D models.

Besides, a higher accuracy on a fixed scan without turntable provides a "0.04" mm accuracy, but with markers, the accuracy is at "0.05" mm. Output file format includes OBJ, PLY, STL, and P3 which are compatible with mainstream software and seamlessly connect to 3D printers for 3D printing. Enhanced handheld HD scan mode with an enlarged scan range will improve capturing efficiency

Advantages

- Easy calibration and software are self-explanatory

Disadvantages

- Too many features and specifications can make it confusing to figure out sometimes

Thunk3D Cooper Desktop 3D Scanner

The scanner is also very versatile to use and features the following specifications, Figure 6.27:

- Industrial CMOS sensor
- High accuracy of 0.04 mm per frame
- structured light Technology
- Free 3D software

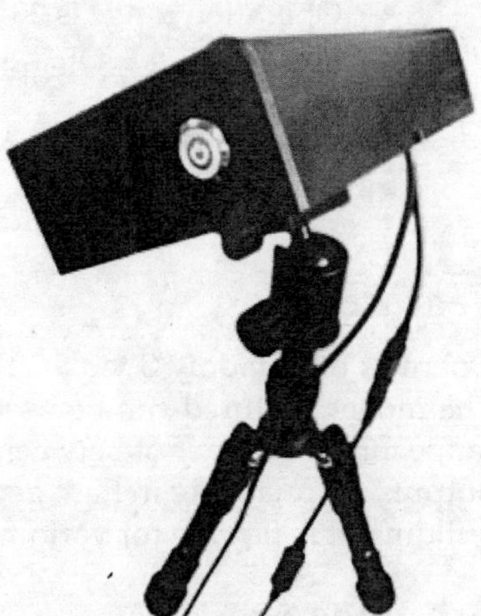
Fig.6.27: Thunk3D Cooper Desktop 3D Scanner

- High integration with a built-in dongle, compact modular and a cable plus a USB
- Compatible with Windows OS

Features

The CMOS sensor from ON-semi improves capturing of details and textured data for export is more accurate. The 1.3 MP-camera captures every dot to every pixel, and the texture data can be easily exported in PLY format. Its compatibility with Windows operating systems is also something to watch. It can work with Windows 7, 8 and 10.

The High accuracy of 0.04 that makes it an ideal fit for industry objects of between 5 and 60 cm. The 3D scanning software is free and simple but can be slightly sophisticated in creating complex models in just a few minutes. The scanner also has 1280 x1080 LED surfaced structured light that uses Texas instruments raster for enhanced core functions. The turntable alignment can also be used for various objects with various shapes.

Advantages

- It's direct plug and play, does not need any calibration
- Results are high quality and highly accurate, a good fit for precise applications like prototyping
- They have efficient tech support on standby to help you in case of any queries or assistance

Disadvantages

- A high-tech tool for industrial work, not suitable for amateurs

3D Systems SENSE2 3D Scanner

3D Systems SENSE2 3D SCANNER

This scanner has also several favorable specs to use, Figure 6.28. The following is a simple break down.

- Operation range is 20- 60 cm
- Field of view 45 degree
- Interface USB
- Wight 1.1 pounds
- Compatible with window 8 or later Operating system

Fig.6.28: 3D Systems SENSE2 3D SCANNER

Features

Operates on window 8 via a USB, it is provided with a competent handheld capturing solution. The model captured can be used for archival or asset creation for digital content. The operating range for the scanner of between 20 and 60 cm gives full- color images with a 45-degree view. The outputs are relatively in high geometry with good texture quality facilitated by the sense active 3D building. It is flexible for various applications

Advantages

- Price function balance

Disadvantages

- Requires Practice to reach good results

LEICA 3D DISTO

Leica is well known not only in the world of cameras but also of industrial metrology. The 3D Disto scanner has had some upgrades in recent years that make it a smart choice for every application, Figure 6.29.

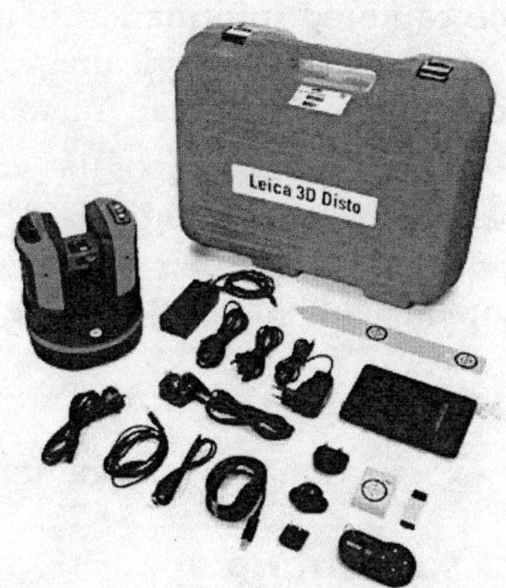

Fig.6.29: Leica 3D Disto

Specifications

- Upgraded Leica 3D DISTO software
- BIM building information modeling exports data
- The horizontal and vertical turning axes are Motorized
- OS; Windows 10
- Interface USB Wi-Fi stick

Features

DISTO can capture as much measurement as possible on any job and export them easily to CAD. It also offers excellent visualization. The upgraded software is now usable with smart windows devices and displays results in 3D. The software also makes it easier to check the measured points that are captured as Built measurements.

The provided interface facilitates the direct acquiring of DXF data into windows devices and uses the DISTO software for laying out all details including the anchor points, verifying critical construction locations. The Kit comes with 3D DISTO, GZM3 target plate, target pad, windows software, remote control, cables charging accessories and an offset tool.

Advantages

- Very accurate in measurement, measure once cut once
- Downloadable software

Disadvantages

- Works with windows tablet only
- Very Expensive

Naked 3D Body Scanner

The scanner has several essential specs, stated as follows, Figure 6.30:

- Technology – naked three, Intel Real sense, depth sensors and the mirror, embedded 3D depth sensors
- Takes 15 second for 360-degree capture
- OS compatibility; iOS and Android app
- Built-in feedback
- Wi-Fi and Bluetooth connectivity
- High processor power

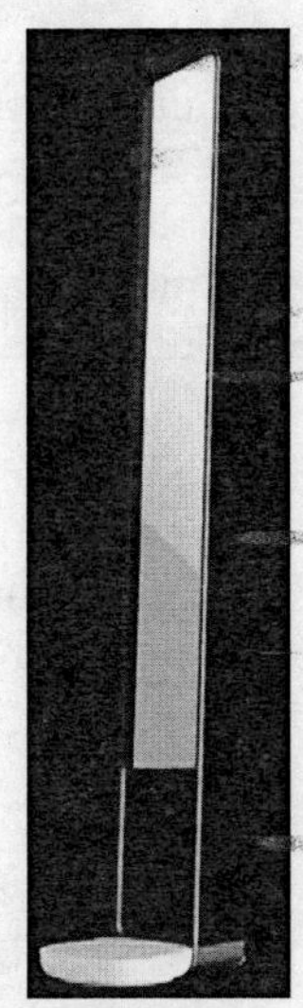

Fig.6.30: Naked 3D Body Scanner

Features

The three Intel Real sense depth sensors will scan the object and capture all the details. For instance, the captured information is not only physical but also other health Metrix including body fat percentage, fat mass, lean mass, circumference measurements. The built-in feedback comes with a scale alignment and an interactive light ring which combines to help in achieving the perfect scan.

For proper sync of the scan to the smart phone, the Bluetooth and Wi-Fi chips automatically handle those requirements. The scanner has high processor power that will easily convert 4GB of data into 3D. There is also a weighing scale for the scanner. Also, it rotates and works well on both smooth and hard surfaces, Figure 6.30.

Advantages

- Suitable for monitoring the body's health

Disadvantages

- Will need some skill to operate
- Expensive

Summary of 3D Scanning Current Technology

3D scanning is usually the vital part of the 3D printing. It is the silent partner, which impacts directly the quality of the 3D print. However, the 3D scanning technology has received less attention than 3D printing, while they are synonymous and inseparable. The 3D Scanner resolution, its ability to acquire the data, stores them, and faithfully transmits the information provides high integrity of the 3D print entity.

IDEAL METHODS TO SCAN OBJECTS FOR 3D PRINTING

It is conceivable to capture present objects into the computer so they can be manipulated or merely re-created using a 3D printer. This is predominantly valuable in the case of artwork or other exceptional formations that could not then be designed easily in a computer model.

The Glen Rose dinosaur track, for instance, came from a laser scan of the original fossilized impression, which was utilized to generate a digital copy of the track that can be shared without jeopardy to the original, Figure 6.31.

Fig.6.31: Glen Rose Dinosaur Track

Optical scanning captures only the surface shape of an object, but it is likely to use ultrasound imaging or CT (Computer Tomography) scan data to generate models of internal structures as well. Scientists have recently generated a model of the first exposed full skeleton of a living animal, for instance, by 3D-printing the bone Skeleton acquired from a CT scan of the subject.

Comparable data is being utilized to recreate the facial features of mummified remains in Egypt and of the newly exposed remains of King Richard III. Using CT scans and a stereolithographic system, scientists at the University of Dundee were capable to print King Richard III's skull into compact form, re-generating what this long-dead past monarch viewed like in life.

Primary 3D capturing systems depend on a probe that touched the printed object at many different positions, describing a ***"point cloud"*** around the object's outline to express its basic geometry, which is then packed in with superior detail as the scanner measures finer points between the original markers. These arrangements are still utilized in machinery analysis and other durable environments.

More current scanners incorporate light from lasers or ***structured light*** — projections that measure the distance from the camera to diverse parts of an object, so there is no jeopardy or harm to the object under examination from the touching points of the scanner. A handheld self-contained scanner delivered by "Creaform" was utilized to scan a human face.

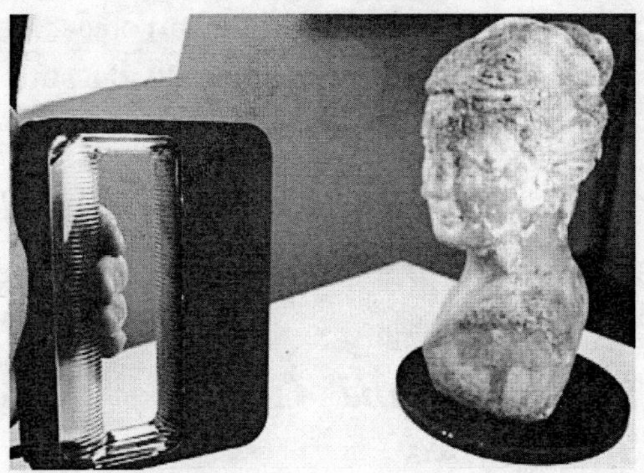

Fig.6.32: Restoring and Preserving Artifacts

Integrated with software on a computer, this structured-light scanner can shape a 3D model from repetitive measurements of an object's surface structure as the scanner is merely swayed directly above an object of interest, Figure 6.32.

Optical scanners can have trouble scanning highly reflective surfaces or scanning objects lacking in detailed topographies. While a mirrored surface would seem as just a longer path to whatever is mirrored, a large sphere would show same to the scanner from one point to another; the software would fail stitching the numerous diverse angles together to generate an entire model.

When scanning bulky objects with limited features, it is possible to help the scanner by ascribing small reflective dots to the object in various positions; the scanner can use the dots to compute the positioning of numerous parts of the scan.

Commercial 3D scanners deliver very high-resolution models of scanned objects; such devices can be as small as a handheld scanner or can contain bigger, more intricate systems that map many angles at the same time.

Scanners can image the inside surface of pipes, map out mineshafts and subterranean caverns, or even scan whole build sites for large build, utilizing laser tools comparable to "RADAR" called "LiDAR." Such systems are employed to map mining processes to compute ore subtraction, or in surveying to generate digital topography maps.

Commercial systems such as "Creaform, FARO, Artec, XYZ/RGB," and a multitude of other options deliver very high-resolution object models appropriate for engineering applications and manufacturing. Nevertheless, a home user can use low-cost lower-resolution scanners — like the Kinect video game controller — to model objects for 3D printing. Composed with software such as SCENECT, ReconstructMe, or Microsoft Fusion, the Kinect game controller can be used to create scanned 3D models at home.

CAD DESIGN TO 3D PRINTING

A Simple Step to CAD

Additive manufacturing permits the generation of complex internal structures rather than simple, solid-block components — a progress over traditional CAD tools which cannot always generate models to compete the true capabilities of 3D printers. Novel tools are becoming accessible in specific fields, as in the case of medical implants that must be tailored to fit an individual's bone structure though also permitting intricate lattices and other forms of fabricated non-solid models.

Steps to Print a Simple Cube on a 3D Printer

The novice user may attempt to print cubes to test all types of configurations, including materials and settings. This is an efficient way to check out one's capabilities and gradually improve upon. It does not take too long time to print, and may highlight the user's settings, such as "Infill" and "Solid Layers" change the way an object looks.

The novice user may download a simple 20mm x 20mm x 10mm height from "**Thingiverse.**"

In order to 3D print, consider the follow steps stated below:

(1) Load The Cube Object into "Slic3r."

Select an Infill of 20% and 3 solid top and bottom layers, Figure 6.33.

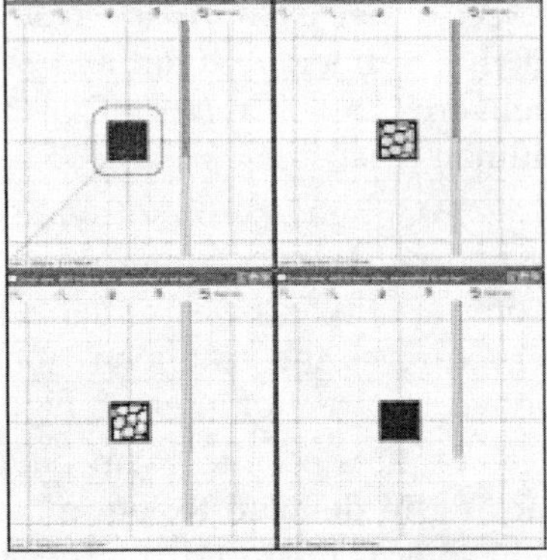

Fig.6.33: Load The Cube Object into "Slic3r."

(2) Export the G-code - Load G-code into "Pronterface."

"Pronterface" demonstrates the object loaded in the central of the virtual print-bed. The user may click the object, and scroll through the many layers. Above, showed in Figure 6.33, the user can view the base layer with a ring around it.

This ring is not part of the object being printed; it is an additional illustration by "Slic3r," as the first part to be printed, which primes the plastic flow of the extruder before the object begins to print. The central sections show the hexagonal infill and -finally the solid top surfaces. The user is now prepared to print the first object.

(3) Set the Print-Bed & Hot-End to the Required Temperatures for the Printing Material - Allow Printer to Heat-up.

If the user forgot to follow this point, the G-code will do it for you.

Fig.6.34: Extrude Some Material to Ensure Printing

(4) Manually Extrude Some Material.

This is achieved to ensure that the printing process is working correctly as planned and to home the printer axis, Figure 6.34.

(5) When Reached Temperature - Press Print.

After a minor delay whereas the G-code verifies and steadies the temperatures the print head travels to the central and begins printing the cube, usually with a border outline to start the stream of plastic. Though the first border is being extruded, check to view whether the polymer is sticking; ensure that the print head is not rubbing through the surface, and is not too distant.

One of the difficult settings is the correct distance of hot-end nozzle over the print-bed. Carefully observe the printer completed two perimeters. The infill must look solid. The first layer must not have any gaps between the individual extruded lines.

If the user does not see a good bond of the plastic, the user must stop the print and adjust the nozzle-head an increment of distance closer. If material does not flow smoothly material is likely squashed out and the nozzle being dragged through the plastic. The extruder is a too close. Move the nozzle-head slightly away from the build bed.

(6) When Cube is Finished, Measure to Confirm Dimension.

If it has printed significantly larger or smaller — it is probably set incorrectly for the Z and Y axes. Work out the new value by performing the same calculation for the extruder. Be sure to mark and remember the orientation of the cube when it is printed:

- The distance from front to back of the cube on the build-plate is the Y measurement.
- The distance from left to right is the X axis.

Novice users quite frequently find that the Z axis is more likely to be set incorrectly. This is due to many 3D printers use similar belt-and-pulley drives for X and Y, however, different for the Z axis. Many 3D printers may use a range of threaded rods, belts, or lead-screws for the Z-axis motion.

(7) Check the Cube Corners 90-Degree Angles.

If they corners are not 90°, user must ensure the X-axis carriage is aligned straight across the moving Y axis.

The novice use should by now be able to print more objects and begin exploring the competence of the 3D printer. It takes consumes more time to learn the diverse speeds, temperatures, and settings needed for printing dissimilar objects. The best way is by Experimentation, and Trial and Error are the best way to learn many factors that influence different 3D printers.

Preparing Future 3D Designers

3D printers can contribute in the training of future operators and designers who may form the next generation. Designer best exploit the capabilities of 3D printers, through innovative thinking. In addition to learning new software tools, tomorrow's designers may take advantage of 3D already printed items to augment their out-of-date settings, such as historical designs to capture interest in architecture and designs.

Future designers of bio-printers may consider recent advancement in 3D printers to formulate their designs before a living organism is required for their research and growth into tissue and organ substitutes or advanced medical implants.

Whether planning more intricate designs or creative arrangements, 3D printers can aid to progress prototypes of new designs and ideas with rapidly modified changes possible before delivery.

Designers can generate 3D objects from their electronic designs in a matter of hours, so that objects can be verified and adjusted to fit their users' preferences. Many times, they can be utilized to generate particular tools that replace costly commercial alternatives, illustrating the benefit of new manufacturing techniques.

Additive manufacturing will touch almost every trade, creating new jobs and improving our capability to clothe, feed, and house the growing global population using 3D printers.

REFERENCES

1. E. Charbon, M. Fisburn, R. Walker, R. Henderson, C. Niclass, SPAD-Based Sensors.
2. D. Van Nieuwenhove, W. Van der Tempel, R. Grrotjans, M. Kuijk, "Time-of-flight Optical Ranging Sensor based on a Current Assisted Photonic Demodulator", Proceedings Symposium, IEEE/LEOS Benelux Chapter, 2006, Eindhoven.
3. SoftKinetic DepthSense™ sensors, http://www.softkinetic.com/enus/products/depthsensesensors.aspx.-
4. J. Geng, "Structured-Light 3D Surface Imaging: a Tutorial", Advances in Optics and Photonics 3, 128-160 (2011).
5. M. Hansard, S. Lee, O. Choi, R. Horaud, "Time-of-Flight Cameras: Principles, Methods and Applications", Springer Brief in Computer Science, 2012.
6. A.P.P Jongenelen, D. G. Bailey, A.D. Payne, A.A. Dorrington, D.A. Carnegie, Analysis of Errors in ToF Range Imaging with Dual-Frequency Modulation, IEEE Transactions on Instrumentation and Measurement, Vo. 60, No. 5, May 2011, SSN 0018-9456.
7. Lopes, E. MacDonald, R. Wicker, "Integrating Stereolithography and Direct Print for 3D Structural Electronics Fabrication", Journal of Rapid Prototyping, Issue 2, Volume 18 (2012).
8. Soloman, Sabrie "Sensors Handbook, 2nd Edition," McGraw-Hill Publishing Co., NY – ISBN 978-0-07-160570-0
9. Soloman, Sabrie "Sensors and Control Systems in Manufacturing 2nd Edition," McGraw-Hill Publishing Co., NY – ISBN 0-07-059626-3
10. Soloman, Sabrie "Affordable Automation," McGraw-Hill Publishing Co., NY – ISBN 0-07-059633-6
11. Soloman, Sabrie "Sensors Handbook, 1st Edition," McGraw-Hill Publishing Co., NY – ISBN 0-07-059630-1
12. Soloman, Sabrie "Sensors and Control Systems in Manufacturing 1st Edition," McGraw-Hill Publishing Co., NY – ISBN 978-07-10572-4.

7 Evolving 3D Printing

INTRODUCTION

This chapter contains highly advanced topics the reader may find. However, easy to follow, and step-by-step explanations are furnished. Nonetheless, reliable comprehensive and simplistic explanations to some contents of this chapter have been provided in previous chapters. Consequently, once the reader is acquainted with the preceding materials, the flow of information herewith is distinctive and well-focused, illuminating a clear insight in the mind of the reader's intellect to grasp the most efficient manners to gain wisdom to enthusiastically apply the evolution of 3D printing technology.

Abundant prosperities and technological advancements were attained in Europe and the United States when the industrial revolution swept through their hemisphere during the 18th and the 19th centuries. Iron and textiles were wings in the revolution lifter, up by the other wing of the development of the steam engine.

Similar prosperities took place on the account of the second industrial revolution immediately before World War I. This revolution is contributed to the invention of the internal combustion engine, the telephone, light bulb, and the phonograph.

More prosperities were achieved thanks to the industrial evolution of the digital innovation began in 1980s whereby the fast advancement of technology was contributed to the migration of analog electronic, and mechanical devices to the nowadays digital explosive technology. These advancements encompassed the personal computer, the internet, and information and communications technology (ICT), robotics, artificial intelligence, nanotechnology, quantum computing, and biotechnology.

The current industrial revolution formed over the powerful digital explosion is the innovation of 3D printing through additive manufacturing technology. Presently, 3D printing technology contains an industrial revolution in itself. The 3D printing is a technology which makes it possible to build real objects from virtual 3D objects.

The 3D Printing Revolution in Progress – The Development of Physical Goods

The 3D printing is wisely called additive manufacturing, which is a process of making three-dimensional solid objects using a digital file. This is achieved by "cutting" the virtual object in 2D slices and printing the real object slice by slice. Slices are printed on top of each other and since each slice has a given thickness (e.g. 0.1 mm); the real object gains volume every time a slice is added. 3D printing is changing the way physical goods are developed, produced and distributed across all industries. Products that took weeks and months to design, prototype and manufacture now take a matter of minutes, with manufacturers enjoying a shorter time to market and lower cost base. 3D printing is portrayed as a manufacturing revolution. The question is not if but when companies need to consider 3D printing. Digital fabrication will allow individuals to design and produce tangible objects on demand wherever and whenever they need it. The revolution is not *"additive versus subtractive"* manufacturing; it is the ability to turn data into entities and entities into data!

Advanced Consolidated Nomenclatures

Triangle and polygon: A polygon is a closed geometric figure whose sides are simple line segments. Each corner of a polygon where two sides intersect is called a vertex of the polygon. For example, a triangle is a polygon with 3 sides. There are also three vertices, one at each point. This is the simplest polygon, because construct one with just 1 or 2 sides can't form a polygon. A polygon may be identified by the number of sides. A polygon with 4 sides is called a quadrilateral. A polygon of 5 sides is called a pentagon. A polygon of 8 sides is called an octagon. A polygon of 10 sides is called a decagon. Triangles are polygons, but polygons aren't necessarily triangles.

Watertight — A 3D object model is watertight when the printer can tell the inside from the outside of the model. If you filled up the model with water, no water would drain out.

subD — A subdivision (subD) surface is used in polygon modelling software to provide rounded edges to otherwise square surfaces, for example, a cube with rounded edges. The subD surface allows a specified set of edges to be subdivided into additional polygons to provide the rounded edge.

Slicing — The process of converting (slicing) a 3D model computer file into layers and then translate each layer into machine code used by the 3D printer to build the three-dimensional object.

NURB — Non-Rational Uniform Splines (NURBs) are a type of shape where the edges of the shape are splines (curves) controlled and adjusted by handles visible on-screen. Dragging a handle will change the length and position of the curved spline edge.

PLA — Polylactic acid (PLA) is a thermoplastic used in 3D printing as an alternative to ABS. Because it is made from sugars, PLA when heated can give off a sweet smell.

Gantry — Larger non-desktop 3D printers use a steel frame made of connected beams to suspend the print head and other parts of the printer as objects are created.

Polygon Modelling Software — Polygon modelling software creates models by connecting and combining polygon shapes at their edges. Polygon models rarely, if ever, use NURBs for curved edges. Instead, polygons are subdivided into new polygons to create the appearance of smooth round edges.

Triangle Geometry — the STL (stereolithography) file format describes 3D models as a set of triangles. Models also can be described with polygons or quad geometry.

Plastic and Resin – Plastic is a form of resin. Natural resins are thick, sticky organic liquids that are insoluble in water. The sap of pine and other coniferous trees, often called pitch, is a resin. The materials, which are called plastics (polystyrene, ABS, acrylics, polyethylene, etc.) are synthetic resins in the form of long-chain polymers derived from (usually) petroleum. Resins are more of a natural origin directly derived from plant oozes whereas plastics are of a synthetic polymeric nature.

Quad Geometry — Quad geometry is used to describe 3D models as four-sided polygons, for example, in polygon modelling and sculpting software. Printing a 3D model requires converting quad geometry to the triangle geometry used by the STL (stereolithography) file format.

Build Volume — the maximum length, depth, and height a 3D printer can use to build an object. It is called Print Volume

Build Platform — in non-resin 3D printing, the flat level surface used as a base to create a three-dimensional object. Resin 3D printing uses a tank filled with resin instead of a build platform.

Sculpting Software — sculpting programs can be used to create 3D models. They are polygon modelling software optimized to handle large numbers of polygons, for example, to create human shapes and round edges.

.STL — The STL (stereolithography) file format describes a 3D model as a set of triangles. STL files are converted to G-code which 3D printers use to print objects. Most 3D modelling software lets you save or export files in STL format. Otherwise, files are saved in OBJ (object) format and netfabb software is used to create the STL file from the OBJ format.

.OBJ — The OBJ file format is a common 3D model format used by modelling software. For 3D printing, OBJ files must be converted to STL file format and then G-code for printing.

G-code — a programming language used by machine tools, including 3D printers. A 3D model must be saved as an STL file format before it can be converted to G-code for printing.

Extruder — A nozzle that takes thermoplastic wire or other material and deposits the material at a precise location with the print head.

3DP = 3 Dimensional Printing

3D Printing Glossary – *Also, described in previous chapters*:

3D — having three dimensions: length (width), depth, and height.

Additive — in 3D printing, the process of building an object by adding layer upon layer.

AM = additive manufacturing = 3DP

ABS — Acrylonitrile butadiene styrene (ABS) is one type of thermoplastic used in 3D printing. It becomes soft and pliable when heated.

CAD — Computer-Aided Design (CAD) software is used to create complete models, usually buildings structures and mechanical parts. CAD models are more precise than other types of models. CAD models also use "NURBs" to define curved edges. Polygon modelling and sculpture modelling software also are used to create 3D models for printing.

Manufacturing –The making of things, usually on a large scale.

Prototype – A first or early model of some device, system or product that still needs to be perfected.

Subtractive — Creating an object by removing (subtracting) parts of a single block of material.

Thermoplastic — Plastic that becomes pliable upon heating then solid when cooled.

FDM — Fused deposition modelling (FDM) is a process to use a fused material to deposit layer upon layer and create a three-dimensional object. FDM is trademarked. An alternative un-trademarked term is Fused Filament Fabrication (FFF).

Stereolithography — an additive 3D printing process building layer upon layer by curing a resin with a laser.

Three-dimensional (3D) printer – A machine that takes instructions from a computer program on where to lay down successive layers of some raw material to create a three-dimensional object.

Three-dimensional (3D) printing: The creation of a three-dimensional object with a machine that follows instructions from a computer program. The computer instructs the printer where to lay down successive layers of some raw material, which can be plastic, metals, food or even living cells. 3-D printing is also called additive manufacturing.

Abbreviations and Synonyms

FDM	=	Fused Deposition Modelling
		Same as = **PJP** = plastic jet printing
		Same as = **FFM** = fused filament method
		Same as = **FFF** = fused filament fabrication
SLA	=	Stereolithography
CAD	=	Computer-aided design
STL file	=	Stereolithography file or Standard Triangle Language file or Standard Tessellation Language file
AMF file	=	Additive Manufacturing Format file
3MF file	=	3D Manufacturing Format file
SLS	=	Selective Laser Sintering
DLP	=	Digital Light Processing
CLIP	=	Continuous Liquid Interface Production;
SLM	=	Selective Laser Melting
EBM	=	Electron Beam Melting
ABS	=	acrylonitrile butadiene styrene
PLA	=	polylactic acid or polylactide
CT	=	computed tomography
MRI	=	magnetic resonance imaging
PDMS	=	Polydimethylsiloxane
PE	=	Polyethylene
PP	=	Polypropylene
PVC	=	Polyvinyl chloride
PS	=	Polystyrene
PET	=	Polyethylene terephthalate
PUR	=	Polyurethane
PVA	=	Polyvinyl alcohol
HDPE	=	High-density polyethylene
PETT	=	Polyethylene terephthalate
RP	=	Rapid Prototyping
IoT	=	internet of things

NON TRADITIONAL MANUFACTURING INNOVATIVE TECHNIQUE

Once again, reminding the reader; 3D printing or additive manufacturing is a process of making a three-dimensional solid object of virtually any shape from a digital model. Successive layers of material are laid down in different shapes. Traditional machining techniques rely on the removal of material by methods such as cutting or drilling whereas 3D printing layers are added successively. Thus it uses a layering technique where an object is constructed layer by layer until the complete object is manufactured. In this way, 3D printing moves away from the mass production line to a one-off customizable production. Any object can be made from a house to a bar of chocolate.

The Initial 3D Printer

The initial 3D printers were used in the 1980s where a pattern submerged in a liquid polymer would be traced by a computer. The traced pattern hardened into a layer, thanks to the laser, and that was how an object can be built out of plastic. Since then substantial progress has been made in additive manufacturing, such that material extrusion is now used. By this method, an object is built out of matter that is pushed from a mechanical head like the way an inkjet printers extrudes ink onto paper. Interestingly, the cost of acquiring 3D printers has been decreasing with the advancement of technology. Domestic usage of 3D printers has been on the rise with the average cost ranging from a few hundreds of dollars going up. However, one major drawback is that it requires expertise to print 3D objects. In fact, it requires a competent person to make both the digital file and the final printing.

Commercial Usage

Commercial usage of 3D printers has been on the increase too in sectors such as the automotive industry and aero-space engineering. Spare parts, for example, are being made in the automotive and aero-space industry leading to improved economies of scale. 3D printing is changing how the production line in industries works which made some analysts to dub the emergence of 3D printers as another Industrial revolution. 3D printing has also had a considerable usage in the field of medicine; from the field of Bionics, to Prosthetics to Digital Dentistry. Inevitably, this is positively affecting and changing every aspect of medicine. Most of the work is still in its exploratory phase but pundits predict medicine of tomorrow is going to be revolutionized by the integration of 3D printing as a tool.

Classrooms/Homes - Converting 2D to 3D printing

Printers have become so commonplace that they are found in many classrooms and homes. Today's models are relatively inexpensive and can even print photograph-quality pictures in color. For years, printers have been limited to two dimensions. The vast change, took place as the 3D printing was invented. With the regular printers, the image, text, or other graphics is printed on a white of paper. Also, the file required for printing diagrams gives the imagery in 2-dimension. But for 3D printers, a 3-dimensional representation is needed. For this purpose, the use CAD (Computer Aided Design) diagrams became eminent. These diagrams are produced in CAD tools. For example, software like AutoCAD 3D or Microsoft Visio or printer customized software could be helpful in making these diagrams. If the user has no knowledge to render these diagrams, a considerable library of thousands of such 3D designs are available on the internet for free.

Solid Imaging – The Leap of Stereolithography

A 3D printer is not anything like a regular printer, however. Instead of printing a simple text document on a single piece of paper, 3D printers can "print" a real three-dimensional object! The technology behind 3D printing has actually been around for several years. Back in the mid-1980s, a solid imaging or stereolithography was discovered. The discovery set the stage for the forward leap of this technology to take place in subsequent years.

CAD -The Life Beginning of 3D Printer

The process of 3D printing starts with a design created with computer-aided design (CAD) or animation modelling software. The design could be for a scale model of a new engine part, a simple toy action figure, or even a prosthetic limb. Once a digital blueprint is created, it can send it to a

3D printer. Unlike an inkjet printer that sprays ink onto a piece of paper, 3D printers use similar mechanical print heads to spray or squeeze raw material, including rubber, plastic, paper, metal, or other composite materials, onto a platform.

3D printing is an additive manufacturing process. Therefore, it creates a three-dimensional object in layers from the bottom up. For example, to build a simple plastic toy action figure, a 3D printer would heat thin filaments of plastic and squeeze it onto a platform layer after layer, slowly building the three-dimensional toy from the bottom up. To heat the materials they extrude, 3D printers often use lasers or similar devices. 3D printers can even create metal objects by heating metal dust with lasers.

The amount of time it takes to "print" a three-dimensional object can vary from a few hours to several days, depending upon how complex it is. 3D printing has become particularly popular in manufacturing sectors, where new prototypes can be 3D printed in a fraction of the time — and at a fraction of the cost — of making traditional wood or metal prototypes. 3D printing technology has thus streamlined and economized the manufacturing process in many industries.

The Future of 3D Printing

The future appears bright for 3D printing technology. There are particularly-exciting possibilities for the medical field, where 3D printing may be used to create everything from hearing aids to prosthetic limbs. NASA has even begun testing 3D printers that could be used on space missions to allow astronauts to print things they need on demand!

The Concept of 3D Printing

If an observer looks closely (with a microscope) at a page of text from a home printer, the letters do not only stain the paper, they are actually sitting slightly on top of the surface of the page. In theory, if printing the character over that same page a few thousand times, eventually the ink would build up enough layers on top of each other to create a solid 3D model of each letter. That idea of building a physical form out of minuscular layers is how the first 3D printers worked, Figure 7.1.

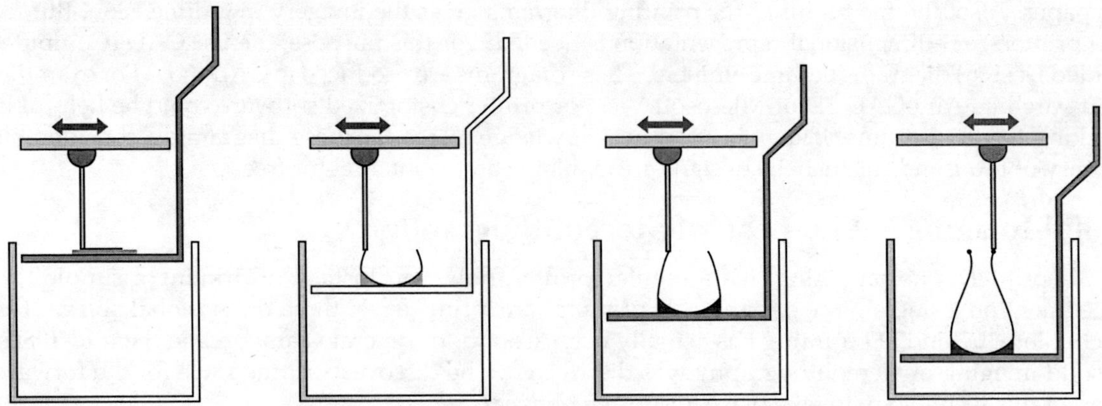

Fig.7.1: Building a Physical Form out of Minuscular Layers for 3D Model

Digital Material Convergence

The 21st century is a witness to escalating movement towards the materialization of digital information that encompasses and crosses multiple areas, including recent innovations in ubiquitous and physical computing, open source hardware, and digital manufacturing, in concert with ongoing advances in social networking and online communities. In the spirit of other forms of media merging, earlier unseen amalgamations between digital and physical production are developing that encourage unparalleled control over material and figurative landscapes. Progressively, groups holding various levels of technical expertise are capable to concurrently make and share both things ("material") and information ("immaterial") through newly conceived digitally arbitrated practices. Technologies such as computer numerically controlled (CNC) mills, laser shears and engravers, as well as '3D printers' that were once the limited domain of large industry are now roaming from the factory to the desktop. It is easy to detect how readily other forms of digital expansion such as document publishing, digital photo management and Web site formation progressed from expert to lay people. It can be predictable that a similar change will occur with digitally designed and physically fabricated media as well. Revealing of the increasing porosity between the digital and the material, an insurgency in desktop digital construction is at least moderately underway.

3D Printing - High-Impact Implementation

Three-dimensional (3D) printing technology, also called additive manufacturing (AM), has lately come into the focus because of its latent high-impact implementation in applications extending from personal tools to aerospace equipment. Other terms utilized as replacements, hyponyms, and hypernyms to additive manufacturing comprise layered fabrication, desktop manufacturing, rapid manufacturing, an heir to rapid prototyping, and on-demand manufacturing, which echoes on-demand printing. Even though 3D printing technology has only recently arisen as a impassioned topic, its recent broad-minded history indicates positive impression to near and far future. Since then, new and extensive applications and markets for 3D printers have performed rapidly, especially with the expiration of a number of core 3D printing patents owned by large number of corporations worldwide.

Developers can easily build or modify 3D printers by themselves or take advantage of the rapidly rising availability of low-cost 3D printers. The recent obtainability of highly able 3D design software and 3D design websites permits the distribution of user-created free 3D digital design files or models, leading to more contact to 3D printers and additional propagation of 3D printing technology.

As 3D printing becomes other than just a simple production procedure, it has come to sustain a merging of technologies and applications such as sports equipment, food packaging, and costume jewelry, as well as products in the high tech fields of aerospace, medicine, architecture, education, automotive industry, military support, and others. At the 2016 New York Fashion Week, two unique 3D printed costumes were unveiled. These masterworks were fashioned through partnership between fashion designers and the 3D printing company. The intricate designs (e.g., mixing a diversity of dovetailing weaves, bio-mimicking natural animal feels, and cutting-edge material, e.g., nano-enhanced elastomeric 3D printing material, delivering the dresses durability and flexibility.

Regenerative Medicine

The extent of regenerative medicine has also accomplished inspiring applications within the 3D printing field. Medicine has positively utilized 3D printing technology to construct living organs and tissue, together with muscle structures, and bone and ear tissue. These bioprinted body parts are able of creating functional substitute tissue.

National Aeronautic Space Agency NASA

NASA has also been applying 3D printing techniques and 3D printers to advance materials that permit astronauts to restore or substitute crucial parts and build structures in space. NASA recently cooperated with many scientists and research engineers to manufacture a reproduction of a moon rock utilizing raw lunar regolith simulant and 3D laser printing technology. The assembly of modular structure materials utilizing giant 3D printers for use in the housing industry has extended significant interest, particularly for poorer countries, during natural disasters, or unexpected emergencies. Some 3D companies have prospered in building houses or bridges with cement, sand, or concrete materials.

3D Printing Unique Qualities

3D printing possesses unique qualities described as follows:

- Objects are shaped by adding or depositing layers of material, not subtracting or cutting out pieces from a block of material.
- Objects are shaped by adding layers, thus, the computer file with particulars about the model must be transformed into slices allowing the printer to generate layer by layer.
- Printing a three-dimensional object may take hours or days to finish, contingent upon the difficulty and size of the object.
- The cost is founded on materials utilized, amongst other factors. In difference, subtractive technology is cut out pieces to generate the necessary object, which is paid for the total original piece.

Digital Inputs Creates Vertical Axis Manufacturing - 3D Models

3D Printing is the process of making a physical object from a three-dimensional digital model, characteristically by laying down many thin layers of a material in series. "3D printing" or "Additive Manufacturing" receipts digital input in the form of Computer Aided Design (CAD) model and forms solid, three dimensional parts through an additive, layer by layer process. 3D printing refers to processes used to generate a three-dimensional object in which layers of material are shaped under computer control to produce an object. The technology of printing physical 3D objects from digital data using vertical axis manufacturing procedure is called "additive manufacturing." The quantity of material, the rotation of the platform and the design of the object is controlled in a computer controlled environment.

Object Shapes and Geometries Standards

Objects can be of virtually any shape or geometry and are shaped utilizing digital model information from a 3D model or another digital data source such as an Additive Manufacturing File (AMF) file. Additive Manufacturing File Format (AMF) is an open standard for defining objects for additive manufacturing processes such as 3D printing. The authorized "ISO/ASTM 52915:2013" standard is an XML-based setup designed to permit any computer-aided design software to define the form and structure of any 3D object to be manufactured on any 3D printer. Contrasting its predecessor STL format, AMF has built-in support for color, materials, lattices, and arrangements. This technology permits for intricate shape, size and dimensions which are not easily or economically obtainable utilizing customary manufacturing tooling techniques. 3D printing services permits low cost production of highly precise, low volume products or mass markets quicker. Eliminate unwanted limitations to the designs, produce prototypes as one unit or pre-built unit assembly to help in mass fabrication.

Printing Materials & Forms

At a high level, a 3D printer takes material; usually polymers wrapped on a reel, heats the material, and then fuses the material layer by layer to build a three-dimensional object. The material is extruded, or forced out and deposited. Accuracy is vital for 3D printers. The print head, pulleys, and extruder hardware must be precisely aligned, as well as software dimensions translated to the precise real world dimensions used by the printer hardware.

There are at minimum two kinds of additive 3D printing, resin and non-resin. Printing objects with resin utilizes laser technology to harden layer by layer to generate objects. It frequently is very costly and utilizes light and chemicals to cure the object. Though, the consequences are more precise than non-resin printing. Typically, 3D printing is referred to a non-resin printing. Instead of resin, lasers, and chemicals; non-resin printing uses polymer, food, ceramics, or other material and adds layer upon layer to generate an object.

Reduce Time of Prototype Creation

The invention of 3D printing concentrates on two issues: decrease time to get the first form of a product and unshackle many constraints that are not possible with traditional production methods. For instance, with 3D printing, it is likely to print intricate geometric figures and meshing parts that require no assembly. It is also likely to yield single objects, in small amounts, at low cost and rapid delivery. This technology also aids in the decrease of production-related material waste. 3D Printing can yield diverse objects deprived of generating explicit tooling or even utilizing numerous tools. This is how 3D Printing aids increasing flexibility in the manufacture flow and aids dropping industrial expenses.

On-demand and Customization

Subsequently, there's no necessity to build devoted production line, it benefits to save time: 3D Printing allows to transforming faster and mechanize faster. Meanwhile, 3D Printing makes a copy of the 3D files one by one, so economies of scale can't be comprehended when the same file is produced several times: this is clearly different from series industrial methods, which target for producing millions of units of the identical objects. On the contrary, 3D Printing is the ideal technique for on-demand and customization requirements. Lastly, 3D Printing provides to everyone the power to manufacture objects only when they are wanted or anticipated. It is an effective procedure for the inventive designers who wish to examine or confirm a novel concept. Therefore, 3D Printing allows innovative entrepreneurs to yield and market their creations by avoiding classical edition and distribution networks. Also, 3D Printing is valuable to replicate objects that are no longer on the market or outdated, for instance in order to overhaul an old machine for which spare parts are obsolete.

SHORTER LEAD TIME AND DESIGN FREEDOM

3D printing brings a revolutionary approach to manufacturing through two key advantages:

Lead Time

The way industrialists have come up with prototyping for the past three decades might even be reflected as a luxury when related to prototyping prior to 3D printing. Today, 3D printing an initial phase design and re-printing it overnight are possible and inexpensive thanks to rapid prototyping or 3D printing platforms like "PolyJet" and "Stereolithography." 3D printing a concluding product in just one to two days is possible with several 3D printing technologies, such as Laser

Sintering, Fused Deposition Modeling and Direct Metal Laser Sintering. Nevertheless, previous to these quick-turn prototyping and production manufacturing processes, conveying an idea into physicality was a complex and costly procedure and there frequently was no simple substitute or time to re-prototype regularly or make multiple design changes.

Design Freedom

Conventionally, designers and engineers have depended heavily on the engineering process to dictate the end design. Complicated conventional manufacturing procedures like CNC machining have intrinsic strict restrictions on assembly rules, manufacturability and overall possibility. Avoiding the design practices for these conventional manufacturing processes openly consequences in improved cost and labor. Nonetheless, observing the design rules of the past unavoidably consequences in repressed innovation development. Additive manufacturing, or 3D printing, has unlocked new potentials previously inconceivable to designers and engineers as it does not depend on the same design and manufacturing restraints as conservative manufacturing. As a result of 3D printing, free-flowing, organic and intricate designs are flawlessly performed while upholding strength in ways impossible via any other manufacturing procedure.

An excellent case example of the design freedom of 3D printing was conceived by NASA. NASA's Marshall Space Flight Center. They were able to transform a part that previously contained in excess of 150 parts and, through 3D printing, consolidated the whole design into one continuous unit! Design freedom in 3D printing is considered "zero-cost" due to the layering process. Design features are seamlessly integrated within each cross-section as the part builds; eliminating the need for tooling, labor intensive assembly, and reducing time and part count to result in significant cost savings.

3D PRINTING POSSIBILITIES

The possibilities of 3D printers are numerous, as they are becoming a commonplace tool in fields like engineering, product design, manufacturing and architecture. Typical examples of their uses are illustrated as follows:

Custom models

Create customized products that are completely tailored to the needs for size, and shape. Making parts that may not be possible with any other type of technology.

Fast prototyping

3D printing makes it possible to quickly create a model or prototype, helping engineers, designers and companies to get iterative feedback on their designs in a short time.

Complex geometry

Models that are difficult to envisage can be created in a 3D printer with ease. These models are vital at educating others to the geometric complexity in a favorable and interesting manner.

3D Printing - Transforming Technology

Transforming technology ass a feature of the 3D printing revolution, summarized as follows:

1. 3D printing technology is increasingly allowing localized production of industrial and commercial goods.

2. It allows for intricately customized prints from specialized organs to custom houses.
3. It allows for rapid prototyping of nanotechnology, thus opening opportunities for significant technological advances.
4. It allows for computer-based goal-directed design allowing for unlimited geometric and material complexity, Figure 7.2.

Categories of 3D printing Applications

1. Manufacturing applications include rapid prototyping, mass customization of products, and rapid manufacturing.
2. Industrial applications include manufacture of industrial parts from aerospace to automotive.
3. Tissue fabrication, implants, prosthetic limb manufacture, and tablet/capsule manufacture mostly make up medical applications.
4. Socio-cultural applications include artwork, fashion, preservation of cultural heritage etc.

Steps to Create 3D Printing Model

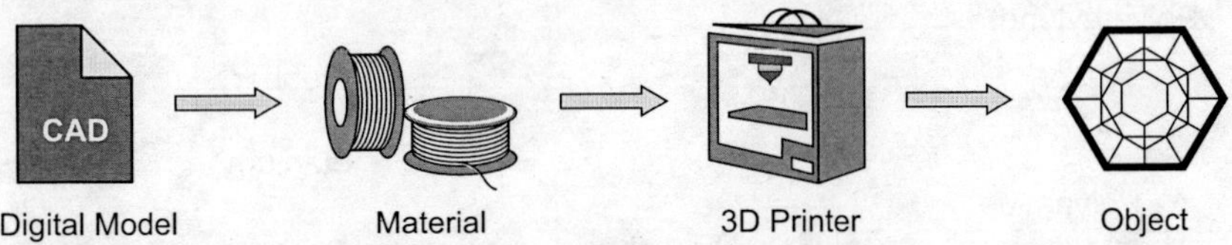

Fig.7.2: Innovative 3D Printing Processes

Printing a three-dimensional model is more complex than opening up Microsoft Word and printing to 2D printer. At a high level, to generte a 3D model follow the steps stated as follows:

- Find a model or create a model with "Blendr," "SketchUp," or "OpenSCAD" software.
- Prepare the model for printing, for example, ensuring the model is watertight with no gaps.
- Create an "STL file" which converts the model into code the 3D printer understands, using "Netfabb," "Meshlab," or similar software.
- Print the 3D model, adjusting the printer as needed. 3D Printing uses software that slices the 3D model into layers, (0.1mm thick or less in most cases. Each layer is then traced onto the build plate by the printer, once the pattern is completed, the build plate is lowered and the next layer is added on top of the previous one.

All of these steps or more may be essential contingent on the availability of a 3D printer or use a printing service, Figure 7.3.

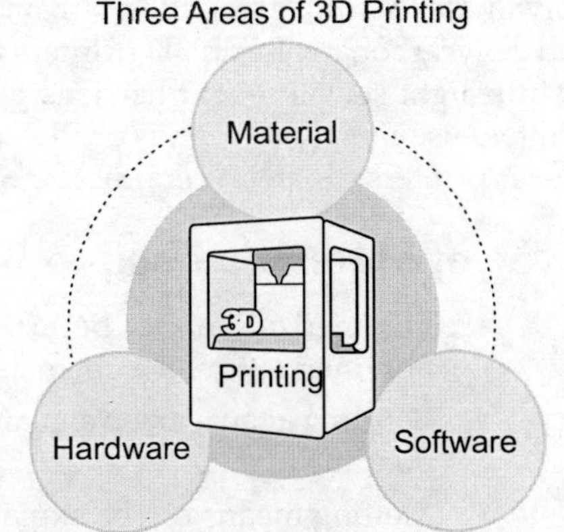

Fig.7.3: The Three Fundamental Areas in 3D Printing

With 3D printing, an object can be made by using one of the following methods:

- Extruding,
- Sintering
- Curing.

Extruding includes using continuous filament of a thermoplastic material as the feedstock. Sintering is an additive manufacturing technique that utilizes a laser as a power source. Curing utilizes a liquid resin as the feed material and the object is constructed layer-by-layer and cured; this procedure progresses slowly.

Materials used in 3D Printing

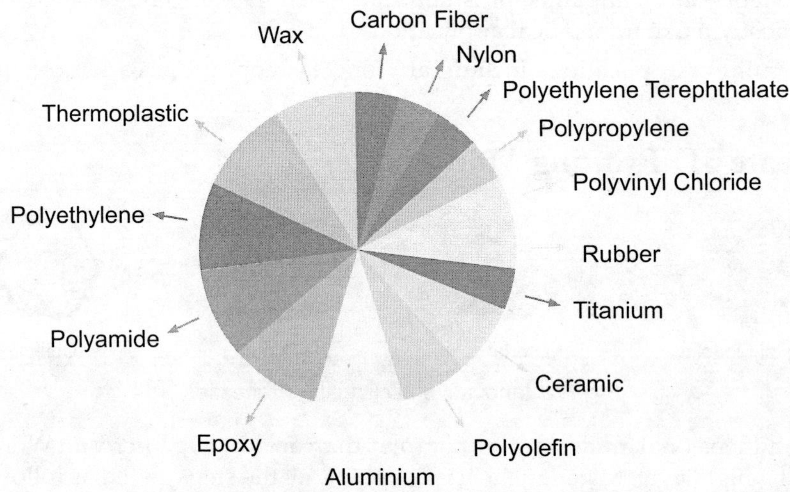

Fig.7.4: Different Materials Used in 3D Printing

Many different materials can be used for 3D printing, such as ABS/PLA plastic, Figure 7.4, polyamide (nylon), glass filled polyamide, silver, titanium, steel, wax, photopolymers (UV resins) and polycarbonate. Resin 3D printers use custom formulated plastic resin poured into a tank. The resin is light sensitive and hardens when laser light heats a spot within the resin. Non-resin 3D printers use mostly ABS and/or PLA plastic which becomes pliable when heated then solid. And newer printers can print ceramics or with food materials

Type of Materials Used to Build Parts:

- Polymerization can be explained that parts are constructed via a UV-light actuated polymerization of a chemically sensitive liquid material.
- Bonding agent can be explained that powder material is pasted together through a liquid bonding agent.
- Melting means can be explained that material is heated and melted together.

Contingent upon the part properties needed, additive manufacturing technologies are either used to directly yield the parts or through subsidiary procedures in combination with customary manufacturing techniques. Contingent on the needs, these technologies are either utilized alone or can be added to other manufacturing methods:

Direct Processes

One stage process: The part is directly created with the additive manufacturing machine. There is post- applied in order to advance tolerances or surface finish.

Multi-stage process: A "green" part is created with the additive manufacturing technology which is then going into additional procedures. For instance, binder jetting is utilized to yield green metal parts, where metal powder is bonded together, which are then sent into a sinter procedure and are infiltrated subsequently.

Indirect Processes

Additive manufacturing can be combined with classical fabrication. For instance, master patterns are made with 3D printing technologies, which are then utilized for investment casting of finished parts.

Software Process Chain

3D printing starts with a digital file derived from computer aided design (CAD) software. Once a design is completed, it must then be exported as a standard tessellation language (STL) file, which is the file translated into triangulated surfaces and vertices. The STL file then has to be sliced into hundreds – sometimes thousands – of 2-D layers. A 3D printer then reads the 2-D layers as building blocks, which it layers one atop the other, thus forming a three dimensional object. All design files, regardless of the 3D printing technology, are sliced into layers before printing. Layer thickness – the size of each discrete layer of the sliced design – is considered partially by the technology, partially by the material, and partially by the anticipated resolution and the project timeline; thicker layers associates to faster builds, thinner layers associate to finer resolution, less visible layer lines and therefore less rigorous post-processing work. After a part is sliced, it is oriented for construct.

Part Orientation

Orientation denotes to how and which direction a part is located on the 3D printing construct platform. For instance, a part may be oriented at an angle, lying flat, or standing vertical. Comparable to CNC machining, orientation factors into the outcome of surfaces and specifics on a 3D printed part. As 3D printing constructs one 2-D layer at a time, the discrete lines appear as corrugated surfaces on parts. Downward facing surfaces typically disclose more layer lines. Specific construct orientations are better for curved or square structures, while delicate structures require special attention.

Technologies with higher occurrences of warp, or material distortion, must account for big flat surfaces during construct orientation. It is vital to consider these influences because how a part is oriented governs where supports are supplemented, or required within the construct. Supports are a essential factor for 3D printing, and can affect material finish and accuracy of a 3D printed part.

Part Supports

Most 3D printing procedures need support structures to act as "scaffolding" for topographies that can't be built above open air, such as overhangs, undercuts, holes, cavities, etc. Where supports are essential mainly be contingent on the material, build procedure, 3D print technology, and build resolution, layer thickness, among other factors. Support structures are typically made utilizing the same or comparable material as the final build and are removed after the model cures.

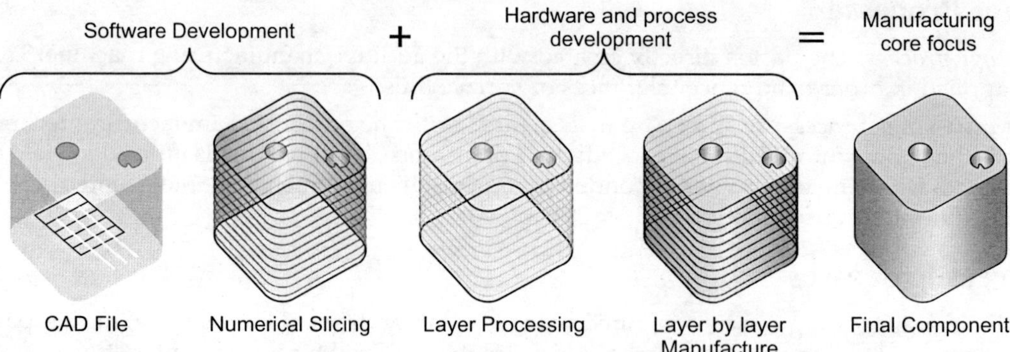

Fig.7.5: Sequential 3D Printing Processes

No matter which method a 3-D printer utilizes, the general printing procedure is usually the same, Figure 7.5. There are following eight steps in the common AM process, which are stated as follows:

Step 1: CAD

Yield a 3-D model utilizing computer-aided design (CAD) software. The software may deliver some clue as to the structural integrity that could be predictable in the finished product, too, utilizing scientific data about selected materials to generate virtual simulations of how the object will perform under specific circumstances.

Step 2: Conversion to STL

Transform the CAD drawing to the STL format. STL, which is an abbreviation for standard tessellation language, is a file format established for 3D Systems by its stereolithography apparatus (SLA) machines. Most 3-D printers can use STL files in as well as to some exclusive file types such as ZPR, and ObjDF.

Step 3: Transfer to AM Machine and STL File Manipulation

An operator copies the STL file to the computer that controls the 3-D printer. There, the user can specify the size and orientation for printing. This is comparable to the method one would set up a 2-D printout to print 2-sided or in landscape against portrait orientation.

Step 4: Machine Setup

Each machine has its individual necessities for how to formulate for a new print job. This comprises replenishing the polymers, binders and other consumables the printer will utilize. Also, it shelters adding a tray to function as a basis or accumulating the material to construct temporary water-soluble supports.

Step 5: Build

Let the machine does its job; the construct procedure is typically automatic. Each layer is typically about 0.1 mm thick; however it can be much thinner or thicker. Depending on the object's size, the machine and the materials utilized, this procedure could take hours or even days to finish. Be sure to check on the machine every so often to ensure there are no faults.

Step 6: Removal

Eliminate the printed object, or multiple objects in some instances, from the machine. Be sure to take any safety safeguards to avoid harm such as wearing shielding gloves from hot surfaces or toxic chemicals.

Step 7: Processing

Many 3-D printers will necessitate some amount of post-processing for the printed object. This could contain brushing off any residual powder or washing the printed object to remove water-soluble supports. The new print may be delicate during this step since some materials necessitate time to cure, so carefulness must be implemented to ensure that it does not breakdown or crumbled.

Step 8: Application

Make use of the newly printed objects or objects.

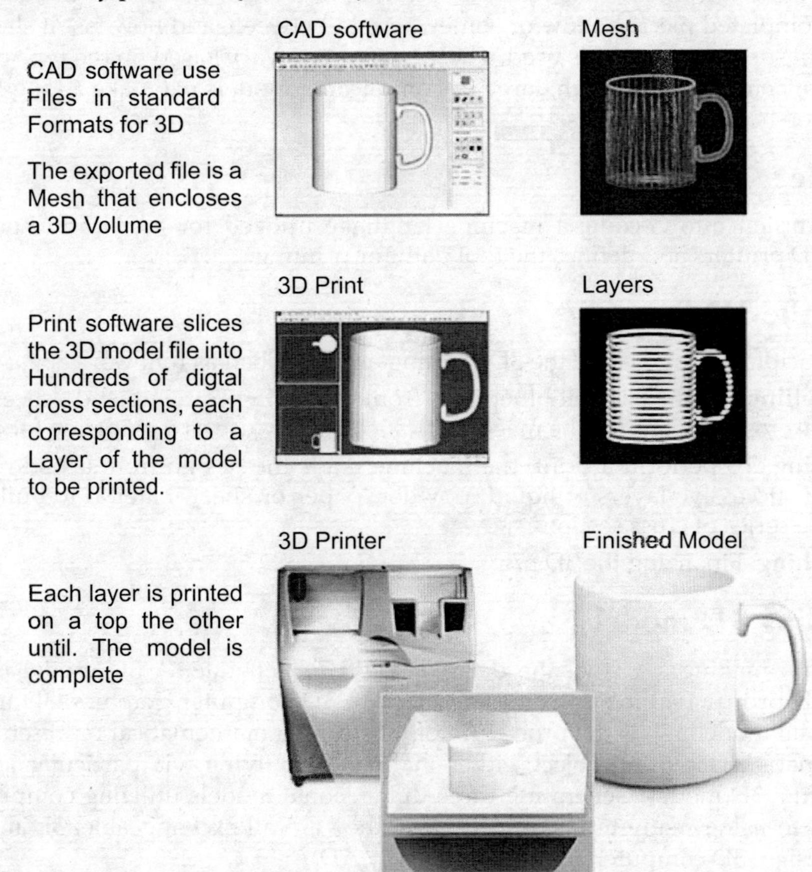

Fig.7.6: 3D Printing integration of Digital and Mechanical Operation

THE DIGITAL MODEL

The digital model is characteristically generated in either CAD (computer aided design) which is generally utilized as mechanical parts and precision are required or a polygon modeler, which are normally utilized for simulation and game assets, Figure 7.6.

Examples of CAD programs are AutoCAD, SolidWorks and 123D Design. Some popular polygon modelers are" Maya," "3ds Max" and "Blender."

Slicing the "STL," File onto Layers

The model can even be generated by 3D scanning a actual object, but irrespective of how it is assimilated; the model is ultimately exported as an "STL," standard tessellation language, file, which defines the surface geometry of the model in triangles, as preliminary described in chapter 5. The STL should always be treated through a mesh repair program such as "netfabb" meanwhile during building it is very easy to generate a mesh that may not print correctly due to defects in the mesh. As soon as it is repaired, the STL file has to be 'sliced' into the layers that will be drawn by the printer, one at a time, layer-by-layer on top of each other until the object is concluded.

The STL is conveyed into a slicing program where stipulating the thickness of each layer must take place. The thickness is the resolution in microns. Also, stipulating the use of material as either solid completed piece, hollow or somewhere in-between, and how fast it should print as well as structural support material is used. These parameters vary based on the type of the printer and the software being used. Though, large or complicated models may take a long time to slice, predominantly on home use printers.

The G-Code

The slices are turned into G-code; a machine language utilized for fabrrication on both CNC machines and 3D printers and defines the tool path for printing.

Processing the 3D Printing

The general principles of processing the 3D printing are described as follows:

1. **Modelling**: It takes virtual blueprints from modelling software and "slices" them into digital cross-sections for the machine to successively use as a guideline for printing.
2. **Printing**: To perform a print, the machine reads the design from a ".stl file" and lays down successive layers of liquid, powder, paper or sheet material to build the model from a series of cross section.
3. **Finishing**: Finalizing the 3D print.

Understanding 3D model

It all begins with making a virtual the desired object to generate. A 3D model is the virtual design that a 3D printer requires to print an object. In 3D computer graphics, 3D modelling, or three-dimensional modelling, is the procedure of emerging a mathematical representation of any three-dimensional surface of an object, either inanimate or living, via particular software. The product is called a 3D model. Generating three-dimensional models utilizing computer software can be referred to using many terms, nonetheless, they are all exchangeable: 3D modelling, 3D drawing, 3D design, 3D computer aided design (3D CAD).

Scanning a Model

One may own a 3D file by downloading it, 3D scanning an object, by a plain digital camera and photogrammetry software, employing a 3D designer, or designing it with 3D modelling software. The 3D printed models, which are generated by CAD end in reduced errors, and can be modified before printing, permitting confirmation in the design of the object before it is printed. Once a 3D model is generated, the next step is to prepare the model to make it 3D printable.

Computer-Aided Design (CAD) Software - The Initial Processing Step

Creating a digital model is the first step in the additive manufacturing procedure. The most collective method for manufacturing a digital model is computer aided design (CAD). There are numerous numbers of free and professional CAD programs that are well-matched with additive manufacture. Commercial software such as CAD software AutoCAD and Pro Engineer, software packages Rhino, Maya, and SolidWorks are all satisfactory for designing 3D models.

A novice in 3D design may try some of 3D modelling software which can be downloaded for free; for example "SketchUp," "3Dtin," "Blender," "OpenSCAD" and "Tinkercad." When designing a part to be 3D printed the following issues must be considered stated as follows:

1. The part must be a solid, that is, not just a surface; it must have a real volume.
2. Creating very small, or delicate features may not be printed properly, this depends greatly on the type of 3D printer.
3. Parts with overhanging features must be structurally supported to print properly. This should be taken into account, as the structural support of the model will be removed and the model will be cleaned. This may not be an issue unless the part is delicate or fragile and that it may break.
4. Be sure to calibrate the 3D printer before using it, as it is essential to ensure that the part sticks properly to the build plate. If it does not, at some point the part may come loose and ruin the entire print.
5. Some considerations should be given to the orientation of the part, since some printers are more precise on the X and Y axes, then the X and Z axis.

3D Scanners Process

3D scanning is a procedure of collecting digital data on the figure and appearance of a real object, generating a digital model based on it. All the digital manufacture, CNC machine tools and CAD software programs are means to turn bits into atoms. Is there any reverse concept of turning atoms into bits? It's undoubtedly, difficult to draw 3D objects with little knowledge on a screen unless proficiency is acquired first in "AutoCAD" or "Solidworks." It is advisable to start with a similar model that already exists, and then modifies it. This process is called 'reality capture'. An object can be scanned creating a "point cloud" of dots, which define the topography of its surface. Then software turns these cloud points into a mesh of polygons. They are similar to "wireframes" that make up the characters in computer-animated films, which can be manipulated, and modified on screen. A commercial 3D scanner may be purchased, which can perform the animation with Lasers to trace over an object, and cameras to capture the positions of points on object surface. However, there are inexpensive ways to achieve the same result. Autodesk offers a free online service called "123D Catch," which allows uploading regular photographs of an object, taken from all angles, and then cloud-based software will turn it into a 3D object ready to be modified and printed on a 3D printer.

3D Printing without Human Interface

Some 3D printers have a built-in 3D scanner and computer, making it an inspired powerhouse. The scanner may be utilized to scan any item and then slice and adapt the subsequent image with the built-in computer. This kind of equipped 3D printer can print a 3D copy without human interface.

3D Printing Pen

3D Printing pens are recent developments in modelling as a result of advancements in 3D designing technology. The most favorable 3D drawing pen enables users to make drawings in three dimensions which give clear and concise designs, and representations of initial abstract ideas possessed only by the user.

The pen has a heated plastic component to form a liquid, which is used to make drawings when expelled from the nozzle of the pen. The printing has a low power requirement. The pen size and weight enable working with ease by a single hand. These pens may offer endless possibilities to the passionate hobbyists or entrepreneurs of making their 3D modelling and drawing ideas real. This is a new technology, and is still in its infancy. There are few individuals and organizations yet understand it fully.

Parametric & Explicit Modelling

There are two distinct methods for defining 3D geometry.

a. The parametric approach

b. The explicit approach

The parametric method is characteristically utilized by engineers and architects because the file encompasses more precise dimensions, relationships, and can include a design history. The clear method is characteristically used by artists and manufacturing designers because it can be more flexible. Unambiguous modelling permits modifications to be made on the fly and small modifications are not essentially related to any other defined point of the geometry.

3D Printing Geometry Restrictions

The size and geometric limitations must be measured when shaping whether a model is appropriate for 3D printing. These type of attentions apply to a wide variety of industries whereby use CAD in design and are aiming to acquire a design 3D printed from basic component modelling to intricate architectural models.

Physical Size

The table below identifies the build volume of a variety of different printing technologies. It is vital to note that as build volume upsurges more industrial printing technologies are characteristically used, and thereby increasing the whole price of the print.

Build volumes for a range of 3D printing technologies, Table 7.1:

Table 7.1: 3D Printing Technology at Corresponding Construct-Volume

Technology	Maximum build volume
Desktop FDM	220 mm x 223 mm x 205 mm
SLA	145 mm x 145 mm x 175 mm
SLA	145 mm x 145 mm x 175 mm
Industrial FDM	914 mm x 610 mm x 914 mm

Element Thickness

Often engineering designers or game-designers may produce fundamentals within a design that have an infinitesimal thickness, hair, capes, sails etc.. These topographies are difficult to 3D print unless they meet a smallest printable feature size. These necessities vary by printing the preferred technology, the smallest possible feature size being 0.3mm printed with SLA technology. All basics within a model must fulfill with these feature constraints to successfully print.

Water-Tightness

Any models that are envisioned to be utilized for 3D printing should be totally manifold - watertight. Each edge on the model should have precisely 2 polygons joined to it and contain no holes. Models that are not manifold - watertight might become misconstrued by the software that produces the instructions for the 3D printer's slicer software. This may end in the object having unreliable layers, holes or cause the model to be unprintable. Water-tightness is often connected to the thickness of essentials in a model with very thin structures typically causing a problem. Non-manifold issues are frequently not observable at the modelling phase. The simplest method of checking whether a model is water-tight is to upload the model to an analyzer program, such as "Netfabb" or "Meshmixer." These programs sense model landscapes that will cause issues at the 3D printing phase and provide repair choices, often without influencing the general aesthetics of the model.

Curved Surfaces

Most CAD modelling programs such as "Solidworks" and "Fusion360" use non-uniform normal basis spline, "NURBS" to show surfaces of a model. When transferring the file to ".STL" for 3D printing it is significant that an acceptable number of polygons are utilized to characterize a surface to guarantee it will print with a smooth appearance. If the number of polygons is representative the surface sufficiently, the edges connecting discrete polygons, fundamentally appearing as flat planes, will be observable in the final 3D print. When printing in high-detail 3D printing technologies such as "SLA," "DLP" and "PolyJet" these polygons are frequently observable. This effect is more prominent with large scaled models, which larger than 300 mm^3, where the polygon consequence becomes more visible on curved surfaces.

The mainstream of modelling programs export solid models with a satisfactory number of polygons ensuing in the 3D print that is satisfactory for most applications. If a sophisticated polygon count is essential to progress 3D print surface finish, the analyzer programs, such as "Netfabb" or "Meshmixer" allow this parameter to be diverse.

Printing Hollow Part

When designing for the 3D printing, the structural reliability of the part must be considered. If overlooked it may breakdown while the part is being constructed. Therefore, build internal supports that will be utilized to build upon, and adjust the internal dimensions so that there would be a more helpful structure on the interior permitting it to continue hollow and retain the external features of a hollow part to a more buildable position.

There is always an essential guide line with 3D printing stating: If success is accomplished on the first try, it was either a simple structure or by a likelihood pure chance. The most significant part of any print is the first few layers. If the construct is on a good solid foundation for the 3D print, then it will at least forecast its success. If error happened in the first few layers the consequence will not be of any better. Operator must abandon the print before further waste of material and time.

File Types Used in 3D Printing

When designing 3D models, and 3D printing, it is expected to encounter a variety of file types stated as follows:

A. Some file types are related to the design software,
B. some are associated with 3D scanners, and
C. other file types are proprietary to certain brands of printers.

With many diverse files, keeping track of what files are utilized and for what procedures can be puzzling. While there are many diverse file types that can be used in 3D printing and modelling, the following is a list of some common file extensions shown in 3D printing:

- **OBJ**: OBJ is an open file format that characterizes 3D geometry. It is a shared file utilized in 3D printing since several design software transfers to OBJ and most printers receive OBJ as a printable file.
- **STL**: STL files are a typical file type that interfaces between Computer Aided Design (CAD) software, and 3D printers. Majority of CAD software transfers to an STL file and majority 3D printers receive STL files. ".STL" format was generated support a layering procedure for one extruder. As printers advanced, other formats, both open-source and exclusive, were developed.
- **VRML**: VRML, or WRL files are usually utilized as a 3D model has color and anticipated to assignment that specific color to the print.
- **X3G** : X3G is the file kind that interfaces with the "MakerBot" 3D printer. "MakerWare" and "ReplicatorG" will yield an "X3G" file.
- **PLY**: PLY files are generally scanner- created models and will have to be imported into some form of design software before transfer to a 3D printer to print.
- **FBX**: FBX is a file exclusive format to Autodesk. It is utilized to competently interchange data between "Autodesk" programs including "3ds Max," "Autodesk" and "Maya."
- **3MF**: 3D manufacturing format – In 2015, the 3D Manufacturing Format ".3MF" was introduced. Established and supported by the 3MF Consortium, the purpose is to standardize 3D print files.

STL file

STL, Stereolithography is a file format built-in to the stereolithography CAD software generated by 3D Systems. STL has some after-the-fact backronyms such as "Standard Triangle Language" and "Standard Tessellation Language." This file format is sustained by many other software packages; it is extensively utilized for rapid prototyping, 3D printing and computer-aided manufacturing CAM.

STL files define merely the surface geometry of a three-dimensional entity deprived of any demonstration of color, texture or other collective CAD model characteristics. The STL format stipulates both ASCII and binary demonstrations, Figure 7.7. Binary files are more common, since they are more compacted. STL file encompasses data labelling the layout of a three-dimensional object. These files are typically created by a computer-aided design (CAD) program. ".STL" is the file extension of the STL file format. The STL file format is the most frequently utilized file format for 3D printing. When utilized in combination with a 3D slicer, it permits a computer to connect with 3D printer hardware.

Graphic representation of information is an .STL file. The object, shown on the left, was created on a CAD program and was subsequently saved as an .STL file. The graphical information displayed in the .STL file is shown on the right of the same subject. Notice that the surface of the is triangulated. The spatial coordinates of the triangle vertices are stored in the .STL file, and that information is transmitted to the printer for fabrication.

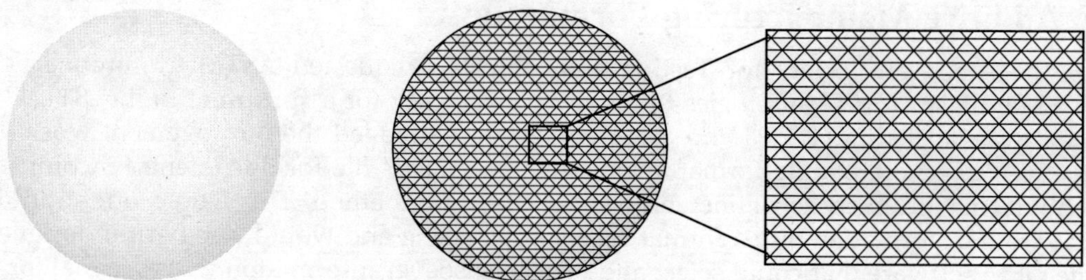

Fig.7.7: .STL Information

In distinct terms, the STL file format utilizes a sequence of linked triangles to regenerate the surface geometry of a solid model. For a rudimentary model, its surfaces can be signified utilizing a few triangles. For greater resolution models, more triangles are necessary to replicate the surface of the model. The extra triangles that make up a model, is indicating that the larger the file size and the more comprehensive the object.

The STL file format can describe intricate shapes, mathematically, such as polyhedrons with any polygonal facets. In run-through, it is typically utilized to define the layout of triangles in a virtual space. Each triangle facet is defined by:

1. a *perpendicular direction* and
2. *three points* which represent the corners of the triangle.

An STL file delivers a comprehensive listing of the x, y and z coordinates of these corners and perpendiculars. The STL file format made it likely to transmit three-dimensional CAD models to 3D Systems' very first commercial 3D printers, the Stereolithography Apparatus. Since this original release, the technical provisions of the STL file format have continued virtually unchanged.

STL file is not 3D printable

Merely a 3D design that's explicitly made for 3D printing is 3D printable. The STL file is only the vessel for the data, not an assurance that it is printable. 3D models appropriate for 3D printing reqiure to have a minimum wall thickness and a "watertight" surface geometry to be 3D printable. Even if it is seen on a computer screen, it's beyond a miracle to print something with a wall thickness of zero. There is likewise the attention of overhanging elements on the model. When downloading an STL file that was not initially generated by the user, it is important to confirm that it is indeed 3D printable. This may save time, and filament material.

While .STL controls the primary routine, the .AMF, Additivc Manufacturing Format, and ".3MF," "3D Manufacturing Format" file formats are creeping to take esteemed position.

3MF file

3D Manufacturing Format or 3MF is a file format established and issued by the 3MF Consortium. 3MF is an XML-based data format designed for utilizing additive manufacturing, as well as information about materials, colors, and other information that cannot be characterized in the STL format. As of today, CAD software related companies such as "Autodesk," "Dassault Systems" and "Netfabb" are portion of the 3MF Consortium. Other firms in the 3MF Consortium are "Microsoft," for Operating system support, "SLM" and "HP," whilst "Shapeways" are also encompassed to give insight from a 3D Printing background. Other key players in the 3D printing and additive manufacturing business, such as "Materialize," "3D Systems," "Siemens PLM Software" and "Stratasys" has lately joined the consortium.

AMF, Additive Manufacturing Format file

In 2016, the American Society for Testing and Materials requested ASTM Committee F42 on Additive Manufacturing Technologies to define the necessity for a substitute to the .STL file. At the time it was labeled "STL 2.0" and a design subcommittee led the improvement work. Their objective was to generate a file format analogous to a ".PDF" file for documents, encompassing as much information likely to define an object in the same method. This would offer additional data prior to transformation into a format for manufacturing and would also permit designers to generate OEM software that could select and take whatsoever information was essential for their 3D printing hardware to function. The subcommittee established the ".AMF" format for technology objectivity, effortlessness, scalability, presentation, and both future and backward compatibility. Its first reiteration was established in 2017 and consequently accepted in 2018 in conjunction with the International Standards Organization (ISO).

AMF is an XML-based open format that delivers comprehensive information through a chain of command of five main elements: object, material, texture, constellation, and metadata. This delivers the foundation for information on a model's shape, composition, color, materials, and geometry. It also presents the impression of print constellations, permitting information on several objects to be situated and organized together. This would end in enlarged packing effectiveness or for utilization with big arrays of equal objects that would be 3D printed on a single construct tray. Among .AMF's added features: XML-based for easy reading, writing, and treating with as much information as possible to define an object, its materials, and other manufacturing structures. Whereas beginning off slow, acclimation to .AMF has matured since its starter. CAD vendors initially took a "wait and see" attitude to the matter, favoring to integrate .AMF yielding to customer request.

Growth has been continuing since and ultimately increased a position with larger presentation such as SOLIDWORKS and Autodesk's suite of products. Stratasys also gestured its cooperation, predicting the potential benefits in .AMF and remain to support it today. 3D printing service benefactors such as Shapeways and Materialise added .AMF to their listing of recognized formats while also backing resources to its growth. Due to the myriad of mutually supporting technology necessities these companies require, .AMF was ready to work with 3D printers on the market today. Furthermore, adaptation and generation tools have been established to aid users migrate older .STL files to .AMF, Table 7.2.

Table 7.2: Comparison between STL and MF File Format

Comparison of STL & AMF – 10cm Sphere - Solid Edge		
	STL Format - Binary	**AMF Format - Curved Triangles**
Precision	10 micron	10 micron
Number of Mesh Triangles	49,500	320
File Size	2400k	10k

Printing

Prior to printing a 3D model from an STL file, it is a requirement first to examine for errors. Most CAD presentations yield errors in output STL files: holes, faces "normals," self-intersections, noise shells or manifold errors. A step in the STL creation known as "repair" fixes such glitches in the initial model. Mostly STLs that have been fashioned from a model gained over 3D scanning frequently have more of these errors. This is attributable to how 3D scanning works-as it is frequently by point to point achievement, rebuilding will embrace errors in most circumstances. Once finished, the STL file necessities to be treated by a piece of software named a "slicer," which translates the model into a series of thin layers and produces a G-code file encompassing commands tailored to a precise kind of 3D printer. This G-code file can then be printed with 3D printing client software (which loads the G-code, and utilizes it to guide the 3D printer during the 3D printing procedure.

G-code

Digital depictions of "0's" and "1's" are complex, which is extremely adequate for computer code. As soon as a 3D model is designed or simply downloaded, the file, typically have extensions such as 3MF, STL, OBJ, PLY, etc., must be transformed into the G-code. G-code is a numerical control computer language utilized mostly for computer aided manufacturing, both subtractive and additive manufacturing. It is a language which guides a machine how to travel. Without G-code there would be no method for the computer to communicate where to deposit, cure or sinter a material throughout the manufacturing procedure.

Programs such as "Slic3r" are essential in order to translate 3D model files into G-code. Once the G-code is generated it can be sent to the 3D printer, as long as a blueprint as to what its next several thousand moves will involve. These steps all add up to the complete manufacturing of a physical object. There are other computer languages in the market today, and possibly many will ultimately increase popularity, but for now G-code is by far the most significant one. G-code is a numerical control "NC programming language." It is utilized in computer-aided manufacturing (CAM) to control automated machine tools, comprising CNC machines and 3D printers. The slicer program also permits the developer to customize the construct parameters together with support, layer height, and part orientation. The G-code communicates with the computerized machine tools – such as Laser systems – what to achieve, and in what way to achieve it. For instance, how fast to travel, what path to take, etc. The majority of the 3D Printer software are written in "Python," "Java" or" C++."

M-code

M-code is for various functions; it can control auxiliary tasks such as coolant but can also achieve machine specific tasks.

Printer Resolution

Printer resolution defines layer thickness and X-Y resolution in dots per inch, dpi or micrometers, µm. Characteristic layer thickness is around 100 µm, although some machines can print layers as thin as 16 µm. 3D printing machines often comprise of many small and intricate parts so correct maintenance and calibration is critical to produce accurate prints.

Raw Materials

The raw materials utilized in additive manufacturing frequently have a partial shelf life and need cautious treatment. While some procedures suggest the capability to recycle surplus build material, frequent recycle can cause a drop in material characteristics if not substituted frequently. Most additive manufacturing machines do not require to be examined after the print has begun. The machine will shadow an automated procedure and issues commonly only ascend once the machine emptied of material or there is a fault in the software. Building of a model with modern methods can take anywhere from several hours to several days, contingent upon the technique utilized and the size and intricacy of the model. Additive systems can characteristically decrease this time to a few hours, although it differs extensively contingent upon the kind of machine utilized and the size and number of models being shaped concurrently. Traditional methods similar to injection molding can be less costly for fabricating polymer goods in high quantities, but additive manufacturing can be quicker, more flexible and less costly when creating relatively small numbers of parts. 3D printers provide developers and concept development teams the capability to yield parts and concept models utilizing a desktop size printer. Apparently inconsistent, more complex objects can be inexpensive for 3D printing production than less intricate objects.

Prints Removal

Many of additive manufacturing technologies, the removal of the print is as simple as extrication the printed part from the construct platform. For further more industrial 3D printing means the removal of a print is a highly technical procedure concerning exact withdrawal of the print while it is still enclosed in the construct material or attached to the construct plate. These means require intricate removal measures and highly expert machine operators along with safety equipment and controlled surroundings.

Post Processing and Finishing

3-D printers abolish the need for further machining or subtracting procedures such as cutting and grinding; the finishing product is constructed in three dimensions deprived of waste. Though the printer-produced resolution is adequate for numerous applications, printing a slightly large version of the wanted object in standard resolution and then take away material with a higher-resolution subtractive procedure can attain superior accuracy. Some printable polymers such as ABS, permit the surface quality to be smoothed and better-quality utilizing chemical vapor procedures based on acetone or comparable thinners. Some additive manufacturing methods are proficient of utilizing numerous materials in the course of building parts. These methods are capable to print in numerous colors and color mixtures concurrently, and would not essentially need painting. Some printing methods need internal supports to be constructed for overhanging features during building. These supports must be mechanically detached or dissolved upon conclusion of the print.

All of the commercialized metal 3D printers require cutting the metal element off the metal substrate after deposition. A novel procedure for the GMAW 3D printing permits for substrate surface alterations to eliminate aluminum or steel. Post processing procedures again vary by printer technology. SLA needs a element to cure under UV before handling, metal parts often require to be stress relieved in an oven while FDM parts can be held at once. For technologies that utilize support, this is also detached at the post processing phase. Most 3D printing materials are capable to be sanded and other post processing methods as well as tumbling, high pressure air cleaning, polishing and coloring are applied to prepare a print for end use.

X, Y & Z Axes

3D Printers utilize a Laser or extruder, the material output part of the printer that travels along an X, Y and Z axis to construct an object in three dimensions, where consecutive layers of material are placed down in diverse shapes. 3D printers use a diversity of very dissimilar kinds of additive manufacturing technologies, but they all share one essential thing in common: they generate a three dimensional object by constructing it layer by consecutive layer, until the whole object is finished. It is much like printing in 2D on a sheet of paper, but with an added third dimension: "UP" the Z-axis. Each of these printed layers is a thinly-sliced, horizontal cross-section of the ultimate object. One may envisage a multi-layer cake; with the baker laying down each layer one at a time pending the whole cake is shaped. 3D printing is rather alike, but more exact than 3D baking. In the 2D world, a sheet of printed paper output from a printer was "designed" on the computer in a program such as "Microsoft Word™." The file, which is word document, comprises the orders that direct the printer to print. In the 3D world, a 3D printer also desires to have commands for what to print. Also, it needs a file. The file, Computer Aided Design (CAD) file is generated with the use of a 3D modelling program, or from beginning with a 3D model generated by a 3D scanner. Either method, the program generates a file that is directed to the 3D printer. Along the way, software slices the design into hundreds, or more likely thousands, of horizontal layers. These layers will be printed one atop the other until the 3D object is completed.

"U Controller" of 3D Printer

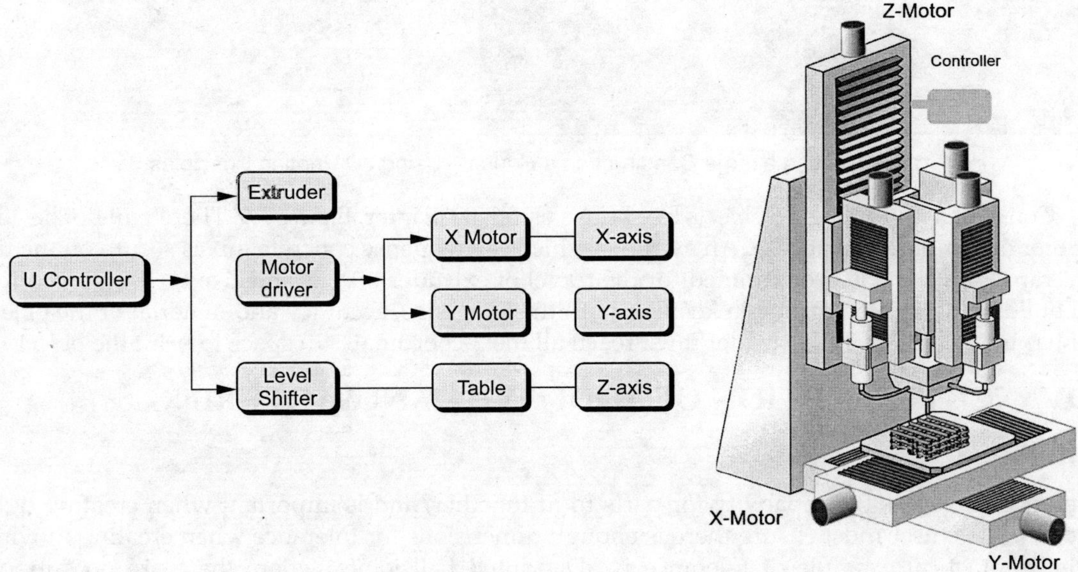

Fig.7.8: Schematic Diagram Representing the U Controller

Figure 7.8, illustrates the build of a typical 3D printer. The print table is the platform where the object for printing has been located. It delivers the plain support for fabricating objects layer by layer. The extruder is the most significant part of a 3D-Printer. As the extruders in the typical paper printers, also this extruder is utilized to dispense ink for printing. The travel of extruder in many dimensions generates the 3D print. For printing a 3D object, the extruder has to access X, Y and Z coordinates. For attaining this, many methods are utilized consistent with the printer requirement essential for numerous applications. If the 3D-Printer is a desktop printer, the Z axis travel of the extruder can be evaded and that purpose can be moved to the print table. This will evade difficulty in 3D printing as well as time consumption. When the STL file is input to the printer, the microcontroller excerpts each layer from it and also excerpts each line section from each layer. Then it provides controls to the travel of the extruder at necessary rate.

The X-direction travel of extruder is made likely by the X-motor. As the X-motor rotates the shaft also rotates, and the extruder trvels in the X-direction. The Y-direction travel of extruder is made likely by the Y-motor. As the Y-motor rotates, the shaft also rotates and the extruder travels in the Y-direction. The Z-direction travel is made by the print table. In the instance of desktop printers, the printing ink is typically plastic wire that has been melted by the extruder at the time of printing. Though printing, the plastic wire will melt at the travel it falls down to the printing table.

The original "MakerBot" is one of the simplest 3D printers. It has just four motors: the x, y, and z, along with a fourth motor to drive the ABS plastic filament, or other material, through a heater to melt it and then onto the construct platform to make the object.

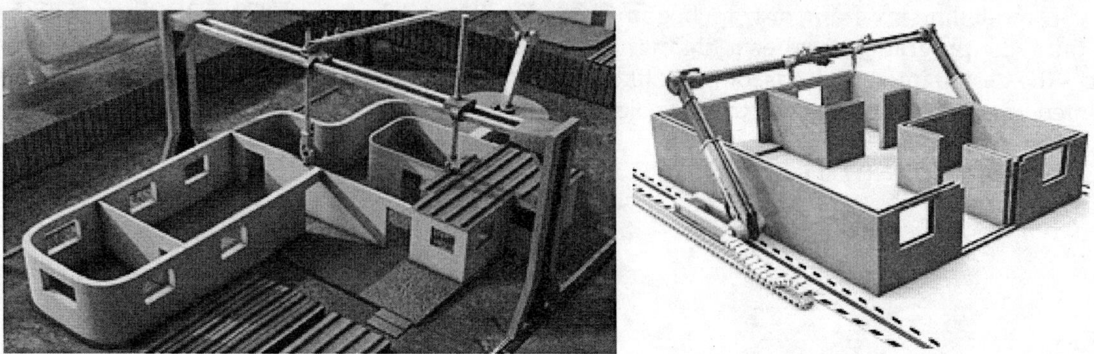

Fig.7.9: Printing a large Construct as a House – Using 3D Printing Principles

Consider printing bigger objects like house using 3D printer, Figure 7.9. There will not be any X motor or Y motor in that case. An extruder which can dispense concrete mix is secure on the tip of a crane. The crane is programmed for the travel of extruder in X, Y and Z axis. The perception and build of 3D printer change in keeping with the kind, size, accuracy and material of the object that has to be printed. The extruder must reach all the 3-coordinates in space to print the object.

ADVANCE PERCEPTION OF ADDITIVE MANUFACTURING

Tolerance

Tolerance illustrates the capability for parts to fit together, and is important when creating tight fit joints. Also, user must ensure there is enough dimensions for tolerance when creating moving parts. A practical example of tolerance is 3D printed ball joints, where there are no support structures in place between the socket and the ball. The ball joint comes fresh off the print bed, already articulated and fully rotating.

Maximum Size

If the model falls outside the printable area of the 3D printer, the developer must consider breaking it down into smaller pieces that may be printed separately, or scale the size of the model to fit the working envelope.

Colors

Color is critical in the appearance and aesthetics of a 3D printed object; it has a significant impact on bringing out finer details to complement the final design. When evaluating the range of colors for a 3D print job, or conducting a demonstration, the developer ought to browse through a color swatch book. Alternatively, the user may lug several spools of filament to select the most appealing color. It is possible to print the basic blocks of color with the shade labelled on the filament. When printing, ensure aligning the correct color filament spool with the intended nozzle.

Infill

The belief of infill is that, for those inner parts of a model, which may not be visible once the print job is finished, the infill doesn't essentially require being 100% solid. As an alternative, it can be a basic honeycomb or diamond grid to deliver mechanical integrity. The advantage of this value is two-fold. First, it avoids waste of time when printing an object. Secondly, it decreases the quantity of filament being utilized.

Over Hang

Developer is required not permit the model to have too several over hangs deprived of structural supports underneath them. An arranged design constituent located at less than a 45 degree angle must be reinforced. Any element more than 45 degrees does not need support.

Bridging

Bridging is understandable. Developer must print "bridges" crossways unfilled space between two solid shapes. The longer the space the bridge has to traverse; the more expected it will sag down as the fused filament cools. This drawback can possibly be alleviated when allowing for the orientation of the model while being printed. For instance, consider a small rectangular table for a doll's house. Printing it standing — as you would normally picture it — will cause bridging snags when the printer begins manufacturing the surface of the table. But printing it upside down with legs in the air shouldn't cause any difficulties.

Supports

Considering the preceding notion, another remedy for bridging is to embrace detachable supports, or rafts in the print job. These may not necessitate widespread particulars as the main body of the object. They can be cut off effortlessly after the print is completed.

XYZ Axes

The distinct axes — X, Y and Z — are essential to 3D modelling. Synchronizing the three together, over the commands confined in a STL file, is in what way a 3D printer plots a point in space when manufacturing a physical object. This model can be attached to the 3D printer for appropriate reference, or perhaps buttressed up together with the computer monitor when working with a CAD program.

Polygons

The resolution of the digital model is a distinct thought from physical print surroundings like layer height. This is precisely about the number of polygons utilized to erect a shape. The low poly mesh model is a simple design encompassed of expressively fewer polygons than the high poly mesh model. The consequences are comprehensible, where the surface of the high poly model is smoother and more meticulous. The efficiencies of selecting for a low-poly model is increased at the design phase, in file-size, and sometimes, but not essentially always, at the printing phase. Low poly models are particularly common with the 3D model public, permitting developer to render familiar objects in an artistic way utilizing a sophisticated arrangement of polygons, Figure 7.10.

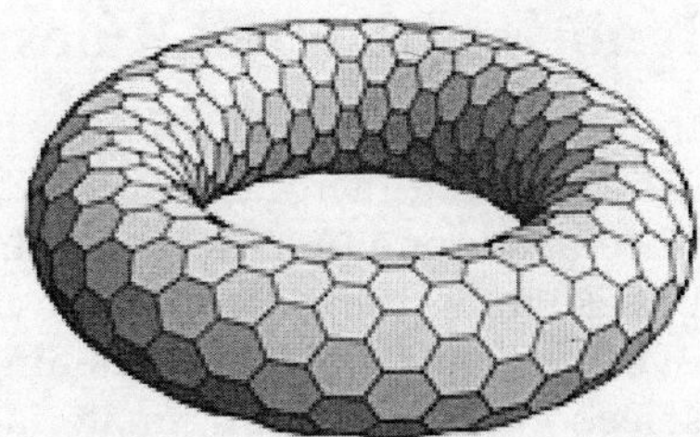

Fig.7.10: Creating a Torus with a Hexagonal Mesh for 3D-Printing

Example:

Create designs on the surface of 3D objects use a hexagonal mesh on a torus, Figure 7.10. Consider a hexagonal mesh that has a specific thickness, so that it would be 3D-printable. Devise a method to generate the hexagonal design on the surface, and the procedure of plotting it onto the torus.

Assumption:

As we also require "nn" to be even (or else the tiling does not fit properly on the torus),

- we can let n=2kn=2k and this gives us k3 – √=rmk3=rm;
- thus for a given rr we should try to choose k, mk, m as the nearest integers satisfying this equation.
- This gives us a very nearly angle-preserving tiling.
- with r=23 – √r=23, we can choose m=11m=11, n=44n=44 to get something that looks like this:o
- hexTile[n_, m_] :=
- With[{hex = Polygon[Table[{Cos[2 Pi k/6] + #, Sin[2 Pi k/6] + #2}, {k, 6}]] &},
- Table[hex[3 i + 3 ((-1)^j + 1)/4, Sqrt[3]/2 j], {i, n}, {j, m}] /.
- {x_?NumericQ, y_?NumericQ} :> 2 π {x/(3 m), 2 y/(n Sqrt[3])}
-]
- ht = With[{torus = {Cos[#] (3 + Cos[#2]), Sin[#] (3 + Cos[#2]), Sin[#2]} &},
- Graphics3D[hexTile[20, 20] /. Polygon[l_List] :> Polygon[torus @@@ l], Boxed -> False]
-]

Calibrating 3D Printers

Fundamentally, the developer desires to guarantee that Printer's Head is at X0, Y0, and Z0. A piece of paper is barely fitted between the extruder nozzle and printer bed.

Avoid Mistakes in 3D Printing Design

There are diversities of software to utilize, variations of prints in diverse materials, and variations of printers of diverse technologies. It appeared impossible to advance an ideal design an ideal 3D model for an ideal 3D printing. Thus, an easy guide may help the user to evade drawbacks by avoiding past faults learnt when turning a 3D model into 3D print. The following statements are a gathering of faults learned:

Ignoring Material Rules

Individually and each printing material is dissimilar. Materials can be stiff or ductile, flexible or solid, smooth or rough, and heavy or light. Consequently, an object should may preferably be developed for a precise material. If the user wishes to print a 3D model in Ceramics, there will be precise material-related design references that essential to take into account such as supporting overhanging parts, strengthening elements that are sticking out, and rounding off corners. The selection of the printing material pre-governs some of the elementary design rules that must be perceived.

Ignoring Printing Technology

It's not only the rudimentary chemical features of the printing materials that are diverse, but also the technologies that are utilized for printing each of these materials. The best case of this is meshing parts: In materials like ABS, Polyamide, Alumide, or Rubberlike, which can print meshing parts, while in others, such as Gold, Silver, Bronze, or Resin may not be feasible. This is not due to the material itself, but the technology, which is utilized for printing each of these materials. ABS is utilized in Fused Deposition Modeling, filament-based, with an additional nozzle for material support, while Polyamide, Alumide, and Rubberlike are utilized in Laser Sintering, powder-based. Also,, precious metals are uses in lost wax molding, founded on a 3D print in wax and a mold, although Resin utilized in Stereolithography, liquid polymer-based. This may seem confusing. Nevertheless, the essential matter to consider is the followings:

a. Must not consider that Stainless Steel and Silver will have comparable features simply as they are both metals.
b. Diverse technologies are employed to 3D print them; therefore, some design characteristics will differ.
c. Materials that use the identical technology such as Gold, Silver, Bronze, and Brass (lost wax casting) are probable to share comparable design necessities.
d. Also, consider the utilization of diverse printers and printing technologies, even for diverse printing sizes.

Ignoring Wall Thickness – Eminent Failure

The difficulties happened from wall thickness are by far, the most typical motives that many 3D models are not printable. In some circumstances, wall thickness is too thin. Walls that are too thin will lead the small parts on the model to be non-printable or very brittle, which could breakdown easily. In other instances, walls that are too thick create too much inner stresses and may cause models to crash or break.

Ignoring File Resolution

The most common file format for 3D printing is STL, "Standard Triangle Language," which offers the guarantee that the design will be interpreted into triangles in a 3D space. Most 3D modeling software has the selection to transfer the design to an STL file and set the desired resolution.

a. ***Low-resolution STL file***: It's significant to be mindful that a poor-quality will not deliver a good print. Low-resolution shows that the triangles in the STL file are big and the surface of the print will not be smooth. It will lead to a "pixelated" print.
b. ***Very high-resolution STL file***: A file with a resolution that is too high will make the file too big that may be impossible to execute. Also, it may encompass an risky level of details that the 3D printers cannot print.

When transferring a file in most 3D modeling software, a user-query will be elevated to describe the tolerance for the export. This tolerance is described as the maximum distance between the original shape and the STL mesh prior to transferring.

i. Choose 0.01 mm for a good export. Exporting with a tolerance smaller than 0.01 mm does not is not representative, as the 3D printers cannot print at this level of detail.
ii. When exporting with a tolerance larger than 0.01 mm, triangles might become noticeable in the 3D print.

Ignoring Software Guidelines

Contingent upon the software to utilize:

a. Shells may require to be combined together,
b. Models may require to be made watertight,

c. Wall thicknesses may require to be implemented, and
d. Printing sizes may require being set.

The developer must always put in the upper most of mind that *"each and every software is different."*

Five Axes Technology of 3D printing

Five axes technology of 3D printing refers to the utilize of five-axes technology to 3D print objects from numerous directions. Nearly all 3D printers nowadays can be defined as three-axes machines, which print parts one horizontal layer at a time. The blemished problem with this approach has caused flaws at the lines of adhesion between layers. Combining a trunnion table into a 3D printer as its print bed evaded such difficulty. This incorporation allowed the machine to print parts as a concurrent five-axes machine. Therefore this permitted the device to print parts that were three to five times stronger with 25 percent less material than the same parts printed in the predictable three-axis fashion.

Different Types of 3D Printing Technologies

Not all 3D printers utilize the same technology. There are numbers of means to print retaining available additive manufacturing, opposing mainly in the way layers are built to create the final object. Some approaches utilize melting material to yield the layers. Selective Laser Sintering, SLS and Fused Deposition Modelling, FDM are the most shared technologies utilizing this method of 3D printing.

Additional technique is utilizing curing a photo-reactive resin with a UV laser or comparable power source one layer at a time. The most collective technology utilizing this method is called Stereolithography, SLA. The American Society for Testing and Materials, ASTM group "ASTM F42 – Additive Manufacturing", since 2010, established a set of principles that categorize the Additive Manufacturing procedures into 7 categories rendering to Standard Terminology for Additive Manufacturing Technologies, Figure 7.11. These seven procedures are stated as follows:

1. Vat Photopolymerisation
    a. Stereolithography, SLA
    b. Digital Light Processing, DLP
    c. Continuous Liquid Interface Production, CLIP
2. Material Jetting
3. Binder Jetting
4. Material Extrusion
    a. Fused Deposition Modeling, FDM
    b. Fused Filament Fabrication, FFF –Different types of FFF 3D Printers
5. Powder Bed Fusion –Selective Laser Sintering, SLS
6. Sheet Lamination
7. Directed Energy Deposition

Additive Manufacturing Monogram

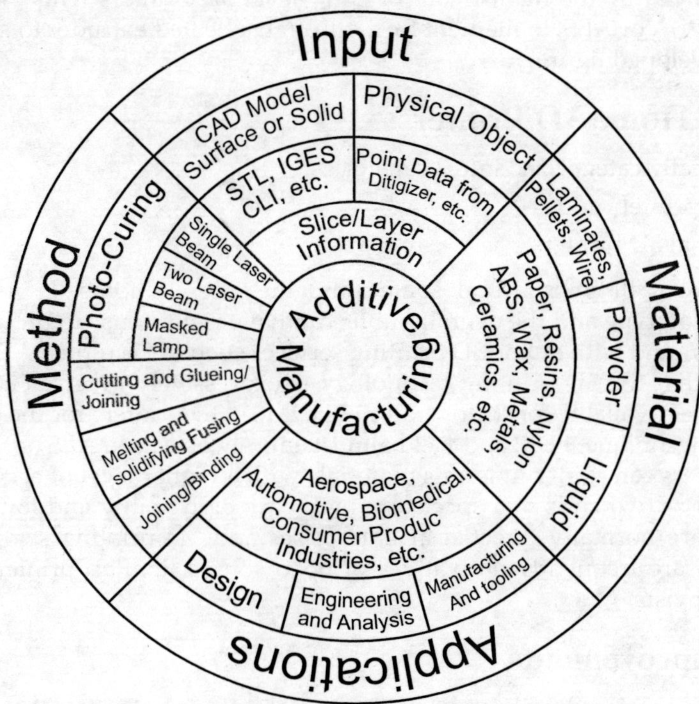

Fig.7.11: Additive Manufacturing Monogram

Business 3D Printers – Beyond Prototypes

Summary

As the name suggests, a 3D printer permits designer to print an object as 3-Dimensional CAD, Computer Aided Design images. It is an inventive technology that helps businesses cut costs and grow new ways of production. Though 3D printers have been in utilization for more than 30 years now, they have been in advance prominence only recently. Easiness of use and nearly limitless customization potentials has caused in 3D printers becoming more and more common as design tools. Originally, 3D printers were utilized as a rapid prototyping answer to make one or two quick physical models providing the designers a possibility to remedy the flaws and revise the product. Prototyping still is the main usage of 3D printers but the advances in technology have made it potential to utilize this 3D printing further than prototypes, investigation and development stages to incorporate it into the manufacturing stage, Figure 7.11.

The different kinds of 3D printers, –each employs a diverse technology that process different materials in diverse way, Table 7.3. It is significant to comprehend that one of the most basic restrictions of 3D printing — in terms of materials and applications — is that there is no 'one answer fits all'. For instance some of 3D printer's operates on powdered materials, as well as nylon, plastic, ceramic, and metal, which use a light-heat source to sinter, melt, or fuse layers of the powder formed in defined shapes. Others handle polymer resin materials and yet again use a light/laser to harden the resin in ultra-thin layers. Jetting of fine precipitations is additional 3D printing procedure, suggestive of 2D inkjet printing, but with greater materials to ink and a binder to fix the layers. Possibly the greatest mutual and easily familiar process is deposition, and this is the process employed by the mainstream of entry-level 3D printers. This procedure extrudes plastics, frequently PLA or ABS, in filament form through a heated extruder to arrange layers and generate the pre-developed figure.

Professional & Home 3D Printers

3D Printers are typically categorized into 2 categories:

- home 3D printers and
- professional 3D printers.

This merit is not at all times factual: some companies possess home 3D printers for the very initial phases of a prototype and the overall public can have right to use to professional-grade 3D printers in "FabLabs" and with online 3D printing services such as "Sculpteo." Though, developer must be conscious that the 3D printing technology is expressively diverse between specialized printer and 3D printer available for the overall public. Most 3D printers for the overall public use filament deposition technique, FDM and yield almost entirely objects in ABS or PLA plastic, while, specialized 3D printers can print numerous materials with a higher level of accuracy. Today's 3D printers are focused at two ends of a spectrum: high cost–high ability and low cost–low ability. High-end printers are normally directed at enterprises and 3D printing service bureaus; low-end printers, which are frequently derivatives of open source RepRap printers, are directed at consumers and hobbyists.

3D Printers Improvements

The printers must be enhanced in three areas to take hold of the potentials that exist further than today's predominant utilization, Table 7.3:

1. **Performance**: Advance key routine features, such as speed, resolution, process, simplicity, utilization, reliability, and consistencies.
2. **Multi-material Ability and Multiplicity**: Integrate numerous kinds of materials, as well as the capability to combine materials though printing a mere object.
3. **Completed Products**: Deliver the capability to print fully working and vigorous systems that integrate many modules, such as implanted sensors, batteries, electronics, micro-electromechanical systems (MEMS), and others.

Table 7.3: Available Types of 3D Printers

Type	Technologies	Materials
Extrusion	Fused deposition modeling (FDM)	Thermoplastics (e.g. PLA, ABS), Eutectic metals, edible materials
Granular	Direct metal laser sintering (DMLS)	Almost any metal alloy
	Electron beam melting (EBM)	Titanium alloys
	Selective heat sintering (SHS)	Thermoplastic powder
	Selective laser sintering (SLS)	Thermoplastic, metal powder, ceramic powders
	Powder bed and inkjet head 3D printing, Plaster-based 3D printing (PP)	Plaster

Laminated	Laminated object manufacturing (LOM)	Paper, metal foil, plastic film
Light polymerized	Stereolithography (SLA)	Photopolymer
	Digital Light Processing (DLP)	liquid resin

World Favorable 3D Printers

3D printer manufacturers are continually broadcasting new models that cost less, print faster and yield larger objects than ever before. The resolution of the present printers is around

- 328 x 328 x 606 DPI (xyz), at
- 656 x 656 x 800 DPI (xyz), in ultra-HD resolution.
- The accuracy is 0.025 mm – 0.05 mm per (2.54 cm).
- The model size is up to 737 mm x 1257 mm x 1504 mm.

Selecting the most suitable 3D printer, designer may consider to the follow recommendations:

Printer Type

While there are many numbers of 3D printing technologies, as stated previously – ***"Different Types of 3D Printing Technologies,"*** still the two most common main types of 3D printers are:

- FDM, Fused deposition Modelling, and
- SLA, Stereo Lithography.

FDM printers operate by melting a plastic filament in a traveling print-head to shape the model. SLA printers utilize an ultraviolet, UV Laser to harden a resin, focusing the Laser to generate the solid model. FDM printers are usually less costly, and easier to utilize, while SLA models, such as the "XYZprinting Nobel 1.0," are dropping their price.

Printing Materials

Whichever type of printer the consumer may select, consideration must be highly regarded to the kind of material it can utilize to print. The filament material utilized by FDM printers, such as the "LulzBot TAZ 6" is obtainable in numbers of diverse materials, such as PLA, a brittle, biodegradable material, ABS, the similar plastic utilized in Lego blocks, nylon, TPE, a soft, rubberlike material and "HDPE" light, tough polystyrene. Numerous of these materials, particularly PLA and ABS, are obtainable in big diversities of colors. Filaments found in two sizes: 1.75 mm, and 3 mm, is not interchangeable. SLA printers have less choices than their FDM matching part, but printers such as the "Form 2" can utilize resins that yield models extending from very stiff to flexible and elastic. The suggested printers are capable to use an extensive variety of materials, each of which has its individual strengths and weaknesses. "HDPE," for instance, is light and hard, but not appropriate for food use, while nylon is food-safe. Some printers permit only the use of authorized materials or materials formed by the same manufacturer that made the printer. Consequently, those kinds of 3D printers are similar to customary paper printers: The producers suggest the hardware inexpensively, and then make the revenue back on the consumables. The top budget 3D printer, the "da Vinci Mini," only works with PLA filament from manufacturer "XYZprinting," for instance; though "XYZ's filament" charges about the similar as most third-party materials. Other 3D printers have no limitations on the kind or source of the material.

Print Volume

All printers have restrictions on the size of the 3D print they can yield. That restriction is described by the size of the print bed and how far the printer can travel the print-head. This is typically calculated in cubic inches or cubic mm. Consumer may also give consideration to each of the discrete dimensions, which govern the maximum size 3D print the device can generate. If a printer such as the "LulzBot Mini" has a print volume of 223 cubic inches, 6.2 x 6 x 6 inches, it can print objects that are up to just less than 6 inches high, wide and deep.

Print Speed and Quality

3D printing is a slow task. At present, there's no innovative means to improve it. The user should expect a 7- to 10 –cm model to typically take between 6 and 12 hours to print, depending on the print quality selected. This is contributed to the nature of 3D printing. The print is constructed in layers. The thicker these layers are, the quicker the print is produced but the lower the print quality, as the layers become more visible. Consequently, there is a trade-off between print speed and print quality. The favorable printers will allow determining the manner the user requires producing prints quickly or more slowly but at higher quality. Most printers provide a wide range of quality settings, from fast, of low quality, to slow but of higher quality.

Printing Time

Time consumed in printing depends on several factors. When utilizing FDM technology, 3D printing speed is affected by print nozzle's temperature, materials to heat, filament thickness, layer thickness, and the number of supports required. For SLA and Polyjet or Multijet technologies, the speed factors include Laser power and size of the drop and materials. Meanwhile, for powder sintering technologies, the factors include Laser power, layer thickness, and material. Aside from these factors, the most significant is the part's orientation and the amount of material to print. User can orient the model in a particular way to minimize the part's height and reduce the printing time. Also, the user may consider that a large 3D model takes longer to print than a lightweight wire structure.

PHASES OF RAPID PROTOTYPING TO HOME FABRICATION

The acceptance of 3D printing technologies has happened in four consecutive phases

1. Rapid prototyping,
2. Rapid tooling,
3. Digital manufacturing, and
4. Home fabrication

These parallel to a diverse level of contribution of 3D printing in the production process. The implementation of 3D printing technologies is essentially a "multi-layered" implementation procedure that parallels to different usages. The main cause relays both to the technology itself, specifically the materials employed, and to the cost of practice. The 3D printing manufacturing by now attends four separate market segments, all of which are at dissimilar stages of maturation. As illustrated in figure, 7.12 below, the most established market for 3D printing encompasses the production of prototypes. This is shadowed by rapid use of 3D printers to manufacture the molds and other tooling from which final products are completed. After this comes *direct digital manufacturing*, DDM — or the use of 3D printers to make final products or parts thereof. Finally, a novel market for individual home manufacture is merged.

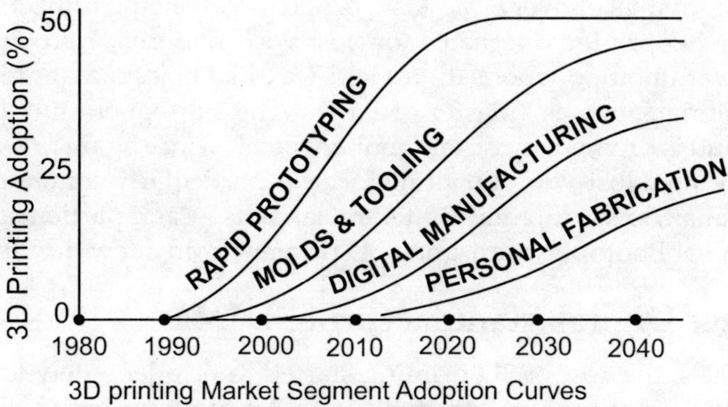

Fig.7.12: 3D Printing Market Segments

Chronicle - Rapid Prototyping

The first 3D printing technologies were stereolithography, selective laser sintering, fused deposition modelling, and laminated object manufacturing appeared in the late 1980s. Operational application began in the early 1990s. At the time, only plastics could be used. The level of details and quality of finished products were low, which produced rough surface texture and looked dull objects. Printing was slow, expensive and restricted to small objects. Consequently, the first application of 3D printing technologies was rapid prototyping, i.e., the capability to rapidly build plastic models of objects. Though rapid prototyping was the original effort in 3D printing, owing to its low cost, it was mostly utilized by big companies. Progressive decline in price led to a broader acceptance of the technology. Today, all 3D printers cost are in the range of the $1000–4000. Prototyping quality has considerably better-quality. At present, upper choice printers cost may extent above $250,000. Lucrative-end 3D printers are capable to construct multi-materials of completely working prototypes in one run. There are even additional expensive 3D printers founded on material jetting, or powder bed fusion than may yield extremely detailed, working prototypes out of a extensive variety of materials, and progressively in full color. There is already a 3D printer — the "DragonFly 2020" — that can 3D print prototype circuit boards.

Chronicle Printing - Molds and Tooling

In the second half of 1990s, the arrival of 3D printers by means of heat-resistant polymers and metal alloys activated the second phase of acceptance of 3D printing, namely, rapid tooling. The fastest rising area of 3D printing implementations is the additive manufacture of production tools. Conservatively, the molds, patterns, jigs, fixtures and other tooling utilized to make final products has been hand made at a substantial cost. The molds were utilized to mold plastic parts, which frequently costs ten thousands of dollars. Presently, in difference, it is now likely to utilize 3D printers to yield low-run injection molding masters for a small portion of the conservtive cost. The influence of the 3D printing cost on conservative production cost is dramatic. Also, through 'binder jetting', it is likely to directly 3D print sand cast cores and molds. Traditional sand casting needs wooden patterns to be fashioned, around which casting sand is filled. The wooden pattern then requires being detached from the subsequent mold or core, which is not always a simple or positive job, before molten metal is dispensed to cast the final part. Using 3D printers, such as the "S-Max+ from ExOne" to fabricate sand cast molds and cores, businesses can spare themselves the cost of fashioning wooden pattern masters. They can also cast more complicated parts from sand molds that would be impossible to style by traditional means. Once again time and cost benefits can be important. 3D printing sand cast molds and cores can save weeks from production times, as well as bringing cost savings of over 80%. Also, progressively being utilized in traditional metal casting procedures are 3D printed alternates for 'lost wax' patterns. Here 3D printers based on material extrusion or vat Photopolymerisation yield wax-like objects that are later enclosed by a traditional casting material, such as "plaster of Paris." The subsequent mold is then heated, the wax-substitute melts and is drained away, and molten metal is once again presented into the mold. The utilization of 3D printed lost-wax-substitute forms is particularly common in jewelry or other businesses that make small, low-run, complicated metal parts — and particularly as it is now likely to buy high-resolution vat Photopolymerisation 3D printers from a few hundred dollars.

Chronicle - Direct Digital Manufacturing DDM

In the latter part of 2000s, the cost of 3D printing started to be adequately low, with satisfactory quality to start directly fabricating final products with 3D printers. The speed, quality, accuracy and material characteristics have advanced to a degree that 3D printed parts can be completed for final utilization. This led to the third wave of acceptance, commonly referred to as "Direct Digital Manufacturing - DDM, or simply Direct Manufacturing, which infers a completely digital production procedure, with end-products directly manufactured using digital (CAD) models and 3D printers, without molds, casts or machining.

As stated by the American Society of Manufacturing Engineers, direct digital manufacturing - DDM denotes to the procedure of proceeding directly from an electronic digital depiction of a part to the final product through additive manufacturing. On occasion, some such final products may be molds or other items of tooling. Though, in manufacturing subdivisions counting space, aerospace, automotive manufacture, healthcare, toy making, art-and-crafts, developer of goods and fashion, the market is initially witnessing the implementations of 3D printing to directly manufacture final products or parts thereof. The motive of the DDM Revolution are innumerable, and comprise the capability to utilize 3D printing to enhance the geometry of finishing parts, which frequently save raw materials in the process to make parts with geometries that could not be generated through traditional means. Also, through DDM low-run can be made with highly made-to-order products that would be not cost affordable to manufacture without 3D printing.

Improve Sustainability

3D printing is likewise a technology that can aid to recover sustainability via augmented localization. Presently, many of the companies controlling DDM are in the space and aerospace segment. "SpaceX" has now flown one its "Falcon 9" rockets with a 3D printed main oxidizer valve, and is utilizing powder bed fusion to 3D print the engine chambers for its approaching "Crew Dragon" manned space capsule. In more conventional aviation, Airbus is now including 3D printed mechanisms into its aircraft, with its latest "A350 XWB" encompassing over 1,000 3D printed mechanisms. In the meantime, GE Aviation's new "CFM LEAP" jet engine integrates several 3D printed parts. Most remarkably, these comprise one element that would be impossible to manufacture utilizing conventional manufacturing techniques. Over in the automotive segment, 3D printing is now extensively engaged to manufacture parts for F1 racing cars. More shockingly, a company called Local Motors is aiming to unveil a vehicle called the "LD3M Swim" that will be around 75% 3D printed by volume.

Optimize Component Geometries

In the space, aerospace and automotive subdivisions the key rewards of 3D printing are to enhance module geometries, to save costly materials, and to decrease the cost of low-run manufacture. In difference, in the healthcare subdivision a more significant advantage is customization. At present, at least in the United States, 3D printing is allowing 'digital dentistry', with dental applications starting to be regularly 3D printed based on patient scans. Hearing aid casing production is also being converted by 3D printing. 3D printing has commenced to be utilized in other production of larger custom prosthetics, as well as in medical and surgical planning. Presently, other noticeable entrepreneur of Direct Digital Manufacturing includes manufacturers of jewelry, designer's goods and toys.

The Infancy of Home Personal 3D Printer

The fourth stage of adoption is the home fabrication, which is sweeping the Western World. It involves end-users to manufacture objects directly using 3D printing equipment they may have at home. It is observed that few consumers own 3D printers, and those who may possess 3D printers are mainly hobbyist and engineering students. However, a limited and slow adoption is expected, as prices are still high and the technology yet has some inhibitions.

While there is a large consensus about the value and the potential of 3D printing technologies in general, there has been much debate about 'home manufacturing' revolution at the horizon. The cost may be the major factor, as well as on quality produced. Also, the home demands to manufacture an item, as a quantity of one, at a time on a spasmodic interval, made the possession of home 3D printer unattractive. To this known factor, the very similar arguments were made in the past in regard to technologies, such as personal computers or the Internet, which are now in every home. This, in turn, is the very argument used by those who believe in a widespread home adoption of 3D printers.

3D printing, as a disruptive technology, follows the same adoption pattern as other past andd current disruptive technologies. In some respects, the market for personal 3D printing is now growing very rapidly, with over a million personal 3D printers are likely sold annually by 2020. However, the types of object that currently may be fabricated on personal hardware are limited. Despite fulfilling the dreams of some enthusiasts, personal 3D printers are therefore a product category in an early stage of development. Hardware fabrication allowing mainstream consumers to manufacture a wide range of products is unlikely to arrive on the market for another decade. The manufacturers who are content with producing objects through thermoplastics, or thermoplastic composites are satisfied with the revenues acquired in the past few years. Interestingly enough, nowadays there are many 3D object repositories are providing substantial access to a wealth of free or for-a-free 3D models for download-and-print. Astonishingly, numerous developers discovered the ease to visit a website such as "Thingiverse" to download a additional part for an automobile dashboard, or a novel lens cap for a camera. These substances can then be printed out at serviceable excellence on a local 3D printer.

Even beforehand individual 3D manufacturers capable to print intricate products become obtainable, 3D printing is just beginning to enable a new age of individual manufacture in which anyone can own their digital designs turned into material certainty, and even provided for profit-making sale. Authorizing this revolution are multitudes of 3D printing bureau services that permit anyone to upload or else obtain a digital file that is then 3D printed. As most bureaus work at scale, they can pay for to acquire and operate a extensive variety of 3D printers, therefore permitting persons to manufacture a far broader variety of objects in wider variety of materials than conceivable on domestic and customer hardware.

Bureaus such as "Shapeways" and "i.materialise" moreover allow persons who have uploaded 3D models to open an online store, from which their goods can be sold. This permits anyone to launch industrial business deprived of any venture in stock or construction tooling. Somewhat, the 3D printing bureau takes the client's order, yields the product, ships it, and delivers a revenue share to the designer.

The Table 7.4, below summarizes the diverse adoption phases of 3D Printing technologies. It is vital to observe that each new phase does not make the preceding one 'obsolete.' Instead, spreads it to 3D Printing utilized for rapid prototyping. Also, the fourth stage of home-based use manufacture, spreads the role of 3D printing beyond manufacturing, as home-based printers permit to use 3D printing as a ways of product delivery.

Table 7.4: Time-Line Adoption of 3D Printing

Adoption stage	Started	Design	Tooling	Manufacturing	Distribution
Rapid prototyping	Early 1990s	✓			
Rapid tooling	Late 1990s	✓	✓		
Direct manufacturing	Late 2000s	✓	✓	✓	
Home fabrication	Early 2010s	✓	✓	✓	✓

In addition to these four distinct phases, an intermediate phase between direct manufacturing and home-based printing has begun to transpire. Home-grown manufacture denotes to direct manufacturing that is conducted not at home, but at a native print-shop. The appearance of this intermediary phase is not astonishing, as it allows bridging the gap between prime of life technology and infant technology with large installed consumer base. This is in fact comparable to "Internet cafes" or 2D "print-shops." It remains to be seen if the latter, such local facilities will vanish as home adoption increases or if, instead, the environment of 3D printing technologies is such that they are here to stay.

The Inception of The 3D Printing Techniques

The original 3D printing technologies first came to fruition and appeared visible in the late 1980's, at which time they were named "Rapid Prototyping - RP" Technologies, Table 7.5. The name was a derivative from the rapid progressions to infancy after their inception in 80's. The technology has become effective method in creating prototypes for simplistic products. The first patent application for "RP" technology was submitted by a Dr Kodama, in Japan, in May 1980. He was the original to define a layer by layer methodology for manufacturing, fashioning an ancestor for SLA: a photosensitive resin was polymerized by an UV light. Regrettably, he did not submit the patent prerequisite before the deadline. Four years advanced, a French team of engineers was engrossed in the "Stereolithography" but ceased to advance and abandoned due to a shortage of interest and lacked business perspective. In the meantime, Charles Hull was also attentive in the technology and submitted the formal first patent for "Stereolithography – SLA" in 1986. He established the "3D Systems Corporation," and a year later, uncovered the "SLA-1."

In 1988, at the University of Texas, Carl Deckard conveyed a patent for the "SLS technology," another 3D printing method in which powder grains are fused together locally by a laser.

Meanwhile, Scott Crump, a co-founder of Stratasys Inc. submitted a patent for "Fused Deposition Modelling - FDM": the third of the foremost 3D printing technologies. Within less than a decade, the three foremost technologies of 3D printing were patented and 3D printing was formally conceived!

Milestones

1992 – "3D Systems" formed the first SLA - Stereolithographic Device. The procedure incorporated a UV laser solidifying photopolymer, a liquid with viscosity and color comparable to honey. It made 3D objects, layer by layer. The first machine was defective, but it was promising, as it can manufacture highly intricate parts.

1999 – 3D Printing added new improvements in medicine through engineered organs. The scientists at the "Wake Forest Institute" for "Regenerative online Pharmacy Medicine" engineered the first lab-grown organ utilizing a 3D printer. They utilized a artificial scaffold layered with the patient's individual cells. The procedure had little to no risk of refusal because it is made with the patient's cell. This cemented the method to emerging new strategies for engineering organs.

2002 – The scientists fashioned a mini-functional kidney that was able of filtering blood and creating diluted urine in an animal. This improvement had led to greater goals of creating organs and tissues utilizing 3D printing technology.

2005 – Dr. Adrian Bowyer established "RepRap," an open-source inventiveness to generate a 3D printer that can print most of its own mechanisms. The goal of the project was to allocate inexpensive "RepRap" units so persons can generate products on their own.

2006 – The first SLS – "Selective Laser Sintering" - machine was created. SLS machine utilizes a Laser to fuse materials and construct 3D products. This improvement cemented the path for mass customization and on-demand fabrication of industrial parts and prostheses. In the same year, a machine able of printing multiple materials, counting elastomers and polymers was generated. The said machine also allows a single part to be made with diverse densities or material characteristics.

2008 – The first "Self-Replicating Printer" originated to life that permitted users to make more printers for family members or friends. In the same year, the first being with a 3D-printed artificial leg walked. All components of the leg, knee, foot and socket were printed in the similar intricate build deprived of any assembly required.

2009 – Do-It-Yourself, "DIY 3D Printer" kits hit the market. On the same year, Dr. Gabor Forgacs utilized a 3D bioprinter to print the first blood vessel.

2010 – Engineers at the University of Southampton, England, designed and fly the world's first 3D printed airplane. The airplane was only constructed in seven days with a budget of £5,000.

2011 – Urbee, the world's first 3D-printed automobile was completed. It is sleek, environmental-friendly and is designed to be fuel-efficient and economical. In the same year, 3D printing services offer 14K gold and sterling silver as materials, a less costly option for jewelry inventors.

2012 – Medical valium professional from the Netherlands employed a 3D printer to generate a customized 3D prosthetic lower jaw that was embedded into an 83-year old woman suffering from a chronic bone infection.

2014 — NASA brought a 3D printer in space to style the first 3D printed object off of the earth.

2015 – Dr. Sabrie Soloman --- The Chairman & CEO of American SensoR$_x$, Inc., successfully incorporated his "Near Infrared – NIR" Patented Technology, "The SpectrR$_x$™" into 3D Printer, printing patient custom-made pharmaceutical Tablets and Capsules immediately after accurately measuring the active ingredients by weight.

As technology matured, some authors had started to venture that 3D printing could support in sustainable improvement world.

3D Printing – Unique Manufacturing Techniques

3D printing is a substitute to traditional manufacturing method of metal casting technique. If the object was poured into a mold, it must utilize casting tools. Casting is one of the first manufacturing methods known to humankind. It actually dates back 5,000 B.C. The most famous bronze casting took place during "King Solomon Dynasty (970-931 BC), Figure 7.13.

Huram placed these two bronze columns in front of the entrance of the Temple: the one on the south side was named Jachin (Establish) and the one on the north was named Boaz (Strength). 1Kings:7-21

The bronze casting did not gain popularity until 800 B.C., when it became commonly used during king Solomon reign in old Israel. The original casting process involved molten metal poured into a mold. Die casting is a manufacturing process that can produce geometrically intricate metal parts through the utilization of reusable molds, called dies. The die casting procedure includes the utilization of a furnace, metal, die casting machine, and die. The metal, characteristically a non-ferrous alloy such as aluminum or zinc, is melted in the furnace and then inserted into the dies in the die casting machine.

The Resin casting, a prevalent hobbyist technique today, includes the utilization of liquid synthetic resin mixed with a hardening catalyst and dispensed into a mold. The casting process starts by making the product model with the correct dimensions in a casting material, which is often silicon. Next, a two-part resin is put together and dispensed into the mold. The chemicals in the resin cure. Once the resin hardens, the final model is generated. 3D printing is progressively an alternative to casting.

The anticipated object is printed in a small container filled with synthetic resin. The resin has a very distinct characteristic. It hardens exactly where it is flood-lit with focused beams of light. Layer for layer, the synthetic resin is exposed at precisely the right spots. When one layer hardens, the next layer can be bonded to it, until the object is finished. Intricate geometrical objects can be shaped with an intricate inner build, which could not be generated by means of casting practices.

3D printing is also utilized to create tooling molds and dies, as well as patterns for castings. Moreover the actual mold or the model to create the mold can be formed more rapidly and economically than with conventional approaches.

Injection Molding and 3D Printing

The two procedures are completely diverse. In injection molding it is melting a molten material, and injecting it into a big pre-designed cavity inside a steel tool, analogous to a steel box, with an internal figure that equals the skeleton of the intended part. Injection molding machines are developed to keep the plastic melted while enroot to the tool cavity and then to quickly cool it once injected so that another cycle can be achieved.

3D printing and injection molding are not challenging, but rather complementary means of fabrication. An industry may utilize "SLS" for rapid prototyping and low-volume manufacture, and then shift to injection molding once the volume of parts is raised above a positive threshold. The capacity to produce a "just-in-time" number of parts is where 3D printing shines. Compare that with injection molding, which is characteristically utilized in mass production – i.e. producing hundreds, if not thousands, of parts economically.

Different Fabrication Processes

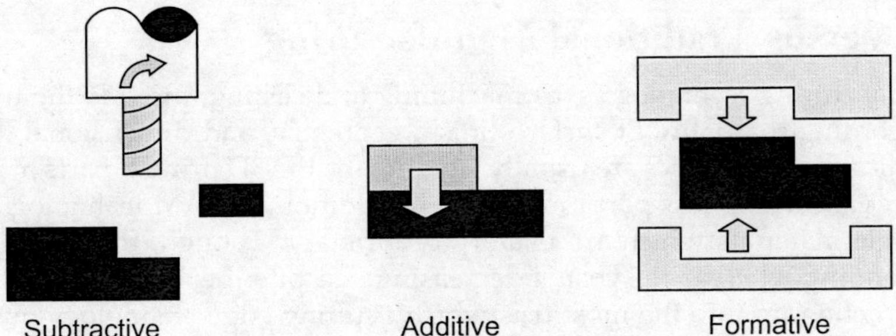

Fig.7.14: Types of Manufacturing Processes

Subtractive Manufacturing

Typical manufacturing techniques are known as 'Subtractive Manufacturing' because the process is one of removing material from a preformed block. Processes such as Milling and Cutting are subtractive manufacturing techniques. This type of process creates much too much waste, since the material that is cut off cannot be re-used and sent discarded as scrap, Figure 7.14.

Additive Manufacturing – The True Potential

Additive manufacturing has the potential to basically change the manufacture and delivery of goods. Contrasting conventional or subtractive manufacturing processes, such as drilling, which create a part by cutting away material, additive manufacturing shapes a part using a layer-by-layer process. 3D Printing removes such waste since the material is placed in the location that it is required only, while the rest will be left out as empty space.

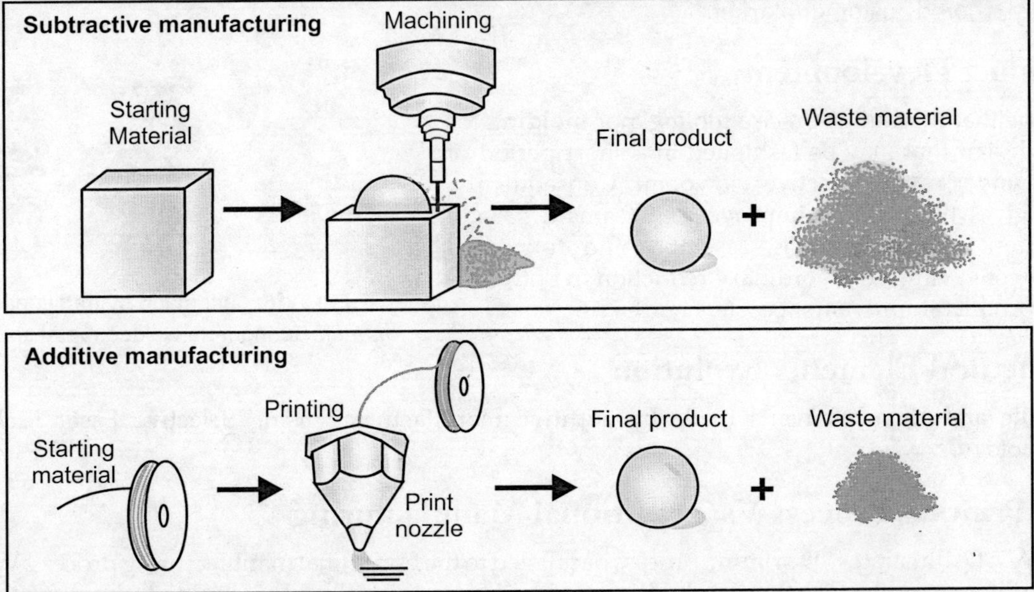

Fig.7.15: Representations of Subtractive and Additive Manufacturing Technologies

3D Printing Versus Traditional Manufacturing

Traditional manufacturing technologies, as machining and casting, provide the uppermost parts quality level, according to quality of surface finish, geometry, and dimensional toleance, Figure 7.15. The mechanical characteristics are usually superior, as the 3D printed parts may be not totally filled by the material causing less part toughness. In addition, the AM technologies have a short variety of available materials which are usually proprietary. Though, the materials diversity is unceasingly expanding whereas the cost is decreasing. Despite the stated disadvantages, the 3D printing is constructing up into the most trustworthy cutting-edge technology businesses within trustworthy segments such as aerospace, automotive and medicine. This is conceivable due to AM does present substantial prospects, resulting from its excellent returns. Some of the advantages are listed as follows.

Unlimited Designs

AM makes likely the manufacturing of any design irrespective the design intricacy and permitting the manufacture of combined components. This is exceptionally significant towards a extensive open-mind design, revolutionary shapes and lighter parts, deprived of bearing in mind manufacturing restrictions related to machining, molding. The 3D printing technology is considered "complexity is free", as complexity has no an influence in the manufacturing cost.

Flexible Supply Chains

Production by AM does not require any distinct tooling neither molding, so changing the manufacture to a completely diverse object is totally simple and instant. Redesigning phases has no impact in the manufacture price. Accordingly, it is the ultimate manufacturing technology for both prototyping and low volume manufacture batch, as low as one if necessary. This is very helpful product progress and low-to-medium batch production. Also, 3D printing is an ideal supporter for highly tailored customer market.

Product Development

AM neither it requires precise tooling, nor molding for any design that may be fashioned in a short period of time under a cost effective viewpoint. Consequently, AM gives the product improvement teams a speedy iteration between designs, assembly and functional tests, conveying a extraordinary reduction in both time and product improvement costs, Figure 7.16 .

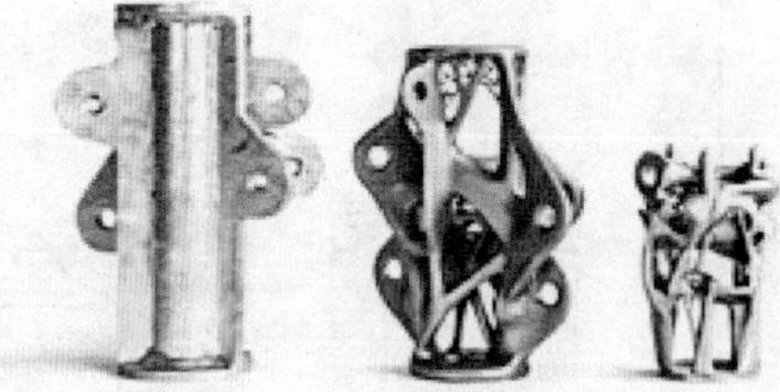

Fig.7.16: Left: made by traditional manufacturing from welded metal sheets.

Structural Elements Evolution

Middle and right: elements made by additive manufacturing using Selective Laser Melting technology.

3D Printing Process Vs Traditional Manufacturing

Figure 7.17, illustrates 3D printing process paralleled to the traditional manufacturing process. When 3D printing manufacturing technology is compared to traditional manufacturing technologies such as molding, machining, and drilling, 3D printing is measured as efficient technology in materials, utilizing up to 90% of materials. The traditional manufacturing techniques, such as milling machine, drill press, and lathe machine are essential to be attendant by the operator. The work piece requires to setup in the exact location, measuring, and machining by the user, which presents an additional human miscalculation into the manufacture of the part. In difference, 3D printing is a hands-off manufacturing procedure; just by clicking a button, whatsoever the complexity of the design is, the design is totally completed. The principles of using 3D printing over traditional manufacturing make it appropriate for various surroundings requiring the following conditions:

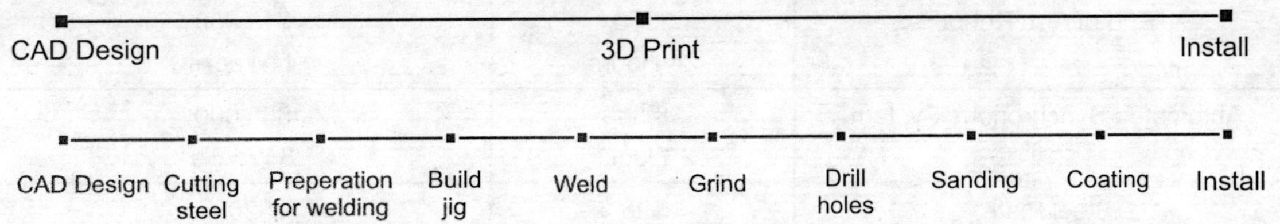

Fig.7.17: Manufacturing Procedures of 3D Printing Vs Traditional Manufacturing

1. Rapid prototyping
2. Mass customization
3. Complex geometries stated as follows
   i. Cannot be manufactured by any other method
   ii. Improved material property, e.g. strength, elasticity, transparency
   iii. Cost-effectively to manufacture with 3D printing.

Limitations.3D Printing

1. 3D printers may devote a number of hours instead of minutes to finish a piece and therefore do not advance themselves to mass production for certain applications, particularly in the promising surgical cases.
2. The partial selection of 3D printing material, particularly those believed appropriate for medical practice, such as biocompatible, serializable, high strength, multi-color, and inexpensive, may delay wider application.
3. The size of the work pieces anticipated for printing may restrict the applications, as printers capable of making bigger prints are more costly with less options obtainable.

Bypass Supply Chain

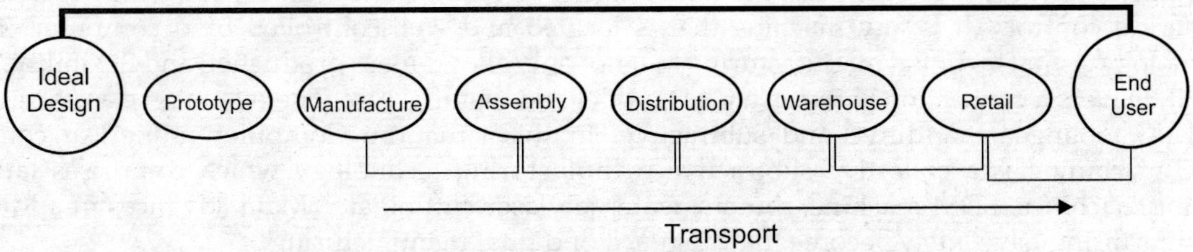

Fig.7.18: 3D Printing Bypasses Traditional Supply Chain

The low cost 3D printing enables anyone with a digital design to bypass the traditional supply chain and manufacture a product themselves, Figure 7.18.

Less Time and Cheaper Product:

Table 7.6: Cost Comparison of FDM – 3D Printing Vs traditional Manufacturing Method

Part Tool	FDM	Alternative Method
End of Arm Robot	$600 24 Hours	$10,000 44Weeks
Automated Synchronous System	$8,800 2 Weeks	$50,000 8 Weeks
Steel Plate	$20 2 Weeks	$200 2 Weeks

Table 7.6 shows the benefits of "Fused Deposition Modeling, FDM," - 3D Printing Compared to traditional manufacturing method.

3D Printing Vs. Rapid Prototyping

Prototype is the initial model of a mechanism, assembly or product that still requires to be manufactured. Rapid prototyping denotes to the fast fabrication of prototypes for diverse goals. In countless cases, such rapid prototypes are attained by means of high speed computer numerical control, CNC machining or utilizing rapid method copying procedures, after locating initial models.

The RP procedures were initially considered as a fast and more cost-effective technique for generating prototypes for product improvement within industry. Today rapid prototyping is one of several applications under the 3D printing additive manufacturing shelter. There are numerous motives that companies utilize rapid prototyping. Primarily, the novel project can be touched or viewed as it would appear on an exhibition. Secondly, the form, fit, function, and aesthetics of an object can be verified and viewed. Also, if there are any faults in the CAD file, they can be discovered in the rapid-prototyped work piece.

CNC Machining – The Reverse of 3D Printing

There are many diverse tooling procedures that digital manufacturing uses. Though, every digital manufacturing procedure includes the utilization of CNC machines. CNC stands for "computer numerical control." It is any machine that is located and well-controlled by a computer. This technology is vital in digital manufacturing as it not only allows mass production and flexibility, but also it delivers a connection between a CAD model and manufacture. The two main classifications of CNC tooling are additive and subtractive. In this situation, Computer numerical control CNC machining is a collective subtractive manufacturing technology which comprises lathes, milling machines, EDM machine, routers, water-jet, laser cutting etc. Main advances in additive manufacturing have lately become the vanguard of digital manufacturing.

CNC routing is fundamentally the opposite of 3D printing. As an alternative of using a computer to control the moving-mechanism and print-head that layers material in 3D, CNC routing uses a drilling tool to cut materials. This is essentially the variance between making a statuette out of clay and carving it from marble, only in this instance, there's a machine achieving it as an alternative of a human.

CNC Routers

While CAD software operates with CNC routers and 3D printers, the two tools use numerous approaches to generate outputs. In fact, CNC routers use a technique that operates contradictory of the method 3D printers operate. CNC machines begin with a block of material and cut away until nothing left excluding the projected product. Instead, 3D printers begin from naught and add layers of a particular material to shape a build that take up the shape of the product design. This additive process styles 3D printers flexible enough to generate a diversity of outputs, restricted only by the ability of the printer. Albeit the subtractive process of CNC technology frequently harvests faster results than the additive process of 3D printers, both methods to production have rewards that make each type of machine compatible to accomplish divergent goals.

CNC – 3D Printing Comparison of Properties

Table 7.7, below compares the main manufacturing properties to be considered when making a part by CNC machining or 3D printing.

Table 7.7: Properties Differences between CNC Vs 3D Printing

Property	CNC	3D Printing - AM
Material	Mainly utilized for machining metals. Can also be utilized for machining softwoods and hardwoods, thermoplastics, acrylic, modeling foams and machining wax. Needs diverse cutting tools for each material.	Predominately polymeric plastics with some technologies capable to yield parts from metals, ceramics, wax, sand and composites.
Speed	CNC machines are capable to take away material at abundantly faster rate than AM is capable to build it. They commonly need a substantial amount of procedure planning and setup, mainly when multiple machining steps are required. Parts often necessitate relocation or transfer.	Can yield a part in single step, which is independent from other manufacturing phases other than post processing. Also delivers batch manufacturing for some process (most notably SLS and metal printing).
Complexity	Undercuts, tool access, internal structures and clearances are all restrictions that must be well thought-out. A sound consideration of the machining procedure, the order a part will be manufactured and part orientation is essential.	Intricate designs can be created using AM in a single step with very little procedure planning. Consider how to appropriately orientate a part, feature size limitations and physical construct size are commonly the foremost restrictions.
Accuracy	Accuracy is described by the tool geometry. Because all tools are rotated, internal corners at machined with a radius. Structures smaller than the tool size can be produced resulting in walls with a thickness smaller than the tool diameter.	Minimum feature size is usually ruled by the diameter of the material delivery mechanism (e.g. the nozzle for FDM or jets for material jetting) or the diameter of the energy transferring component of the machine (e.g. sintering laser or UV light

	Delivers superior surface quality when equated to the best outputs 3D printers can produce.	source). FDM printers yield parts with a layer height of 100 – 200 microns while material jetting printers can print at resolutions as low as 16 microns.
Geometry	CNC machines depend on a point to point machining procedure following a programmed tool path to remove material. There are consequently limitations on the surfaces a CNC machine can reach without requiring to manipulate a part. Can be utilized to machine very large and very small parts.	Features that are not linked to the model or have nothing below to brace them need additional support material to be printed. This upsurges the cost and time to complete a print.
Programming	Needs an expert operator or engineer to consider tool choice, spindle speed, approach position and angle and cutting path. These issues all greatly influencing the final part quality and construct time.	Once a model has been uploaded and the orientation, layer height and support locations are carefully chosen most AM machines can yield a complete part without any human intervention.

Additive Manufacturing is the best suitable technology for intricate and complicated design as well as the manufacture of prototypes for fit, form, and function validation. The variety of part materials is more restricted than CNC machining, and frequently the surface finish and dimensional accuracy are not as good as those accomplished by a CNC machine. A number of 3D printing technologies provide batch manufacturing. CNC machines are perfect for simple geometric designs made from traditional materials that need high accuracy and surface finish. The need for skilled CAM design as parts become more intricate can upsurge cost and lead time. CNC machines frequently necessitate more human interface than 3D printers. The burdens for quality and speed to yield parts be contingent heavily on operator interface.

3D printing and CNC are totally in divergence in almost all aspects as defined as follows.

1. 3D printing included of additive manufacturing, where material is supplementary layer by layer to gain the final workpiece, while CNC machine is traditional manufacturing, where material is removed to attain the final workpiece. 3D printing is adding manufacturing process, while CNC is a subtraction method.
2. As 3D printing is adding of material, no waste of material will be left, while in the CNC the wastage is substantial.
3. *"G-code consists of G- and M-commands that have an assigned movement or action."* The 3D printing G & M codes are generated automatically when slicing and object. However, CNC a manual entry of G & M codes must be entered manually. 3D printing saves substantial time.
4. 3D printers have only one pattern program regardless of size, shape and material of workpiece to print. The CNC has diverse programs (G & M codes) has to be written for different size, and shape of objects.
5. CNC necessitates tools to achieve diverse all processes, while 3D printing does not.

Transforming Businesses

Additive manufacturing may transform businesses and entire industries. There are several benefits of the technology that make it a revolutionary production method stated as follows:

1. ***Efficient use of resources***: Additive manufacturing requires fewer processing steps, less assembly, and less energy. It also minimizes waste.
2. *Small-lot production*: Additive manufacturing eliminates the need to produce parts or products in large quantities. As a result there is no inventory or stock cost.
3. *Rapid manufacturing*: The technology allows makers to move directly from design to manufacturing. It dramatically reduces cycle times since it doesn't require a tooling process.
4. *Agile manufacturing*: Additive technology can create spare parts on demand and thus simplifies supply chains and logistics.
5. *Reverse engineering*: Parts for legacy systems can be easily replaced with the help of a scanner and a 3D printer or other additive manufacturing technology.

Additive manufacturing is not replacing traditional manufacturing, but complementing it and creating new markets within the industry. It is capable of making parts that cannot be manufactured by traditional processes and enables new technologies that are lightweight, customizable and complex.

REFERENCES

1. Lopes, A.J., Navarrete, M., Medina, F., Palmer, J., MacDonald, E. and Wicker, R.B. (2006), "Expanding rapid 960 prototyping for electronic systems integration of arbitrary form", Proceedings of the 17th Annual Solid Freeform Fabrication Symposium, University of Texas at Austin, Austin, TX.
2. M. Navarrete, A.J. Lopes, J. Acuna, R. Estrada, E. MacDonald, J.A. Palmer, and R.B. Wicker "Integrated Layered Manufacturing of a Novel Wireless Motion Sensor with GPS," Proceedings from the 18th Annual Solid Freeform Fabrication Symposium, University of Texas at Austin, August 06-08, 2007.
3. David Espalin, Danny W. Muse, Frank Medina, Eric MacDonald, Ryan B. Wicker, (forthcoming) "3D Printing Multi-Functionality: Structures with Electronics", International Journal of Advance Manufacturing Technology
4. Douglas Brooks, "Fusing Current: When Traces Melt Without a Trace", December 1998
5. Soloman, Sabrie "Sensors Handbook, 2nd Edition," McGraw-Hill Publishing Co., NY – ISBN 978-0-07-160570-0
6. Soloman, Sabrie "Sensors and Control Systems in Manufacturing 2nd Edition," McGraw-Hill Publishing Co., NY – ISBN 0-07-059626-3
7. Soloman, Sabrie "Affordable Automation," McGraw-Hill Publishing Co., NY – ISBN 0-07-059633-6
8. Soloman, Sabrie "Sensors Handbook, 1st Edition," McGraw-Hill Publishing Co., NY – ISBN 0-07-059630-1
9. Soloman, Sabrie "Sensors and Control Systems in Manufacturing 1st Edition," McGraw-Hill Publishing Co., NY – ISBN 978-07-10572-4

8

3D Printing & Design in Healthcare, Food, Fashion & Environmental Safety

HEALTHCARE

The state of current 3D printing technology has set a prosperous and secure position in the history of manufacturing technology. This prominent position is highlighted by Today 7,000,000 working 3D printers acquiring US$3 billion in various industries, laboratories, academic and research institutions, medical and dental offices, as well as Universities and homes. Only two years earlier less than half of this quantity were in operation worldwide, Figure 8.1, illustrates the global 3D printing revenues among large public companies.

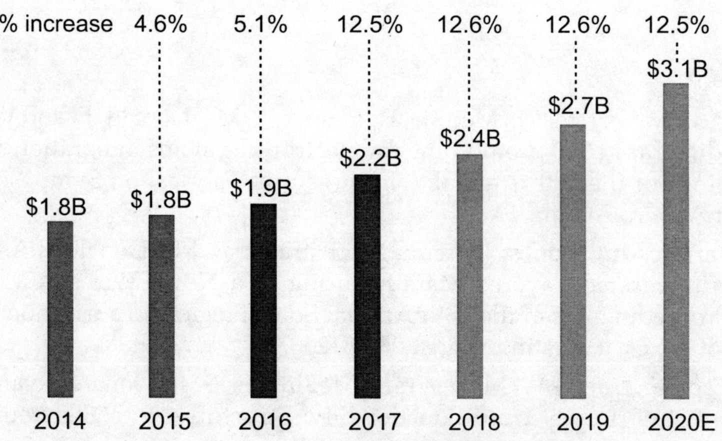

Fig.8.1: Growth in the 3D Printing Market

The progress of current 3D printers' capabilities is equally impressive when one assesses the expected breakthrough in the speed and the qualitative characteristics within the short and long range plans.

3D PRINTING IN HEALTH CARE

Bone Reconstructive Surgery

In March 2014, surgeons used 3D printed parts to rebuild the face of a patient who had been seriously injured in an accident. Surgical uses of 3D printing therapies started in the mid-1990s with functional modelling for bone reconstructive surgery planning. Patient-corresponding implants were a natural addition, leading to truly personalized grafts that fit each exclusive individual, Figure 8.2. Virtual planning of surgery and guidance using 3D printed, personalized instruments have been applied to many areas of surgery involving total joint replacement and "craniomaxillofacial reconstruction" with steady, Figure 8.3.

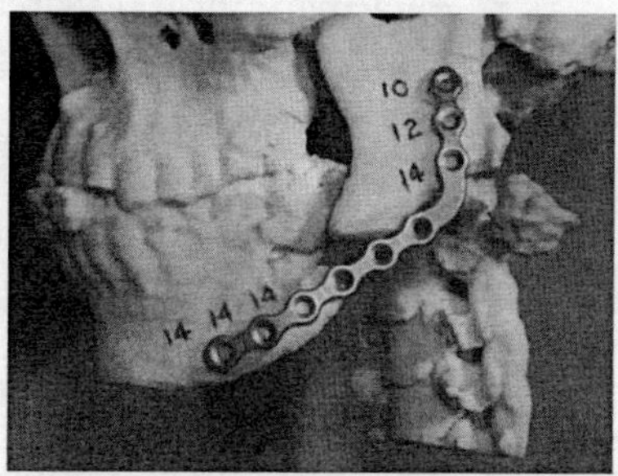

Fig.8.2: Stereolithographic Model of Craniofacial Region - Pre-bent Reconstruction.

One instance of this is the bioresorbable trachial splint to treat newborns with "tracheobronchomalacia," Figure 8.4. The use of additive manufacturing for serialized manufacture of orthopedic grafts, utilizing metals, is also increasing due to the capability to efficiently generate permeable surface structures that enable "osseointegration".

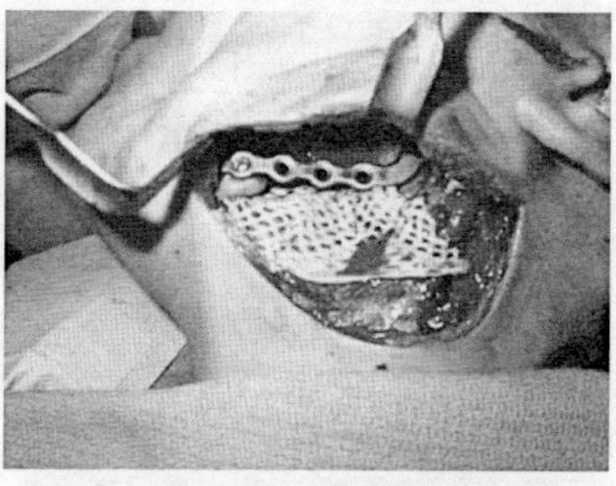

Fig.8.3: Intraoperative Placement of the Acellular Dermal Matrix Mesh after Fixation of the 3D Printed Reconstruction Plate

The hearing aid and dental industries are expected to be the largest area of future development using the custom 3D printing technology, Figure 8.5. 99% of the hearing aids worldwide are manufactured through 3D printing.

Living Tissues

As of 2012, 3D bio-printing technology has been implemented by many biotechnology organizations and academia for potential utilization in tissue engineering implementations in which organs and body parts are constructed using inkjet methods. In this procedure, layers of living cells are deposited onto a gel environment or sugar matrix and slowly constructed up to form 3D structures containing vascular systems.

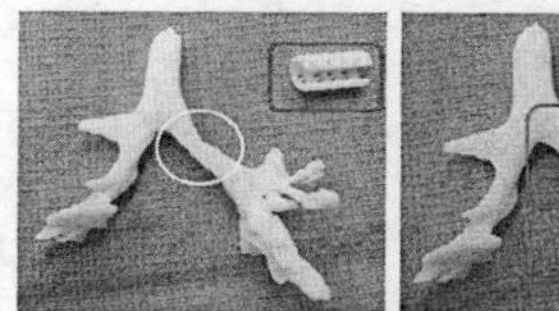

Fig.8.4: 3D Printed Bioresorbable Trachial splint for a Newly Born Baby

Lately, a heart-on-chip has been fashioned which matches characteristics of cells. A non-profit medical center is utilizing 3D printing to further its mission of genegrating the next innovation of medical technology to treat vascular illness, such as heart attack and stroke. They have established a way to 3D-print brain arteries to better comprehend brain aneurysms and stroke.

Medical Applications

Medical implementations for 3D printing are increasing rapidly and are predictable to transform health care. The genuine and potential medical applications can be organized into several broad categories, including:

Fig.8.5: Hearing Aids Used Stereolithography 3D Printing

- tissue and organ fabrication;
- creation of customized prosthetics, implants, anatomical models; and
- pharmaceutical research regarding drug dosage forms, delivery, and discovery.

Recently scientists define the utilization of 3D printing to yield bones reconstrruction, ears, exoskeletons, windpipes, a jaw bone, eyeglasses, cell cultures, stem cells, blood vessels, vascular networks, tissues, and organs, as well as novel dosage forms and drug delivery devices.

The implementation of 3D printing in medicine can deliver many adavnatages, comprising:

- the customization and personalization of medical products, drugs, and equipment;
- cost-effectiveness; increased productivity; the democratization of design and manufacturing; and
- enhanced collaboration.

However, it should be warned that notwithstanding recent substantial and exhilarating medical improvements concerning 3D printing, distinguished scientific and regulatory challenges continue, and the most transformative implementations for this technology will require time to advance.

The categorization of healthcare-applications of 3D printing can be stated as follows:

1. Tissue and organ manufacture
2. Medical implants comprising tailored prosthetics and grafts

3D printing founded on imaging data from CT/MRI

1. Functional Models for Surgical Research and Planning
2. Drug quantity forms, delivery and discovery.
3. Dentistry

The categorization of healthcare application is detailed according to the current state of their descending level of importance worldwide. The list of importance may change in near future.

Bioprinting Tissues and Organs

Bioprinting - Heart on a Chip

In October 2016, Harvard researchers 3D-printed the world's first heart-on-a-chip with incorporated sensors. The heart-on-a-chip is manufactured completely utilizing multi-material 3D printing in a single automated process, assimilating six custom printing inks at micrometer resolution, Figure 8.6.

Fig.8.6: Heart-on-a-Chip, Made Entirely Using Multi-material 3D Printing in a Single Automated Procedure

Courtesy: Harvard University

Tissue or organ failure owing to old-age, heart ailments, accidents, and birth defects is a serious medical problem. Current cure for organ failure depend on typically on organ transplants from living or deceased donors. Though, there is a chronic shortage of human organs obtainable for transplant. In 2009, 154,324 patients in the U.S. were waiting for an organ. Only 27,996 of them, 18% were donated organ transplant, and 8,863, 25 per day deceased while on the waiting list. As of early 2014, about 120,000 people in the U.S. were pending an organ transplant. Organ transplant surgery and continuation is also costly, costing more than $300 billion in 2012. An added problem is that organ transplantation embroils the often problematic charge of finding a donor who is a tissue match. This difficulty could possibly be abolished by utilizing cells cultured from the organ transplant patient's individual body to construct a substitute organ. This would lessen the jeopardy of tissue rejection, as well as the requirement to take lifelong immunosuppressant.

Solution - Organ Donor Shortage

Rehabilitations centered on tissue engineering and regenerative medicine are being followed as a possible solution for the organ donor unavailability. The traditional tissue engineering approach is to isolate stem cells from small tissue samples, blend them with growth factors, grow them in the laboratory, and seed the cells onto scaffolds that direct cell proliferation and differentiation into operative tissues.

While still in its infancy, 3D bioprinting bargains supplementary important benefits outside this traditional regenerative technique, which fundamentally delivers scaffold support only, such as, highly detailed cell assignment and extraordinary digital regulator of speed, resolution, cell concentration, drop volume, and diameter of printed cells. Organ 3D printing takes benefit of 3D printing technology to yield cells, biomaterials, and cell-laden biomaterials separately or in tandem, layer by layer, directly generating 3D tissue-like builds. Numerous materials are obtainable to construct the scaffolds, contingent on the anticipated strength, porosity, and kind of tissue, with hydrogels usually well thought-out to be most appropriate for generating soft tissues.

Droplets of Living Cells

Although 3D bioprinting systems can be laser-based, inkjet-based, or extrusion-based, inkjet-based bioprinting is most shared. This technique deposits "bioink," droplets of living cells or biomaterials, onto a substrate according to digital commands to duplicate human tissues or organs. Several print-heads can be utilized to deposit diverse cell kinds (organ-specific, blood vessel, and muscle cells), a essential characteristics for manufacturing complete hetero-cellular tissues and organs.

A procedure for bioprinting organs has appeared as follows:

1. Generate a blueprint of an organ with its vascular construction;
2. Create a bioprinting procedure plan;
3. Segregate stem cells;
4. Differentiate the stem cells into organ-specific cells;
5. Formulate bioink reservoirs with organ-specific cells, blood vessel cells, and support medium and load them into the printer;
6. Bioprint; and
7. Place the bioprinted organ in a bioreactor prior to transplantation.

In bioprinting, there are three main types of printers that have been utilized.

These are stated as follows:

- Inkjet:
- Laser-assisted, and

Extrusion Printers:. Utilized in bioprinting for rapid and far-reaching products. One kind of inkjet printer, named drop-on-demand inkjet printer, prints materials in precise quantities, reducing cost and unused.

Laser printers have also been active in the cell printing procedure, in which laser drive is utilized to stimulate the cells in a specific array, if 3-D control of the cellular setting.

Bioprinting 2D and 3D Tissues

Utilizing inkjet bioprinting schemes, scientists effectively printed skin and cartilage replacements, 2D tissues. In the meantime, other scientists achieved to utilize CT imageries of a patient's airway to laser bioprint a bioresorbable tracheal splint, 3D tissue. This splint was then positively embedded in a baby with tracheobronchomalacia, Figure 8.7.

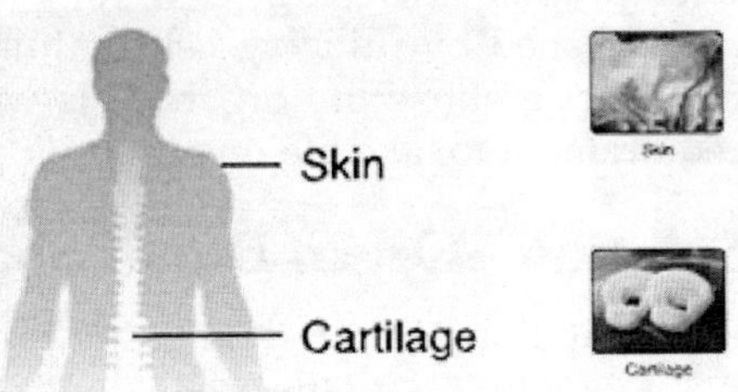

Fig.8.7: 2D Tissue

While tissue and organ bioprinting is still in its early stages, many investigations have delivered evidence of perception.

Scientists have utilized 3D printers to generate a knee meniscus, heart valve, spinal disk, other kinds of cartilage and bone, and a synthetic ear. Researchers implemented inkjet 3D printing technology to restoration human articular cartilage. They have utilized 3D bioprinting technology to deposit diverse cells inside many biocompatible hydrogels to yield an simulated liver. Doctors issued a case study in the New England Journal of Medicine recording that utilization of a 3D printer and CT imageries of a patient's airway permitted them to manufacture a exactly demonstrated, bioresorbable tracheal splint that was surgically embedded in a baby with tracheobronchomalacia, Figure 8.8. The baby recovered, and full resorption of the splint has to occur successfully.

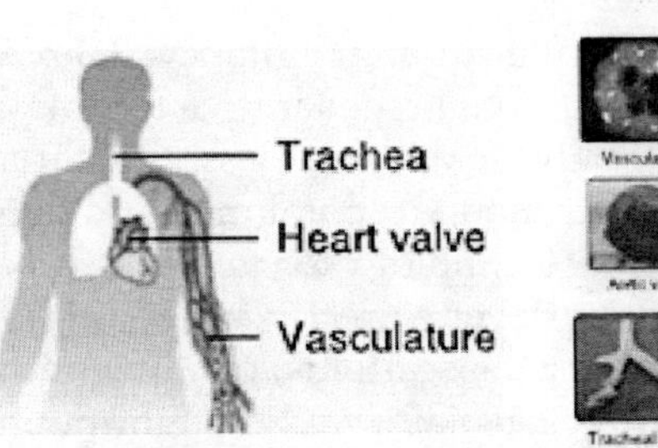

Fig.8.8: Hollow Tube

Many biotech corporations have concentrated on generating tissues and organs for medical investigations. It may be likely to quickly monitor new possible healing drugs on patient tissue, significantly reducing investigation costs and time. Researchers at "Organovo" corporation are emerging bands of printed liver tissue for this strategy; they believe the material will be progressive enough to utilize in studying new drug behaviors. Other biologists are occupied with methods to produce whole human organs that can be utilized for understanding phases of drug discovery. An organ generated from a patient's individual stem cells could similarly be utilized to monitor drug behaviors to regulate if a drug will be useful for that individual.

Bioink Components

Bioprinting utilizes a computer-controlled 3D printing means to precisely deposit cells and biomaterials into the simulated organs. Bio-printing characteristically utilzes two inks. One is the biological material and the other is hydrogel that delivers the surroundings somewhere the tissue and cells cultivate.

Essentially, when a tissue design is designated, scientists produce "bio-ink" from the cells. Stem cells can familiarize effortlessly to tissues, so they are an striking choice for bioprinting diverse organs and bones. The cells are layered between water-based layers until the tissue is constructed. That hydrogel in between layers is occasionally utilized to seal spaces in the tissue or as supports to the 3D printed tissue. Collagen is additional material utilized to fuse the cells combined. This layer-by-layer method is comparable to the common 3D printing procedure, where products are constructed from the ground up.

Hydrogels for Bioprinting

Hydrogels have been distinct as two- or multi-constituent systems containing of a 3D set-up of polymer chains and water that plugs the space between macro-molecules. Contingent upon the characteristics of the polymer or polymers utilized, as well as on the nature and density of the network joints, such build in symmetry can comprise several volumes of water; characteristically in the swollen state, the mass fraction of water in a hydrogel is abundant than the mass fraction of polymer.

In preparation, to accomplish accurate swelling, it is normal to utilize artificial polymers that are water-soluble when in non-cross-linked form. Hydrogels come out as a type of biomaterial with good biocompatibility and now they are extensively utilized as the cell-laden materials for bioprinting. As the bio well-matched materials, hydrogels have been extensively utilized in 3D bioprinting/organ printing to load cell for tissue engineering. They have been extensively utilized as cell transporters and scaffolds in tissue engineering due to their physical resemblances to the natural extracellular matrix. Hydrogel alginates have developed as one of the most regularly employed materials in organ printing investigation, as they are extremely custom tailor, and can be fine-tuned to mimic certain motorized and biological characteristic of normal tissue. The capability of hydrogels to be custom-made to precise requirements permits them to be utilized as malleable scaffold materials that are matched for a diversity of tissue or organ assemblies and physiological circumstances. Amid these, intelligent hydrogels denote to a collection of hydrogels that is receptive to numerous peripheral incentives such as pH, temperature, light, electric, and magnetic field. Combining the possible of 3D printing and intelligent hydrogels is a stimulating new example in the manufacture of a efficient 3D tissue.

Generating materials that will embellish in the human body is a complicated procedure. The bioink utilized in the organ printing procedure is very more intricate than normal printer ink. There are numerous dependent variables centered on the quantified organ and patient with regards to exclusive cells, biomaterials, and biochemical signals. The difficulty of these constituents permits for effective organ manufacture. In order for the manufactured organ to endure inside the body, biochemical and physical cues must be preserved that endorse cell endurance. Biochemical cues are related to development and adhesion factors, signaling proteins, etc. The physical characteristics are constituents outside the cell and fluid found inside the cell.

Functional Bioprinted Scaffold

There are two classifications that make up the bioink process, practical scaffold and scaffold-free. Practical scaffold utilizes biomaterials that may or may not have cells as the actual ink, while scaffold free utilizes exclusively cells. The biomaterials utilized in the usefulness of scaffold procedure come in a variation of dissimilar materials and size. From hydrogels to metal grafts, single to multiple nanometers in size – materials generated through these phases create the extracellular (outside the cell) matrix. Through these phases essentials such as dimensions, internal geometry, anti-degeneration measures, biocompatibility, as well as others must be highly considered.

i. Bioink Spherid Printed on layer of Bio paper Gel

ii. Additional Layers Printed to Build Object

iii. Bioink Spherpid Fused together – Biopaper Dissolves

iv. Final Living Tissue

Fig.8.9: Hydrogel 3D Printing Steps Forming Living Tissue

Bioprinting Approach

Scientists in the Biological Tissue Engineering field have established methods to yield living organs that are built with the appropriate biological and mechanical characteristics. 3D bioprinting is established on three main approaches:

- Biomimicry,
- Autonomous Self-assembly and
- Mini-tissue building blocks.

The first method of bioprinting is entitled biomimicry. The main objective of this method is to create fabricated structures that are identical to the natural structure that is found in the tissues and organs in the human body. The second approach of bioprinting is autonomous self-assembly. This method depends on the physical procedure of embryonic organ growth. The third method of bioprinting is a mixture of both the biomimicry and self-assembly methods, which is entitled mini tissues. Organs and tissues are constructed from very small functional components. Mini-tissue approach takes these small pieces and manufacture and organize them into greater framework.

SUCCESSFUL IMPLANTS OF PRINTED ORGANS

Bioprinting of Human Ear

Scientists at the have effectively printed bone, muscle, and ear structures and imbedded them into rats. The novel tissues have been acknowledged by the immune system and have full-grown in size and produced their own networks of blood vessels. Human sized organs are now being printed. The completely purposeful human organs are at hand, Figure 8.10.

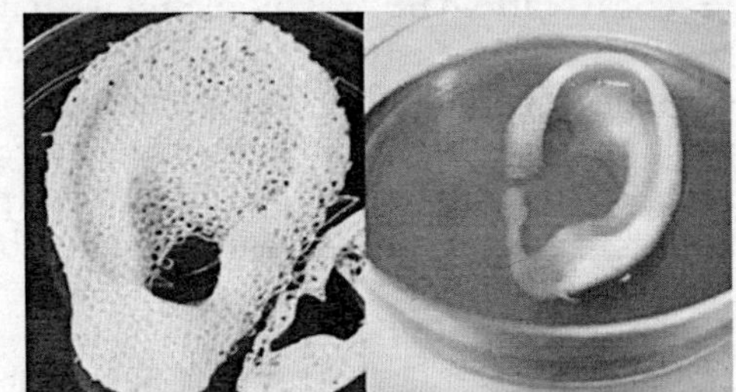

Fig.8.10: 3D Printing Human Ear with Dissolvable Matrix Scaffold

Tissues Assimilate with Bone Structures

Recently, scientists successfully created a titanium jaw graft for an 83-year-old woman, Figure 8.9. Titanium was particularly most appropriate because of its light weight and non-interference with the immune system. Titanium did not fully combine with existing bone. However, Japanese scientists concentrating in biomedical 3D printing of ceramics have effectively printed tissue that has fully assimilated with current bone structures. The new 3D printing material is calcium phosphate which is the main composite in natural bone. Several months later, the grafts accomplished utilizing this novel 3D printing ceramics technology merged with the natural bone, Figure 8.11.

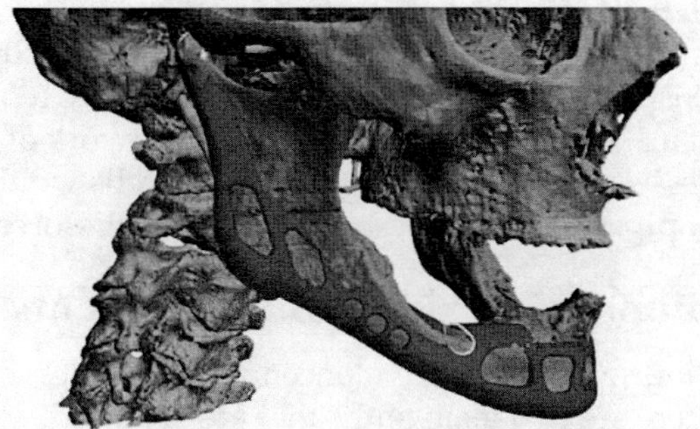

Fig.8.11: 3D Printed A Jaw Implant for 83 Year Old
Courtesy: Innovation Technology

3D Bioprinting Kidneys

The kidneys are one of the toughest organs to reconstruct—if not the toughest. This is owing to the huge amount of nephrons confined in each kidney and to the nephrons' complex construction. Scientists lately advanced forwards generating an simulated kidney that could one day substitute biological donor kidneys. Utilizing 3D printing, scientists were able to reproduce the tubule element of the kidney's nephrons and provide it a vascular network for blood flow. The 3D printed tubules were capable to remain active for over two months, Figure 8.12.

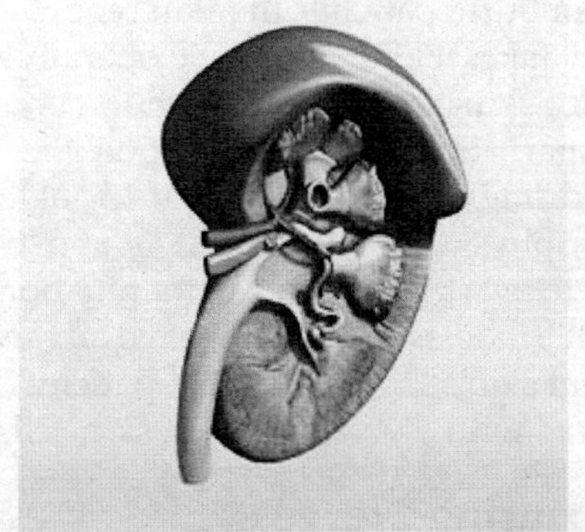

Fig.8.12: 3D Printing Kidney Research
Courtesy: Oreganovo

Bioprinting must have more advancement time before being capable to generate a fully-efficient organ. Though, the technology has other valuable implementations. Bioprinted nephron tubules could be utilized for drug toxicity evaluation, serving govern biological kidneys' capability to filter specific chemicals, or they could be combined into existing dialysis approaches to enable the technique more comparable to the method our bodies anticipated.

3D Printed Sugar Network - Artificial Liver

Scientists are a step nearer to generating a artificial liver, after a US team generated a pattern for blood vessels to propagate into, utilizing sugar. They have long been investigating with the 3D printing of cells and blood vessels, arranging up tissue structure layer by layer with synthetic cells. The synthetically engineered cells often expire before the tissue is generated, Figure 8.13. The technology, in which a 3D printer utilizes sugar as its construction material will one day be utilized for transplants.

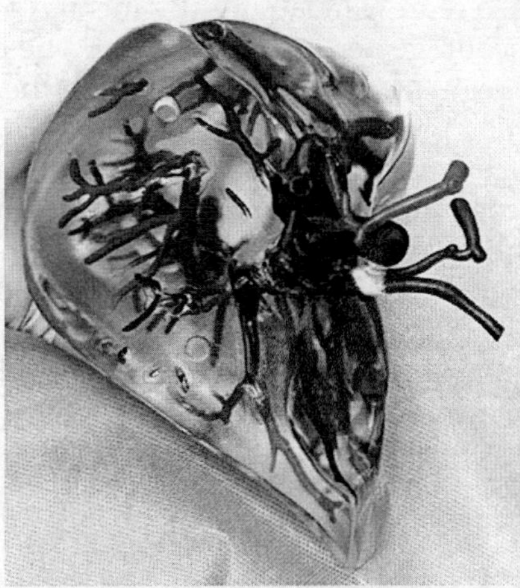

Fig.8.13: 3D Printing Research - Artificial Liver Network
Courtesy: Health Essential

The task in comprehending how to propagate large simulated tissue is how to retain all the cells alive in the engineered tissues, because when cells are put together, they feed nutrients and oxygen from adjacent cells and become choking and dying. Sugar can be dissolved away in the existence of living tissue; it is responsive to biological tissue. Scientists presented that the utilization of a 3D printer to print an random network of vessels for any tissue figure or any network of blood vessels, and then setting them with cells would generate the organ. And when the build is shaped of pipes-to-be and tissue, the sugar is dissolved away utilizing water.

Bioprinting Skin – Burnt Patients

Bioprinting is mostly concentrated on the progress of 3D printed of skin and tissues. Currently, there are two main paths that are explored.

The first path: It is one is the possibility to print skin directly on the body and

The second path: It is to 3D print tissues that could actually be implemented in a human body.

For burnt patients, the skin renovation is a vital subject. The surgery takes a long time to be operative and flourish. Typically, a part of skins is detached from the body and positioned on the burnt part. Still, contingent upon the injury significance, this method might not be sufficient to shelter the wound. Not only from a physical viewpoint, it is above all a problem for their health to avert them from contamination and fluid and heat loss. Several Schools of Medicine have advanced 3D printers that can 3D print skin straight on burnt patients. The ink of the printer is prepared with diverse kinds of skin cells, Figure 8.14.

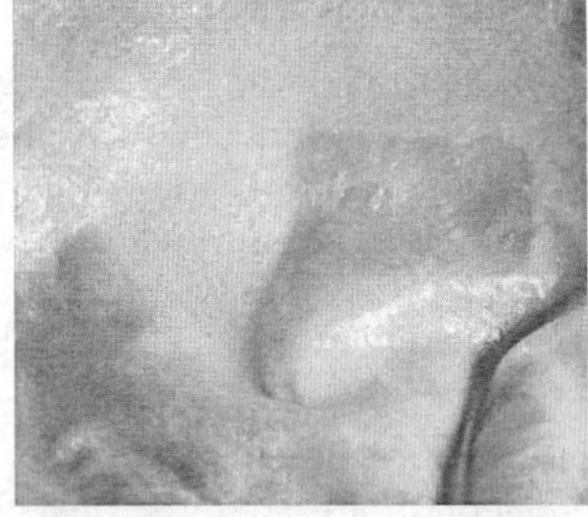

Fig.8.14: 3D Tissue Printing Skin - Transplanted to Burnt Patient

Courtesy: Linköping Burns Treatment Centre – Linköping University - Sweden

Bioprinting Nerves

Every year, 250,000 Americans lose sensation or mobility because of nerve impairment from diseases such as diabetes or injuries from accidents. Scientists discovered a method to excite those nerves to regrow in lab rats. By means of a 3D printer to generate divided tubes of silicon to excite and encourage nerve development. The procedure originated by scanning the rats' sciatic nerves, which regulate their back legs. From the scan, he 3D printed silicon tubes of the precise size and branching pattern. Scientists cut off a half-inch length of each rat's sciatic nerve, and then stitched the silicon guide onto the ends of the nerve endings. At first, the rats tripped as predictable. But the nerves were capable to develop. Within three months, the rats ran about normally. An artifact of printing—long parallel channels at the length of the tubes—delivered further direction to the developing nerves. It is since revealed that biomolecules can be placed down, such as proteins inside the channels as part of the printing procedure to guide nerve development. These improvements may pave the way toward stimulus of intricate nerve injuries for faster recuperation.

3D Printed Robotic Skin

A technique was developed by scientists and engineers to print directly 3D print tactile sensors onto prosthetic limbs or surgical robots, Figure 8.15.

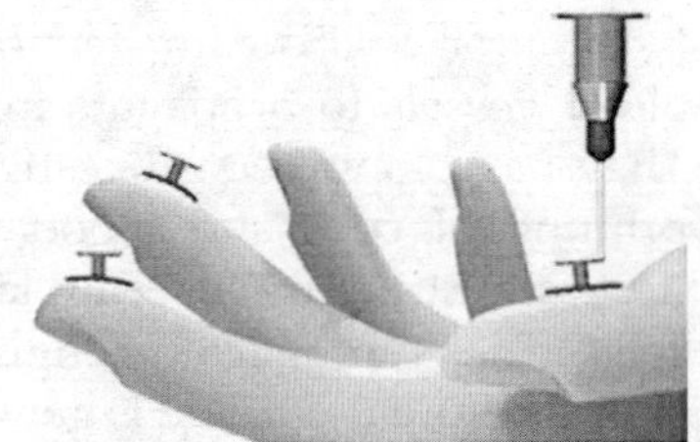

Fig.8.15: 3D Printing Tactile (Touch) Sensors Directly on Robotic Skin/Hand

Courtesy: University of Minnesota

U.S. scientists have constructed a "revolutionary" procedure for 3D printing stretchable electronic sensory instruments that could provide robots the capability to sense their distance increments. The detection, is a foremost step onward in printing electronics on actual human skin.. The novel procedure is utilizing a one-of-a-kind 3D printer construct in the lab. The multifunctional 3D printer is furnished with four nozzles, to print diverse types of dedicated "ink" that make up the layers of the instrument. These flexible 3D printed sensors can expand up to three times their initial size, rendering to the investigation.

This expandable electronic textile material has functional uses. Applying this kind of "bionic/robotic skin" on surgical robots would give doctors the capability to essentially sense during minimally invasive surgeries, which would cause surgery to be easy instead of using advanced cameras. These sensors could also brand it easier for other robots to stroll and interrelate with their surroundings. Scientists trust the novel 3D printing method could also be utilized to print electronics openly on actual human skin. This eventual wearable technology may ultimately be utilized for health observation or by militaries in the field to sense dangerous chemicals or explosives.

This is a fresh method to 3D printing of electronics that could take numerous instructions from health intensive care to energy production to chemical detection.

3D Printed Ovaries to Give Birth

The generation of simulated ovaries for humans is a step closer after birth of vigorous pups from mice provided 'ovarian bio-prosthesis, Figure 8.16. Infertile mice have given birth to healthy pups after their fertility has been restored with ovary inserts made with a 3D printer. Scientists formed the artificial ovaries by printing porous scaffolds from a gelatin ink and providing them with follicles, the manicure, fluid-property pouches that hold immature egg cells. In tests on mice that had one ovary surgically detached, researchers established that the implantations attached to the blood supply within a week and continued on to discharge eggs naturally through the pores constructed into the gelatin structures. The work provides a step nearer making simulated ovaries for young women, whose reproductive systems have been injured by *cancer treatments*, leaving them infertile or with hormone disparities that necessitate them to take regular hormone-boosting drugs.

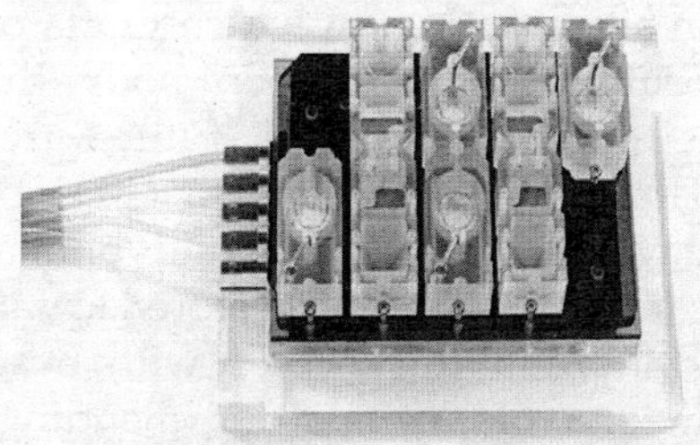

Fig.8.16: 3D Printed Ovaries – Step Closer Aiding Pregnancy

Courtesy: Warren Centre

3D Printed Vascular Networks in Ischemia

Scientists indicated they have established 3D-printed coverings permeated with cells that deliver a fresh technique to developing healthy blood vessels to treat ischemia. Their study combined 3D-printed vascular networks to guide healing angiogenesis in ischemia, Figure 8.17.

Therapeutic angiogenesis, when development factors are inoculated to boost new vessels to produce, is a hopeful trial method to treat ischemia. But in trial, the new subdivisions that sprout form a unsystematic and winding network that looks similar sort of a hairball and does not permit blood to stream efficiently through it. Researchers developed two coverings with endothelial cells—one where the cells were restructured into a precise architecture, and another where the cells were simply inoculated short of any organizational structure. In vivo results established the coverings with pre-organized structure reproduced a marked development in decreasing the occurrence of ischemia, while the coverings with no organization caused in the "hairball" situation.

This preclinical work offers a fresh method to guide improved blood flow to precise areas of the body. The increased blood nourishment delivers valued oxygen to heal and functionally preserve essential organs such as the heart and limbs.

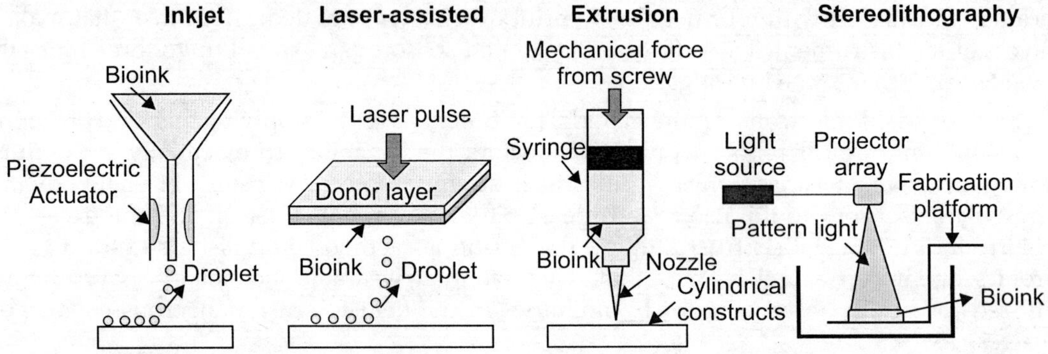

Fig.8.17: 3D Printed Vascular Networks to Direct Therapeutic Angiogenesis in Ischemia

In Situ Printing

In situ printing, in which grafts or alive organs are printed in the human body during procedures, is another expected future trend, Figure 8.18. Through utilization of 3D bioprinting, cells, development factors, and biomaterial scaffolding can be deposited to restore lacerations of numerous kinds and thicknesses with exact digital control. In situ bioprinting for mending exterior organs, such as skin, has previously taken place. In one case, a 3D printer was utilized to seal a skin lesion with keratinocytes and fibroblasts, in stratified zones through the wound bed. This method could perhaps progress to utilize for in situ reparation of partly injured, diseased, or failing internal organs. A handheld 3D printer for utilization in situ for direct tissue restoration is an expected modernization in this area. Progressions in robotic bioprinters and robot-assisted operation may also be essential to the development of this technology.

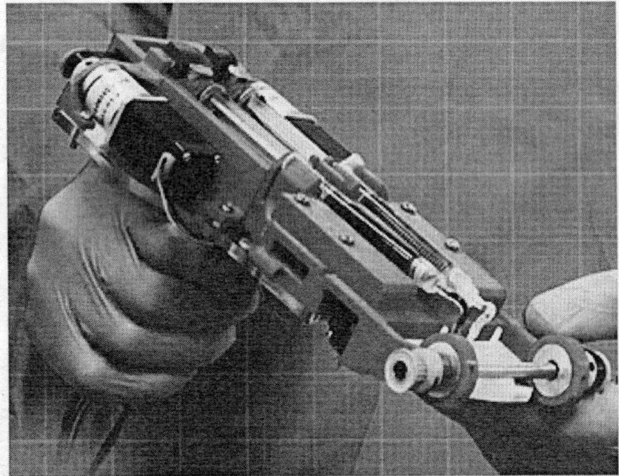

Fig.8.18: In Situ Portable 3D Skin Printer

Courtesy: University of Toronto - Researchers have created the first handheld 'skin printer' that forms tissues in situ for application to wounds.

Magnetic 3D Bioprinting

Magnetic 3D bioprinting is a practice that allows biocompatible magnetic ***nanoparticles*** to print cells into 3D build or 3D cell cultures. In this procedure, cells are labeled with magnetic nanoparticles, ***nano-shuttle*** that is utilized to render them magnetic. Once magnetic, these cells can be quickly printed into precise 3D designs utilizing exterior magnetic forces that mimic tissue build and performance. Magnetic 3D bioprinting can be utilized to screen

Fig.8.19: Magnetic Levitation 3D Bioprinting Market – *Space Market Research*

for cardiovascular noxiousness, Figure 8.19, which accounts for 30% of drug extractions. Vascular plane muscle cells are magnetically printed into 3D ring configurations to mimic blood vessels that can contract and dilate. This system could possibly substitute trials utilizing ex vivo tissue, which are expensive and harvest insufficient data per trial. Furthermore, magnetic 3D bioprinting can use human cells to estimate a human in vivo reaction better than with animal experiments. This has been established by the bioassay which associates the advantages of 3D bioprinting in constructing tissue-like structures for investigation with the speed of magnetic printing.

The key objective users for magnetic 3D bioprinting are in the pharmaceutical and Control Research Organization, CRO industries, where this system can be incorporated initially in the drug discovery procedure as a composite screen for toxicity and effectiveness. In the upcoming years, magnetic 3D bioprinting could be implemented to the field of regenerative medicine and organogenesis.

Overall, magnetic 3D bioprinting is an operative tool to generate realistic models of native tissue, Figure 8.19.

3D BIO-PEN WITH BIO-INK

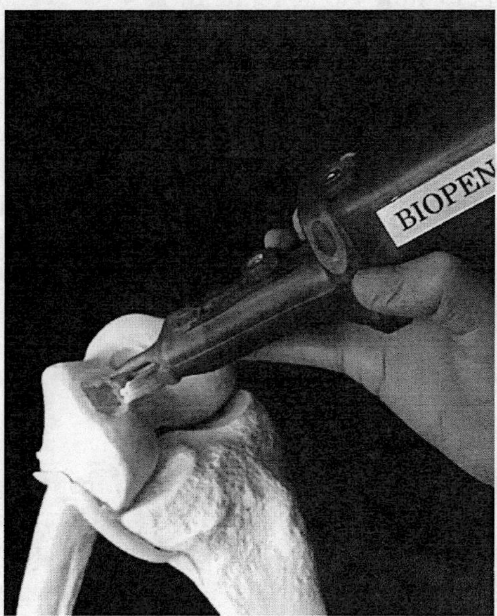

Fig.8.20: Handheld 3D Biopen Prints Human Cartilage

Courtesy: University of Melbourne

Handheld 3D Bio-Printing Pen

It is tailored cartilage grafts prosper in sheep trial, Figure 8.20:

A radical tool for printing living cells in situ during operation inches nearer to human trials. A handheld pen that yields a bio-ink comprising a patient's own cells could soon day be utilized to mend knees injured by osteoarthritis. In a initial studies, the 3D-printing pen created excellent data when verified on sheep. The Biopen is an efficient new tool accessible to doctors. In the trial, cells taken from each sheep were combined with a particularly formulated Bio-ink. During surgery, the tool was then utilized to generate an exact, bespoke graft that substitute cartilage lost through osteoarthritis. The ink performs to guard the cells during the procedure, and then maintain them until they become completely incorporated into the joint. Because they are resulting from patient – or animal – being preserved, there is no risk of rejection and no need for immunosuppressive drugs.

Scientists verified the tool on eight osteoarthritic sheep. Each had one knee cured by the Biopen, and another using current technique. The consequences were excellent, as it was discovered that the Biopen achieved better in terms of quality and features of new cartilage formation.

Obstacle to Bio Printing Organs

In an ordinary organ, every cell is close only a couple of cells of a minor blood vessel and if that cell or cell group is supposed to secrete somewhat substance like insulin, a development factor or digestive enzyme, there is a channel for that as well close. These minute channels and capillary vessels are frequently lesser in diameter than a single cell which makes them exceptionally challenging to engineer by standard technologies, which can be arranged. This is where the collapse usually occurs. Samples can be made in large numbers of cells on flat sheets or a limited layers thick and can utilize abnormal mixtures to generate these vessel beds, put them together has not successful yet. Additionally, the capability to shape small beds has not been positive. Blood vessels are not that different to an oak or maple tree. If one would envisage the heart is the main trunk, and the furthest twigs are the furthest roots. It begins as a single large volume and then continues dividing and splitting and is clear that the volume in the margin is greater than in the principal core, but all the leaves are ultimately fed by a single point where those leaves signify cells.

Too many researchers are attempting to produce organs by the sameness of developing a tree by congregating a pile of leaves and hoping. Vascularization is a big difficulty in the method of 3D printing organs, since they require having a system of arteries, capillaries, and veins that support the system. They must be exit to deliver nutrients and eliminate waste generated by the cells. One choice is to vacate the space in the 3D printed tissue for veins to be included later on in the process, but scientists are now attempting to find out a method to print blood vessels as well.

Medical Devices

3D printers are utilized to fabricate a diversity of medical instruments, comprising those with intricate geometry or characteristics that equal a patient's exclusive anatomy. Some instruments are printed from a normal design to make multiple matching duplicates of the same device. Other devices, called patient-complemented or patient-specific devices, are generated from an exact patient's imaging data.

There are diverse commercially available 3D printed medical devices include the followings:

- Hearing aids and eyewear
- Instrumentation (e.g., guides to assist with proper surgical placement of a device),
- Implants (e.g., cranial plates or hip joints),
- External prostheses (e.g., hands).

Hearing Aid

3D printing has by now had a Changeable consequence on hearing aid manufacturing, Figure 8.5. Today, 99% of hearing aids that suitable into the ear are custom-made utilizing 3D printing. Everybody's ear canal is molded in a different way, and the utilization of 3D printing permits custom-molded devices to be created competently and cost-effectively.

3D Printed Custom Eyewear

Tailored eyewear is more costly than the typical mass-manufactured prescription frames. This is because producing custom tailor eyewear is time-consuming and augments line items to the costs of a discrete frame. For instance, capturing facial dimensions is supplementary to an optician's responsibilities during the custom eyewear process.

Likewise, custom clay patterns augment to the material and time costs. Furthermore, errors at any phase of the procedure are costly and necessitate recalibration of patterns and measurements. 3D printing services, such as "Shapeways" and "iMaterialise," have cut down access to industrial grade printers for normal customers and allowed them to design custom frames for themselves. Though, these benefits are counterbalance by restrictions in completed products. Custom 3D printed eyewear is afflicted by the same drawbacks that distress other 3D printed objects. The final product is restricted in strength, finish and diversity. Selective Laser Sintering or SLS printers, which are characteristically employed to construction custom eyewear, are costly and restricted in their capacity to custom fabrication eyewear utilizing available materials for 3D printer.

Self-Adjustable Glasses

Scientists promoting the extensive prospective of 3D printing lately utilized a "RepRap" printer to construct self-adjustable glasses. The glasses cost just $1 to create, and the lenses can be self-adjusted, serving low-income people to evade the costs normally related to vision alteration. The project validates the humanitarian consequences of 3D-printing. With 90% of visually impaired persons living in countries where remedial eye-care is either unreasonable or impossible, the "Adspec" glasses could have a considerable global influence.

Customized Implants and Prostheses

Grafts and prostheses can be manufactured in almost any conceivable geometry via the conversion of X-ray, MRI, or CT scans into digital ".STL" 3D print files. In this method, 3D printing has been employed positively in the health care segment to allow both typical and intricate customized prosthetic limbs and surgical grafts, occasionally within 24 hours. This method has been employed to construct dental, spinal, and hip inserts. Beforehand, before grafts could be employed clinically, they had to be authenticated, which is very time-consuming.

The aptitude to rapidly yield custom grafts and prostheses resolves a clear and tenacious difficulty in orthopedics, where typical grafts are often not adequate for some patients, mainly in intricate circumstances. Beforehand, surgeons had to achieve bone graft surgeries or utilize scalpels and drills to adjust grafts by shaving pieces of metal and plastic to a wanted figure, size, and fit. This is also true in neurosurgery: Skulls have uneven figures, so it is tough to normalize a cranial graft.

In wounded of head injury, where bone is detached to provide the brain room to swell, the cranial plate that is later tailored must be precise. Although some plates are milled, more and more are shaped employing 3D printers, which enables it much easier to modify the fit and design. There have been numerous commercial and clinical achievements concerning the 3D printing of prostheses and grafts. A team of scientists successfully implanted the first 3D printed titanium mandibular prosthesis. The implant was made by using a laser to successively melt thin layers of titanium powders.

In 2013, "Oxford Performance Materials" was given FDA endorsement for a 3D printed polyetherketoneketone, PEKK skull implant, which was first effectively grafted that year. The company manufactured 3D-printed titanium orthopedic, maxillofacial, spinal, and dental grafts,

Figure 8.21. A structurally correct 3D-printed prosthetic ear able of sensing electromagnetic frequencies has been manufactured utilizing silicon, chondrocytes, and silver nanoparticles. There is a emergent trend in the direction of making 3D-printed grafts out of a diversity of metals and polymers, and more lately grafts have even been printed with living cells.

Fig.8.21: 3D Printed Bionic Ear

Knee Replacement

Printing technology has already transfigured joint substitutes. Knee emergency is a very common practice; there are about six diverse kinds of knees that doctors utilize, with each one the requirement to cut the bone another way. But with 3D printing, doctors are not restricted to those six knees. They can develop one exact to each patient, Figure 8.22. Patients with tailored knees do not have to drop extra inches of bone, in its place the surgeon can amend at the optimal point, which could lead to faster heling times and better performance. Robust, flexible new knee joints imitating bone and cartilage can now be printed with nylon. These surgeries are accessible at top-tier medical services in USA.

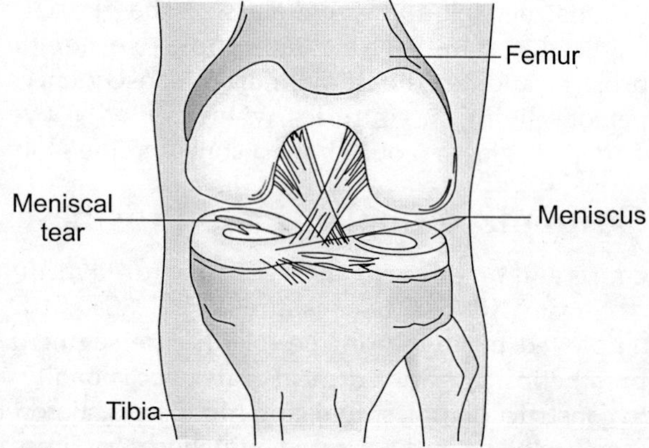

Fig.8.22: Researchers at Columbia University - Medical Center, USA Replaced Damaged Menisci Using 3D Printed Implant Infused with Proteins that Stimulate Stem Cells - Body Regenerated a New Functional Lining

3D Printing - Imaging Data CT/MRI

Graspable (Intelligible) 3D objects bypassed the restrictions of 3D imaginings which can only be projected on flat screens. 3D workpieces can be fashioned established on CT or MRI volumetric medical imageries. By means of dedicated post-handling algorithms, a spatial model can be mined from image data sets and transferred to machine-readable data. That spatial model data is exploited by special printers for creating the final rapid prototype model. The main necessities are isotropic data, most circumstances from CT or MRI, dissection

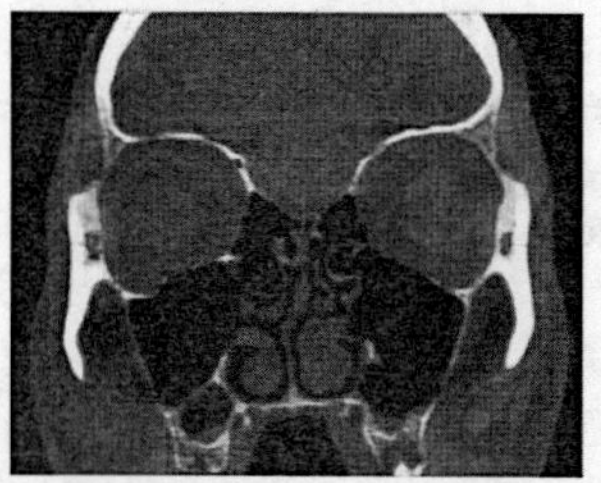

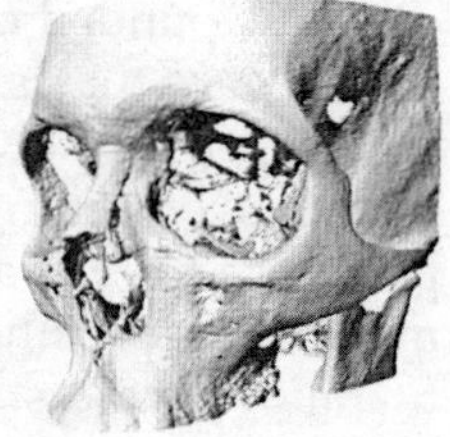

Fig.8.23: Transforming 2D Radiographic Images to 3D for 3D Printing

software, e.g., 3D slicer, capable to transfer a volumetric object as STL file format, slicing software, encompassed with the printer, and entry-level 3D printing, most models utilize fused filament manufacture, FFM, Figure 8.23.

The main implementations fields for 3D printing are urology, cardiology, cardiac surgery, vascular surgery, craniomaxillofacial surgery, neurosurgery, orthopedics, and pulmonology.

Some examples of potential applications of 3D printing in the following medicals implementations:

Cardiac surgery

In the situation of a septal metastasis of a malignant melanoma, a 3D print based on an electrocardiogram-triggered, contrast-enhanced CT scan aided visualization of the tumor, helped facilitate the preoperative planning, and was used for patient education. The segmented tumor matched the resected specimen.

Pulmonology

Anatomic relationship of a small peripheral tumor is found in regard to the bronchi. A hollow, high-resolution resin, 3D print allowed the pulmonologist to better plan the bronchoscopy and was used as a guiding reference for orientation during the procedure. It was also used for training purposes.

Urology

In a case of renal cell cancer, the 3D print-based multiphasic CT scan, hand-colored after printing, boosted visualization of the tumor and facilitated the preoperative planning for the kidney vasculature and the ureter. In another case of renal cell cancer, 3D printing facilitated preoperative planning and permitted nephron-sparing robotic surgery.

Orthopedics

3D-print mirrored the original size of a complex fracture of the right hemi-pelvis. The mirrored print made it possible to use the non-fractured side as a template for the preoperative bending of the osteosynthesis plates, thereby saving time and increasing accuracy.

3D Printed Casts for Broken Bones

It is significantly painful is a broken bone. They take a long time to recover. 3D technology can be convenient in this situation. Doctors can take X-rays or MRI scans, and then build a needed structure that is an exact match for the patient using a 3D printer. This can be helpful for thousands of patients. A cortex exoskeletal cast is one such example. These casts are far better than the traditional casts used for broken bones.

Structured Anatomy

Pro-section of the hand and wrist with 3D images and 3D printed model provides the following observations:

A. Image of CT-scanned pro-section of hand and wrist;

B. The 3D computer image is constructed from the CT data, in this case, exported from the scanner workstation in DICOM format, using image processing software, e.g., Amira, Avizo, Mimics, Simple ware, 3D Slicer, which creates a stereolithography file .STL;

C. Because CT scan does not provide information on color, anatomically realistic colors can be added using a package such as 3D Coat;

D. The colored .STL file can then be printed in full color as a 3D copy of the original pro-section.

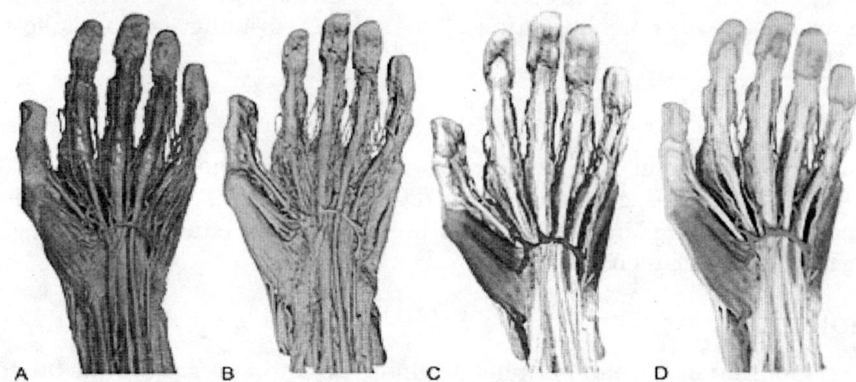

Fig.8.24: 3D Printed Model of Pro-sectioned Hand

The education of anatomy has steadily been the subject of societal disagreement, particularly in the context of employing cadaveric materials in professional medical and associated health professional teaching, Figure 8.24. The reduction in dissection-based teaching in medical and allied health professional training programs has been in part due to the monetary concerns tangled in maintaining legacy programs, accessing human cadavers and concerns with health and safety attentions for students and staff showing to formalin containing embalming fluids. 3D printing permits the formation of reproductions of pro-sected human cadaver and other anatomical examples that remove many of the above issues.

Anatomical Models for Surgical Preparation and Planning

The separate modifications and difficulties of the human body create the deployment of 3D-printed models perfect for surgical groundwork. Having a palpable model of a patient's anatomy obtainable for a physician to learn or utilize to mimic surgery is desirable to relying solely on MRI or CT scans, which aren't as informative meanwhile they are observed in 2D on a flat screen. The use of 3D-printed models for surgical exercise is also desirable to exercise on cadavers, which present problems with respect to obtainability and cost.

Cadavers also often lack the suitable pathology, so they deliver more of a experience in anatomy than a illustration of a surgical patient. Scientists at the National Library of Medicine creäte digital files from clinical data, such as CT scans, that are employed to make special 3D-printed surgical and medical replicas. 3D-printed neuroanatomical replicas can be mainly helpful to neurosurgeons by as long as a depiction of some of the most intricate configurations in the human body. The complicated, sometimes disguised relationships between cranial nerves, vessels, cerebral structures, and skull construction can be problematic to understand grounded solely on radiographic 2D images. Even a minor error in circumnavigating this complex anatomy can have actually shocking significances.

A accurate 3D model reproducing the relationship between a lesion and normal brain constructions can be helpful in deciding the safest surgical passage and can also be valuable for the neurosurgeon to practice perplexing cases. Intricate spinal malformations can also be learnt better through the utilization of a 3D model. High-quality 3D anatomical replicas with the right pathology for teaching doctors in executing colonoscopies are also essential, since colorectal cancer is the second leading reason of cancer-related demises in the U.S.

With the assistance of 3D sculpting skills, surgeon tools are effortlessly customizable in a 3D file, to be printed after. Rendering to the fact that the doctor or hospital possess a 3D printer or use a 3D printing service, tools can be arranged for the surgery in a day and are tailored for the patient's body. Mayo Clinic shaped around 750 surgical case models on its own 3D printers in 2018, has achieved complete cranial reconstructions, detached intricate tumors and positively detached conjoined twins with 3D printed monitors and tools.

Surgical Planning

There are numerous advantages to patients, surgeons, and hospitals in utilizing a 3D-printed model to devise an optimal surgical plan. It is summarized as follows:

- Allows surgeons to try a particular approach or device in a risk-free environment;
- Provides the surgeon with practice time before performing the actual surgery, much like medical simulation;
- Minimizes the time a patient is on the table, being exposed to harmful radiation, as a surgeon tries to figure out the best approach;
- Reduces surgical cost associated with longer surgery;
- Reduces cost associated with incorrect device selection, catheters, wires, or devices types and sizes;
- Determines which are the best tools for the specific patient's case;
- Helps the surgeon choose the appropriately-sized device;
- Identifies complications so they can anticipate them in surgery.

3D Printing Aorta

Figure 8.25, illustrates 3D imagining and 3D rapid prototyping of the "Aorta" in a patient with thoracic aortic aneurysm, arrow. 3D models are supportive in representing intricate pathologies. Vascular surgeons may discover models valuable for assessment of the best treatment plan as well as for exercise. Discrete patient care can help from rapid prototyping by as long as a understandable model, thus serving the patient to comprehend the pathology and to provide knowledgeable consent for surgical involvements sufficiently skilled on models, surgeons feel more self-assured while going to the operating theatre for actual surgery. Moreover, the pre-operative recreation of a exact and intricate surgery to offers a exceptional occasion to employ surgical steps in order to decide the best operating strategy, again growing the surgeon's confidence during the operation.

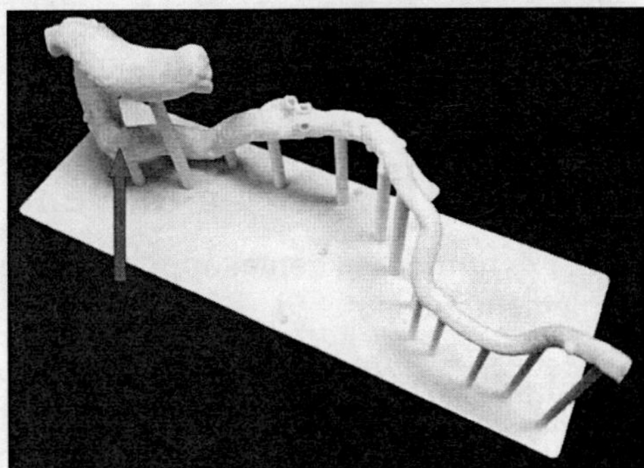

Fig.8.25: 3D printed Model Created by Computed Tomography (CT) - Scan of 89-year-old patient with Severe Aortic Stenosis - Scanned for "Transcatheter Aortic Valve"- Replacement – TAVR Evaluation

Courtesy: Stanford University – Stanford Medicine

Pharmaceutical Discovery - Drug Dosage - Delivery

3D printing technologies are even now being employed in pharmaceutical investigation and manufacture, and they capacity to be transformative. Benefits of 3D printing embrace accurate control of droplet size and amount, high reproducibility, and the capacity to yield dosage forms with intricate drug-release profiles. Intricate drug engineering procedures could also be standardized through utilizing of 3D printing to crreate them simpler and more feasible. 3D printing technology could be very significant in the improvement of tailored medicine, too. The first tablet, capsule, or gel fabricated by 3D printing was accepted by the FDA in August 2015. Binder-jetting into a powder bed of the drug permits very absorbent tablet, capsule, or gel to be manufactured, which allows high drug doses in a single tablet, capsule, or gel which dissolves rapidly and can be dissolved easily. This has been established for Spritam, a reformulation of levetiracetam for the treatment of epilepsy.

Dentistry

Already valued $2.10 Billion in the dental market, 3D printing could become a $3.1 billion industry in this sector by next year 2020, agreeing forecasts in a report released in May 2019. The prevalence of 3D printing in other markets is predictable to mature as well, but improved 3D printers and materials are now being mass-produced precisely for dentists, Figure 8.26. Also it is forecast the sale of 3D printing systems to dental labs will doubled by 2020. 3D printing technology is also predictable to deliver more than 60% of all dental production needs by 2025, and perhaps even more in certain areas such as dental modeling.

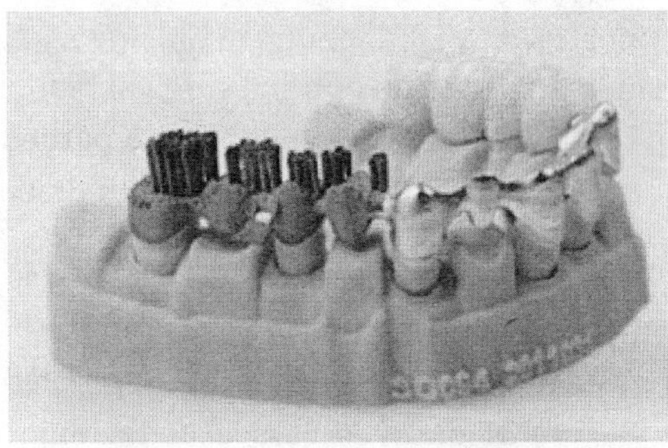

Fig.8.26: 3D Printed a Patient Specific Denture

Dental Implants

Dental industry has been employing synthetic material for dentures, orthodontics, implants, crown, and bridges for numerous years. As these parts are tailor made for each person, the procedure is both time consuming and costly. Direct and indirect 3D printing, specifically printing the actual part or a mold, has been revealed to be a inexpensive and faster substitute to conventional methods.

The Use 3D Printing of Dentistry Technology

Here are numerous dental applications employing 3D printing technology, below are a few a few implementations stated as follows:

Replace or Repair a Damaged Tooth

The dentist scans the patient's mouth with a small size digital wand. This generates a 3D image of the teeth and gums, which is stored as a computer file. Computer Aided Design, CAD software permits the dentist to digitally develop the tooth restoration and print the completed product on a 3D printer.

Create an Orthodontic Model

Pre-3D printer technology comprises having the patient bite down on gooey, painful clay so it could toughen into a mold, which develops the preliminary model for developing a cure for braces or "Invisalign." This is not so with 3D printing. A dentist can utilize the same technology highlighted in the first example to scan the teeth, design an orthodontic appliance and print the end result in-house. "Invisalign" braces are successful commercial use of 3D printing, with 60,000 printed every day. These clear, detachable, 3D-printed orthodontic braces are tailor-made and exclusive to each patient. This product offers a good example of how 3D printing can be utilized competently and beneficially to make single, tailored, intricate items.

Produce Crowns, Bridges, Caps, Dentures asnd More

The identical procedure defined above can be utilized to 3D print all kinds of dental implants. The only change is the exact material employed in the printing procedure.

Construct Surgical Tools

Not only can 3D printers deal with the dental implants themselves, but they can also 3D print the drill fixture required to complete specific dental procedures.

3D Printing in Education

Fig.8.27: 3D Printing - University Curriculum

The capability to yield nearly any workpiece in 3D is self-assured to transform education. As a substitute of utilizing linear, 2D teaching approaches to handover knowledge and teach ideas, one can expedite the education capability by providing hands-on 3D models, Figure 8.27. Perceptions that have been historically problematic to grasp can be directly established with 3D printed visual aids that aid students learn and recollect ideas more effortlessly. Learning through immersive experiences has long been a gratifying approach to education of all categories, and with the advent of 3D printing technology, one can implement this method virtually everywhere.

Interpreting ideas into 3D objects vicissitudes the method one teach as well as the way the students learn. From desktop printers to specialized online services employing high-grade printers, 3D technology is becoming a priceless inventive tool in many courses of study outside 3D computer design. While one may find numerous 3D printers are too expensive for utilization in a classroom environment, even the rudimentary models simplify cooperative learning. Apart from cost, another impediment to 3D printing in schools is the absence of familiarity in most mentors for utilizing this technology or the practices for implementing it to everyday educational use. Hitherto, 3D printers are a likely choice for teaching in STEM curriculum, Figure 8.27. Science, technology, engineering and mathematic, STEM is a term that denotes to the academic disciplines of science, technology, engineering and mathematics. The term is characteristically utilized when addressing education policy and curriculum adoptions in schools to advance competitiveness in science and technology advancement.

The implemented knowledge in designing and generating working objects prepares the students to excel in these areas of study while procuring valuable experience in chasing after prosperous technical careers.

Some of the most favorable uses for 3D printing in education embrace the followings:

- 3D printing and mathematics can work together seamlessly in order to generate precise representations of hard to comprehension mathematical ideas. Some students have struggle understanding formulas and diagrams that occur only on paper and 3D printing can be utilized to help them envision perceptions they might have trouble with.
- 3D printing permits those students who have anxiety envisioning equations, intricate graphs, and multipart mathematical arrangements to practice real-world models. Also, 3D printers convey clarity of methodological aspect to a subject which most students conventionally find out-of-date and boring.

3D printing delivers a brand new approach to procedures of design and graphics. Students no longer have to count on 2D screens and imageries to comprehend 3D models. 3D printing consents design students to discover scaled, sophisticated models they could never formulate in the lab or workshop. The overview of 3D printing to educational surroundings means students can recapitulate through distinctions of their designs and appreciate how they accomplish. Young designers can thrust their creative philosophies as far as the 3D software permits.

Implementing 3D Printing in Science – Technology - Engineering – Mathematics "STEM"

Ways 3D Printing can be used In Education are stated as follows:

1. Engineering design students is capable to print out prototypes
2. Architecture students is capable to print out 3D models of designs
3. History classes is capable to print out historical artifacts for consideration
4. Graphic Design students is capable to print out 3D varieties of their creation
5. Geography students is capable to print out landscape, demographic, or population maps
6. Cooking students is capable to generate molds for food foodstuffs
7. Automotive students is capable to print out auxiliary parts or adapted samples of current parts for testing
8. Chemistry students is capable to print out 3D models of molecules
9. Biology students is capable to print out cells, viruses, organs, and other serious biological relics
10. Math students is capable to print out "problems" to solve in their own learning atmosphere, from scale models to city infrastructural design trials

3D Printing in Education – International Implementations

In 2019, Dimensional Research Institute executed an international survey of over 300 educational institutions concerning the implementation of 3D printing. Of those surveyed, 99% have at least one printer though direct student admission and further expansion remains to be limited due to three key obstacles:

1. **Security & Access**: Guaranteeing safe and protected right of entry to 3D printers.
2. **Cost Management**: Managing continuing costs & mitigating the Return on Investment, ROI.
3. **Incorporating into Curriculum**: Integrating 3D printing ventures into classroom curriculum.

The Benefits of 3D Printing in Education

Numbers of education sectors have received this technology with open arms. This is contributed to the abundant benefits 3D printing can bring, examples of benefits are stated as follows:

- New learning possibilities – In addition to plainly feeing and viewing their designs take wings and come to life, students can touch and interrelate with the completed product and understand it the way not previously conceivable.
- Access to learning materials – If a school or an educator doesn't have right of entry to education materials, they can merely create them.
- Generates excitement and engagement – Somewhat than inactively reading from a textbook or listening to the teacher, students will have a added lively curiosity in 3D printing, as it is very much a hands-on action.
- Inspires imagination and novelty – With right to use to design software and a 3D printer, the potentials of what a student can fashion are endless, which inspires greater advancements.
- Encourages critical thinking and problem solving-skills – In ranges of education where students must workout critical thinking and problem-solving skills, a 3D printer can contribute incalculably.
- Better knowledge retaining rates – People study novel information and knowledge more efficiently if they do somewhat than just seeing or listening.
- Reasonable teaching tool – As contrasting to a few years ago when 3D printers cost thousands of dollars, they are now much more reasonable and inexpensive than alternative education means such as laptops and computers.

While some may feel as though 3D printing is merely a method for students to research and express their inspiration, this technology can be much more practical than that. In fact, education materials fashioned for subjects like mathematics and science could stimulate the next age group of engineers, architects, designers, or any other inspired yet empirical profession. It can also aid students that fight with learning theories and topics from a classic text books but are much more adept when given palpable objects to work with. Consequently, students that were previously disheartened with education could not only be revived by the prospect of learning but also reveal their hidden potential.

Revolution in Science-Technology-Engineering- Mathematics (STEM) Education

Science, Technology, Engineering & Mathematics some writers have appealed that 3D printers give an unparalleled "revolution" in STEM education. The indication for such prerogatives comes from both the low cost capability for rapid prototyping in the classroom by students, but also the manufacture of low-cost high-quality scientific equipment from open hardware designs creating open-source labs. Engineering and design philosophies are discovered as well as architectural forecasting. Students recreate replacements of museum items such as fossils and historical artifacts for education in the classroom without possibly harming sensitive collections.

Other students attentive in graphic designing can build models with intricate working parts easily. 3D printing stretches students a new viewpoint with topographic maps. Science students can investigate cross-sections of internal organs of the human body and other biological samples. And chemistry researchers can discover 3D models of molecules and the association within chemical complexes.

Architecture

Architectural models have long been a principal implementation of 3D printing procedures, for generating precise demonstration models of an architect's dream. 3D printing bids a relatively fast, easy and thriftily viable method of creating thorough models directly from 3D CAD, BIM or other digital data that architects utilize. Many effective architectural firms, now frequently utilize 3D printing (in house or as a service) as a vital part of their workflow for augmented innovation and enhanced communication. More lately some visionary architects are considering 3D printing as a direct building method. Investigation is being showed at a number of organizations on this front, most particularly Loughborough University, Contour Crafting and Universe Architecture. Iron oxide-free Portland cement powder has been utilized to generate architectural constructions up to 9 feet in height.

AUTOMOTIVE INDUSTRY

According to a January 2018 report by the US consultancy Frost & Sullivan, the implementation scope of 3D printing is presently limited to the production of extremely low volume parts and production tooling. Notwithstanding this, the firm predicted that the technology will produce $ 4.3 billion in the automotive industry by 2025, and achieve deeper infiltration in manufacture and the aftermarket. In cars, trucks, and aircraft, AM is start to change both:

(1) Unibody and fuselage design and manufacture and

(2) Powertrain design and manufacture.

General Motors

When General Motors, GM was underway to build the 2014 Chevrolet Malibu, engineers at the company utilized 3D printing to save time essential in prototyping the parts for the vehicle. GM employed stereolithography, dedicated software, math data and laser sintering to construct parts out of liquid resin in order to make developments to the Malibu. General Motors said that the rapid prototyping showed particularly valuable for the floor console, which has smartphone holders for the driver and passenger. The parts were also lighter weight so that it provides the Malibu an better-quality fuel economy. GM utilized 3D printing for the front fascia design to test vehicles in wind tunnels. 3D printing was too employed to sculpt the front-seat back panels.

Ford Motor Company

Ford Motor: Company uses 3D printing to create prototypes of numerous parts in their vehicles, such as cylinder heads, brake rotors, shift knobs and vents. 3D printing was employed at "Ford's Torrence Assembly Plant" for the manufacture of the Explorer and EcoBoost engines.

Urbee 2

The team of engineers is constructing a whole vehicle with 3D printers named the "Urbee 2." The body of the original Urbee was finished through the employment of 3D printing. Though, the "Urbee 2" as a whole will be typically 3D printed, Figure 8.25. Similarly, the "Urbee 2" will also appear more similar a production-ready car associated to the original.

Fig.8.28: 3D Printing Urbee2

Urbee2 is the name of the initial car in the world car mounted employing the technology 3D printing its bodywork and car windows were "printed." The car is fashioned by collaborative work between the US engineering group and the company Stratasys, the manufacturer of Stratasys 3D printers. The Urbee2 it is a hybrid vehicle with innovative look, Figure 8.28. The engineering group will be driving the Urbee 2 from New York to San Francisco upon conclusion and is predictable to use less energy than most other production cars. Their goal is to exploit only 10 U.S. gallons of gas for the whole trip.

3D Printed Bus & Aviation

Like numerous industries, the aerospace industry is progressively implementing 3D printing and rapid prototyping technologies to advance aircraft parts in the pursuit of trimming down engineering and fabrication costs. As a matter of fact, one of the main players in the aerospace domain, "Boeing," already makes use of 3D printing technology comprehensively and printed over 22,000 parts last year. In May 2017 Airbus announced that its new "Airbus A350 XWB" incorporated over 1000 components fabricated by 3D printing, Figure 8.29.

Fig.8.29: 3D Printing Air-Bus

3D printing is also being employed by air forces to print replacement parts for planes. In 2018, a Royal Air Force Eurofighter "Typhoon" fighter-jet took-off with printed parts, Figure 8.30. The United States Air Force has started to work with 3D printers, and the Israeli Air Force has also acquired a 3D printer to print replacement parts.

Fig.8.30: 3D Printing Metal Airbus Wing Bracket – Lighter and Stronger Than Conventional Wing Bracket

In 2019 GE Aviation shown that it had made design for additive manufacturing to construct a helicopter engine with 16 parts instead of 900, with great possibly-impact on decreasing the intricacy of supply chains. Furthermore, General Electric (GE) recently proclaimed an investment of $80 million to 3D print fuel-nozzles for the next-generation LEAP jet engine. The unexpected predisposition of aerospace industry in the direction of additive manufacturing is mostly due to the opportunity to accomplish substantial weight decrease. According to American Airlines, for every pound of weight decreased from the aircraft, the company saves 11,000 gallons of fuel per annum.

3D Printing Advanced Metal Casting

Scientists developed a 3D design utilized to print a polymer airline seat frame. The frame was then treated in ceramic material and heated until the polymer evaporated. The subsequent ceramic mold was then utilized to metal cast the new seat frame. A proof-of-concept project could afford airlines hundreds of millions of dollars.

Autodesk has joined with a Michigan foundry in a 3D printing proof-of-concept project that caused in a novel magnesium commercial airline seat frame so light it could save an airline more than $200 million in fuel charges. Autodesk and Aristo Cast stated that if an aircraft manufacturer such as "Airbus" were to substitute all 615 seats on its "A380 jets" with the novel, lighter magnesium seat frames — across a fleet of 100 planes which characteristically have a 20-year lifespan — the interchange would save $206 million, knowing that the average jet fuel charges in 2019. Along with decreasing fuel costs, the novel seats could also ensure 126,000 fewer tons of C02 emission release if utilized on a single model aircraft.

Unmanned Aerial Systems

Recently, BAE Systems revealed 2,040 aircraft manufacturing conceptions integrating on-board 3D printing to improve Unmanned Aerial Vehicles UAVs. The concept clarifies how an aircraft scrutinizes the disaster and accounts to the mission control where the compulsory engineering data is provide to the on-board printers to print unmanned aerial vehicles rendering to the necessities of the disaster situation. Ultimately, these 3D printed UAVs will accomplish rescue operations or monitor the condition. While this idea is still on the drawing board, BAE Systems has now capitalized $250 million in research and development to guarantee that these concepts can be revolved into a actuality.

Space

NASA engineering technical staff illustrated on inventiveness and forward-thinking technology. About 70 of the parts that brand up the nomad were constructed digitally, directly from computer designs, in the heated chamber of a production-grade "Stratasys 3D Printer." For its 3D-printed components, NASA utilizes ABS, PCABS and polycarbonate materials. FDM, patented by Stratasys, is the single 3D-printing method that sustains production-grade thermoplastics, which are lightweight but tough enough for harsh end-use components.

Fig.8.31: 3D Printer Manufactured Rocket Engine Injector

Recently, NASA's skyrocket engine injector manufactured from a 3D printer succeeded a most important hot fire investigation. In the test, the skyrocket engine injector produced 10 times additional thrust than any injector manufactured from 3D printing in the past, Figure 8.31.

NASA engineering technical staff was talented to construct the injector with selective laser melting with the aim of fusing together nickel-chromium alloy powder. The examined part was the size of an injector utilized in a lesser rocket engine, but the design is comparable to what an injector appears like in a bigger engine. The injector can be scaled to a bigger size when it is fabrication-ready.

Also, NASA is looking forward to utilize 3D printing as a provision for speedy pre-prototyping. "3D printing makes it simpler to seize the dreams of the mission concepts. Using 3D Punters, engineers could gain peer-reviews, different design impressions, and support for final prototype. With open source design improvement, there will be a potential to incorporate multiple ideas from the outside, thereby decreasing the construct time significantly and also reducing costs.

3D Printing in Zero-Gravity

Traveling through space takes a much fuel and preparation, especially dealing with unexpected incidents in space, there may be limited equipment or spare parts available. Transporting items into space is, of course, exceptionally expensive. Astronauts may need to wait for the next mission to obtain the parts they need or resolve the failure or find out alternative solutions to their problems. With 3D printing available however, they can print out parts, tools or other items they need in the event of an emergency. One of the most logical uses for 3D printing is printing parts, tools, and other devices for astronauts while they are in space.

It can also help accelerate the building of parts for the International Space Station. The Zero-G Printer, the first 3D printer designed to operate in zero gravity, was built under a joint partnership between "NASA Marshall Space Flight Center, MSFC, and Made In Space, Inc." In September 2014, "SpaceX" delivered the zero-gravity 3D printer to the International Space Station, ISS.

On December 19, 2014, NASA emailed CAD drawings for a socket wrench to astronauts aboard the ISS, who then printed the tool using its 3D printer. Applications for space offer the ability to print parts or tools on-site, as opposed to using rockets to bring along pre-manufactured items for space missions to human colonies on the moon, Mars, or elsewhere.

On a related note, an engineer won a grant from NASA in 20119 to prototype a machine that will print food that is better than the freeze-dried food astronauts normally eat.

Fire Arms

AM's influence on firearms includes 2D: novel engineering methods for well-known companies, and new potentials for the making of do-it-yourself weapons. Recently, the US-based group "Defense Distributed" revealed strategies to design a traditional polymers 3D printed firearm "that could be downloaded and replicated by anyone with a 3D printer." Afterward "Defense Distributed" declared their strategies, queries were upturned regarding the consequences that 3D printing and extensive consumer-level CNC machining may have on gun control efficiency.

CONSUMER PRODUCTS

Consumer Electronics

Poly-Jet technology can yield models with extraordinarily thin walls — 0.6mm or less — perfect for small tools and electronic instruments densely compacted with miniature components. Smooth finish and realistic colors make these models almost indistinguishable from the finished product.

Sporting Goods

Prototypes frequently necessitate a mixture of tough and flexible materials. A hard shell helmet, and its embellished interior, or a pair of ski goggles with tinted lenses and neoprene over molding. Only Poly-Jet technology can yield prototypes with many materials and colors in a single run, and automated construct. Consequently, it is ideal for sporting goods creators with an aesthetic charm.

"FDM Technology" operates with production-grade thermoplastics to yield components with high impact strength and great toughness. "FDM Technology" is well matched parts that essential to endure tough and frequent functional testing.

Toys

Children may develop their restricted selections of toys when buying toys. Children may imagine the precise color of their toy's hair. Children can modify their own exclusive toy, thanks to the novelties of 3D printing. The toy manufactures is among the largest recipients of the technology. Toys already incline to be small and made out of polymers, enabling it to 3D print printed easily. It is projected the total economic influence of 3D printing technology will be $230-$550 billion per year by 2025, $100-$300 billion of which will be direct consumer goods such as toys.

Jigs and Fixtures

Inkjet-based technology and FDM Technology, both available from "Stratasys," provide rapid and precise techniques to yield and fabricate tools. 3D printing jigs and fixtures with inkjet or FDM can aid decrease the backlog for an in-house machine shop and be utilized as a bridge-to-tooling answer.

ADDITIVE MANUFACTURING

Cloud-Based Additive Manufacturing

Additive manufacturing in engagement with cloud computing technologies permits decentralized and physically independent disseminated fabrication. Cloud-based additive manufacturing denotes to a service-oriented networked manufacturing entity in which service users are capable to construct workpieces through Infrastructure-as-a-Service, "IaaS," Platform-as-a-Service, "PaaS," Hardware-as-a-Service "HaaS," and Software-as-a-Service "SaaS." Distributed manufacturing is supported out by some corporations; there is also a services like "3D Hubs" that place people requiring 3D printing in connected with owners of printers. Some businesses offer on-line 3D printing services to both industrial and individual customers, working from 3D designs uploaded to the business website. 3D-printed designs are either transported to the customer or picked up from the service provider.

Mass Customization

Businesses have created services where users can customize objects using simplified web based customization software, and order the subsequent items as 3D printed exclusive articles. This now permits users to generate custom cases for their mobile phones. "Nokia" has released the 3D designs for its case so that owners can modify their individual case and have it 3D printed.

Rapid Manufacturing

Progresses in RP technology have presented materials that are suitable for final manufacture, which has in turn presented the opportunity of directly manufacturing finished mechanisms. One benefit of 3D printing for rapid manufacturing lies in the comparatively low-cost production of small numbers of workpieces. Rapid manufacturing is a new method of manufacturing and many of its procedures continue unproven. 3D printing is now spreading to the field of rapid manufacturing and was acknowledged as a "next level" technology by many experts in a 2018 report. One of the most encouraging procedures appears to be the variation of selective Laser sintering (SLS), or direct metal laser sintering (DMLS) some of the better-established rapid prototyping approaches.

Rapid Prototyping

Industrial 3D printers have occurred since the early 1980s and have been expoited expansively for rapid prototyping and investigation purposes. These are commonly larger machines that utilize exclusive powdered metals, casting media, e.g. sand, plastics, paper or cartridges, and are employed for rapid prototyping by universities and commercial companies.

Agile Tooling

Agile tooling is describing the procedure of utilizing segmental means to develop tooling that is created by additive manufacturing or 3D printing approaches to permit fast prototyping and answers to tooling and fixture requirements. Agile tooling utilizes a cost effective and high quality technique to rapidly react to client and market requirements. It can be utilized in hydro-forming, stamping, injection molding and other engineering and fabrication procedures.

3D Printed Food

Food is one of essential elements of life which is at the foundation of the pyramid of human essentials. Conveying the food industry to the digital era is one of the indispensable and groundbreaking implantations of 3D printing. Spread over this technology allows fast automated and repeatable procedures, freedom in design, as well as permitting large and easy changeability of the cooking process which can be tailored for each region or individual. Employing robotic layer based food printing schemes permits the recipe of the food to be digitized and saved in order to cook very repeatable and high quality dishes deprived of any margin for operator blunder. Also, the figure and beautification of the food can be tailored based on the customer or the occurrence. Additive manufacturing of food is being advanced by squeezing out food, layer by layer, into 3D substance. A large diversity of foods are applicable candidates, such as chocolate and candy, and flat foods such as crackers, pasta, and pizza.

NASA has carefully provided resourcefulness of the concept, awarding a contract to the "Systems and Materials Research Consultancy" to invetiate the viability of printing food in space. One of the difficulties with food printing is the environment of the consistency of a food. For instance, foods that are not sturdy enough to be filed are not suitable for 3D printing, Figure 8.32.

Fig.8.32: 3D Printing Food in Layers

Construction Industry

In the construction industry, Construction 3D printing can be employed to generate building modules or to 'print' entire buildings. Construction is well-suited to 3D printing as much of the information essential to produce an item will survive as a consequence of the design procedure, and the industry is already skillful in computer aided manufacturing CAD. The recent development of "building information modelling," BIM in specific may enable better exploitation of 3D printing.

Far East Decoration Design Engineering has exploited super large 3D printers to spray a combination of rapid drying cement and salvaged raw materials. This has allowed them to build 10 small demo 'houses' in less than 24 hours. They have recommended that each house can be printed for less than $5,000. Their system constructs blocks off-site by layering the cement mix in a diagonally strengthened arrangement. The blocks are then draw together on site. It is possible to use the procedure to construct larger houses or even skyscrapers in the future. In 2017, they also printed and entire villa and a five-storey apartment building. Scientists at MIT have developed a system that 3D printed the basic structure of an whole building. It is a plan concerning building that could fundamentally decrease the costs of construction, Figure 8.33.

Fig.8.33: MIT 3D Printing machine created the largest 3D printed structure to date - *courtesy neri oxman, steven keating, julian leland, levi cai,.*

3D Concrete Printing Robot

At the Viva Technology conference – 2017, in Paris, France, they provided registered visitors a live demonstration of the 3D Concrete Printing robot.. The host company claims the technology vividly drops the time necessary for the construction procedure by as much as a immense 75%. The concrete is truly a product called CyBe MORTAR which sets in just a matter of minutes, accomplishing its total structural strength in about an hour. Thanks to these characteristics, the robot is capable to extrude layer after layer building the structure in almost any shape physics will permit. Walls are just one of the applications, park benches and even planter boxes are also likely utilizing the 3D concrete printing technology.

The robot can transfer in as many as 6 axes and prints at a rate of 20cm per second. That means every 5 seconds it can cover a 1m. Its arm can cover a range of 2.75m and in some applications, its positioned on a height adjustable. The robot attaches through an Ethernet cable to a computer running dedicated CyBe software. This software can deal with most CAD models and does the structural authentication ahead of time.

Benefits of 3D Printing in Construction

- Lowered materials usage
- Increasing the capacity to design a larger diversity of tailored homes and structures
- Savings of 30%-60% in construction surplus
- Decrease of manufacture time by 50%-70%
- Decrease of construction manual labor costs by 50%-80%

Environmental

In Bahrain, large-scale 3D printing employing a sandstone-like material has been utilized to form unique coral-shaped configurations, which encourage coral polyps to inhabit and restore impaired reefs. These configurations have a much more natural figure than other structures utilized to build synthetic reefs, and, unlike concrete, are neither acid nor alkaline with neutral pH.

SOCIOCULTURAL APPLICATION

Fashion and Digital 3D Clothes

3D printing has arrived to the world of fashion with style designers carry out trial with 3D-printed bikinis, shoes, and dresses. In the world of extraordinary fashion courtiers such as Karl Lagerfeld designs for Chanel, Iris van Herpen and Noa Raviv experimenting with technology from "Stratasys," have engaged, and presented 3d printing in their collections. Assortments from their lines and other formation with 3D printing were displayed at the 2018 Metropolitan Museum of Art Anna Wintour Costume Center, exposition "Manus X Machina".

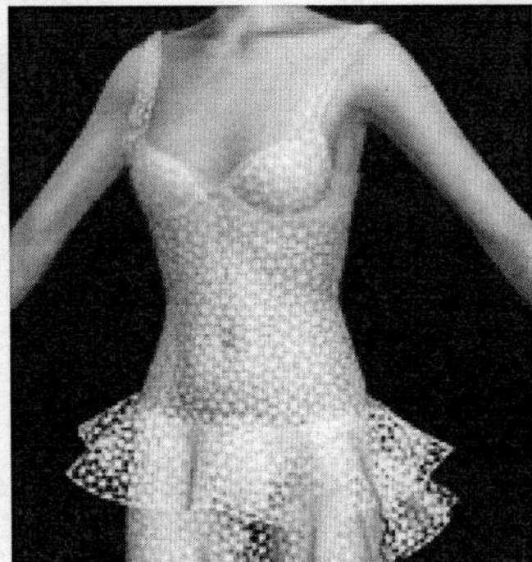

Fig.8.34: 3D Printing in US and European Fashion

3D Printed Fashion Clothes

Fashion apparels are fashioned with diverse types and classes of fabrics, thread and a needle. Frequently buttons or zippers are utilized to permit people to open and close diverse portions of the clothing, Figure 8.34. To create clothes seamstresses draft clothing patterns, cut patterns and sew patterns together to fashion the final vestment. Intricate clothing patterns are often drafted and restyled dozens of times. In some cases it may take a seamstress around 40 hours to cultivate a final clothing pattern, and an extra 60 hours to cut and sew it, Figure 8.34. Conventionally for thousands of years, seamstresses and tailors would sew clothing by hand.

Later in years, people invented sewing machines and transitioned to utilizing sewing machines more frequently instead of sewing by hand.

3D-Printed Fashion Shoes - On Demand

It is not merely bikinis that are being reproduced out by innovative businesses. "Kerrie Luft," a British designer fashioned a complete range of shoes exploiting the technology, which were presented at Selfridges in Oxford Street, London, England in March 2013. She originally commenced utilizing 3D printing to envisage her fashion without having to fashion an expensive mold, which is another advantage that the technology can have for fashion designers. Her outstanding fashions have established to be a ground-breaking.

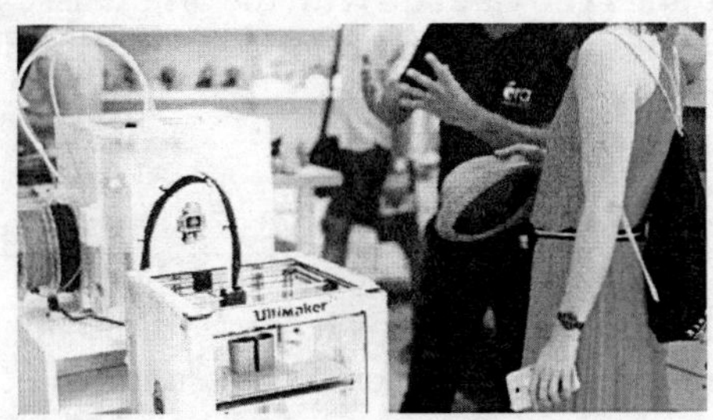

Fig.8.35: 3D Printing Shopper Create Custom Made Shoes

Courtesy: Retailers- Eram and Unistudio.

In commercial production "Nike" is utilizing 3D printing to prototype and fabricating the "Vapor Laser Talon" football shoe for players of American football, and New Balance is 3D manufacturing custom-fit shoes for athletes. Also, Customers have the power to create 3D print of their own high heel designs in the store, Figure 8.35.

Art, Design & Sculpture

Artists and Sculptors are involved with 3D printing in countless of diverse means to discover form and function in ways formerly impossible. Whether purely to discover new unique expression or to study from old masters this is an exceedingly charged sector that is progressively finding novel ways of working with 3D printing and presenting the outcomes to the world. There are plentiful of artists that have now made a name for themselves by working explicitly with 3D modelling, 3D scanning and 3D printing technologies, sample of the artist as listed below:

- Joshua Harker
- Dizingof
- Jessica Rosenkrantz
- Pia Hinze
- Nick Ervinck
- Lionel Dean
- And many others.

The discipline of 3D scanning in combination with 3D printing also carries a novel element to the art world, though, in that artists and scholars now have a established procedure of reproducing the work of past masters and creating precise duplications of antique, and more recent, sculptures for close examination – works of art that they would otherwise never have been capable to interrelate with in person. The work of Cosmo Wenman is predominantly informative in this field.

Modernist Fashion Jewelry

3D printed jewelry is now on the market and it not astonishingly inexpensive. By utilizing 3D printers, jewelry creators can yield complicated designs in half the time, utilizing precious metals as well as polymers and brass. This 3D printed jewelry is an art practice in its own right, generating geometrically modern custom designs that are already rising popular amid consumers around the globe.

Museums

The popular of artworks and sculptures reserved by the Smithsonian Institute are not exposed to the public. Only 2% of its 137 million objects are being revealed at any one time as they are too historically significant or too delicate to be unprotected to the public continually. With 3D printing, museums can compromise scans of their smithereens to the public instead. In fact, not only can they envision these masterworks, they can also touch and get up close to the art or sculpture – things you can't do with the originals. On top of serving reservation great masterpieces, 3D-printed sculptures can aid educate the multitudes about old and modern art.

Cultural Heritage

In the last numbers of years 3D printing has been intensively utilized by in the cultural heritage field for conservation, renovation and dissemination objectives. Several Europeans and North American Museums have acquired 3D printers and vigorously recreate misplaced pieces of their relics. The utilization of 3D scanning technologies permits the reproduction of real objects without the utilization of molding methods that in numerous cases can be more costly, more challenging, or too invasive to be achieved, mainly for valuable artwork or fragile cultural heritage artifacts where direct interaction with the molding materials could damage the original object's surface. While one seldom think about it when visiting a museum, many of those showed pieces are very delicate and require to be handled with exceptional care. Take delicate works of art out to multiple venues, or even a national tour is endangering much to encounter the damage.

The curators from a museum in Salem, MA, selected a 3D printing answer. For a nationwide tour of an exhibition on Thomas Hart Benton, the Peabody Essex Museum generated a 3D printed reproduction of a predominantly fragile clay maquette.

The Natural History of Man, Metropolitan Museum of Art and the British Museum have employed their 3D printers to generate museum mementos that are accessible in the museum shops. Other museums, like the National Museum of Military History, and Varna Historical Museum, have gone additional extent and sell over the online platform liberating digital models of their artifacts, fashioned by utilizing "Artec 3D scanners," in 3D printing friendly file format, which everybody can 3D print at home.

Guitars

Distinct violins, whose form and sound does not wander far from the classic standard, there is a much extra liberty in guitar design and material use. 3D-printed "Gibson Les Paul" is a fully well-designed guitar exhibited at Maker Faire exhibition. Summit is already imagining a future where musicians order tailor-made guitars to yield exclusive sounds.

Gifts

3D printing is becoming widespread in the customizable gifts business, with products such as tailored models of art and dolls, in many figures: in metal or plastic, or as consumable art, such as 3D printed chocolate.

Challenges of 3D Printing

Defending the faith of a newly born novel product has never been free from challenges, and many road blocks. The greater the challenges he greater the reward at thhe end. Below is a few of selected challenges the 3D printing technology and implementations may face:

Creating Valuable 3D Project is not an Easy Task

Though many individuals can create a document for printing, nonetheless, developing 3D data require challenging tasks, such as the utilization of sophisticated design software and/or scanning and digitizing the data. Also, the ability to use these products necessitates training and investment, which is often beyond the reach of the average consumer.

Not all 3D is Created Equal

Just because one may envisage a 3D image on the computer monitor, does not indicate that it may be 3D printed. 3D printers necessitate 3D data to fit in to definite geometric controls as well as to be of a precise format. This indicates there is a necessity for "middle-ware" software systems that take input 3D data and translate it to 3D data that is friendly to be 3D printed. Again, Utilization of these systems necessitates training and investment of time and resources.

Functional Challenge and Automation Challenge

There are other two types of considerable challenges; the first challenge is functional challenge, while the second is an automation challenge. It is recommended, to have the ability to use the 3D-printer to print more realistic objects than just plastic toys. The need to enable these objects to have more functions by incorporating electro-mechanical features, and electronics, such as hardware inserts and electric wires to the objects. Also, adding "InfraStructs-like" technology to make these 3D printers readable. The automation challenge is innovating the technology by speeding up the production throughput process from a small quantity in the 10's to a larger quantity in the thousands if not in the millions.

REFERENCES

1. ining, G., Kowalski, S.,"Statistical Methoods for Engineers", 3rd ed. Boston:Brooks/ Cole, ISBN:978-0-538-73518-6, 2011
2. Cooper, K. G. (2001). Rapid Prototyping Technology: Selection and Application, Marcel Dekker Inc., SAD, New York. ISBN 0-8247-0261-1
3. Gibson, I.L; Rosen, D.W. & Stucker, B. (2010). Additive Manufacturing Technologies: Rapid Prototyping to Direct Digital Manufacturing, Springer, ISBN 978-1-4419-1119-3, New York
4. Godec, D. (2005). Utjecaj hibridnog kalupa na svojstva injekcijski prešanog plastomernog otpreska, D.Sc. Thesis, FSB, Zagreb
5. Soloman, Sabrie "Sensors Handbook, 2nd Edition, McGraw-Hill Publishing Co., NY – ISBN 978-0-07-160570-0
6. Soloman, Sabrie "Sensors and Control Systems in Manufacturing 2nd Edition," McGraw-Hill Publishing Co., NY – ISBN 0-07-059626-3
7. Soloman, Sabrie "Affordable Automation," McGraw-Hill Publishing Co., NY – ISBN 0-07-059633-6
8. Soloman, Sabrie "Sensors Handbook, 1st Edition," McGraw-Hill Publishing Co., NY – ISBN 0-07-059630-1
9. Soloman, Sabrie "Sensors and Control Systems in Manufacturing 1st Edition," McGraw-Hill Publishing Co., NY – ISBN 978-07-10572-4

9 Additive Manufacturing

THE PUSH TO MATURITY

Three decades of struggle to maturity Additive Manufacturing had to face un-avoidable obstacles, discouragements, to gain notable success and assured victory in various facets of 3D printing technology. The limitations of 3D printing did not inhibit the spread of its virtues neither the slow progress restricted to mark victories in what it does best. Additive Manufacturing is here to stay!

In spite of its expensive hardware and its expensive materials leading to expensive parts, still, it is a favorite technology to cut-down the cost of developing novel innovative products and substantially reduces the lead time to market. However, Additive manufacturing, as a young sprouting plant basking in the sunshine of inventive novel products in desperate needs of variations prototypes, still young enough to grasp the other side product mass production demands.

Nonetheless, the writing on the wall predicts the sprouting Additive Manufacturing is poised to maturity as it is in the most caring hands of entrepreneurs, inventors, engineers, scientists, mathematicians, researchers, and even medical doctors, dentists, and surgeons.

At present, the demands are thus making it difficult to compete with mass production. Also, Additive Manufacturing requires its CAD designer to create what the customer has in mind, and can be expensive if the part is very intricate. For now, 3D Printing is not the answer to every type of production method. 3D printing may not replace other manufacturing technology. The 3D printing has not yet revolutionized the mass production of the high volume of the manufacturing sector, to render traditional factories obsolete. 3D printing simply works best in areas where customization is the key, from printing hearing aids to dental implants.

Using a combination of 3D scanning and printing, implants can be customized to specific anatomic circumstances in a way that was not feasible beforehand. However, it is also known that 99% of all manufactured parts are standard and do not require customization. In these cases, 3D printing has to compete with scale-driven manufacturing processes and rather efficient logistics operations. The simple fact of the matter is the economics of 3D printing - only for now - an unfeasible way to produce the vast majority of parts manufactured today. So instead of looking at it as a substitute for existing manufacturing, consider looking to new areas where it can exploit its unique capabilities to complement traditional manufacturing processes till the time of Additive Manufacturing adulthood.

LIMITED PROTOTYPING - LOW VOLUME OF PARTS

3D printers have a limited precision. The nozzle output is about 0.4mm, and trying to print details below 0.1mm is usually worthless. It is very well suited to many objects, but you will have hard times trying to replace the smallest broken nylon gear in your compact camera with a 3D printer. Even the professional or semi-professional selective laser sintering technique will not make it possible sometimes. Precision can be increased by using a smaller nozzle. But you will suffer from excessively long prints. Also a smaller nozzle will make it increasingly difficult to fine-tune the proper printing parameters. The quality and homogeneity of the filament being extruded will also impact the extrusion much more. Nonetheless, while the world is being blessed by its ability to produce the initial ideal parts fundamental to eminent success for a timely high volume product through the phase of traditional manufacturing stage.

Strength

Some 3D printed parts are not as strong as traditionally-manufactured parts. Their layer-by-layer technique of manufacturing is both their strength and weakness. In injection molding, a very even strength across the part is maintained, as the material is of a relatively consistent material structure. In 3D printing, building it in layers — indicates that it has laminate weaknesses as the layers do not bond as well in the Z axis as they do in the X and Y plane. This is comparable to a Lego wall — similar to placing all the bricks on top of each other, and press down: feels strong, but push the wall from the side and it breaks easily.

Limited Printable Material

Printing only in one material, usually is in a plastic. Many items are made up of plastic. Nowadays, most items in our homes are made up from multiple materials, and most of them are both metal and plastic. Those two cannot be made together as their melting temperatures are hundreds, if not thousands of degrees apart. Then, professional printing techniques offer metallic powders to be fused by high power lasers, but the results still may be brittle, or at least expensive. The common PLA becomes soft on as low as 50° C, ~120° F. 3D printed plastic cannot stand more than 200° C, ~400 °F.

Machine Range

Printing 3D print can use several hundreds of different types of materials. There are many different types of 3D printers, all of which have advantages and disadvantages. FDM, the main consumer form of 3D printers, extrudes ABS or PLY in a relatively safe manner, with the materials cooling down quickly, where parts are safe to touch coming off the machine. Resins, though higher detail, are disordered and expensive; powder-based printers are really untidy, and sometimes explosive, never make the 3D printed gun on a powder-based printer; others operate at high temperatures or produce masses of waste. This basically leaves us with FDM, which by the way has the worst surface finish, as the only really suitable technology for the home.

Surface Finish Quality

It is understood the ability to print in plastic, and imagine smooth, soft, and glossy plastic item. The actual result surface is a matt finish with rough layer lines all over. Many companies offer a 'smooth' surface finish, after post-processing the parts, which frequently involves labor, and/or chemicals like acetone.

Printing Speed

Printing is slow! Slicing an object into 0.1mm layers and filling them successively with a 0.4 mm nozzle just take long time. In fact, as of today, printing something quicker than 20 minutes is rare. 3D printing owner experienced hour-long prints, if not day-long. Many sales individual may suggest that 3D printing is quick but items regularly take hours to print, even days. And the longer it takes to print, the more chances are that a failure happens at some point in the process. Unless the user has very specific requirements and pays much for it, a 3D printer is not a productive factory.

This is even true for professional 3D printers, except for some limited fields like in the prosthetic or the space industry, or when deeply mixed material is needed, or when intricate shapes cannot be molded or carved. Past a few items per day, te speed is really not appropriate. In order to speed up the process the layers could be thicker. Once the speed increase, the surface finish quality becomes poor. 3D printing is limited by the chemical properties of materials such as ABS and PLY — these materials can only be extruded at certain speed, and at a specific rate. Beyond the appropriate chemical properties the process fails and starts to destroy the properties of the part. This happens with the well-respected machines for FDM, Fluid Deposition Modeling.

3D Printable Shapes

Home 3D printing relies on a filament pushed through a melting nozzle. Accordingly, it must print on top of solid reliable base, which in turns impacts the design of the object. Consider 3D printing the letter *"M"*. The inside sloped lines on the letter "M" are called "overhang." They require support to be printed as they float in the air. Support structure may be added automatically by the software, but it must be carefully and manually removed afterwards, unless printing is conducted using esoteric and complex dual-head setups with water-soluble PVA plastic for example. As soon as the overhang angle is above 45°, the printer will start pouring plastic on empty space, which generally ruins the print. Hence the 3D object designer must take this into account.

If the part of the object in contact with the bed is too small, as it happens for a sphere, it will most probably dislocate the end of the print, the plastic retracts a little when cooling down. Once again, there is no definitive answer exist to make sure the printed part stay on its the support. As a consequence, designers take this into account and add explicit tabs or "rafts" to counteract the issue. It must be removed manually when the print is completed. However, there are other professional printing technologies that avoid the overhang and dislocating issues, such as those based on fusing the top layer of a powder with a laser, then adding another layer of power and cycling. Some other applications use a liquid polymer that is cured with hard light or lasers. However, none of them made their way to homes yet, as they require untidy powders or polymers with their own sets of limitations.

Reliability and Routine Simplicity

Printing an object requires extensive use of software and algorithms. The need to "slice" the 3D objects into layers printed one of top of each other; it is easy to understand that there are about many parameters at the minimum to be taken into account.

The 3rd Power Law of 3D Printing

While the propaganda highlights the hope of limitless replications of lost components, shoes, body parts, musical instruments, and even guns a key fact is that the 3-D printing may be unfettered by complexity, it is constrained by volume. Everything from cost and time to amount of material increases exponentially, specifically, to the third power. If it is required an object 2 times as big, it will cost $2^3 = 8$ times as much and take 8 times as long to print. If it is required object 3 times as big, it will cost about $3^3 = 27$ times more and takes **27** times longer to print

3D Printing - Immature Technology

Not ever previously consumer has had a technology where we so easily interpret the concepts into a touchable object with little concern to the machinery or talents available. Yet just as the microwave didn't replace all other forms of cooking as originally foretold, 3-D printing will not substitute other manufacturing technologies, let alone industrial-scale ones for a variety of motives.

It will accompaniment them. The truth is that 3-D printing is, still, an immature technology. As soon actual use of 3D printers are taking place for a period of time, the immediate wish for faster construct times, higher quality prints, larger construct envelopes, better and inexpensive materials becomes a demand.

Limited Usability

3D printing an object, requires a CAD model, this is not totally simple, as described below for the following reasons:

Learn CAD

Advisable, but difficult. It needs long time to master the CAD programming and use. Also, learn how to design, and understand tolerance.

3D Scanning

Scan the outer surface of a part, cannot scan the inside of items

Downloading 3D Files

An original 3D model library opens up on the internet. The files are populated with the identical 3D files that are on all the other sites. These files are 99% unmoderated, which may not be assured to be 3D printable. Even the main 3D printing libraries are full of files with errors and/or have been designed with little attention to the abilities and cost of 3D printers.

Photo-based Apps

It is similar to scanning, nonetheless poor.

Exaggerated Expectations

Despite the many potential advantages that 3D printing may provide, expectations of the technology are often exaggerated by the media, governments, and even researchers. This promotes unrealistic projections, especially regarding how soon some of the more exciting possibilities—such as organ printing—will become a reality. Although progress is being made toward these and other goals, they are not expected to happen soon. 3D printing will require vision, money, and time for the technology to evolve into the anticipated applications.

While it is certain that the biomedical sector will be one of the most fertile fields for 3D printing innovations, it is important to appreciate what has already been achieved without expecting that rapid advances toward the most sophisticated applications will occur instantly.

The world has become aware of 3D printing — most world population read about it in the papers, on blogs or seen it on TV. The social acceptance seems to be that, in the future, population will be able to download the products or make them with CAD programs, apps and 3D scanners, then just print them out, either at home, or in localized print shops.

Populations have seen a 3D printed violin and a shoe which actually works, straight out of a printer. But it is a very expensive, high-end printer which uses lasers or resins. These people think that they can create objects as well without much input or training, on a machine which costs $800 or less. 3D printing is severely overhyped. The main issue lies with elevated potentials, build quality, price and usability.

Risks and Dangers of 3D Printing

There are a few issues the users must consider when examining the safety of a 3D printer, stated as follows:

- Electrical safety
- Mechanical safety
- Risk of burns

Electrical Safety

When purchasing a 3D printer, user must check if the power supply is protected against short-circuit, overload, over voltage and over temperature. There should be a thermal fuse mounted in the heat sink to monitor temps and cut off all power to the heater in the event of any over temp condition. Incorporate a smoke sensor that is separately powered from the machine, which cuts off the power to the heater when smoke is detected.

Mechanical Safety

As 3D printers contain many moving parts, industries used NEMA-17 stepper motors, which do not have enough power to cause serious injuries. Though, in search for a safe 3D printer, user must consider that moving gears that are not covered can still form a potential risk. User may wish to limit the search to 3D printers that have a 'closed' design.

Risk of Burns

There is a possible danger of burns, as the print head of majority of 3D printers can range temperatures of up to 280° C, the heated bed of up to 120° C. A 3D printer is considered safe when contact with the nozzle is prohibited by enclosing it with for an aluminum cover. User can obtain burns from touching a heated bed. Therefore, a 3D printer with a heated bed needs to have a closed environment.

Environmental Hazards

3D printers are an environmental hazard due to emitting microscopic particles and chemicals that have been linked to asthma. Several studies have shown that 3D printers produce high amounts of ultrafine particles, UFPs and volatile organic compounds VOCs, while in use and that these particles and vapors are detectable for many hours after the printers have been shut off.

Ultrafine particles, UFPs have been connected to adversative health situations, such as asthma and cardiovascular matters, because they can pass over the lungs and travel to other organs. They can also transmit noxious material into the body, as well as blood and tissue cells. A National Institute for Occupational Safety and Health, "NIOSH" report these discharges peaked a few minutes after printing began and returned to baseline levels 100 minutes after printing finished. The problem was reduced by using manufacturer-supplied shelters and full enclosures, utilizing proper ventilation, protecting workers away from the printer while wearing respirators, turning off the printer if it jammed, and using lower emission printers and filaments.

However it must also be specified that 3D printing radically decreases the wastage of material, resulting in less pollution, and is therefore harmless for environment.

HAZARDS OF PRINTING MATERIALS

Individual 3D printer has been developed to utilize specific sorts of materials. These materials have characteristic hazards and may become more dangerous when they are exposed to the 3D printing process or are unintentionally ignited.

Thermoplastics and Photopolymers

Thermoplastics can be combustible, cause exasperation and skin sensitivity. Some may comprise small quantities of noxious constituents. Photopolymers utilize introduction to UV light to harden during the printing process. These often contain hazardous monomers, such as acrylates. Additionally, UV light is a radioactive hazard, which can cause harm to sight and skin.

Support Structural Materials

The 3D printing procedure frequently exploits a support material to permit for formation of the empty spaces in the manufacturing design. Support materials often encompass dangerous chemicals, such as phenyl phosphates, that are integrated into the thermoplastic acrylic polymer, and thus are hazardous during use and discarding, Figure 9.1.

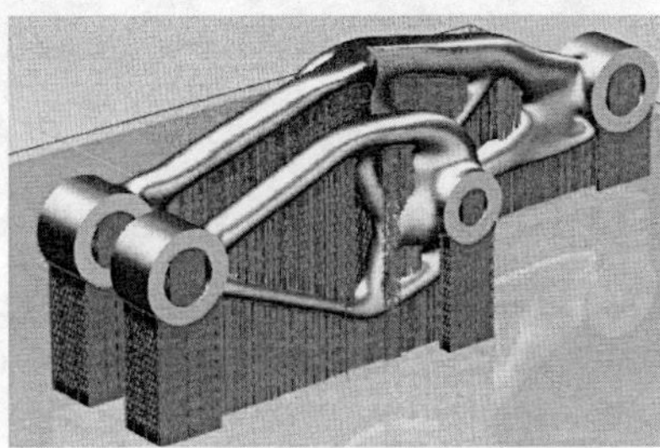

Fig.9.1: 3D Printing Construct Structure Support Material

Metallic Materials

Sensitive and highly flammable powder metals are utilized in the manufacturing 3D-printed metal alloy tools and parts. Divided metal powders, such as titanium and aluminum, can instinctively combust triggering fires, pyrophoric. Remove sources of possible explosion and do not store the powder materials close detonation hazards.

Biological Materials

3D printing has extended to contain printing of biological materials, such as cells for engineered-tissue generation. Possible contamination from exposure to the sprays and the suitable procedure controls required to be deliberated.

Protective Equipment

To evade chemical contact while utilizing the eroding base bath, generate a wall through individual protective equipment as well as a laboratory coat or smock, corrosion-resistant rubber gloves and, at minimum, splash goggles.

A neoprene apron should be worn when pouring big quantities of corrosive materials. A face shield provides extra defense and averts stray splatters from getting to the face. Long pants and shoes that cover and shield the feet should be worn, since splashes often land below the waist.

ABS and PLA Food Safety

The main food safety concern with 3D printers is bacterial growth due to small fissures in printed objects. Items made by 3D printers are porous, allowing bacteria to grow over time and to potentially become a health hazard. Also, children playing with 3D-printed objects run the risk of coming into contact with bacteria growing on the object after it has become soiled.

Washing the object might demonstrate problematic because high dishwasher temperatures can dissolve or warp the item, especially with PLA-printed workpiece. Some filaments, other than ABS and PLA, are food safe, as well as polyethylene terephthalate, PET, which has been FDA permitted for direct food interaction. ABS-created plates, cups, and utensils can transmit hazardous chemical remains, including "1,3-butadiene," a probable human cancer-causing agent and can cause irritation to skin and mucous membranes through interaction with food, Figure 9.2. Also, UFPs and other remains from the 3D printing procedure could still be on the printed material. These levels are minor but should still be of concern to those utilizing 3D-printed material as dinnerware.

Fig.9.2: 3D Prinitng Food – *Food Graade Material Approved by FDA*

3D PRINTING STATE AND FEDERAL LAWS

3D Printing and Crime

3D printing has given rise to safety and security issues that merit serious concern. 3D printers have already been employed for criminal purposes, such as printing illegal items like guns and gun magazines, master keys, and ATM skimmers. These occurrences have highlighted the lack of regulation of 3D printing technology. In theory, 3D printing could also be used to counterfeit substandard medical devices or medications. Although 3D printing should not be banned, its safety over the long term will clearly need to be monitored.

Firearms

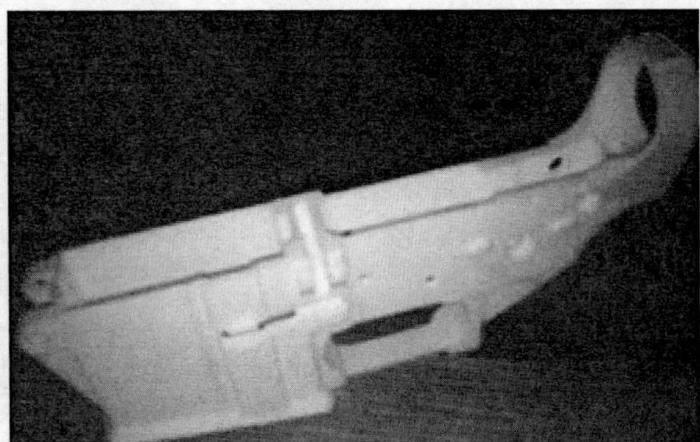

Fig.9.3: 3D Printed Firearm and its Enhanced Magazine

Nearly everybody in USA has received the news about Texas law undergraduate student, who was in headlines in 2013 by 3D printing a practical plastic gun and placing the blueprints on the Internet. The blueprints for the gun were downloaded 100,000 times prior the U.S. law enforcement official forced its removal from the server, Figure 9.3.

In 2015, police authority in Chiloquin, Oregon, made arrests for the unlawful ownership an "AR-15" assault rifle. Its lower receiver—the crucial part to what makes it a firearm—was alleged to have been 3D printed. It has always been lawful for U.S. citizens to make firearms, so long as they were detectable by airport metal sensors or x-ray machines. The gun, though, established that anybody with a 3D printer can print not only a gun, but also an undetectable gun—causing the inescapable, and frightening, knowledge those terrorists and other treacherous persons could reasonably make undetectable guns.

In August 2016, the TSA found a 3D-printed pistol in carry-on luggage at the Reno-Tahoe Airport, Nevada. The gun seems to have been sensed only because it was loaded with live ammunitions. Similarly, albeit U.S. law forbids the sale of guns to sentenced criminals, 3D printing delivers a technique for criminals to make or obtain guns deprived of anyone knowing about it.

This is not only a U.S. serious matter, though. In a police search in Manchester, England, police exposed 3D-printed gun mechanisms, along with a 3D printer. In distinct searches in Brisbane, Australia, and the country's close Gold Coast, police discovered 3D-printed gun mechanisms and a completely operative, loaded 3D-printed gun.

Philadelphia, Pennsylvania, was the first US state to illegalize 3D printed guns. Numerous other local governments have well-thought-out prohibiting 3D printed guns, counting the territory of Queensland, Australia.

California lately delivered a law demanding 3D printed weaponries creators to apply for an authorized serial number. California is also attempting to pass legislature demanding plastic 3D printed guns to encompass at least some metal.

New South Wales, Australia, forbidden digital blueprints for 3D printable guns.

Duplicate Keys

3D printers have shady-side implementations beyond weapons. In developments assuredly being shadowed by the gangland, a German hacker utilized a 3D printer to replicate handcuff keys for high-security handcuffs.

Massachusetts Institute of Technology, MIT students have CT-scanned locks, and then utilized the scans to 3D print master keys. Accomplished lock-pickers 3D printed a bump key, which can be utilized to pick virtually any pin tumbler lock.

In July 2016, hackers provided 3D-printable blueprint for the master luggage keys utilize by the U.S. "Transportation Security Administration."

3D Printers - Credit Card Fraud

In France a dishonest man 3D printed counterfeit fasciae for cash machines, which had replicated the data on customers' ATM cards.

Felons in Sydney, Australia, utilized 3D printers to make accessories for bank machines that scan bank card data from unsuspecting ATM customers. By utilizing 3D printers, the felons can make the scanners look like they are part of the ATM.

Organized crime is hurdling on board with 3D printing, as well. In synchronized break-ins against mobs in Malaga, Spain, and the Bulgarian cities of Sofia, Burgas, and Silistra, police officials' detained equipment utilized to 3D print refined scanner equipment, as well as fake card openings for bank machines.

An illegitimate individual who calls himself "Gripper" makes a scanner by the same name, which sells online.

Illegal Drugs

It may be soon enough for 3D printers to be utilized in printing unlawful drugs. This drug-printing ability would be a advantage to structured law-breaking, which would be capable of printing unlawful drugs at the point-of-demand, in that way removing the capital investment and dangers of shipping and storing big amounts of unlawful drugs. Actually, there is no motive drug dealers will endure to sell drugs when they can sell the digital blueprints in its place, which local dealers or customers can utilize to print their drugs of choice.

Counterfeiting

Additional argument of interest to law enforcement is the utilization of 3D printers for the production of forged products. Criminal mobs could be involved in creating fake products or device parts for making profits, but security specialists are more concerned by the likelihood that those mechanisms could be utilized for sabotage. Counterfeit products could be presented into the supply chain in order to create irregularities and faults in the products designed.

Another possible shady utilization use of 3D printing could lurk governments and monetary organizations universally: *3D-printed currency.* For instance, the U.S. dollar is printed on distinctive linen-like paper with colored threads utilized to guard against counterfeiting; though, this kind of paper may be a good nominee for 3D printing. Paper currency is also printed from plates, which could be 3D printed. Maybe most significant, numerous countries trust on large denomination coin currency, and 3D printers are ideal tools to print counterfeit coins. Cash is not the only thing 3D printers can counterfeit.

Criminals will be capable to print virtually whatever with 3D printers or sell the blueprints for others to do so. They will not be restricted to Rolex watches and YSL handbags—almost any proprietary product might be counterfeited via the utilization of 3D printers, which could print it with or deprived of the brand or print a basic product with a brand name on it. 3D printers could make it much simpler for imaginative counterfeiters to enter the game. Brands are secured by trademarks covering the brand name and, occasionally, the look and feel of the product, such as its figure or color. Brands may also be secured by copyrights and patents; in those circumstances, brand owners will have tools to fight back against the counterfeiters, if they can discover them. This trick will not be essentially diverse from brand counterfeiting today, except that it will be on a considerable and greater scale.

Nuclear Weapon Components

In the article entitled *"Toward the Printed World: Additive Manufacturing and Implications for National Security,"* subject matter specialists specified that 3D printing might make nuclear proliferation tougher to sense because countries and political groups who declined memberships of the "Atomic Club" could utilize 3D printers to make parts for nuclear armaments. In fact, reprobate nations or groups might utilize 3D printers to make products they cannot buy because of international sanctions and restrictions.

As modern and future conflict skilled professionals observed, "3D printing could crack sanctions—which have been a crucial part of foreign policy for a generation or more—into an old-fashioned idea." Russia's nuclear energy corporation, Rosatom, is in progressive phases of applying a plan to develop progressive metal additive manufacturing technology by creating 3D printed portions for the country's nuclear industry.

The Disarming Corruptor

Designed by a university scientist, the Disarming Corruptor camouflages 3D-printable blueprints, permitting them to slipup through filters meant to block them from design repositories. The scientist became exasperated when an online design repository automatically obstructed the upload of his artistic version of Mickey Mouse, and once more when he learned that additional online repository obstructed blueprints for 3D-printable guns. His answer was a software tool that camouflages the blueprint, making it disguisable by software procedures developed to block content that disrupts intellectual property, IP rights or is otherwise unlawful or objectionable. Once disguised, any project can be uploaded to any 3D-printing blueprint repository. The "Disarming Corruptor" also creates a password that can be shared independently of the repository, so that persons who download the project can eliminate the masking and print the object.

The Disarming Corruptor—and other tools like it—may render impotent any efforts to prevent the spread of 3D-printable designs that trespass third-party IP rights or disrupt laws that forbid printing specific types of objects, like guns. Tools like the "Disarming Corruptor" are a 3D-printing criminal's dream, permitting the sharing of designs for any and all products, legal or not.

3D Printing and Cybercrime

Additive manufacturing, AM is one area where cyber risk positions a particularly significant menace. AM, 3D printing, changes the way in which physical products are designed and produced. Designs are generated digitally, and, via printers communication and production lines, can theoretically be manufactured wherever, at whenever by whoever with the means to do so. AM's dependence on digital files and connectivity can also open the procedure up to completely new kinds of cyber threats, from product breakdowns to intellectual property theft and brand jeopardy. Wherever data and information are conveyed, used, or retrieved, corporations must understand in advance that someone somewhere will try to exploit those data and information for private gain or to cause harm or damage.

Understanding who implements cyber-attacks and why they do so can help manufacturers predict conceivable threats. Early on in the digital revolution, incidents of system breaches were not necessarily for militant, but rather caused by people curious to see if and how they could gain access to systems. Many occurrences have been concentrated on accessing what makes the organization competitive; in many cases, companies may be uninformed that their systems have even been penetrated.

Today, however, cyber-attacks can and cause real and sometimes significant harm to businesses and people and particularly to manufacturers. Cyber-attacks can initiate from individuals or organizations that are determined by a variety of reasons, including financial gain, the desire to cause harm to a specific company or companies, or as a means to interrupt society. In dangerous cases, the objective may be to cause threat to life and security.

Manufacturing, principal intimidations can embrace theft of intellectual property by nation-states or competitors, especially those capable to compromise insiders or business partners.

STEPS TOWARDS AM CYBERSECURITY

Secure 3D Printers

Companies who connect 3D printers to their LAN must respect the printer as a node that requires to be secured with satisfactory security software. Otherwise, the printer must be actively detached from the network altogether.

Encrypt Critical Data Elements

For the first time in popular manufacturing, a company's proprietary competitive advantage can be stored as a single computer file that can be stolen, making it crucial for companies to encrypt their CAD files with the strongest algorithms available to prevent cyber theft.

Secure the Cloud

Cloud storage has gained widespread adoption as a cost-saving mechanism. However, CIOs should not trust a cloud provider's security claims blindly. Written security procedures and, if possible, the provider's premises should be reviewed to ensure they are being followed.

Verify Intellectual Property Rights

Product developers often generate their own CAD design files and contract manufacturing companies to yield products for them. Contract manufacturers utilizing 3D printers should confirm that the designs provided do not violate any current patent, copyright or intellectual property laws, as well as secure a hold harmless contract to protect/indemnify them in the occasion of an intellectual property right against the original designer.

Secret Flaws Keep 3D-Printing Safe from Hackers

Mount secret Flaws, and share the "Decoder Ring." Researchers have discovered means for manufacturers utilizing 3D printing to implant hidden flaws in files to prevent intellectual property theft. Since the global supply chain for 3D printing, also called additive manufacturing, AM, needs businesses to share computer aided design, CAD files within the institute or with outside parties via email or cloud, intellectual-property pilfers and criminals have many occasions to filch a manufacturer's design files to yield counterfeit parts.

There are certain deliberately induced flaws can disappear when the part is printed under a very specific set of circumstances. Those without the information required to process the files properly—such as parties with stolen CAD models—would print a part with undesired flaws and, accordingly, lower quality. The purposeful manufacturing flaws established in this work range from two-dimensional features to three-dimensional shapes that can be hidden inside the part.

One CAD model seems to have a sphere inside a rectangular block. Though, the block prints without the spherical piece if the processing is directed under the precise set of parameters, while incorrect processing generates a void in the block. Scientists confirmed that flaws implanted in 3D printed components can be so small that they can evade discovery by commonly utilized imaging methods but can nonetheless disturb the performance. So far, the main ways designers have secured CAD files is with cybersecurity tools such as encryption and password protection and by restraining access to vital files. It is known that "cybersecurity tools can be implemented as usual to make the files and cloud secure; though, in situation the design files are stolen, there is nothing in the designs to discourage printing a high-quality component. The new method is intended to deliver an advantage in this scenario and to make printing high-quality parts from stolen files problematic."

3D PRINTING IN FORENSICS SCIENCE

3D Printing Solve Crimes

Forensic Uses of 3D Printing

The utilization of physical representations in court is a common exercise. Small scale replicas of crime scenes or other objects of evidence have been documented into court to demonstrate jurors where and how a crime may have been committed. Nevertheless, the utilization of 3D printing for fact-finding or court purposes is still fairly new. This may be in part because of a perception of an intricate technology, cost, or simply a lack of understanding of what can be accomplished with 3D printing. It's a wonder why more investigators, attorneys, and expert witnesses haven't seen the advantage of 3D printing for use in court.

3D Fingerprints

Time can often be an influential factor and in many distant areas where resources and equipment may not be readily obtainable, first responders have a chance to acquire photographs of evidence utilizing nothing more than a digital camera. Utilizing innovative photogrammetry software such as "PhotoModeler" Scanner or "3DReality," a dense and precise surface model can be generated. Also, it is significant to learn that the 3D model is a reproduction of the footprint and not a surface that is cast as a "negative." Consequently, the generated digital model can be transformed into a freely satisfactory format for 3D printing and in the lack of more satisfactory casting materials, time, or resources, laser scanning or photogrammetry can demonstrate to be of advantage.

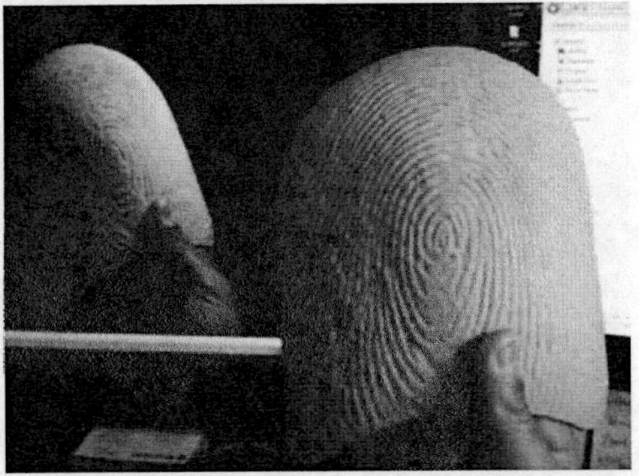

Fig.9.4: 3D Printing Finger Print with Detailed Ridges for Forensic Science

Fingerprint examination permits 3D printing to prove to be useful, Figure 9.4. A small scanning system, such as those created by "FlashScan3D5," permits a suspect's fingerprints to be taken fully in 3D. While fingerprints at crime scenes are usually captured through the utilization of powder and tape, they are ultimately scanned or photographed as a high resolution image. The basis of the prints, i.e. parts of the fingers and palms are all bent and encompass highly thorough ridges and pores. The subsequent 3D model is capable to capture all the ridge detail and can be utilized for investigative contrast determinations. While in court, a fingerprint assessor could use a large replica of a suspect's fingerprint to make identifications and comparisons by color coding particular ridge features, such as islands, crossovers, and bifurcations, and corresponding them to a discovered print at a crime scene. Jurors profit by being able to easily envisage the 3D replica and they have the advantage of haptic insight.

Fingerprints are a good specimen of where a small object can be generated to much larger scale projection to bring out specific details, which would not ne usually easily visible by the naked eye. Fingerprint examiners in training benefit likewise from having the capacity to easily visualize and "feel" an enlarged 3D replica of a person's finger appearances like before making a flat print contrast.

Facial Reconstruction and Identification

When skeletal leftovers are found and the skull is whole, it is conceivable to utilize the skull to gain information about the kind of individual who was discovereed. Gender and race are capable to be identified from various benchmarks on the skull to support with identification of the individual. At the "Central Identification Laboratory" of the Joint "POW/MIA" Accounting Command, JPAC, they have an assignment to recognize the remains of American soldiers from past military battles. Among the lab's tools for forensic identification are multicolor 3D printers. For instance, JPAC prints a model of a skull utilizing digital information from CT scans of the remains. The 3D printed skull is then photographed from multiple angles and superimposed with photographs of known soldiers to measure possible equals, a process called "skull photographic superimposition." Some mention should be made that these methods need not be found from a CT scanner since photogrammetry or other 3D scanning systems are able of capturing the data at diverse levels of detail contingent on the necessity, Figure 9.5.

Fig.9.5: 3D Printing Recreating the Egyptian Queen Nefertiti – *(She was the right hand of Amenhotep IV - later called* Akhenaten, *c.1350-c. 1333 BC)*

3D Printing Identifying Victims

Florida scientists are now utilizing 3D printing to aid crack nine cold cases. For these nine murder cases with nameless victims, the evidence was exceptionally restricted — typically just the body itself. And because forensic sculptors are not given direct access to these skulls, there is much conjecture involved in the process. Researchers at the University of South Florida have scanned and printed 3D printed replicas of the victims' skulls. These are then be given to forensic artists with the National Center for Exploited and Missing Children, who will use the bone structure to fill out the faces with clay, sculpting artists' renderings of the deceased. Using the genuine bone structure as a foundation for the models should make them much more precise. This exclusive collaboration between law enforcement, scientific researchers, and forensic artists marks the formal re-opening of these cases: seven adults, and two children. Shared with new skeletal testing and chemical analysis of the remains, the team aims to recognize the victims, solve the crimes, and give people closure.

Forensic Pathology

A method could be used on the internal skeletal components of a shooting victim which may be impossible to document without an invasive procedure. Small bone fragments, bullet, or pellet particles lodged in clothing or internally would all be disturbed as part of a traditional medical examination. Through the use of CT and MRI scanning, the relative positions of objects remain intact and without further disturbance. The benefit to the investigator or pathologist is that the object is able to be inspected first hand and if created at a one to one scale, measurements, bullet trajectories, and other important data may be recovered.

There is a special need for techniques that allow pathologies to be presented clearly in the courtroom. For medical laymen such as judges, lawyers, and especially relatives of murder victims, the presentation of autopsy photographs can be disturbing, making findings difficult to present this way. From a juror's perspective, being able to see such an object which is much more familiar than trying to interpret radiographs allows for a better spatial appreciation and understanding of injuries. Unlike volume renderings, 3D printed models offer real three-dimensionality as well as a haptic component, which make it easier for medical laymen to understand.

Accident Reconstruction

Vehicle accidents represent an area where much is disputed and litigated on a regular basis. In some instances, the complexity of an accident or a failed component may require more than just photographs to do a proper investigation. As more police agencies adopt laser scanners in their workflow, it is possible to scan vehicles just as they are found at an accident scene to give a more accurate account of how vehicles may have been found in their final rest positions relative to one another or to some other object. Insurance companies do not hold on to "write-offs" indeterminately and they are frequently demolished not long after the event; removing any chance of further examination and all confirmation with it. In the courtroom, a skillful witness can physically hold a model of a crumpled car and point out areas that were of significance. In this method, physical 3D duplications can reserve some of the evidence for future observation by a juror. Crush and extent of destruction can be shown to a juror constructed by the physical replica and it is conceivable to take multiple models of vehicles and demonstrate the engagement between them by only "fitting" the 3D replicas back in alignment with one another.

Structural and Industrial Accidents

Beside the lines of accident reconstruction, there is frequently much consideration and examination into the reason and origin of mechanical and large scale industrial accidents. In the case where a big structural member has failed, 3D Laser Scanner can be used to document crane as in big object disasters. The catastrophe and analysis of the 3D scan information can deliver signs to how and where an vital structural member may have failed. Additionally, being capable to reconstruct the circumstances of the failure utilizing a scaled model is also likely.

3D Printing Reconstructs Crime Scene

3D printing certainly has many army applications, nonetheless the technology can similarly be very valuable in terms of serving out our local law enforcement, from manufacture traffic stops safer and aiding to recognize victims of homicide, to utilizing 3D scanners to construct a simulated reestablishment of crime and accident scenes and constructing digital, 3D printable archives of evidence. Autonomous 3D printing company can provide a complete variety of high-performance desktop 3D printers and materials to its customers, accompanied by training facilities and expert services. There are 3D Printers utilized to support the Police team with its exploratory events.. "3D laser scan technology has been used to yield 3D animations and plans for court use. It was a normal development therefore to be capable to re-erect the crime scenes in solid 3D. This would permit judge and jury to grasp simulated the crime scene in their hands.

An oversized scale 3D printer was essential to produce an advanced prototype and succeeding product for government partners. Other forensic utilizes of 3D printing are far-reaching and are open to creativity.

Some Forensic Uses of 3D Printers

- Printing a scale model of the floor in a home where a crime taken place.
- Reconstructing a physical copy of a firearm discovered at a crime scene.
- Exhibiting bullet trajectories through a 3D scanned object of clothing.
- Constructing a model of a suspect's dentition and presenting how well a bite mark aligns.
- Printing a small scaled model of a buckled building due to a bombing.
- Constructing test pieces of a piece of proof that might be utilized in testing.

While there are limited cases where 3D printing has been implemented for fact-finding or court purposes, the capability to physically reconstruct a piece of evidence is an remarkable methodology.

The variety of articles can be as small as a fingerprint or can be a whole crime scene that is scaled down to just a few feet. As detectives and scientists begin to recognize the advantage of reproducing evidence, they will require to sstart looking at digitizing technologies such as close range laser scanners, structured light scanners, and photogrammetry. Once these technologies have been implemented and additional evidence is captured in 3D, there will very possible be numerous more cases where 3D printing will be applied.

ETHICS AND LEGALITY OF 3D PRINTING

Intellectual Property - IP Concerns

Copyright, patent, and trademark are the three primary types of intellectual property. Generally speaking, copyright covers creative works, patent covers technical works, and trademark covers the ways in which goods are identified in the marketplace. Manufacturing applications of 3D printing have been subject to patent, industrial design, copyright, and trademark law for decades. However, there is limited experience regarding how these laws should apply to the use of 3D printing by individuals to manufacture items for personal use, nonprofit distribution, or commercial sale.

Patents with a finite duration usually provide legal protection for proprietary manufacturing processes, composition of matter, and machines. To sell or distribute a 3D-printed version of a patented item, a person would have to negotiate a license with the patent owner, since distribution of the item without permission would violate patent law. Copyright is also an issue encountered in 3D printing. The fact that copyrights traditionally don't apply to functional objects beyond their aesthetic value may limit the significance in this area. However, that does not mean that concerns about copyrights are inconsequential. In at least one case, a designer filed a copyright takedown notice demanding that a 3D print file repository remove another participant's design because the complainant considered the design to infringe on his copyright. Incredible design and manufacturing capabilities are available to almost anyone right now, thanks to 3D printing services and the designers/makers who populate those services' marketplaces with designs that are creative, original and/or functional. While we wait for the law to catch up to the technologies underlying 3D printing services, it makes sense to consider where the biggest IP risks lie. By simply focusing on original and/or function designs and avoiding well-known trademarks and copyrighted characters, you'll go a long ways toward lessening the likelihood of finding yourself a target of an IP lawsuit. The rapid emergence of 3D printing will create major challenges in relation to intellectual property (IP) theft. Gartner predicts that by 2018, 3D printing will result in the loss of at least $100 billion per year in IP globally.

3D Printing Testing Existing Laws

Multidisciplinary experts at the international law firm lately distributed a comprehensive report testifying the legal difficulties and risks related with the augmented utilization of 3D printers. The areas of legal problems and risk are specified as follows:.

1. Constitutional Trepidations
2. Commercial Lawsuit
3. Secrecy
4. Indemnification
5. Insurance
6. Product Liability

Because 3D printing becomes further conventional, it is only a matter of time before courts are confronted with the dilemma of whether outdated tort liability doctrines will apply to 3D printed products and manufacturing methods, or whether novel laws will require to be shaped. While the explosion of 3D printing may posture new trials to traditional IP prosecution, there will be "affirmative measures" IP owners can take to adapt to its greater role in manufacturing.

Safeguarding endorsement from watchdogs is another significant barrier that may obstruct the prevalent medical implementations of 3D printing. Numerous simple 3D-printed medical objects have obtained the FDA's 510(k) approval. Nevertheless, achieving more challenging FDA regulatory necessities could be an obstacle that may obstruct the accessibility of 3D-printed medical products on a large scale. For instance, the essential for large randomized organized trials, which necessitate time and funding, could cause a wall to the obtainability of 3D-printed drug dosage formulae. In addition, fabricating guidelines and state lawful necessities could enact obstacles regarding the provision of 3D-printed medications. 3D drug printers must also be lawfully defined as manufacturing or compounding apparatus to better control what laws they are subject to. Eventually, the controlling decisions that are made should be founded on sound science and technology. With this objective in mind, the FDA lately generated a working group to measure technical and regulatory concerns regarding 3D printing

As 3D printing technology cultivates in importance while the easiness of conveying such data upsurges, the guideline of such data grants both exclusive opportunities and trials. Nation-states hold vastly diverse controlling frameworks and philosophies on the improvement, formation, and ownership of various goods and products within their borders.

As the continuing 3D-printed weapon dispute proves, current global lawful frameworks are exclusively unequipped to the task of governing the stream of such information. Such openly obtainable designs tend to be un-copyrighted and open-source, rendering them unquestionable to cross-border copyright and intellectual property IP contests. While national regulation—coupled with irregular international cooperation—may deliver a incomplete solution to such apprehensions, a time may rapidly come when countries must essentially consider the legal mechanisms that will bind what many have come to claim the "Fourth Industrial Revolution."

Ethics in 3D Printing

Additive manufacturing has extended broadly over the past three decade, particularly with the accessibility of home 3D printers. In the near future, many objects may be fabricated at home, which elevates two ethical subjects.

First, there are demands of safety. Our current safety guidelines be contingent upon central manufacturing traditions; they will be problematic to enforce on this novel model of fabrication.

Second, there are intellectual property IP concerns. In addition to a 3D scanner, it is presently conceivable to scan objects and print copies; many items are not secure from this by existing intellectual property decrees. Intellectual property depends on the concept of inventiveness, but what counts as inventive changes with the growth of novel technologies. Some of the ethical subjects surrounding bioprinting embrace equal entrée to treatment, medical safety, and the augmentation of human body.

Bioprinting concentrations are on the discrete care rather than developing a general treatment strategy for all patients. Tailored medicine is costly and upsurges the inequality between the rich and poor. Since 3D printing is an discrete treatment, the general public assumes that it may thwart people with financial issues from receiving care. Any novel treatment connecting with 3D bio-printers is chancy and patients must be well well-versed of the health consequences. Bioprinting may be utilized to upsurge human performance, strength, speed, or stamina.

THE INTRICACY OF 3D PRINTING

Complex Geometries

One of the greatest advantages of 3D printing is the potential for producing components and parts with complex designs. Since designs can be made digitally, products do not need to be limited by geometric complexity. This indicates that manufacturers can produce stronger and lighter components, a fact of which the aerospace industry in particular is taking advantage. 3D printing lets designers create complex shapes and parts – many of which cannot be produced by conventional manufacturing methods. Though this benefit has been occupied by developers and artists to inspiring visual influence, it has also made an important impression on industrial implementations, whereby implementations are being advanced to emerge complex mechanisms that are demonstrating to be both lighter and tougher than their forerunners. Distinguished utilizations are evolving in the aerospace division where these subjects are of main prominence. By the natural laws of physics, fabricating through additive techniques means that intricacy doesn't have a price; elegant product designs with intricate design characteristics now cost just as much to yield as simple product designs that follow all the customary rules of conventional manufacturing.

Mass Customization

Along with liberty in complexity comes the choice to tailor a product to nearly any prerequisite. 3D printing technology permits fabricators to build multiple products inside the same construct chamber, each designed to answer to diverse end-user requests. With traditional production approaches, it is less expensive to create and sell products at an reasonable price to the consumer. Alternatively, 3D printing permits for simple customization; one only requires to modifying the design digitally to allow alterations with no added tooling or other costly fabricating procedure essential to yield the final product. Each and every article can be tailored to meet a user's exact requirements deprived of extra manufacturing costs.

Less Tooling

For industrial fabrication, one of the most cost, time, and labor concentrations phases of the product development procedure is the production of the tools. Low to medium volume implementations, industrial 3D printing — or additive manufacturing — can eradicate the necessity for tool production and, consequently, the costs, lead times and labor related with it. This is a favorable proposition that an increasing number or manufacturers are investigating.

Additive manufacturing permits products with intricate designs to be prepared with one machine. This decreases the total number of stages in the operation, as well as the amount of tooling essential at each stage. The final product can be completed sooner and necessitates fewer parts. When metal casting or injection molding of each workpiece of each product needs a new mold, a considerable and rapid upsurge in manufacturing cost. To recover these extra manufacturing costs, most businesses depend on on thousands of the similar item being sold. Instead, since 3D printing is a "single tool" procedure there is no necessity to modify any feature of the procedure and no added costs or lead times are necessary between making an object complex or simple. Eventually, this leads to considerably lower fixed cost.

Environmentally Friendly

With less assembly needing less tooling and containing fewer parts, a 3D printing project generates less total waste. This drops the environmental impact of a 3D printer, particularly when equated to conventional machining approaches. Numerous conventional manufacturing procedures are subtractive: beginning with a chunk of material, cut it, machine it, and mill it until it has been treated as the envisioned design. For several products – such as a bracket for an airplane – it is standard to waste 90% of the raw material during this procedure. Instead, 3D printing is an additive process; creating an object from the raw material layer by layer. Naturally, when a workpiece is fabricated in this way, it only utilizes as much material that is required to construct that particular piece. Furthermore, most of these materials can be reprocessed and repurposed into more 3D printed pieces. Also during an additively manufactured product's operating life, by method of lighter and tougher design that imposes a deceased carbon footprint equaled with traditionally manufactured products.

Superior Creativity

This is one of the best compensations of 3D printers. Different than traditional manufacturing approaches, there is no restriction by the workforce or the machines utilized. A 3D printer allows you generate practically anything imagined. utilize any polymer, metal, ceramic, glass or alloy material preferred, and selected inspired colors, intricate details or whatever else the product needs.

Faster Product Delivery

The 3D printing operation can be accomplished anyplace in the world providing a printer is accessible. The product fabricated locally, where the customers located. For instance, if a product is ordered in England, a 3D printing facility in England can receive the product design, print it out and transport it from there. Then delivery cost is saved as well as the customs fees. The customer received the product much earlier, simply and affordably. 3D printing is presenting good potential in terms of satisfying a local manufacturing model, whereby products are created on demand in the place where they are wanted — removing inventories and unmaintainable logistics for shipping big volumes of products around the world.

Single Step Manufacture

One of the highest concerns for a entrepreneur is how to manufacture a workpiece as proficiently as possible. Most workpieces need a large number of manufacturing stages to yield and the order the stages happen in influences the quality and manufacturability of the design.

Consider a tailored steel bracket that is created through traditional manufacturing approaches. Like additive manufacturing the operation starts with a CAD model. Once the design is completed fabrication starts with the steel being cut to length. The cut lengths are then fastened into position and welded one at a time to form the bracket. Occasionally a jig will require to be created up to safeguard all mechanisms are properly aligned. The welds are then ground to provide a good surface finish. Following, holes are drilled so the bracket can be fixed to the wall. Finally the bracket is sand blasted, primed and painted to improve its aesthetics. Additive manufacturing machines finish a construct in one step with no interaction necessary during the construct phase. For additive manufacturing as soon as the CAD design is completed it can be uploaded to the printer and printed in one step in a couple of hours. The capacity to yield a part in one step greatly decreases the need on different manufacturing streams (machining, welding, and painting) and provides the designer superior control over the finished product.

Unique Efficiency of 3D Printing

The 3D Printing Technology enjoys unparalleled efficiency among manufacturing operations to prototype any product unabated, for the reasons stated as follows:

- Inexpensive customization
- Permits manufacture of more efficient designs — lighter, stronger, less assembly necessary
- One machine, unrestricted product lines
- Very small articles (nano)
- Efficient utilization of raw materials (less waste)
- Pay by weight — intricacy is free
- Batches of one, formed on demand
- Print at point of assembly or depletion
- Manufacturing available to all — lower entry barriers
- New supply chain and retail prospects

WORLDWIDE IMPACT OF 3D PRINTING

Impact on Manufacturing Competitiveness

Additive manufacturing, beginning with today's infancy period, necessitates manufacturing corporations to be flexible, ever-improving users of all accessible technologies to continue competitiveness. Supporters of additive manufacturing also forecast that this arc of technological improvement will stand globalization, as end users will do much of their own fabrication rather than involve in trade to purchase goods from other people and corporations. The real incorporation of the newer additive technologies into commercial manufacture, though, is more a substance of supplementing traditional subtractive methods rather than dislocating them entirely. The insinuations of evolving technologies such as 3D printing on the industrial manufacturing marketplace are frequently hotly discussed. Some authorities in the industry sense that it will be immensely disruptive, while others trust the technology is years away from full capability.

The 3D printing industry is predictable to adapt approximately each industry it embraces, totally disrupting the outdated manufacturing operation. Consequently, the predictable worth of the industry is anticipated to burst out in the near future." As the market expands and the cost of printers drops, it is possibly to give increase to new contestants in traditional markets, and branch innovative novel products, as prototypes for new products will become meaningfully inexpensive and less perilous to fabricate. Hitherto, product improvement and the competitive situation are just two of the possible inferences. This developing technology is also possible to have a substantial impression on how manufacturers do business, precisely as it relays to shifts in material cost, incremental cost calculations, and outdated assembly line and product pricing plans.

Stratasys retailed a total of 129,197 systems worldwide as of March 31, 2017. A worldwide shipment of 3D printers was 219,168 systems in 2018. Rendering to data from CCS Insights, approximately 158,000 3D printers were sold in the global market in 2014. By 2018, that number has climbed 845,000. That progression could signify a thump from $1.6 billion in revenue to $4.8 billion. Roughly 49 percent of the global revenue will come from North America and 3D printers will become progressively ordinary in both the private sector and commercial world.

3D PRINTING IMPACT ON GLOBAL MANUFACTURING

Few examples and predictions of how 3D printing will intersect with modern manufacturing are presented as follows:

Replacement parts

One of the greatest substances about 3D printing is that the technology stretches the capability to print tailor-made designs. This is cherished to substitute parts on existing products. From the armed printing custom parts for trucks and tanks to industries quickly fixing minute parts for a machine, the capacity to rapidly manufacture substitute parts will be valued.

Healthcare

There is no industry has as abundant possibilities as healthcare. There are now sufficiently of stories about 3D printing being utilized to print objects as prosthetic components, surgical replicas, and even living tissue. Though, there is still the teaching and raining of customers and ethical issues to sift through before 3D printing wholly floods modern healthcare, but the possibilities is surely there.

Retail

There is a robust yearning among modern customers to customize the products they purchase. Over the next few years, 3D printing will confront the traditional mass-manufacturing model. While modern manufacturing representations need scale and standardized design to yield high volumes of products, the novel manufacturer is self-assured for one-off or small-batch production, depending on which 3D-printing technology is utilized. The result will be more customization in the products we purchase.

REVOLUTIONIZING MASS MANUFACTURING

Mass production is the substantial challenge in 3D printing, but with the embracing of large-scale printers and quickly developing technology to yield objects faster, the printers will totally disrupt outdated manufacturing in many businesses:

Food

Whatever that happens in liquid or powder form can be 3D printed; naturally, printed food is one of the frequent topic of home discussions.

Military

The machinery for the armed division is often tailored and substitutions must be made rapidly. A 3D gun has previously been printed, so it is only a matter of time before the technology hooks on in this industry.

Electronics

The size, shape, and materials exploited to create electronics make this industry a natural applicant for 3D printing.

Toys

Home 3D printers and open source design will revolution the way children innovate and play.

Automotive

This industry is previously utilizing the technology—Ford reportedly utilizes 3D printing to test parts. High-end and minor auto companies will benefit first, though 3D printing could advance the competence of creating auxiliary parts for any business.

COST OF 3D PRINTING

The cost of manufacture can be broken down into 4 categories; cost of 3D printer, machine operation costs, material cost and labor costs. These categories are discusses as follows:

Cost of Printers

The cost of procurement a 3D printer still does not make its purchase by the average householder feasible. Also, different 3D printers are required in order to print different types of objects. Printers that can manufacture in color are costlier than those that print monochrome objects.

Machine Operation Costs

Most desktop 3D printers utilize the same amount of energy as a laptop computer. More industrial additive manufacturing technologies devour a high amount of energy to yield a single part, though the capability to yield intricate geometries in a single step consequence in higher efficiency and turnaround. By being capable to fabricate products with less overall parts and assembly, 3D printing permits saving the cost of tooling and the hassle of acquiring all the necessary equipment. 3D printing also allows zero-cost complexity. Machine operation costs are typically the lowest contributor to the overall cost of manufacture.

Material Costs

3D printing cuts down on material costs. Instead of using a block of plastic, metal or other material and cutting the product out of it, subtractive manufacturing, it is feasible to use only the materials necessary for the build, additive manufacturing. This not only cuts the upfront costs for materials, it also reduces the funds normally spend on transporting and disposing of that waste. The material cost for additive manufacturing varies significantly by technology. Desktop FDM printers use filament coils that cost around $25 per kg while SLA printing requires resin that retails around $150 per liter. The range of materials available for additive manufacturing makes quantifying a comparison with traditional manufacturing difficult. Nylon powder used in SLS costs around $70 per kg while comparable nylon pellets used in injection molding can be purchased for as little as $2 – $5 per kg. Material costs are the biggest contributor to the cost of a part made via additive manufacturing. Although 3D printing reduces waste of materials, the materials used for 3D printing is costlier, hence net effect is negligible. Also some 3D printing systems generate more waste than the printed part material, and because their material is expensive, that waste comes at a high dollar cost. Support material, shaving uneven surfaces, dissolving chemicals, etc., all contributes to the total cost of 3D printing. These systems result in significantly higher hidden costs, particularly as that waste are abandoned.

Labor Costs

One of the main advantages of 3D printing is the cost of labor. Post processing aside, the majority of 3D printers only requires an operator to press a button. The machine then follows an automated process to produce the part. Compared to traditional manufacturing where highly skilled machinists and operators are typically required, the labor cost for a 3D printer is almost negligible.

HOW TO CUT 3D PRINTING COSTS

Avoid unnecessary 3D printing costs using the following guidelines:

General Guidelines

1. Buy your own 3D Printer
2. Select the appropriate material
3. Use or Modify a free design
4. Use free 3D Modeling Software
5. Take advantage of Non-Traditional shapes
6. Hollow your 3D Model
7. Use an Online 3D Printing Service
8. Use a Price Comparison Service
9. Reduce Labor costs with Looping and Sinter-shelling

Technical Guidelines

1. Select the Appropriate Finish
2. Choose a greater layer thickness & resolution
3. Use Economy Mode
4. Take advantage of Bulk Discounts - Use Batch Control
5. Reduce the size of your Print
6. Join Loyalty - Student Programs

Impact on Economy

The 3D printing industry will have far-reaching effects on the global economy. 3D printing will cause major disruptions in the global economy by 2025. The analysis firm predicts it will bring about new product development cycles as the systems become cheaper. More companies will adopt the technology and product creation will focus on client feedback and customer-centered design. The industry is also reducing the cost of entry into markets, allowing very niche businesses to pop up everywhere. 3D printing will affect the developed and the developing world in different ways. While it is a boon for the developed world due its aging society and shortage of cheap labor for manufacturing, it is a double-edged sword for the developing world. In developing economies, 3D printing, on one hand, will lower manufacturing costs while increasing manufacturing-related unemployment on the other.

Impact on Society

Thirty years into its development, additive manufacturing has become a mainstream manufacturing process. Additive manufacturing build up parts by adding materials one layer at a time based on a computerized 3D solid model. It does not require the use of fixtures, cutting tools, coolants, and other auxiliary resources. It allows design optimization and the producing of customized parts on-demand. Its advantages over conventional manufacturing have captivated the imagination of the public, reflected in recent mainstream publications that call additive manufacturing the fourth industrial revolution.

The societal impact of additive manufacturing

The societal impact of additive manufacturing include following areas:

1. customized healthcare products to improve population health and quality of life,
2. reduced environmental impact for manufacturing sustainability, and
3. simplified supply chain to increase efficiency and responsiveness in demand fulfilment.

3D printing tackled the headlines with the printing of a working gun and will continue to make it increasingly difficult to stop the distribution of illegal objects. It is far easier to find and confiscate a physical object than a computer file. When that file is finally turned into a physical object it only needs to happen immediately before it is used. The game of cat-and-mouse between law enforcement and criminals will move further online. Another implication is that it will take "mass personalization" to new levels. This is where companies are able to personalize matters for people without the large cost of human intervention to do so. Indeed this will probably be the norm, turning individuals into creatives, all able to personalize jewelry, watches, clothes, food, and cutlery and so on before we buy. This will put out of business organizations that do not adapt to the world of mass personalization. Personal interaction will reduce, with fewer of those incidental chats and conversations that are so important in building strong communities. As more and more 3D printers start to enter people's homes, the conventional relationship between the home and the workplace might get further eroded.

Finally it will have a positive effect on the environment. Place of manufacture has become disassociated from place of consumption. Planes, ships and trucks transport millions of goods around the world every day at considerable environment cost. If the two are brought together this would significantly reduce the environmental damage of goods transportation.

3D Printing Impact on Environment

There are two sides of 3D printing impact on environment; one is an eco-friendly, while the other side is not so eco-friendly as one would have hoped it to be. Currently, 3D printing is not what might call eco-friendly. Here are the some disadvantages of this technology stated as follows:

3D Printing Consumes much Energy to Produce Items

While basic items created by 3D printing may use less energy than those manufactured and shipped traditionally, 3D printing remains a slow and less efficient process compared to subtractive production machining. While an injection mold could produce 1,000 objects in an hour, a 3D printer may only manage 100, using the same amount of electricity. Furthermore, industrial 3D printers are frighteningly energy hungry. The industrial 3D printers that use heat or a laser to melt plastic consume between 50 to 100 times more electrical energy than an injection molding making the same object.

The electricity consumption argument is not really applicable to smaller desktop 3D printers, the environmental and economic effects of desktop 3D printing compared with traditional manufacturing show the claim that these printers are energy-hungry is not true. They have the power draw approximately equivalent to a laptop computer. It is worth noting that energy alone is not the full picture. There's more to an overall carbon footprint than electrical machine power. Other considerations include:

1. The manufacturing process
2. Raw materials used
3. Location of the manufacturer
4. Transportation mode used for distribution

The additive manufacturing procedure did not yield the lowermost carbon output at every stage of the supply chain. 3D printing did fare well at the material production stage. The cause for this was due to using less material and created less waste in general. Some 3D models require support materials though, which the user abandons at the end of the printing procedure. In the instance where a printed part utilizes support materials, the wastage then increase. 3D printed products are capable to yield low carbon parts, but only at the material-production stage, not the energy-production stage—yet. Support material waste can be an issue, but it is not the main concern from an environmental position. That apprehension remains with the energy consumption.

3D Printing Reliance on Polymer Materials

An impediment to 3D printing's carbon footprint is with the plastic materials uses. The majority of widely used plastic filaments are eco-unfriendly. Unfortunately, high-end 3D printers used in industries leave behind considerable amounts of plastic by-products. None of these by-products are ecological, hence the current problem. Any object involving plastic is not an heartening feature for the environment, and while scientists are employed on generating plastic polymers, or the "ink," that are biodegradable, the entire procedure still depend on a material that is environmentally harmful. Unsuccessful prints are typically disorganized. It is not yet acknowledged how to recycle them. The two most typical filaments to print with are ABS and PLA. ABS is a thermoplastic that is great for 3D printing because of its strength and robustness. This material is not biodegradable or compostable, but can be recycled in other methods if you want to re-heat the material to use it again in a filament recycler.

Non-biodegradable plastic poses threat to the environment, but these plastics can essentially be recycled into filaments utilized in 3D printing. While machines that can recycle 3D printer plastics do exist, the quality of the material worsens the further times it is recycled and therefore "the more likely it is to fail so people might be less motivated to recycle their material". Some PLA is compostable, though it necessitates a very specific temperature and environment to do so. It is formed from products like cornstarch, sugar cane, and tapioca root so it can be absorbed by microorganisms. Specialists recommend material manufacturers not toss their PLA in a recycling bin as it may biodegrade in the recycling procedure.

The 3D Printing Process Produces Toxic Fumes

Scientists examined the possibly harmful emissions produced by printing with plastics, also called 3D inks. Their discoveries found that the materials seized toxic by-products when heated at high temperatures. This does not pretense any surprise as all types of heated materials yield some level of toxic emissions, as well as organic cooking. In actual fact, the emissions created by heating these plastic materials were on a par with those produced by cooking indoors. When 3D print, it discharges ultra-fine particles into the air which can be detrimental when breathed. Also, heating plastic emits toxic gases which fluctuate depending on the printing material. ABS emits a stronger scent when printed & new PLA materials are recognized to emit very little aroma when heated.

Encouraging Wastefulness

Research demonstrated 3D printers themselves have inadequacies that make them less environmentally friendly. An inkjet 3D printer wastes 40 to 45 percent of its ink. And if a printer isn't turned off or unplugged, it utilizes a disproportionate amount of electricity. As the printers become further available, manufacturers will require to find a means to develop these wasteful matters.

Eco-friendly Impact of 3D Printing

Traditional manufacturing is frequently wasteful and polluted. In numerous methods, 3D printing diminishes that waste and the carbon footprint manufacturing has on the Earth.

Fewer Wasted Material

Merely the raw materials required to generate the object—be it plastic filament, metal powder, or carbon fiber—are utilized. Using biodegradable PLA plastic filament in fused deposition modeling printers like "MakerBot" is a virtuous beginning. PLA is a renewable sort of thermoplastic material. PLA yields significantly less toxic emissions throughout the printing phase than others, but that's not all. There's even fewer waste product with PLA when employed as a filament for 3D printing. It's likewise healthier at the disposal phase at the end of a product's life cycle. This is why it's presently, and will remain to be, the supreme popular selection over other 3D printing materials similar to ABS.

Possibility of Longer Life Spans

Product components can be substituted with 3D printing (or at least, that's the idea for the future), so the whole product doesn't have to be discarded away and substituted each time it became unsuccessful.

Less Transport

Products frequently travel from corner to corner throughout many continents to reach their final destination. With 3D printing, the production and assembly can be as close as possible locally. However, raw materials are the only subject that will ship, and they take up far fewer space.

Fewer Unsold Products

Thousands of plastic lawn chair are fabricated every year – all necessitating energy and materials to yield the chairs– but not all of them are sold, and those that are not sold either end up abandoned in a place like a warehouse, or tossed as garbage. If it were feasible to print a lawn chair on demand, it would eliminate the needless fabrication and save abundant of embedded energy. If a corporation creates a product, that is obsolete or not sold frequently end up heaping up in landfills, 3D printing can recover this because companies can make them as required.

Smarter 3D Printing

Up until the technology advances to print in 3D with a smaller amount of harm to the environment, there are additional many replacements that can aid. One method to decrease the carbon footprint is to use less energy, and the other is through the use of intelligent materials or filaments, suggested as follows:

- Print objects with hollow parts
- Change part orientation during printing
- Print multiple parts at a time on the printer bed

A hollow part will necessitate less material and consequently print quicker. Intricate components may require support materials; nonetheless it will still indicate faster print times and less emissions in general. Placing tall parts on their side is an alternative way to speed up the printing process and decrease or eliminate the need for support materials. And for printers that have a good sized print bed, or print platform, it is likely to print several components in one run.

Use Superior Material

Not all 3D printing materials are equal. Enhanced materials provide the following benefits:

- Decrease resource utilization
- Cut down on waste
- Yield less toxicity
- Reduce energy utilization

Besides PLA, further extensive, eco-friendly 3D printing materials to look forward to the future may include:

- Bio-plastics
- Powders
- Resins
- Acrylates
- Wax

The Future Scope of 3D Printing Technology

Undoubtedly, Additive Manufacturing Technology itself will experience revolutionary developments just as its competing technologies have gone before it. Naturally, the minds, and the charisma of the entrepreneurs who originally invented the technology, are the same minds needed to equip it with additional wings to freely soar higher and further. Accordingly, the expected future development are logically predicted as follows:

1. Complex Engine parts.
2. On demand parts in Space.
3. Aircraft Wings.
4. 3D Printing as a Service (3DPaaS)
5. Printing large volumes economically
6. Expanding the range of printable materials
7. Decreasing the cost of printable materials
8. Using multiple materials in the same printer, including those for printing electronics
9. Printing very large objects
10. Improving durability and quality

3D Printing Trends:

Localizing Production and Industrial Goods

Interestingly enough, Nike could be a pure software company in a decade or even less. If this came to pass, it would be thanks to 3D printing production. In the future, all consumable items we know will be made closer to us. Think through one day at home, one of the wheels on the door handle is broken. Simply, find the part CAD blueprint on the Web. Download it, transform it to tessellated file, print it out at home or at your neighborhood commercial 3D printer, and out comes the substituted part.

Your daughter outgrows her favorite 3D-printed tailored pair of shoes! Discard the timeworn pair in the material recycled bin, and print out a fashionable new pair, fitting a size and half as deemed appropriate. Creating one's own tailored consumer goods, as in the deemed necessary, will brand it much more expedient, inexpensive, and highly efficient to content the household desires and needs.

Direct Digital Manufacturing

However, the financial impact of localized industrial manufacture and commerce will be even more intense. Around the globe, digital 3D printing industrial units of various sizes and capacities will rapidly become intertwined in a global production network. Denoted to as "Direct Digital

Manufacturing," this capability to manufacture modules and finished goods close their point of exploitation will be much more effective than today's global sourcing.

Custom Production Materials

The skill to customize the materials utilized for production—even printing living tissue—will be actually transformative. Doctors are already exploiting 3D printing to generate knee substitutes tailored for each patient. Very soon, the material itself will encompass pain prescription and antibiotics deveeloped to time-release gradually over time, or as needed.

Visionaries are investigating with 3D printing utilizing "Graphene," a carbon constituent that is flexible, see-through, and yet is 100 times stronger than steel. 3D bioprinting has the possibilities to wholly reform healthcare.

Nano-Printing

3D-printed lithium-ion battery operates as a standard battery; apart from that this battery has the dimensions of a single grain of sand! Batteries as these may be exploited to energize minute medical instruments, dense electronics or even minuscule robots. "3D nano-printing" permits for rapid prototyping of micro and nano constructions. Applying 3D printing conceptions to nanotechnology will make nanofabrication much quicker and more efficient, and consequently economically sustainable much sooner than most creative thinkers ever envisioned, Figure 9.6.

Fig.9.6: 3D Nano-Printing Technology

Courtesy – Nano Dimension

Goal-Directed Design

3D printing permits for virtually limitless complexity in dimensions and materials – so much that humans become the preventive influence in captivating full benefit of the potentials. Nonetheless, a computer could routinely create hundreds upon hundreds of differences of a design until it solves a explicit problem. This is the ultimate drive of goal-directed design.

Currently, computer-aided design CAD implements are utilized to manually generate, document and investigate designs. Conversely, with "goal-directed design," begin with explicit objectives for parameters like strength, weight and durability, and then the computer creates hundreds of design options, appraising and recommending the unsurpassed designs for different objectives. Merging 3D printing's limitless shape and material customization with authoritative computing power will lead to designs that no human could perhaps have made-believe. In the future, the greatest designers will not be those who come up with the finest designs; they will be the ones who are talented to ask the computer the unsurpassed questions.

4D Printing - *The Fourth Dimension*

With this novel technology, printing a 3D article that at some future point may spontaneously self-assemble or change configuration when challenged with a deviation in its environment, such as temperature or moisture. ***This is the fourth dimension.***

Visualize a building brick that only extents its full weight and structure when water is added to it, at the precise position where it will be utilized. See in your mind's eye hydraulic pipes that spontaneously mending themselves if they are ever broken.

Be prepared to wear sneakers that transform to running shoes if you initiated running, or grow cleats if you tread on grass, or become waterproof if it starts to rain. Although this technology is still novel it has the abilities to take 3D printing to an utterly new dimension.

Automated Printers

Printers will be progressed to be more computerized and easier to utilize. Current 3D printers achieve numerous jobs autonomously. Nevertheless, some printers at the hobbyist end necessitate that print-heads be cleaned every so often, that printer powder beds be properly leveled, and that a human fiddle and tinker to minimize errors. These printers all need significantly more individual maintenance than people are accustomed to with appliances. The potential to decrease or remove this human element is real and will be a significant factor of revolution over the next few years.

Mechanizing the features that were originated many of the collective errors and consistency worries, such as support structure creation, part orientation, and others, would very likely advance the ease of utilization in hobbyist printers. For instance, a print run can be squandered if the construct platform is not level. Many printers, such as those from "Robox," "XYZprinting," and "MakerBot," include auto-leveling where the printer calibrates itself to the platform. Predictable in the future is an audible response system that provides real-time monitoring of the printing operation, that senses flaws or nonconformity from the design (as quantified in a 3D model created by a CAD tool), and that permits suitable interference. Together, such characteristics will possibly progress the dependability and repeatability of the printing operation.

Multi-Material Printer

Most printers operate with only one kind of material—polymer, metal, ceramic, wood, or a biological material. To generate additional suitable products and increase the market, 3D printers will require processing multiple material kinds within a single construct cycle. Numerous features, typically connected to materials themselves, brand this prerequisite challenging. For instance, most processes are constructed around an ideal material that reacts to a narrow choice of temperature inputs or UV light frequency. Utilizing heat or light, printers frequently liquefy or solidify substances to influence the material into specific forms. The features that brand this operation work eliminate many other prospective materials—at least at the present level of sophistication.

The chase of multi-material ability will serve particular printing approaches over others. FFF printing has high prospective to house multiple materials without significantly extending the existing technology, as printing heads can be added to accommodate other plastics. Multi-head printers are obtainable from "Hyrel 3D," "XYZprinting," and "MakerBot" for a few than thousand dollars. For multi-material printing, inkjet-like technology such as "Voxeljet" is the current and the future.

Approaches such as selective laser sintering SLS and others use inkjet technology. This technology can deal with multiple materials within a variety that can be brought as a powdered "base," as it already utilizes multiple print-heads. Consequently, objects or assemblies created from diverse materials can be printed in a single print run. Nowadays, this technology is available at the high end from "Voxeljet," "Stratasys," "3D Systems," and others. Today, multi-material printers operate for a single family of materials—plastics, for example—and are mainly utilized for prototyping so developers can check form, function, fit, and feel.

Developments are still required to syndicate diverse families of materials, such as metals and plastics, in a single print cycle. "Stratasys," "Objet500 Connex" printer provisions multi-material and multi-color 3D printing. A printed workpiece can have as many as 14 different material properties and 10 color palettes. Scientists effectively implement more efficient operation to the manufacture of heterogeneous substances, which include diverse materials that cure at different rates. This novel 3D printing operation will permit for dental and robotics models, for instance, to be manufactured more cost- and time-efficiently than ever before. Multi-material printers are commercially obtainable from Stratasys, Objet Connex. Though, only restricted materials, photo-curable resins, can be utilized since liquid resins require passing through small nozzles. This method may increase the selections of base materials that are utilized in multi-material printing.

Printing Complete Systems

Farther out is the capacity to print whole systems or subsystems. Emerging multi-material competences will aid, subsequently most completed products are created from more than one material. Though, challenges encompass to the skill to embed obects such as sensors, electronics, and batteries, so all can be printed in one construct. R&D energies are under way in a many of areas, comprising materials, printing approaches, and combining additive and traditional approaches of manufacturing. The important materials science challenge is to improve inks that can be the foundation for printing diverse kinds of products, be they sensors, electronics, or batteries. For instance, "Xerox PARC" is evolving inks so circuits, antennas, and RFID (Radio Frequency ID) tags can be printed and implemented directly to a product. Likewise, scientists advanced the rudimentary building block of minuscule lithium-ion batteries as inks that can be printed. The future of additive manufacturing is not restricted to inanimate substances. Bio-inks were advanced to create living tissues. Scientists utilize several print heads and the tailored inks to create complex living tissues, complete with minuscule blood vessels. Some pharmaceutical corporations are already utilizing 3D printed tissue for testing drugs. Bio-printing characteristically utilizes two inks. One is the biological material and the other is hydrogel that delivers the environment where the tissue and cells nurture. The innovation to add blood vessels was the improvement of a third ink that has an uncommon property: it melts as it cools, not as it warms. This property permitted scientists to print a connected network of filaments and then melt them by cooling the material. The liquid is drawn off out to generate a network of hollow tubes, or vessels, inside the tissue.

Such formations are conceivable only with 3D printing, creating novel potentials beyond traditional manufacturing.

Another Way to 3D Print Blood Vessels

Scientists' first bioprinted a 3D blood vessel pattern consisted of agarose fibers, which was then sheltered in a hydrogel precursor material. This material was then photo crosslinked and the blood vessel pattern within removed, resulting in perusable micro-channel with several architectural structures. The team was capable to embed this functional microchannel within several diverse hydrogel materials, and they established that these networks could advance transport and cellular viability within tissue constructs. But possibly more prominently, the team presented that monolayers of endothelial cells magnificently shaped within the micro-channels, Figure 9.7.

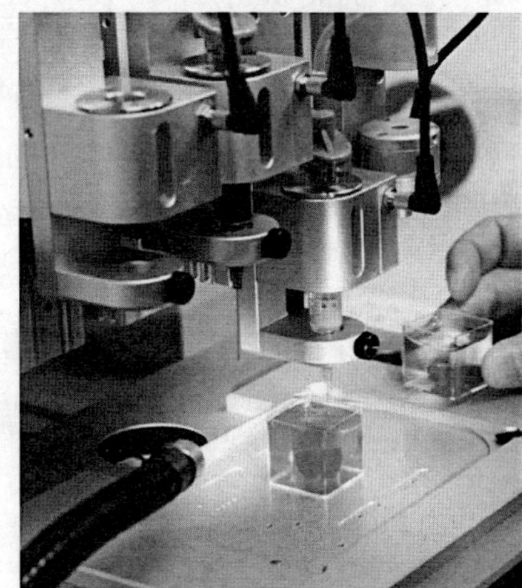

Fig.9.7: The First 3D-Printed Vascularized Engineered Heart

Courtesy: Tel Aviv University, April 15, 2019. Photo by FLASH90

FUTURE OF 3D PRINTING MATERIALS

3D Printing Transparent Fused Silica

Glass is one of the most significant high-performance materials utilized for technical exploration, in industry and in society, mostly due to its supreme optical transparency, exceptional mechanical, chemical and thermal resistance as well as its thermal and electrical insulating characteristics. Nevertheless, silica and especially high-purity silica such as fused silica glass are extremely difficult to shape, needful high-temperature melting and casting procedures for macroscopic objects or hazardous chemicals for microscopic characterisstics. These disadvantages have made glasses unreachable to modern manufacturing technologies such as 3D printing, 3D printing. Utilizing a casting nanocomposite, user generates transparent fused silica glass parts using stereolithography 3D printers at resolutions of a few tens of microns. The procedure employs a photo-curable silica nanocomposite that is 3D printed and transformed to high-quality fused silica glass over heat treatment. The printed fused silica glass is non-porous, with the optical transparency of industrial fused silica glass, and has a smooth surface with a roughness of a few nanos. By doping with metal salts, colored glasses can be formed. This work broadens the selection of materials for 3D printing, allowing the generation of random macro, and microstructures in fused silica glass for numerous implementations in both manufacturing and academia.

3D Printing Removes the Limitations of Gravity

The possibility for 3D printing to transform manufacturing is dumbfounding—if the technology can bypass a few restrictions. Scientists advanced a novel method to both speed up the 3D printing operation, and free it from the limitations forced by gravity. 3D printing includes slowly constructing up a workpiece utilizing thousands of thin layers of extruded molten plastic that can take hours, even days, to finish, contingent on object being printed. The soft feature of the material being employed, which consumes a few moments to cool and solidify, likewise means that models have to be developed and strengthened with temporary structures to account for the force of gravity. One can't 3D print anything that just suspends in mid-air, it will simply downfall before it becomes solid. So the scientists at MIT, are developing a novel 3D printing operation that takes place inside a vat filled with a thick gel suspension that fundamentally negates the properties of gravity. The gel suspension delivers a continuous support for the liquid material as it's being extruded. So as an alternative of a nozzle restricted to moving in just two directions, it's capable to extrude materials in 3D. This permits more intricate workpiece to be printed deprived of the necessity for additional supports, and at a significantly quicker pace. The work that has taken 50 hours to print using a traditional 3D printer is taken a mere 10 minutes using their novel rapid liquid printing operation.

The novel operation permits more than just molten plastic to be utilized as the printing material. Rubber and foam in fluid states can also be extruded, with the gel itself helping as an immediate chemical hardening agent so that workpieces can be removed as soon as the printing operation is over. For now, the MIT lab has been working to generate some complicated structure to prove just how intricate of a 3D-printed workpiece can be fashioned. But without the restrictions of gravity, one might envisage whole machines ultimately being 3D-printed in a single pass, as well as gears, wiring, and other moving components, deprived of necessitating the assembly of hundreds of diverse parts afterwards.

Faster 3D Printing

Manufacture time and the intricacy of multi-material objects has been a obstacle to extensive exploit of 3D printing. Even for simple products, 3D printing still consumes too long—typically hours and sometimes days. 3D printing consumes too much time to finish a job thereby restricts possible materials, upsurges that chance of introducing faults, and decreases an object's eventual strength. Incremental developments as well as novel approaches that have the likelihood for an order of magnitude transformation will aid printers meet the trials for superior speed. There are many methods to advance speed by utilizing higher-quality parts and by optimizing the designs and the movement increments of the lasers. For example, "Form 1+," a stereolithography printer from "Formlabs," employs Lasers that are four times more powerful to print up to 50 percent faster than the previous generation printer "Form 1." Most of today's printers employ a single print-head to deposit material. Adding additional print-heads that print simultaneously can upsurge speed by depositing material quicker whereas incorporating multiple materials or multiple colors of the similar material. Multiple heads can also brand several copies of the same design in the time it consumes to print one. With such novelty, print speed can upsurge more or less linearly as the number of heads surges. At the hobbyist end, "Robox" markets a dual nozzle printer that the company can print three times faster than single nozzle printers. Speed is especially a challenge once printing bigger objects. Big objects require more material to be flowing through the printer nozzle, which normally has a set rate for processing material. Various 3D corporations are currently developing large-scale additive manufacturing systems.

Their design will adapt larger nozzles for quicker polymer deposition, high-speed laser cutters that deal with work areas in feet rather than inches, and high-speed driving motors to accelerate the rate at which printer heads are moved into position. The consequence will be a proficient system of printing polymer components as much as 10 times larger, and at speeds 200 to 500 times faster than present additive machines.

To regulate the movement of the printer head, 3D printers employ diverse methods or configurations. Cartesian printers, which move a print-head in 2D on a plane, are the popular configuration today. "Deltabot" printers, also called Delta-robot printers, use parallelograms in the arms similar to a robot. The Delta printers will basically control all the Cartesian printers, because they have some momentous benefits, one of which is speed, Delta configuration permits for higher speed, since the print heads are lighter and they utilize smaller paths from one point to another.

Scientists developed enhanced mask-image-projection-based stereolithography (MIP-SL) to radically speed up the manufacture of consistent 3D objects. In the MIP-SL operation, a 3D digital model of a workpiece is sliced by a set of horizontal planes and each slice is transformed into a 2D mask image. The mask image is then projected onto a photo curable liquid resin surface and light is projected onto the resin to cure it in the profile of the related layer. The research team also established a two-way movement design for bottom-up projection so that the resin could be rapidly extended into even thin layers. Consequently, manufacture time was reduced from hours to a few minutes.

"ORGANOGEL" MATERIAL AIDS SCIENTISTS IN 3D PRINT FUTURE MEDICAL IMPLANTS

Researchers have revealed that suspending delicate 3D-printed structures in a Jell-O-like goo though the liquid ink hardens retains them from distorting or sagging. This could ultimately advance the fabrication of objects like distinct medical implants — but for now, the novel procedure is still in its proof-of-concept phase. 3D printing is a thin, hollow delicate workpiece — like a substitute windpipe, the 3D printer places down layers of a material as silicone until sufficient builds up to create a tube. But there is lag time between the printer first shoots out the liquid ink, and that ink hardens. This presents a problem to keep the structure from breaking up or bending before it is completely solidifies. Scientists advanced a new material called an "Organogel" that can sustain the liquid ink as it solidifies. It is made out of soggy, microscopic, chemical balls that are packed together in mineral oil. The gel is essentially a microscopic chemical ball pit, where soggy balls of gel are packed together in oil.

3D Printing Carbon Nanotubes

Scientists are testing with striking materials, such as carbon nanotubes, that provide an understanding to the scope of this technology. Carbon nanotubes were first created in the early 1990s. They are tubes of carbon formed on the nanoscale. Although they are thousands of times smaller than a human hair, they are, in fact too tough, outstanding conductors of heat and electricity. Therefore, they have become in high demand for scientists. Many researchers are investigating the combination of the tubes into 3D printed implementations, or 3D printing the tubes themselves. Scientists are occupied with 3D printed carbon nanotubes for the advancement of flexible electronic devices and wearable technology. Researchers have established a novel technology to 3D print highly conductive multiwall carbon nanotubes, or MWNT, using liquid ink.

The technique offers researchers additional control over the 3D printing operation and therefore the capacity to generate more intricate nanostructures in forms and configurations that were not thinkable before. This unlocks up new potentials for wearable electronics, which are becoming increasingly lesser but are still hulking than many people would want. Many 3D printers can print electrical circuits, creating intricate electronics from scratch. Scientists are likewise at work concerning 3D printing at the molecular scale.

Bioprinting in Water

New exploration into materials demonstrated a novel method with implementations for 3D bioprinting. Materials science researchers have established a new family of photo initiators for utilization in "digital light processing (DLP)." These additives, that establish quick solidification of a liquid material, generate quicker responses when subjected to light. By 3D printing in water, it also unlocks up the DLP technique to medical applications, leading toward a viable reaction for patient specific implants and tissues. When working with living cells, hydrogels and bio scaffolds are characteristically utilized as support material to cultivate tissue. As such, there is a mounting volume of 3D bioprinting investigation regarding the optimal environment and materials for cell development. With this facts in mind, it is clearly understood the reasons why water may be an excellent setting to 3D print a structure for medical exploitation. As debatably "the most versatile" 3D printing method in terms of design flexibility and speed scientists focus on Photopolymerisation for 3D printing in water. The obstacle of 3D printing in water is finding an initiator, i.e. the active ingredient that responds upon exposure to light, that likewise dissolves at the right instant.

Most photo-initiators are used up by irradiation in the 3D printing operation. By dissimilarity, the key to rapid 3D printing initiators is in their capacity to divide water, and absorb oxygen molecules that characteristically hinder the performance of the operation. The particles added as the photo-initiator in this case are semi-conductive metal hybrid nanoparticles (HNPs), and are employed to generate high-resolution 3D workpieces on a sub-microscopic scale. The degree of polymerization in material having HNPs is considerably quicker than light-restive material utilized deprived of the particles.

The Future of 3D Printing

The future of 3D printing will escort the next industrial revolution. Fabrication will be disrupted as it will drive the technology from an analogue, mass manufacturing technology, to a digital technology, 3D printing. Consequently, it will become promising to acquire right of entry to the capability to transform products. It will democratize manufacturing. This will be further catalyzed by the fact that the technology of industrial 3D printers endures to upsurge in quality, speed of manufacture, and plummet in cost. This is deep in a diversity of means, as stated below:

Product Realism

Since 3D manufacturing allows for unique approach for designs and products to be printed, as additional products being shaped, in lesser quantities, which is difficult through the procedure of traditional manufacturing. 3D design is carrying the creator spirit to the overall population, allowing them to yield products might have else only made-believe — from iPhone personal cases to jewelry to miniaturization, and beyond.

Digital Manufacturing

3D printing will be close to our native residence again. While mass-used products, paper napkins, and toilet papers will remain to be traditionally manufactured, more defective products can be redeveloped and tweaked based on customer feedback.

Manufacturers and Users

3D printing generates substantial prospects for co-creation between corporations and their clienteles. Co-creation and mass-customization are two very significant trajectories of user improvement, which is, itself, a serious foundation of essential modernization. Co-creation can happen at diverse phases of the manufacturing operation: design phase, co-design, fabrication phase, co-manufacturing, and distribution phase. Additionally, co-creation can too be accomplished between discrete customer, offering increase to "communities of creation" or "communities of co-design".

Patents

Patented custom design products will remain to be nearer to their creators by understanding the benefits of 3D printing creators' art. Brands can leverage the inventiveness of creators to make astonishing products, making it a normal additional revenue stream.

Altering Business Philosophy

Additive manufacturing is not only influencing the "Creator Movement," it is also altering the method enterprises and government agencies do business. Businesses are adapting to additive manufacturing to construct components that weren't likely before Federal agencies are discovering new ways to utilize the technology to better meet their missions.

Sharing 3D Printing in Health

The U.S. Department of Health and Human Services founded the "NIH 3D Print Exchange" to improve on sharing biomedical 3D-printable projects throughout the medical communal although NASA is investigating how 3D printing operates in space. However, this is just the tip of the iceberg to additive manufacturing's possibilities. For fabricators, additive manufacturing will allow a extensive variety of new product developments that can upsurge industry competitiveness, decrease industry energy depletion and aid propagate the clean energy economy.

"America Makes" – Fund

From the "Aiding Fund," namely "America Makes," a public-private government initiative program designed to mark the U.S. the frontrunner in 3D printing, to instituting the "Manufacturing Demonstration Facility" at Oak Ridge Lab, the Energy Department is offering businesses with admission to 3D printing technologies and taming them — and upcoming engineers — regarding the technology's potentials. To ensure the technology changes forward, the Department's National Labs are partnering with industry to create new 3D printing technology.

Novel 3D Printing Materials

Lawrence Livermore National Lab, USA, lately proclaimed an association to develop novel 3D printing materials, hardware and software, and Oak Ridge National Lab is joining to develop a novel marketable additive manufacturing system that is 200 to 500 times quicker and could print plastic components 10 times superior than today's industrial 3D printers. As the prices drip and the technology becomes quicker and more accurate, 3D printing is self-assured to transform the technique businesses and consumers reflect upon manufacturing — much identical to the method the first computers led to the rapid entrance to knowledge that we now take for granted.

PROSPERITY & ADVANCEMENTS – MATERIAL SCIENCE

Digital Material

Digital material merging is revealing of the cumulative porosity amid the digital and the material. Materialization of digital information includes and joints multiple categories, comprising physical computing, open source hardware, and digital manufacturing. 3D printing is additive kind of digital manufacturing; though computer numerical control CNC machining is a collective subtractive manufacturing technology which embraces lathes, milling machines, electrical discharge machining (EDM) machine, routers, water-jet, laser cutting etc.

Analog Manufacturing

Manufacturing nowadays still remains analog; though the designs are digital, the operations are not. 3D printing technology turns data into objects and objects into data. Manufacturing will be disrupted as it will go from analogue operations to digital operations in 3D printing.

Digital Manufacturing

3D printing is a technology which allows it probable to construct real articles from computer-generated 3D objects. 3D Printing is the process of creating a physical object from a 3D digital model, characteristically by laying down many thin layers of a material in sequence subsequent commands from a computer program. This is done by "cutting" the computer-generated 3D object in 2D slices and printing the real object slice by slice. Slices are printed on top of each other, and meanwhile each slice has a assumed thickness (e.g. 0.1 mm); the real object increases volume every time a slice is added. The quantity of material, the rotation of the platform and the design of the object is managed in a computer controlled environment. The size of each discrete layer of the sliced design – is resolute partially by technology, partially by material, and partially by desired resolution and the project timeline; thicker layers associates to quicker constructs, thinner layers associate to finer resolution. There is a trade-off between print speed and print quality.

Digital 3D Modeling

3D modelling is the procedure of generating a 3D depiction of any surface or object by operating polygons, e.g. triangle, edges, and vertices in virtual 3D space. The product is called a 3D model. 3D model, 3D drawing and 3D design are all identical. A 3D model is the computer-generated design that a 3D printer requires to print an object. Once can get 3D model by downloading it, 3D scanning an object, by a basic digital camera and photogrammetry software, engaging a 3D designer, or designing it with 3D modelling software. The greatest collective technique for creating a digital 3D model is through computer aided design CAD software. When a 3D design is finished, it must then be transferred as a standard triangle language STL file, meaning the file is interpreted into triangulated surfaces and vertices. STL file format utilizes a series of interconnected triangles to regenerate the surface geometry of a solid model. STL files are a typical file type that interfaces between Computer Aided Design CAD software and 3D printers. Most CAD software exports to an STL file and most 3D printers accept STL files. Once completed, the STL file requires to be treated by a piece of software called a "slicer," which translates the model into a series of thin layers and yields a G-code file comprising commands tailored to a precise kind of 3D printer. In other words, STL file is sliced into hundreds – sometimes thousands – of 2D layers. A 3D printer then reads the 2D layers as constructing blocks which it layers one atop the other, thus creating a 3D object by utilizing G-code which tells computer guided machine tools like extruder, motor drivers of xyz axis and laser systems what to do and how to do it. While STL is te primary command, the AMF (Additive Manufacturing Format) and 3MF (3D Manufacturing Format) files are rival for that position.

Water Tight 3D Print

Individual 3D design that's precisely generated for 3D printing is 3D printable. 3D models appropriate for 3D printing essential to have a smallest wall thickness and a "watertight" surface configurations to be 3D printable. Print volume must be within the capability of a printer and satisfactory number of polygons, e.g. triangles should be utilized to signify a surface to safeguard it will print with a smooth form. Always follow material and software rules and use printing technology that is well matched to print a specific.

Printing Materials

Numerous diverse materials can be utilized for 3D printing, such as ABS/PLA plastic, polyamide - nylon, silver, titanium, steel, wax, carbon, photopolymers, UV resins and polycarbonate. Resin 3D printers, e.g. SLA utilize custom made plastic resin emptied into a tank wherein UV Laser is concentrated to cause solidification of resin. Non-resin 3D printers, e.g. FDM utilize mostly ABS and/or PLA polymer which converts to flexible when heated then solid. And newer printers can print ceramics, graphene, cellulose, food materials and even living cells bio-ink.

No One Size - Fits All

Opposing to common credence, there isn't a one-size-fits-all method to 3D printing. Diverse kinds of 3D printers engage diverse technologies that operate diverse materials in dissimilar ways. The most basic restriction of 3D printing — in terms of materials and implementations — is that there is no 'one answer fits all'. What is common between all 3D technologies and all 3D printers is the mutual 3D procedure i.e. 3D model to STL file to slicing and g-code. The chief changes among diverse technologies are in the method layers are deposited to generate objects and in the materials that are utilized. Each technique has its own advantages and disadvantages.

Print in 4D – Time, Pressure, Temperature, Wind, Water, and Light

In 3D printing a structure is constructed out of many layers of material in 3D; in 4D printing, the additional measurement is time. Printed 4D objects self-assemble and modify figuration when they are detached from the 3D printer due to chemical reaction in the distinct materials utilized in the manufacturing procedure. 4D printing permits a printed construction to alter its form or function with time in reply to stimuli such as pressure, temperature, wind, water, and light. 4D printing permits the fabrication of objects that convert with time. This fundamentally means that objects can be "programmed" to perform in particular way deprived of the utilization of intricate electronics.

5D Printing – Magnetic Field

5D printing is the utilization of magnetic fields throughout the procedure of orienting elements in the deposited medium (inside the material being deposited) in a layered manufacturing of 3D printing procedure. As well 3 dimensions of 3D printing, other 2 dimensions are local control of arrangement and particle orientation. 5D printing can generate multifunctional shape-changing soft devices that could be utilized to generate mechanical means for connecting parts in the human body such, as tendons and muscles, or for careful pick-and-place systems in soft robotics. Additional view of 5D printing refers to 3D printing using a 5-axis 3D printer.

IoT – Internet of Things

3D-printing and ***IoT*** are corresponding technologies. IoT develops the 3D-printing technology although 3D-printing aids manufacture novel and more effective linked devices.

Internet connectivity to 3D printer can start a 3D print remotely; look at live video of a print procedure to distinguish any problematic in printing; and providing 3D printing quicker data recovery through the internet.

3D printing will not to transform the manufacturing sector, rendering traditional factories obsolete. In general terms, traditional manufacturing technologies, such as machining and casting, deliver the uppermost parts excellence quality level, according to surface finish and geometrical and dimensional accuracy. Though, 3D printing has excellent benefits. When associated to traditional manufacturing technologies, 3D printing is reflected a well-organized technology in the area of materials utilizing up to 90% of materials. Traditional manufacturing approaches require work piece to be aligned, measured, and machined by humans, which may present human mistakes into the creation of the part. In difference, 3D printing is a hands-off manufacturing procedure. 3D printing is more appropriate than traditional manufacturing for rapid prototyping, mass customization and intricate geometries. Furthermore 3D printing avoids supply chain of traditional manufacturing. 3D printing carries a new method to manufacturing through smaller lead time, design freedom and mass customization. Nevertheless, we know that 99% of all manufactured parts are standard and do not need customization. 3D printing is not substituting traditional manufacturing, but supplementing it and generating new markets within the industry. It is accomplished of creating parts that cannot be fabricated by traditional procedures.

Phases of Printing Technologies

The implementation of 3D printing technologies has happened in four successive phases namely

i. rapid prototyping,

ii. rapid tooling,

iii. digital manufacturing, and

iv. home manufacture conforming to different usages;

and each new stage does not brand the previous one 'obsolete', but in its place spreads it, e.g., 3D Printing is still being utilized for rapid prototyping. Progressively 3D-printed objects are being created as completed items, rather than as reproductions or prototypes. 3D printing is journey from a period of trialing to maturing. Today more than two-thirds of American manufacturers are utilizing 3D printing in some form or other.

Over 500,000 3D printers were sold in the global market in 2018. A universal shipment of 3D printers was 489,168 units in 2017, rendering to Gartner. By 2019, that number is predictable to balloon to 1,900,000.

3D printing market could produce to $30 billion by 2022 and could have an financial impact of up to $550 billion a year by 2025. Home-based 3D printing could advance into a $70 billion industry per year by 2030. Western countries (North America and Europe) account for more than two-thirds (68%) of the 3D printing market and Asia Pacific accounts for 27%.

Home-based 3D printing embraces custom jewelry, domestic belongings, minor fixes to household & appliances, dolls, and tools of whatever size, figure, or color. Commercial 3D printing embraces fabricating, industrial design, motorized industry, aerospace engineering, military, medicine, dentistry, architecture, construction, education, sports equipment, toys, food packaging, fashion & clothing, sculpture & museum, jewelry, footwear and others.

The preference of aerospace industry towards 3D printing is mainly due to the option of accomplishing substantial weight decrease, and for every pound of weight detached from the aircraft; the airline saves 11,000 gallons of fuel yearly. Beside the reduced fuel costs, it could also mean decreased CO_2 emissions.

3D Printing Transforming Healthcare

Medical implementations for 3D printing are increasing fast and are anticipated to transform health care. Numerous available appraisals define the utilization of 3D printing to yield bones, ears, exoskeletons, windpipes, a jaw bone, eyewear, hearing aids, cell cultures, stem cells, blood vessels, vascular networks, tissues, and organs, as well as novel dosage forms and drug delivery mechanisms. 99% of hearing aids globally are fabricated uttilizing 3D printing. Hundreds of thousands of individuals have been fitted with 3D-printed orthopedic implants from hip replacement, knee replacement to titanium jawbones, as well as numerous prosthetics. An indescribable number have helped from more precise surgery executed utilizing 3D-printed surgical supplements.

The Creation of Digital Industry

3D printing is allowing 'digital dentistry', with dental applications beginning to be habitually 3D printed founded on patient scans. Already value $780 million in the dental market, 3D printing could develop a $3.1 billion industry in this segment by 2020. 3D printing can substitute or mend an injured tooth; generate an orthodontic pattern; yield crowns, bridges, caps and dentures; and create surgical tools. Not only can 3D printers grip the dental implants themselves, but they can also 3D print the drill supports required to complete certain dental events.

Transforming Education

The capability to yield virtually any object in 3D is self-assured to transform education and ideas that have been historically problematic to understanding can be openly established with 3D printed visual aids that aid students study and recollect ideas more effortlessly. 3D printers provide an unparalleled "revolution" in education.

Cost-benefit 3D printing is the best solution when one requires printing a small number of intricate objects, but it becomes costly to print a large number of simple objects when equated to traditional manufacturing methods. Furthermore, the 3D printing becomes unbeneficial when printing large size objects. The cost of a 3D printed large object is significantly higher than if it had been traditionally manufactured.

REFERENCES

1. Kunwoo, L. (1999). Principles of CAD/CAM/CAE Systems, Addison – Wesley Longman Inc., Massachusetts, ISBN 0-201-38036-6

2. Lim, T.; Corney, J. R.; Ritchie, J. M. & Davies, J. B. C. (2002). RPBlox – a novel approach towards rapid prototyping, Third National Conference on Rapid Prototyping, Tooling and Manufacturing, 20-21. June 2002, Buckinghamshire, ISBN 1-86058-374-1, Rennie, A.
3. E.W.; Jacobson, D. M. & Bocking, C. E. (Ed.), pp 1-8, Publishing Limited, London
4. Liou, F.W. (2008). Rapid Prototyping and Engineering applications: A Toolbox for Prototype Development, Taylor & Francis Group, ISBN 978-0-8493-3409-2, New York.
5. Soloman, Sabrie "Sensors Handbook, 2nd Edition," McGraw-Hill Publishing Co., NY – ISBN 978-0-07-160570-0
6. Soloman, Sabrie "Sensors and Control Systems in Manufacturing 2nd Edition," McGraw-Hill Publishing Co., NY – ISBN 0-07-059626-3
7. Soloman, Sabrie "Affordable Automation," McGraw-Hill Publishing Co., NY – ISBN 0-07-059633-6
8. Soloman, Sabrie "Sensors Handbook, 1st Edition," McGraw-Hill Publishing Co., NY – ISBN 0-07-059630-1
9. Soloman, Sabrie "Sensors and Control Systems in Manufacturing 1st Edition," McGraw-Hill Publishing Co., NY – ISBN 978-07-10572-4

10 The Impact of Additive Manufacturing on the Integral Economy of an Enterprise

OBJECTIVE

Additive manufacturing technology unlocks new visions for the economy of an enterprise, a society, and even for a nation. It can enable the production of robust light-weight products for diverse industries from aerospace to medicine. It permits novel designs to become matchless rare products that were not likely with preceding manufacturing methods. It may transform medicine and biology with bio-manufacturing. This technology has the possibility to upsurge the well-being of a nation's citizens. Additive manufacturing may progress worldwide energy effectiveness in ground, sea and air.

Though, the acceptance and circulation of this new technology is not immediate. With any new technology, new ideals, knowledge, and organization are essential to enable its practice. Establishments such as the "National Institute of Standards and Technology," in USA and in any country can permit the progress of these items; thus, it is significant to comprehend the costs and profits of the additive manufacturing industry. This chapter studies the costs of additive manufacturing and pursues its inventive implementations to classify areas where it maintains a cost benefit and identify possible areas of technological advancements, and numerous means of cost reductions.

There are three primary techniques to manufacture a product depending on the volume required. These techniques may employ one of the following technologies:

1. Subtractive manufacturing - CNC machine parts,
2. Injection molding - designed to pre-produce prototypes, or
3. Additive manufacturing - 3D printing

It is critical to comprehend the dissimilarities between these three kinds of product development approaches to select the most appropriate technology for the intended project. Also, it is critical to determine the validity of utilizing the 3D printing technology if suited.

SUBTRACTIVE MANUFACTURING

Subtractive manufacturing is a product improvement procedure creating 3D objects by uninterruptedly removing material through cutting from a solid block of material such as metal. This can be achieved either manually or automatically with a CNC machine.

Injection Molding

Injection molding is a manufacturing procedure for creating products in great volumes. It can be utilized to do early product prototypes, but is typically employed in mass production procedures, whereas a production part may be generated thousands if not millions of times in sequence.

Additive Manufacturing Value

3D printing - additive manufacturing, produces 3D solid items from a digital computer file. The printing occurs in an additive process, where a solid object is generated through the consecutive layering of material. There are an extensive variety of materials to select from countless lists of polymers and metal. The process begins with the generation of a 3D digital file such as CAD file. The 3D digital file is then directed to a 3D printer for printing using a simple print command.

Once, we understand the value of subtractive traditional technology, one may then evaluate the benefits of 3D printing described as follows:

Corporate Benefits of 3D Printing Technology

Many companies have encompassed 3D printing. Others have strategies to present the technology and substitute conventional subtractive manufacturing. Actually, many investigations showed that over 70 percent manufacturers have now implemented 3D printing. Furthermore, over 50 percent expect 3D printing technology will be employed for high volume product improvement, and about 22 percent others assessed that its effects on supply chains will turn out to be even more disruptive.

In 2018 it is estimated 31 percent a yearly progress in the 3D printing industry between 2019 and 2023. In 2014, the 3D print industry created nearly $4 billion worldwide revenue. In 2016, over 275,000 3D printers were sold worldwide. Predictable progress confirms that 3D printing will eventually create more than $41 billion in revenue across the globe.

Utilizing 3D printing technology may be responsible for many benefits both for people and industries. Various substantial advantages that 3D printing technology delivers:

Faster Production

3D printing is faster than traditional manufacturing counting injection molds and subtractive production. Consider the speediness of a sport car as opposed to the speed of a horse cart. Equally both will arrive their destination in different time. Nonetheless, the time variance is considerably vast. Similarly, starting by a prototype to reach end-product, 3D printing provides fast verification and test of product design ideas very quickly.

Speedy design and prototype production leads to additional time to reiterate the design and improve the prototype, and find product market sufficiently suitable before competitors. 3D printing production may take few hours. In opposition, evaluating ideas and designs with conventional manufacturing approaches may involve days, if not some weeks.

Easily Accessible

3D printing has been since several decades but it actually did not take wings until 2010. The outburst of 3D printing benefits has transformed the 3D printing technology to be easier to use in particularly when innovative software and hardware were implemented to consumers. Consequently, more competitions have moved in the 3D printing arena. It is easy to acquire the technology, and incorporate it in a short time, probably days into the production implementation.

Better Quality

Subtractive technology may consequence inferior or inaccurate designs, resulting in poor quality prototypes. Similarly, Subtractive or injection molds may sometimes be wasteful of parts due to inappropriate design or material. Designer may not be completely certain of part to fit, form, and function always. 3D printing enables the designer to prototype the intended part to evaluate and test leading to a better part design, and assured performance.

Tangible Design and Product Testing

If *"seeing is believing!-"* 3D printing technology is the perfect object image to see, touch, and feel! It is the wings to lift the well sought product, after laboring and toiling in several design iterations to bring the novel product to be a successful implementation. Now it is promising to become familiar with the product prototype and physically test it to find the flaws in the design. If a flaw is detected, the designer can easily modify the CAD file and print out a new unit.

Product Development Cost-effectiveness

The main role in developing a prototype-cost is determining the labor cost, which plays a fundamental role in defining the necessary budget. Traditional prototyping practices including production runs are very costly as they require extensive human labor. Labor cost is very high with conventional subtractive manufacturing. Experienced machine operators and technicians are required to deal with product phases of development and production. Also, production machines such as CNC are expensive machinery. The 3D printing, however, the labor-hours cost is substantially lower than that of the 3D printing technology.

Customization Freedom

Subtractive manufacturing techniques are well suited to produce mass production of identical copies of the same product. Any design change necessary to incorporate to the existing part will render all the mass produced parts obsolete. The subtractive technology does not have the freedom to change easily, as the case of the competing technology of the 3D printing. Obviously, modifying each unique design within the traditional technology is extremely hard.

3D printing allows for boundless personalization, which enables substantial simplicity to accommodate personal modifications requested by customers. This characteristic is vitally valuable in dentistry applications.

Unlimited Shapes and Geometry

Tradition approaches of manufacturing depend on molds and subtractive technologies to create the preferred configurations. Designing geometrically intricate figures can be difficult and costly with this traditional technology. 3D printing takes on this challenge with effortlessness creating limitless shapes and geometries as long as the appropriate materials are provided.

Implementation Assorted Raw Materials

Traditional subtractive or injection mold manufacturing demands that the designers carefully consider the type of material to be used in each step of the product manufacturing operation. Mass Production requires extensive study of the material properties, shelf-life, and its structural alloys, as well as the chemicals involved in affecting part performance. This consideration is costly when employing traditional technologies. 3D printing technology is effortless to accommodate the blending of new chemicals as raw materials, including silica and carbon fiber enforcement. Also, a diverse range of raw material are currently in use 3D printing technology, such as paper, metal, ceramics, biomaterial, silver, and gold.

Minimum Waste Materials and Production

Subtractive technology using CNC and injection molding misuse the natural resources by generating much unwanted material waste to be discarded. Both require the removal of materials from solid blocks. 3D printing technology utilizes the material resources wisely, as it only uses material that is needed to create the desired part. Furthermore, 3D printing technology can reuse the left over materials after printing. Accordingly, additive manufacturing creates little waste, and offers the corporation some cost savings.

Minimal Risk

It is critical before capitalizing in a costly hard-molding tool, a prudent design engineer must find economical way to test and verify the realistic performance of a new component or assembly. 3D printing technology empowers product designers to substantiate product prototypes prior to investing manufacturing investments that may lead to a disappointing manufacturing endeavor.

Thanks to 3D printing technology it supports the corporation to rapidly bring a concept into assured reality. Consequently, products may be invented and constructed fast and economically. Undoubtedly, additive manufacturing technology will continue to be the front runner to transform every industry, altering the manner we work and live henceforth.

Justification to Incorporate 3D Printing in Industries

The economics of 3D printing deliver justification of the reasons to incorporate the 3D printing in many industries and consumers. Companies, continuously attempt to discover various means to decrease manufacturing costs. Consumers would favor to purchase the best quality product at the minimum price. 3D printing has the prospective to decrease costs of manufacturing products that are of equivalent or better quality. As further development is prepared in 3D printers and more performance of 3D printers is attained, 3D printers will be capable to manufacture products more accurately and with fewer waste of material. As industry understands the advantage of this technology, the larger request for these machines will lead to economies of scale for the production of 3D printers and consequently eventually permit these machines to be more inexpensive to both industry and discrete consumers.

Consumers may opt to acquire a 3D printer to generate household and personal products rather than buying them at a store because of cost and individual inclination.

A life-cycle economic analysis investigation by the "Department of Material Science & Engineering at Michigan Technological University" on 3D printers established that self-replicating rapid prototypes 3D printers, even with traditional suppositions, could save the average U. S. domestic use hundreds to thousands of dollars per annum in circumvented acquisitions by printing commercial products in their own home. The investigation was directed with 20 items that would be obtained by a household. 3D printer would deliver a profit on investment between 40% to 200%. With enhancements in reliability, progress in open source designs and a fall in costs, 3D printers will develop a mass-market generated product.

One of the other economic rewards of 3D printing is their capability to decrease costs from the elimination of intermediaries such as wholesalers or retailers in business transactions. By decreasing the parties involved in the production of a product, the general price of that product is dropped. The consumer can also be more participant in the product's development. 3D printers can facilitate rapid manufacturing of products at any site, which can lessen shipping costs for that product, and offer an invaluable advantage of expediency or minimized waiting for the specific product.

Drawbacks of 3D Printing Technology

As learnt from the historical trends of developing any new technology, 3D printing is following the same path in disrupting the manufacturing fields by radically changing product development. The 3D printing - additive manufacturing, encompasses the revolutionary means to construct materials together, layer after layer to create objects in the precise likeness of a 3D digital model.

As it is well described, 3D printing contrasts from conventional subtractive manufacturing, which encompasses removing by cutting away unwanted material from a block of solid material.

Advantageously, the 3D printing method removes several steps utilized in conventional manufacturing, and simplifies the manufacture of complex mechanical components. These characteristics have conceived substantial achievement in the expanses of rapid prototyping and tool improvement.

Accordingly, 3D printing technology has unlocked novel potentials for industrial businesses by empowering more rapidly product design, customization, cost reduction, palpable product testing, medical implantations and much more. The steadfast progress of 3D printing is progressively becoming pertinent in medical and dental industries, where customization is indispensable.

QUESTIONING– ADDITIVE MANUFACTURING- FRIEND OR FOE?

Exercise Prudence - Be Aware!

Nevertheless, 3D printing technology has a questionable sideways. It is not at all times the ideal selection for product improvement for manufacturing a product. 3D printers are still actually produced a byproduct fumes and chemicals considered unsafe, and relatively energy wasteful. Furthermore, their financial justification, political encounters, societal impact, and environmental influence have not been extensively deliberated.

While the potential advantages of the 3D printing technology are significant, the followings represent drawbacks that may influence the selection decision and may, in the unlikely events cast a dim shadow on the advantages:

Energy Consumption

Rendering to research conducted by Loughborough University, 3D printers consume almost 50 to 100 times more energy than traditional injection molding technology. This takes place when heating plastic to the melting point with heat filament or lasers device. In 2018, investigations at "The Environmentally Benign Manufacturing," conducted by experienced scientists dedicated to study the environmental influences associated with product manufacturing. The data exhibited that "direct laser metal deposition" - 3D printing technology uses about 100 times as much electrical energy, as traditional manufacturing. If 3D printing technology is considered, it is well suited batch production arrangements. It is exceptionally wasteful when considering utilizing the technology for mass production.

Printing Technology is Expensive

The equipment, material, and energy of the 3D printing technology are costly. The cost of the 3D printing' industrial-grade is very costly. It may cost few or several hundreds of thousands of dollars, which makes the initial capital investment exorbitant when utilizing this high end application. The capital investment equipment of low end applications starts in the tens of thousands of dollars. It may progressively advance to be as high as hundreds of thousands of dollars or even more. Also, the material employed in commercial grade 3D printers is expensive when compared to product materials utilized in traditional manufacturing.

Limited Materials

The 3D printing is considered a revolutionary technology to manufacture parts and assemblies; however, it has some limitations with respect to the available material as well as on the condition of the presented material to fully utilize. Presently, there are substantial research activities being conducted to develop a new 3D printing materials that are dense enough comparable to traditional materials in traditional manufacturing. For instance, the usual preferred printing material of choice is plastic (polymer). Polymers is typically favored to use for its ease of melting and depositing down in melted layers, and its characteristic for quick solidifications to form the final product. Nonetheless, polymer may differ in strength capability and may not be the best for many components. Several corporations suggest metal as a material, but final product components are frequently not entirely dense. Many futuristic specific materials counting glass and gold are being utilized but are up till now to be implemented.

User-friendly Retractions

While substantial Improvements have been achieved and it is becoming easier to 3D print day-by-day, but it is still a challenge to employ. 3D printers are often requiring high-voltage power supplies, demands specialized equipment, which makes the printers difficult to employ and manage. Sometimes, the excitement and hyped potential about 3D printing technology have come across as easy to use sounded more useful than they really are. The naked truth is 3D printers consume too much energy and several types have low resolution, and may not even connect to Wi-Fi as proclaimed.

Harmful Emissions

3D printers utilized in walled spaces such as homes can produce possibly toxic emissions and cancer-causing carcinogenic molecules according to academics at the "Illinois Institute of Technology." Their 2016 investigation presented that 3D desktop computers might release large quantities of ultrafine molecules and some hazardous unstable organic mixtures for the duration of printing. The printers radiated 20 billion ultrafine particles per minute utilizing PLA filament, and the ABS radiated up to 200 billion particles per minute. The system radiated contaminations that are similar to burning a cigarette, and may settle in the bloodstream or lungs posing health risks including cancer and other diseases.

Much Reliance on Polymers

Plastic filament made of plastic is widespread and cheap to be used in 3D printers. While employing raw plastic decreases waste generation, the machines still leave unutilized excess plastic in the print beds. PLA is biodegradable; however, ABS filament is still the furthermost frequently used kind of plastic. The left over plastic ends up in landfills harmfully disturbing the environment. Additionally, plastic restricts the type of products that can be generated from the material. Upcoming 3D printers may require other materials such as metal, as many 3D printers are presently employ metals, or carbon composites to ensure reliable part strength and performance.

Printers are Slow

Though 3D printers are inexhaustible for mass customization, they are slow to manufacture many objects. Contingent upon printer dimensions and resolution, it may involve several hours to a few days to print. The more intricate configurations the design requires for product development the slower the printers are. Corporations that obtained orders to customize and develop 3D prints utilizing a variety of products can consume up several weeks to print contingent upon the materials used.

Production of Dangerous Weaponry

With 3D printers, it is simple to generate 3D daggers, guns, explosives, and any other hazardous items. Offenders, criminals, extremists and terrorists can, consequently, print such weaponries lacking being noticed. More criminal groups have already employed 3D printing technology to generate card readers for bank machines. Sadly, 3D technology may become more user-friendly and cost-effective, and it is conceivable that design and production of abandoned weaponry will upsurge.

Copyright Infringements

Forging and counterfeiting are the utmost shortcomings of 3D printing. Anybody with a product blueprint may counterfeit products very rapidly. Patent defilements will progressively develop more common, and recognizing counterfeited items will become practically unmanageable. As 3D printing technology progresses, patents, and copyright owners will have a tougher time shielding their rights and corporations manufacturing unique products will be considerably affected.

Manufacturing Job-Losses

3D printing technology may create product designs and prototypes in a matter of hours as it utilizes only one sole step. It removes several stages that are employed in traditional manufacturing. Accordingly, it does not necessitate increase in labor cost. Similarly, embracing 3D printing may cause reduction manufacturing jobs. For countries that trust on a large number of low talent jobs, the weakening in manufacturing jobs could intensely affect the economy. It is probable that robotics will have a much greater influence in this regards.

In several industries, 3D printing delivers myriad advantages. Nevertheless, it may not replace conventional manufacturing. 3D Printing is still a developing technology with some drawbacks that require to be carefully assessed when choosing a product improvement technique. Manufacturers and product designers consequently require study it as a procedure to supplement convention manufacturing. They may utilize its exceptional abilities to advance product design and manufacture wholly novel products that could not be then created.

Manufacturing Statistics

The Optimisms of Economical Facts of 3D Printing & Manufacturing Technology summarized as follows:

1. 3D printing is just one subsection of the manufacturing industry.
2. Rendering to a various reports by US Commerce, the 3D printing industry is predictable to be valued $32.78 billion by 2023.
3. 3D printing jobs are on the increase.

Global Manufacturers

Manufacturers large and small play a significant part in the U.S. economy; rendering to the United States Census Bureau, manufacturers are the nation's fourth-largest employer, and ship several trillions of dollars in goods per annum. It may be a large automotive enterprise manufacturing vehicles or an institution with less than 50 employees. Manufacturers are vital to the country's global success. However, many societies have misunderstandings about the manufacturing jobs are undesirable jobs and offers low-paying compensations. Other countries may be discouraged to compete against USA.

In actuality, the United States' manufacturers are continuing to be strong and provide an extensive variety of well-paying jobs. The followings are the manufacturing statistics and facts:

- U.S. Manufacturing has a positive influence on global economy.
- The United States' accurate accounting and honest reporting put it in a prominent top place worldwide!

- Other competing nations may wish to compete using shady practices by leaning into pirating the intellectual proprietary – IP, instead of equipping their citizens with the power of innovation and free thinking. Eventually, such improper and illegal practices will sink their national economy into disarray.

According to the 2019 Global Manufacturing Competitiveness Index (GMCI), the United States' manufacturing industry currently is holding the top global position, in particularly, after the oil and gas industry were open for fracturing, enabling USA to possess the greatest oil and reserve worldwide, by passing even the largest oil producer in the Middle East. Also, GMCI predicted that Germany, and Japan would hold an economical immediate position behind USA, while USA is predicted to retain its ranking position in 2020. However, if China were to rid its policy of pirating the Intellectual Properties of other nations, it may one day, hold a prominent position among other well technologically developed nations. Interestingly enough, India is the most prominent fearsome competitor worldwide, as it is poised to bypass China due to the positive tenacities of their advanced intellectual academic and technical institutions. It is not a secret that India is the pool of talents to feed the hungry giant of USA's technological inventiveness.

The USA is the largest manufacturer responsible for 20% of world manufacturing activity. If the U.S. manufacturing sector were its own economy, it would the world, with $3.2 trillion of value added in 2018. The USA manufacturing sector has a significant impact on domestic economy in addition to its global impact. The manufacturing industry offers many opportunities domestically, including economic growth in a variety of areas presented as follows:

a. $1.00 spent in manufacturing creates $1.40 for the U.S. economy. The manufacturing comprises nearly 12% of US GDP
b. The **GDP** figure in the second quarter of **2019** was $5.7 trillion; **United States** is the world's leading economy with regard to **GDP.** The **United States GDP** is projected to trend around $21.2 USD Trillion in 2020, according to US econometric models.
c. The largest numbers of establishments in manufacturing are located in the private industry.
d. During the first half of 2016, manufacturers provided 65% more in benefits per hour of employee work than employers in the service industry.
e. 71% of workers in the manufacturing industry are men; however, the conversation around recruiting more women is growing.
    i. Ways to get more women into manufacturing
    ii. Manufacturing leaders are looking to recruit more women

Environmentally-Friendly Practices

The manufacturing industry remains to make advances and expand the way it does business, counting the execution of more and dedication to growth. In 2015 a study from the "National Bureau of Economic Research" states that between 1990 and 2008 air pollution emissions from U.S. manufacturing fell by 60% in spite of a considerable upsurge in U.S. manufacturing production. Manufacturing statistical data accumulated in 2012 provided astounding data in 2019 stating that modern manufacturing in USA, including 3D printing has grown over 1,000% ever since. The global aerospace and defense industry is likely to experience stronger growth in 2020. Following multiple years is predicted of positive but a subdued growth rate. However, the report forecasts the sector revenues will likely grow by about 4.0 percent in 2021.

Finding Return-On-Investment (ROI)

Calculating the Return-On-Investment - ROI of the 3D printer is to first compute the cost the material of the printed part. "Markforged's Eiger" software delivers material approximations to make simpler the process of computing material cost per part. As an example, examination was led on a carbon fiber reinforced "brake lever" printed on the "Mark Two" 3D Printer. It utilizes 13.16 cm^3 of Onyx and 17.30 cm^3 of carbon fiber. The internal view of the given object sample part.

The cost and volume per reel of each material can be set up on the materials page the specific polymer, and with this information one may find the price per cm^3:

For Onyx, reel currently cost $190 for 800 cc, or $0.24/ cm^3.

For Carbon Fiber, reel currently cost $150 for 50 cm^3, or $3.00/ cm^3.

With the following equation the material cost can be calculated.

(Cost per cm^3 of plastic) * (Volume of plastic) + (Cost per cm^3 of fiber) *(Volume of fiber)

= Total Cost ($0.24 / cm^3)(15.87 cm^3) + ($3/ cm^3)(18.34 cm^3)

= $3.81 + $55.02

= $58.83

The total cost of this brake lever is $58.83.

Majority of that expense is derived from continuous carbon fiber, which strengthens this brake lever to be as sturdy as metal — a reasonable price for such material composition to provide high strength. However, not all parts require to be reinforced with carbon fiber to accomplish that strength. Efficient fiber arrangement can decrease costs and print time. Also, diverse material mixtures will achieve also diverse outcomes.

One of the valued benefits of a "Markforged" printer is its adaptability. With diverse fibers, one can accomplish diverse material characteristics. For instance, fiberglass, which is also a continuous fibers option, provides a strong and robust material. However, it is a heavier metal and more brittle. Fiberglass is a cost effective option at $1.50/ cm^3 if a high strength-to-weight ratio is not essential.

Accordingly, recalculating the cost using fiberglass, the saving yields the following data:

($0.24 / cm^3)(14.07 cm^3) + ($1.50/ cm^3)(13.61 cm^3)

= $3.38 + $20.42

= $23.80* $35.03 is saved with continuous fiberglass reinforcement and similar strength.

Though comprehending the strength properties that the part requires, one may utilize the continuous fibers efficiently to reduce cost. Parts may not need to be filled with carbon fiber to encounter the required essential strength for the product. The geometry may be utilized to enable loading situations to optimize fiber direction all over the part. For instance, one needs to avoid bending in one plane. Utilizing a sandwich arranged panel instead of filling every layer with fiber. This technique permits further cost reduction for obtaining robust, aesthetically appealing, and accurate parts.

Determining Costs of Alternate Production

Presently, since the final cost of $55.06 is calculated, a cost comparison can be made against other manufacturing approaches. This "brake lever" is typically CNC machined. For volume fewer than 100, per unit cost is $195.95. For volume over 100, per unit cost is $117.11 and additional decreases in bulk volume. These costs may differ contingent upon the machining service or technique used. For instance, if this part were cast, the preliminary cost would be very high to create the mold, but the cost per part once the mold is made would be small. If this part made in house, the cost would not only contain material costs but add company's time spent to the part.

By equaling performance of the two methods, it is directly clear that 3D printing is the lesser amount of costly choice. The variance in price, CNC manufacturing cost per part subtracting the printing cost per part, is $137.12. Though, there is a more to consider. The 3D printer has an original overhead cost, while the CNC machining service operates on a per-part basis. This is where many developers face difficulty in justifying the cost of a 3D printer. The original price of our "Mark Two Printer" is $13,499, whereas CNC machining a part is a few hundred dollars each time.

Finding the Break Even Point

The $137.12 you might be saving with each part, tool, or product sent to an outside machine shop, one may venture calculating how many parts it takes to *"break even."* This leads to how many parts it takes to fabricate for the cost of the machine.

Presume only printing this type of one part!

The system will pay for itself in 98 parts; which is:

$13,499/$137.12 = 98.45

Verified Return on Investment

The distinctive worth of the 3D printing is to reduce the overhead fabricating costs essential for all part and every design change. The same "brake lever" utilized previously has a tooling cost of $2,865 for an injection mold prototype. If any modifications are necessary to the design, the cost requirement must be included with each design modification.

The price of a 3D printer is paid once, while tooling and prototyping costs must be considered with every design. The overhead costs collectively with fabrication lead-time may postpone the product development cycle, which makes the investment worth considering.

	Cost	Time
Machine Shop - Aluminum	$400.00	420 hours
Printed on "Markforged"	1.59	1.92 hours

Calculating Your ROI

ROI can be computed with facts about overhead and tooling costs from other manufacturing means and the cost of products the 3D printer may fabricate. This is a task related to the number of fabricated prints produced — if the printer used to print only one part, it is not worth the investment. However, the more printed parts, with each new design, the more time efficient and cost effective the justification becomes.

Through an overhead cost is $2,865 and an individual part cost is $3.46, and then each design iteration of the part effectually costs $2869.46 to manufacture a prototype. Knowing these cost data the ROI can be calculated with the following equation:

ROI = 100% * (Gain From Investment – Cost of Investment) / Cost of Investment

The investment gain is recognizing the money saved using additive manufacturing over injection molding. Usually, there are three different products that undoubtedly will experience three iterations. This equates to 9 parts each with their own unique mold or machining process.

The injection mold quote given above will be utilized to calculate the benefits from the investment. The cost of investment is the cost of the 3D printer:

Gain from Investment = 9 * ($2,869.46) – 9 * ($58.83) = $25,295.67

The ROI can now be calculated: 100% * ($25295.67 – $13,499) / $13,499 = 87.3%

This ROI value is the net return contributed to the investment of the printer. accordingly, to 9 product prototypes, not only has the 3D printer paid for itself and broken even. Additionally, the 3D printer investment has returned **87.6%** of the cost on top of that as well. The power of additive manufacturing is manifested in this high return on investment and in the versatility of prototyping and iterations. The cost and time deprived of a rapid prototyping solution will accumulate, to substantial costs. Utilizing the 3D printer provides rapid turnaround time, and a hands-off process. Additive manufacturing can restructure product improvement cycles by accelerating prototyping cost effectively. This provides developers opportunities to iterate on the products they are develop, creating their workflows more creative and bring the product to market rapidly.

3D Printer Achieves Greater Benefits

The 3D printer achieves a greater benefit as it breaks past prototyping. It 3D manufactures parts sufficiently strong for functional use. Continuous compound fiber reinforcement provides the ability ease to manufacture expert, high tensile strength parts. This improves functional prototypes to validate design configurations and loading circumstances to parts that can be rapidly transported to customers. The 3D printer provides the additional advantage of removing many overhead manufacturing costs and time-delay by 3D printing parts that withstand a machine shop environment. Several tooling, fixtures, and work-holding jaws may be easily developed and 3D printed without extensive lead time and overhead cost. Its speed and quality shorten the product improvement cycle, permitting manufacturing aesthetically appealing, and high-performance parts in less than a day. Many Instances testify that 3D printers offer substantial flexibility to yield parts with the adequate tensile strength and quality desired to prosper the technology at a reasonable speed and cost.

THE JUSTIFICATION CHALLENGE

Justifying A 3D Printer Investment for Rapid Prototyping

The significance of 3D printers for rapid prototyping to inventors, developers and product administrators, is incontestable. Thus far notwithstanding the clear importance, it may be indistinct how to persuade organization sides and accounting divisions that the profits defend the capital venture.

The important benefit of 3D printing is that it permits product improvement teams to simply, rapidly and cost efficiently produce models, prototypes, and forms. Autonomous of design difficulty, parts can be formed in hours or days instead of weeks. That benefit harvests the profits of advancing the product improvement procedure and increasing the possibility of prototyping work. Though, these aids may be diluted by the two-fold contest they may create.

Prototyping Expense

Aimed at numerous corporations, prototyping is an expense, not a capital venture, for which there is no reasonable budget. With this attitude, cumulatively the volume of prototyping work means only that increase expense payments. This interprets to an adverse influence on the bottom line that develops hard rationalization. On the way to justify 3D printing founded on printing more prototypes, the worth of prototyping must be calculated in real, concrete ways.

Similarly, a defense founded on speed necessity also is entrenched in tangible improvements that influenced by the creation the prototyping procedure sooner. Though "time is money" is an oft-stated cliché, the direct link between time and money can be problematic to institute, at least in terms that are undeniable in a monetary justification. Deprived of the time-to-money association, 3D printing justifications may be challenging.

These circumstances strengthen many to validate 3D printing on the puniest value intention, saving money by replacing it for present prototyping procedures. This method works, and many corporations have positively employed it for justification. Though, doing so disregards the worth of endorsing modification inside the product development cycle; variations such as implementation more design iterations, prototyping initial and frequently, or making the unreasonable probable. Without these benefits deteriorates the justification, which creates more difficulty to make the case for a capital venture in 3D printing.

The following strategies and guidelines gained by those that have been effective in justifying 3D printer acquisitions are presented below.

Presenting Your Case for Capital Budget

When the cost of novel equipment surpasses the authorization of a director, finance will be derived from a ***capital expenditure budget*** that higher administrator approves. This leads the proposal to be one of many rival for restricted capitals assigned through the company. To demonstrate that 3D printing is the prudent investment amid the other applications, shape a business event that clearly proves the worth of the projected capital expenditure to executive.

The purpose is to corroborate a buying choice by converting profits into substantial, tangible consequences. Once well composed, it demonstrates that the capital expenditure application bids a strong return with controllable risk. It becomes Convincing when fashioned with the decision-makers in mind. Mindful of the attendees emphases the business situation on the decision makers 'sensitive matters, and it makes it easier to embrace the right data.

If the application is organized to justify a precise, directed application, it develops a utilize-case. Though it reports a more restricted possibility than the business case and is frequently accepted at a departmental or regional level, the use case arrangement and details are the similar as that in a business situation.

Together the business-case and use-case cover three parts:

1. Executive summary,
2. Situational analysis, and
3. Financial justification.

The Executive Summary

The executive summary is a one-page, or smaller, outline of the situational examination and economic justification.

The Situational Analysis

The situation analysis designates the present condition, projected resolution, alternate results and related risks.

The Financial Justification

The financial justification, which is the emphasis of this argument, grants the expected return on the capital expenditure investment concluding an examination of the financial outflow and the projected financial advantages. When collective, the financial inflow and outflow yield the financial indicators that quantity the worth of the capital investment.

Whereas constructing the economic justification, pursue direction from the finance department. It will contribute in choosing the correct presentation events, such as return on investment (ROI) or payback period. Also, it may aid data collection and computation approaches, as well as direction on items such as "hurdle" rates, which are the bottom onset for which an asset will be reflected.

Justification Strategies

To shape a captivating justification that avoids the trials of monetizing the gains of 3D printing, those that have prospered in procurement endorsements provide four strategies. Utilized unaccompanied or in combination, these methods deliver a robust basis upon which to shape the economic case for an asset in a 3D printer.

Build on a Tiered Approach

As formerly stated, the easiest and best normally used way of justifying 3D printing is to compute the savings when it is a replacement for prototyping work that is presently being complete with conventional manufacturing methods or done through third-party 3D printing service benefactors. This is the preliminary point and the basis for any justification. Nevertheless, it produces the bottommost general worth to the corporation meanwhile it only reflects the reduction in prototyping expenditure for work that is previously being completed. If bigger economic returns are desirable to gain endorsement of the capital expenditure proposal, transfer to the subsequent tier of 3D printing paybacks, Escalation. This category is founded on doing more of the same kind of work that was encompassed in the replacement tier. The speed, efficiency, and ability of 3D printing eliminate the obstacles of time, cost, and effort when creating prototypes. So, 3D printing makes it simple and useful to yield more prototypes, which convey more worth.

The paybacks of supplementing prototyping work are defendable. Nonetheless, the task is computing a quantifiable and acknowledged worth that consequences from the prototypes. To do so, the decision makers need trust that there is a relationship between the action and the expected consequence. Deprived of this, doing more prototypes could be regarded as just an upsurge in expenses.

The third and last tier is extension. Now the justification is constructed upon doing something in the product improvement procedure that is not a present activity. Extension is possibly to convey the most important worth subsequently it is altering and refining procedures, but it is the toughest to calculate. Since the activity is novel for the corporation, the influence can only be forecast, which makes its value susceptible to debate and doubt. "Target Cost Reduction Revenue" progress is the final purpose of any corporate action, and it is attained by either cumulative sales or lessening expenses.

Nevertheless, when justification is considered, the most influential method is to report the cost side of the calculation. There are many details to emphasize on cost discount in a justification, but the most important is that expenditures are a present element while better-quality sales are a predicted option. Present and quantifiable, a discount in expense is more concrete and secure than a forecast of augmented sales. Moreover, cost containment is typically an importance for all corporations, and consequently, a top-of-mind subject for executive. Address present problems instead of struggling to generate awareness in a possible advantage of 3D printing, aim the justification to solving current difficulties. The strategy is to leverage what company executive has already recognized as fact. If not offered as a resolution to a current problem, executive may envision the proposal as just extra additional expense or be hesitant of the return on its investment.

There are two advantages to this approach:

1. it addresses a current "pain" that the company seeks to eliminate and
2. that problem is likely to have a cost associated with it.

If 3D printing can be presented to be the best option to remove the challenge, the savings are then a portion of its return on investment. For instance, if mold amend is a recurrent matter that has management's consideration and management has recognized a related cost, the advantage of many design iterations, made likely by 3D printing, changes from a suitable benefit to a real-world cost-saving quantity. For the justification, the economic worth of rework decrease has already been recognized — it is the cost that executive has connected with the problem. Since that amount is management's, it is a fact that desires no further confirmation. Control collective funds to avoid the corporate case and executive management endorsement, study selecting well-known budgets from several departments.

When the purchase and operating expenses are divided across departments, the decision making can change from upper corporate executive to managers and directors. The additional advantage of this method is that these decision makers are more expected to recognize the worth of 3D printing. Being near to the day-today actions, directors within the product improvement teams characteristically gain, comprehend, and desire 3D printing's less-tangible compensations. Instead of demanding proof of the technology's influence, the manager simply requires to elect if the budget can back the purchase.

CONSTRUCTING THE FINANCIAL JUSTIFICATION

The procedure of constructing a financial justification has only three stages:

1. Determine the financial benefit
2. Compile the initial and ongoing expense
3. Calculate the investment's performance.

Considering the strategies that have been discussed, start with the monetization of 3D printing's benefits.

Step 1: Calculate the Value:

The value is the economic advantage that consequences from the capital expenditure before the asset expenditure and continuing costs are deducted. It is the profit potential for the corporation, division or subdivision made by expenditure decreases, income growths or a mixture of both that consequence from a 3D printing investment. According to the earlier discussion justification strategies, there are 3 likely worth classifications:

1. Substitution,
2. Augmentation, and
3. Extension.

In that sequence, both the worth and trouble in showing the value of moving from low to high is vital.

Substitution:

There are two foundations of constituents for the justification: outsourced parts from contractors and those made in-house. Collect data for all insourced and outsourced items that could be affected to the suggested 3D printer, counting both the parts that are presently 3D printed and those that are fabricated, molded, cast, shaped and hand-fabricated.

Commence by gathering historical data for any models, prototypes, patterns and tools that are characteristic of the components the novel 3D printer may yield. Use a 12- to 36-month look-back period. For objects in this time extent, collect cost facts, process data and part configurations. Structure from this historical data, development the 3D printing assignment for the near term, which is characteristically three to five years. Approximate if the identical or similar components may need prototyping and couple that with corporate forecasts connected to modifications in R&D expenditure, rate of novel product improvement and product combination.

This appraisal delivers a starting point of all possible part nominees. To ensure the cost justification moderately easy and fairly high level, evaluate the part nominee to define general classifications to which arithmetic mean may be useful. For instance, study the amount of polymer parts, simple sheet metal components, complex machined parts and bulky cast metal parts. Divide these classifications with other qualifiers such as size. For each of the classifications, define if they are appropriate candidates for 3D printing. Lastly, analyze the parts within each classification to define the percentage that will work on the new 3D printer. This delivers a gross approximation of the quantity of parts and their size. This data will be utilized to define the savings possibly, and later, the expenditure to 3D prints them.

Consequently, it is time to compute the real cost of all of these prototype parts when manufactured with traditional manufacturing procedures or by intermediary third party 3D printing services. For subcontracted work, utilize invoices to define average costs for each classification of parts. Ensure to take account of all expenditures, such as:

1. Part cost
2. Engineering charges
3. Labor charges
4. Expedite fees
5. Shipping/handling charges
6. Taxes

Benefits and efficiency improvements that happen when subcontracted work is transported in-house may also be encompassed. Consider, though, that the labor-oriented items are correlated to contest unless they are end results in staffing decreases or less new hires. If not directly involved in the financial justification, position them somewhere else in the business case since they are profits of in-sourcing prototype improvement. Contain labor estimates for:

1. Engineering documentation and detailed drawings
2. Solicitation of quotes
3. Placing orders
4. Creating purchase orders
5. Managing accounts payable
6. Managing the project
7. Inspecting incoming parts
8. Maintaining and protecting confidential information

If counting the savings on in-house work that will be moved to the 3D printer, a cost estimate for these parts must be generated. For large businesses, internal cross-charges make the calculations easy. Accounts of the interdepartmental responsibilities record the expense of these parts. If cross-charges are not utilized, pursue advice from the company's cost accountants. They will be able to develop a cost approximation procedure.

In a cost justification questionnaire (Table 10.1), enter the sum of these costs in the first-year column for return (value). For subsequent years, put on a multiplier to the first-year value that echoes expected alterations to the capacity of prototyping work. Considering that this value does not echo the net return since it eliminates the cost of 3D printing the parts, which will be computed in the expenses section of the justification.

Augmentation

If larger economic returns are required to find endorsement of the capital expenditure proposal, move to the next tier of profits from 3D printing. This classification puts a dollar quantity on the characteristic value of emerging concept models, prototypes, and prototype tooling. For "Augmentation," the business case will demonstration that 3D printing allows the manufacture of more of these product improvement benefits, and therefore, brings more financial value.

Expanding the possibility of prototyping work will embrace reflection of exploiting both more iterations of parts that are presently prototyped, as recognized in the earlier tier, as well as allowing more parts and products to be manufactured. The latter may be defensible both by the speed/efficiency of 3D printing and the ability of creating difficult items rapidly. For ease, utilize the classification method detailed in Substitution to document the nominees and the predictable savings. To transfer this added work from the expenditure column to a financial gain, define the value of the work, from the decision makers' viewpoint. As considered in justification strategies, start by endeavoring to relate prototyping activity to a continuing, top-of-mind difficult within the organization. If not capable to do so, chose benefits that will echo and can be measured in financial terms.

Table 10.1: Cost Justification Worksheet – 5-Year Long Range Plan

	YEAR 0	YEAR 1	YEAR 2	YEAR 3	YEAR 4	YEAR 5
Capital equipment	(- -)					
Operating expense	(- -)					
Total investment	(- -)					
Ongoing expenses		($- -)	($- -)	($- -)	($- -)	($- -)
Return (value)		75,000	75,000	75,000	75,000	75,000
Total	(- -)	75,000	75,000	75,000	75,000	75,000
Cumulative total	(- -)	75,000	150,000	225,000	300,000	375,000

In view of the prototyping volume upsurge compute the expected yearly return founded on the recognized values. Enhance this total to the value for Substitutions displayed to the cost justification questionnaire in Table 10.1

Extension:

The third tier, "Extension," includes modification inside the product improvement procedure. It is as in the notion to **Augmentation**, but it applies to the events that are not being achieved. With a modification to the procedure, the economic benefits can be fairly substantial. Nevertheless, it may also be more intricate to create the economic value since there is no preference. When counting **Extension** requests in a justification look for openings allowed by the speed, cost, efficiency and flexibility of 3D printing. Fundamentally, these openings will be the claims that are unreasonable, or simply ignored, when restricted to traditional manufacturing procedures. In other arguments, pursue the claims that can't be justified—due to time, cost or effort—if using cutting or molding. The openings to use **Extension** extend all activities through the whole product improvement cycle.

Nonetheless, they can be classified in three loads:

1. Different stages of product improvement – For instance, early concept models or late-phase fabrication appraisal prototypes
2. Different product/component kinds – For instance, highly difficult parts or multi-piece assemblies that that are too expensive when utilized as primary form/fit prototypes

3. Different applications – For instance, practical prototypes for initial programs while injection molds are being made as with **Augmentation**, utilize the simplification method of part-type classification discussed in the **Substitution** tier and relate the activity to problems that management seeks to overcome.

Step 2: Calculate the Total Expense

The investment element of an economic justification contains all expenditures to obtain the equipment, get it up and running, and operational. There are two spending classifications: preliminary investment and continuing expense. For 3D printing, the preliminary investment is an up-front computation with simply distinct expenditures. The continuing, yearly expenditures are a more problematic to compute since they are reliant on how many components, and what type, are prepared.

For the initial investment, items to include are:

1. System price
2. Auxiliary equipment and software costs
3. Setting up and training costs
4. IT expenses: – Networks, data storage and computers
5. Facility alterations
6. Shipping expenditure

Continuing expenses may include:

1. Maintenance agreements
2. Routine maintenance costs
3. Materials
4. Other consumables: Cleaning solutions, build platforms and sandpaper
5. Labor, Direct labor for machine operation, maintenance, and part finishing
6. Facility charges

In both classifications of expenses, embrace only the incremental costs for objects such as labor, IT expenses and facility burdens. This is the variance between present expenditures and those sustained after system purchase. For instance, if no personnel will be added to upkeep the 3D printing operation, there will be no labor costs registered even if direct labor will be required.

Declaring a precise valuation of the expenses, the 3D printer supplier will provide much of the data, so a trustworthy business affiliation is vital. This is particularly accurate for continuing expenditures. Deprived of hands-on experience, it may not be likely to approximate construction time, output, capacity use and material use to define the working expense.

To compute these items, provide the vendor with data on the parts encompassed in the value section of the justification. With a reasonably detailed account of the parts — size, configuration, and quantity — the vendor will be capable to approximate the cost of materials and connected working expenses. Also, ask the vendor to approximate the total run time in order to endorse that the quantity of parts utilized in the justification does not surpass the 3D printer's capability.

Table 10.2: Cost justification worksheet with investment, costs and returns.

	YEAR 0	YEAR 1	YEAR 2	YEAR 3	YEAR 4	YEAR 5
Capital equipment	($50,000)					
Operating expense	($5,000)					

Total investment	($55,000)					
Ongoing expenses		($34,000)	($34,000)	($34,000)	($34,000)	($34,000)
Return (value)		$75,000	$75,000	$75,000	$75,000	$75,000
Total	($55,000)	$41,000	$41,000	$41,000	$41,000	$41,000
Cumulative total	($55,000)	($14,000)	$27,000	$68,000	$109,000	$150,000

For the predictable machine exploitation, the vendor can also approximate the direct labor required to arrange builds, run the machine and post process parts. As formerly considered, embrace direct labor only if it is incremental upsurge. Consequently, if the present employees can absorb the vendor-estimated upsurge in labor hours, eliminate them from the cost justification.

Combine all components of the original investment and continuing expenditure and enter them in the year one column in the cost justification questionnaire worksheet (Table 10.2). Do the same for years two through five, utilizing the same multiplier that was applied to the annual returns for those years.

Outside a breakdown by expenditure classification, no other aspect should be offered in the business case. Nevertheless, it is critical to document all computations, expectations and detailed expenditures for reference. There will be queries, so be arranged to answer them with backup, well documented information.

Step 3: Calculate the Return on Investment - ROI

The notable work is finished. The economic information that were assimilated together can be utilized to create any corporate-required metric that shows the worth of the capital expense such as return on investment, ROI, payback period, net present value, NPV or internal rate of return, IRR.

Efficient Time Utilization

3D printing's utmost advantage is making discrete parts rapidly, autonomous of design complications. That speed delivers rapid reaction on the first prototype and the capability to modify the design and speedily re-manufacture the part. As an alternative of waiting days or weeks for a CNC-machined prototype, a 3D printer can manufacture the part overnight. In a fast-paced, pressure-filled business atmosphere, it is clear that decreasing delivery by days is exceptionally valuable. This potential is what attracts numerous to the technology, but time can be very difficult to compute in an economic justification. One strategy to create the time-and-money link is to reflect past incidences where squandered deadlines caused in additional expenditure. For instance, chart the expedite charges that have been funded when projects are at risk of being late or deadlines are fast-tracked. Though the amount of these charges may not signify a substantial share of the 3D printer's acquisition price, they will still aid as confirmation that time has a very actual financial value.

Squandered Deadlines

Allowing for squandered deadlines, could be the substantial cause of real cost is the total expense of a late delivery instigated by delays in product development. For instance, if prototypes for a design team are late, the corporation will have expenditures for a meeting that at no time takes place. Allowing for the investment in labor to prepare the occasion, organizer arrangements, cancellation charges for the venue, and all other obligations, the consequence can be fairly big. Spread missed deadlines out to product launches and the drawback can be huge. If these types of proceedings have occurred in the recent past and have the likelihood of reoccurring, they are excruciating token of the value of time, if not tangible measures for a economic justification.

Decrease Time-to-Market

The reverse of **missed** deadlines is fast-tracked procedures, which may yield productivity improvements in product improvement or overall time-to-market decreases. The advantage of bringing more rapidly can be the most problematic to measure in a tangible way that is undeniable. If desired for the justification, the economic value of a day saved must be strongminded, and the decision makers must trust in the link between 3D printing and the time saved. If time cannot be interpreted to a financial value, think through considering it in the business event as an additional advantage on top of the economic rewards. Even if it cannot be monetized, it is a substantial and exceptional advantage.

The rewards of applying 3D printing are substantial, as 3D printing permits product development teams to effortlessly, rapidly, and cost effectively yield models, prototypes, and patterns. Parts can be manufactured in hours or days rather than weeks.

Indispensable Manufacturing Evolution

Over the past three decades, 3D printing has advanced a reputation as an procedure for prototype parts. Generate a CAD model of a design; forward it to company's 3D printer and a 3D reproduction will be prepared in hours. However these parts are frequently little more than abstract of show-and-tell models, not sufficiently tough for long-term utilization and in some cases disposed to dilapidation by sunlight.

The gusts of fabrication are starting to change, nevertheless, and industrial-grade 3D printing, or additive manufacturing, is nowadays intruding on machining, injection molding, and other traditional manufacturing procedures.

The Case for 3D-Printed Production Parts

Production is a significant phase in a typical manufacturing set-up, while prototyping is producing part for testing its form, fit, and function. The tested part is usually fabricated by CNC or injection molded pending increase in quantities to value investment in high-volume production tooling. Contingent on the part and its proposed use, it is likely to print parts that will achieve every need as well, and in some cases better than traditionally created parts, and do so cost-effectively enough that downstream investment is delayed or even needless.

- One vital instance is **GE Aviation**'s utilization of 3D-printed fuel nozzles in its LEAP engine, a change that saves aircraft proprietors about $3 million per plane yearly.
- In the medical-space, Johnson & Johnson is one of several contractors employing 3D printing for modified surgical tools and patient-specific grafts, as long as a better consequence for patient and surgeon alike.
- Oreck leveraged 3D printing to decrease the cost of its vacuum cleaner assembly pallets by 65 percent.
- Audio manufacturer Soundz utilized it to decrease tweeter grill lead-times from months to days.
- BMW utilized it to decrease the mass of handheld assembly tools by 72 percent. Deprived of 3D printing, all of these parts would have necessitated long-lasting, costly fabricating or injection molding operations to create.

It is recognized, 3D printing may not contest with high-volume production approaches in the widely held of applications, and perhaps would not be for some time to come. Though, it is significant to consider the readiness of these and factually thousands of other corporations to hold additive manufacturing technology for what it is—an operational way to yield accuracy components, which are frequently as proficient as their machined or injection-molded matching part when considering manufacturing end-use production parts.

Conservative Thinking

Production of parts in large volumes determines the choice of the manufacturing technology to create the parts with specific criteria described as follows:

1. Time to produce the product batch
2. Cost of producing the batch
3. The demand of quality to produce the part

A number of plastic mold manufacturers provide fast injection molding services that permit for economic part production at practically any quantity. Very similar is CNC machining services are cost-effective for high volume production with specific material strength. DMLS and Desktop Metal are excellent choice for prototypes and very low volume production. Considering that 3D printers are "complexity-free," while injection molding, machining, casting, and other traditional manufacturing processes, complexity are inherit characteristics of their technology.

Paradoxically, one of the major costs in 3D printing is the raw material. Manufacturing authorities reach a decision, though, that metal powder "supply and demand" is experiencing substantial changes as 3D printing come to be more widespread. Reputable numbers of corporations including BASF, Dow Chemical, HP, are grasping for market share of plastic and metal supplies alike. Undoubtedly, this will lower prices for consumers of corresponding 3D-printed products.

3D PRINTING TECHNOLOGY IMPACT ON SOCIETY AND ECONOMY

The progression of 3D printing technology and its uses have substantial bearing on various social, economic, geopolitical, security and environmental significances. The paramount importance to the survivability of 3D printing technology is price, technical specifications, and its advantages and limitations. The past decades the world experienced the personal computer and internet which had significant impact on our lives. In a similar manner, it is eminent the impact of 3-D printing will progressively upsurge, leading to important conversions, redefining our daily life, economy and society.

3D printing technology has been considerably developed and has progressed into a valuable tool for investigators, manufacturers, designers, engineers, medical doctors, surgeons, dentists, educators and scientists. Additive manufacturing is founded on generating materials and objects, starting from a digital model, by means of an additive procedure of layering, in a consecutive way. Most of the conventional manufacturing procedures are founded on subtractive methods: beginning from a block having an original shape, the material is removed till the wanted figure is achieved. Dissimilar to the above-stated method, the 3D printing is founded on accumulating consecutive material layers in order to gain the wanted form. Since 1984, when the first 3D printer was introduced, the technology has advanced and these machines have developed more and more valuable, while their price points dropped, thus becoming more inexpensive.

Values of Rapid Prototyping Materials

The main factors of the 3D printing technologies and materials are the subject of importance to equip the user to use the technology acquiring vital benefits to the users' industrial developmental needs. To serve the users' 3D printing technological needs synopses of the most valuable 3D printing technologies, highlighting the ebbs and valleys are od each provided as follows: .

Synopsis of 3D Printing Head Substance:

The inkjet head and the powder bed 3D printers sprinkling a preliminary thin layer of powder with fine binder droplets. Then, a roller is employed in order to spread and compress a fresh layer of powder. In the end, an object submerged in powder layers bound together is attained. If it is essential, the utilized binder could be tinted in order to gain a colored concluding object. After the printing, one may too utilize treatments for improving the material's tensile strength, with super glue or for decreasing the color waning, with UV protectants. The final object is ended from more different basic materials, having diverse chemical and physical characteristics, therefore being a compound material.

The powder bed and inkjet head 3D printers are also valuable in generating objects utilizing ceramic powder. The printed objects are then exposed to heat treatment for drying and varnishing, thus refining the material's strength and feature.

Synopsis of the Stereolithography (SLA)

The stereolithography (SLA) is an additive manufacturing technology that employs a liquid photopolymer – resin, and an ultraviolet laser light, in order to gain consecutive objects' layers. With the intention of attaining a layer, the laser draws on the resin a 2D path, therefore reaching a cross section of the final object. The gained layer is then exposed to ultraviolet laser light, curing and hardening the layer with the preceding ones. Through this technology, one may gain very smooth final objects.

Synopsis of Selective Laser Sintering (SLS)

Additional 3D printing technology, the selective laser sintering (SLS) liquefies and fuses fine element layers of powdered materials like plastic - often nylon or metal, using a powerful laser beam. The laser crosses over a powder surface and after the achievement of a printed layer, the plastic powder is spread over. When the laser crosses over this new layer the powder elements melt, interfusing each other and also with the previous layer. The SLS technology is valuable when printing difficult objects having intricate details.

Synopsis of Photopolymer Jetting

The photopolymer jetting technology spreads droplets of resin, utilizing small jets distributed by movable heads like those of an inkjet printer. After scattering the droplets, the resin is hardened through an ultraviolet light. If it is essential, one may too print a support material that embraces the droplets and is detached in the final steps of the printing. This technology is valuable when one has to gain models having very fine details or smooth surfaces, utilizing several materials.

Direct Metal Laser Sintering (DMLS)

The direct metal laser sintering (DMLS) 3D printing technology utilizes a laser with the intention of fusing elements of metal powder, e.g. titanium. This technology is comparable to the SLS technology that prints plastic materials. The DMLS has the drawbacks of high costs and needs precise design procedures.

Additional 3D printing technique contains in the direct metal printing method that creates metallic models utilizing powder elements - mainly stainless steel. The technique consists of some steps. In the first step, the designed object is printed, utilizing an inkjet procedure, in a bed of fine stainless steel powder. The plastic binder is burned out utilizing a heat treatment, though the steels elements are fused together. Then, the unfilled cavities inside the model are filled utilizing melted bronze. In the end, the final printed product containing a permeable steel material having bronze filled permeability can be gold - or other metals platted.

The indirect printing approaches are those founded on generating models or molds that can be further employed in generating metal objects, founded on conventional methods.

3D Printing Advantages and Limitations

In order to continue to examine the influence of 3D printing technology on the society and economy, the following analysis presents the vital advantages and restrictions. However, the greatest significant advantages of 3D printing are described as follows:

Advantages

1. Additive manufacturing provides the opportunity of generating, in a short timeframe, intricate and complex 3D objects, with sufficient specifics, from diverse materials. Through 3D printing, the customer has the option to generate complex objects, shapes and figures that are exceptionally difficult to be attained through any other existing technology.
2. A highly significant benefit of generating objects employing 3D printing technology as an alternative of conventional fabricating approaches is the waste decrease. As the building material is added layer above layer, the waste is virtually naught and throughout the production, it is utilized exclusively the material required for obtaining the final object. In the conventional fabricating procedures, founded on subtractive methods, the final product is fabricated by means of cutting or drilling an original object, consequently leading to a considerable forfeiture of material.
3. One could effortlessly print insignificant mobile parts of the final object.
4. The product's digital design could be referred over the Internet at the customer's site, anywhere it can be printed.
5. The customers likewise have the option of printing items at distant sites considering the element the Internet is currently prevalent and in some countries is even a legal right of the populations.
6. More or less of the materials utilized in 3D printing have enhanced characteristics in terms of tensile strength and deliver an extensive variety of greater final details, likened to the materials employed when fabricating objects over conventional technologies.
7. As the additive manufacturing is a computer-controlled method, it decreases the needed total of human communication and necessitates a low level of know-how for the worker. Additionally, the procedure guarantees that the final product signifies a flawless 3D type of the digital design, without the mistakes that could have shown when employing other current technologies. As the AM decreases the surplus in the manufacturing process, it could help resolving threatening difficulties of the humanity such as the depletion of the building material resources, the energy depletion and the ecofriendly protection.
8. Employing the 3D printing technology one may yield difficult designs valuable in numerous industries: fashion, medicine, arts, jewelry, computer industry, telecommunications, transports etc. AM has led to astonishing progress in medicine, being capable of saving lives, lowering health's care costs and improving the human life's quality. For instance, scientists have successfully generated a valuable 3D printer in generating artificial limbs, organs and tissues. Primarily, it is fashioned a 3D model of the final object using a computer tomography - CT or magnetic resonance imaging - MRI. Utilizing 3D figures, the organic material is printed and later is imbedded in the patient's body.

a. The scientists of "Regenerative Medicine" have effectively formed an abridged size practical human kidney.
b. Additional stimulating event is the one of an eagle's bill that, subsequently being destroyed by a pilferer, it has been positively substituted by a prosthetic one, constructed from titanium utilizing a 3D printer.
c. A highly valuable application of the 3D printing is the generating a Robotic Exoskeleton, employing metal and rubber bands. This means is beneficial for serving patients - especially young people having underdeveloped limbs, as it provides them the option of executing sufficient arm mobility's, permitting individual customization and fine-tuning.
d. Additional significant novelty that uses the additive manufacturing was the improvement of a 3D printer capable to print tissues. One of their greatest significant accomplishments was to print in 30 minutes a blood vessel needing the length of 50 mm and the diameter of 1 mm.
e. Additional 3D printing invention is the custom exteriors that cover prosthetic limbs, therefore gaining a normal figure and feature. The technology employs a 3D scanner -- founded on the gained imageries, the covering is designed and printed employing several materials.
f. Additional use of the additive manufacturing was substituting a woman's jawbone that had to be detached - due to dire disease, with a printed one. Consequently, to obtain the digital model, the corporation has utilized a computed tomography of the patient and at that time has printed the auxiliary utilizing titanium and a ceramic covering.
g. Many investigators have manufactured custom workshop equipment on a lesser scale. Employing particular Computer Aided Design (CAD) software and a 3D printer, scientists were capable to print modified tools employing a polymer gel along with chemical reagents. This can be predominantly beneficial in the pharmaceutical industry.

9. The publicity of the 3D printing tools can be competently attained utilizing the World Wide Web, as these tools are directed to technologically advanced users and therefore, it is not essential to lead costly marketing campaigns.

Drawbacks and Restrictions

Similar to any other technology, 3D printing has a sequence of drawbacks and restrictions that presently hinder a large-scale growth of this technology. The foremost drawbacks and restrictions of 3D printing are stated as follows:

1. The absence of lawmaking and rules concerning the 3D printing.
   a. For instance, there may be printed guns -and this has previously occurred, weapons, portions for aircrafts, military portions, counterfeit parts for commercial or defense operations - designed for disruption, medications or chemical weaponries.
   b. Furthermore, all of these may be attained with comfort, at cheap costs and very rapidly.
   c. Also, weaponries could be easily camouflaged in non- dangerous products. Therefore, 3D printing may develop a likely hazard when utilized by lawbreakers or forgers. Currently, the politicians are mainly attentive in amendable the firearms and more commonly the 3D printed products, but not the 3D manufacturing tools.

d. Even if many representatives endorse, sustenance and observe to the preceding stated approach, added opinion articulated by the representatives considers that the assertion and registering of 3D printing tools become obligatory and also to limit the blueprints' distribution. A part of the 3D printers' manufacturers consider mitigating these dangers and consequently they presented software restrictions on matters that may be printed.

2. Additional central drawback of 3D printers is the element that young people may print out unsafe objects. So as to avoid this, one may use software restrictions and parent control.
3. A main drawback of 3D printing is its high price. At the genuine price of the tool and materials, the 3D printing is the unsurpassed solution when one needs to print a small quantity of difficult and complex objects, but it becomes costly to print a large quantity of modest objects, when associated to conventional fabricating methods. Furthermore, the 3D printing turns out to be nonprofit when printing big size objects. The cost of a 3D printed big object is considerably higher than if it had been conventionally manufactured.
4. As a result of the material costs - particularly concerning the molds, the additive manufacturing is not at all times the top technical selection, most of the molds' materials may deteriorate over time.
5. Occasionally, the 3D printed objects' construction quality is inferior to conventional fabrication. Although the additive manufacturing may print objects needing intricate designs, the end product may occasionally have faults that might influence not only the object's design, but also its performance and reliability.

WORLDWIDE POSITIVE INFLUENCE:

Furthermore to the previous, additional significant element that must be considered when examining the impact of 3D printing on human life is the influence that the extensive dispersion of this technology has on the worldwide economy and on the labor force necessities. Accordingly, the greatest significant matters valuable to be well-thought-out are stated as follows:

1. The option of fabricating products as needed and at diverse sites than when employing conventional practices, could decrease real financial disproportions and may adapt the present ladder of the financial controls.
2. As the additive manufacturing is a computer-controlled method, it decreases the essential expanse of human labor and therefore it may cause substantial decreases in labor force necessities concerning the production, product distribution and fabricating jobs for export businesses, as the AM technique permits fabricating products as needed and nearer to the consumer's site.
3. Alternatively, the 3D printing technology's growth and dispersal will result in generating new entrepreneurs, occupations and businesses related to: the production of the 3D printers, supplies, materials and printing cartridges; the products' engineering and design; the software industry. Furthermore, the AM technology may use inexpensive reprocessed materials. Therefore, the costs of costly imports may be decreased or eliminated altogether.
4. The additive manufacturing's growth will also influence the import of the building materials, as it utilizes diverse materials than other methods, some of which may be nearby brought, deprived of imports.

SELECTING THE 3D PRINTER

In current years, 3D printer tools became inexpensive, improved, more beneficial and more significant with each passing day. Their rising position and the benefits accessible have made them to become progressively prevalent. From a historical viewpoint, meanwhile the advent of the 3D printers in 2007, a real 3D printer evolution has taking place. The evolution surveyed by a sequence of innovations in 2011, and numerous additional devices sustaining the customers' rising needs till 2018.

The greatest significant principles that a customer has to consider when selecting a 3D printer are stated as follows:

1. The printer and its tools must be reasonable for everyday use.
2. The 3D printer's hardware and software must be reasonably priced and friendly to the customer, without having an advanced training in software.
3. The 3D printer must yield objects that meet the customer's necessities and requirements, in a short timeframe.
4. The 3D printer device's dimensions must gauge to the customer's allocation that may be either an extensive corporation's allocation, or a restricted office allocation.
5. An actual evolution in the 3D printer improvement happening by generating a printer to reproduce itself. Its novel design allowed the 3D printer to print its individual parts, so that reproduce itself.
6. A developed model of printers was presented in 2010 at an inexpensive price to print objects up to 200 mm × 200 mm × 110 mm at a comparative high speed.
7. A lesser and inexpensive style was introduced in 2012 with a condensed size, and better movability, as it is one of the smallest printers accessible on the market nowadays, with an strategic price, which delivers better accuracy and prints rapidly of condensed object up to 140 mm × 140 mm × 110 mm.
8. A better-quality model that prints the smallest objects from all the 3D printing devices, the maximum printing dimensions being 120 mm x 120 mm x 115 mm. Additional creation of 3D printers, sprang in 2012, provides a greater printing proportions, 225 mm ×145 mm × 150 mm. Also, it is used as a home assembly kit, for home utilization. The model provides the option of employing twin extruders, being the only cheapest 3D printing device that contains this as a typical option. Consequently, it is likely to utilize this device in order to print objects employing two diverse filament colors. It also has the benefits of rapid printing and does not need early calibration.
9. Another 3D actual printing device is distinguished as an easy system to assemble, to use and to calibrate, with a long-term reliability. It is characterized by a print-bed that moves along two axes, while it lowers along the third one, as printing the object. It offers a printing size of up to 127 mm × 127 mm × 127 mm. It uses precision linear guides and rails along the axes, while its screws are Teflon-coated.
10. One of the most rapid 3D printers provides a great printing space: 210 mm × 210 mm × 220 mm. and does not contain a heated print-bed, consequently the printing materials are restricted, but utilizes a thermal print-head advancing at a very high speed along two axes. It provides high precision, prints with high speed, at outstanding quality, contributing the largest obtainable printing area, necessitating a low upkeep effort, but it is very costly and difficult to assemble.

11. Perceptibly, in reaching the decision to buy a 3D printer, the vital features that one should consider are the reason for which the printer is employed and the financial element, such as the printer's cost the printing material's cost, and the energy cost. Equaling the technical conditions of the 3D printing devices might help in this matter.

SELECTING THE 3D PRINTING – METHODOLOGY

This section concentrate on the costs of additive manufacturing; nevertheless, countless of the benefits of additive manufacturing may lie in developments of the finished goods. Consequently, there is some notion on the product enhancements that result from additive manufacturing technologies. The learning of different 3D printing technologies offers an outline of the processes and materials used in additive manufacturing. It also delivers the basics on additive manufacturing costs and classifying them by their process and material combination, and the scrutiny of the costs and benefits of additive manufacturing. The cost influence is broken into poorly-structured costs, well-structured costs, and product enhancements and quality. Also, the cost models used to survey additive manufacturing, and the inclinations in implementation and implementation of additive manufacturing.

SELECTING THE 3D PRINTING – PROCEDURES AND MATERIALS

There are a number of additive manufacturing procedures; though, at first glimpse it may seem that there are more categories than in actuality. Many businesses have fashioned exclusive system and material names in order to distinguish themselves, which has fashioned some confusion. Luckily, there has been some exertion to classify the procedures and materials by means of standard methods.

The classification and descriptions of procedures and materials below depend on ASTM International Standards. The total global revenue from additive manufacturing system sales was $502.5 million with U.S. revenue projected at $323.6 million.

These systems are branded into various different procedures. ASTM International Committee on Additive Manufacturing Technologies has advanced standard expressions. Provided below are the groups and adapted descriptions from the ASTM standard:

DEFINITIONS PROVIDED BY ASTM

Binder Jetting:

This method utilizes liquid bonding agent placed utilizing an inkjet-print head to connect powder materials in a powder bed.

Directed Energy Deposition:

This procedure utilizes thermal energy, characteristically from a laser to fuse materials by melting them as they are dropped.

Material Extrusion:

These machines enforce material, naturally a thermoplastic filament, through an orifice onto a platform that transfers in horizontal and vertical directions.

Material Jetting:

This procedure, characteristically, exploits a moving inkjet-print head to deposit material through a build area.

Powder Bed Fusion:

This procedure uses thermal energy from a laser or electron beam to selectively fuse powder in a powder bed.

Sheet Lamination:

This process uses sheets of material bonded to form a 3D object.

Vat Photopolymerisation:

These machines selectively cure a liquid photopolymer in a vat utilizing light.

Materials almost $327.1 million was disbursed globally on materials for additive manufacturing in 2011. There are two main kinds of materials: plastics and metals. There are also ceramics, composites, and other materials that are utilized as well, but are not as collective.

3D PRINTING MATERIALS CLASSIFICATIONS

The materials may be characterized into eight classifications:

1. Polymers and polymer blends
2. Composites
3. Metals
4. Graded/hybrid metals
5. Ceramics
6. Investment casting patterns
7. Sand molds and cores
8. Paper

Various procedures advance themselves to specific materials. Table 10.3 provides varieties of additive manufacturing procedures and their matching materials.

Also, the vital 3D printing technologies are selected and sated as follows:

1. Material extrusion
2. Material jetting
3. Binder jetting
4. Vat Photopolymerisation
5. Sheet lamination
6. Powder bed fusion
7. Directed energy deposition

Table 10.3: Additive Manufacturing Procedure and Material Arrangements

	Material extrusion	Material jetting	Biner jetting	Vatphotopoly -merization	Sheetlamination	Powder bed -fusion	Directed energy deposition
Polymers and polymer blends	x	x	x	x	x	x	
Composites		x	x	x		x	
Metals		x	x		x	x	x
Graded/hy-brid metals					x		x
Ceramics			x	x		x	
Investment casting patterns		x	x	x		x	
Sand molds and cores	x		x			x	
Paper					x		

ADDITIVE MANUFACTURING COST

There are two main vital categories for inspecting additive manufacturing costs. The first is to associate additive manufacturing processes to other conventional procedures such as injection molding and machining. The drive of these types of investigations is to govern the environments additive manufacturing is cost effective.

The second classification includes identifying resource use at several stages in the additive manufacturing procedure. The goal of this kind of analysis is to recognize when and where resources are being expended and whether there can be a decrease in resource use. In Table 10.3 the areas that are possible denoted with "x" to indicate possible combinations.

Additive Manufacturing Costs and Benefits

The costs of production can be characterized in two ways.

The first encompasses those costs that are "well-structured" such as labor, material, and machine costs.

The second encompasses "ill-structured costs" such as those related with build failure, machine setup, and inventory.

There inclines to be more emphasis on well-structured costs of additive manufacturing than ill-structured costs; though, some of the more important advantages and cost savings in additive manufacturing may be concealed in the ill-structured costs. Furthermore, bearing in mind additive manufacturing in the situation of lean production might be beneficial. A key perception of lean manufacturing is the detection of waste, which is categorized into seven categories:

Overproduction:

It happens when more is manufactured than is currently essential by customers

Transportation:

Transportation does not cause any modification to the product and is a source of jeopardy to the product

Rework/Defects:

Rejected flaws consequence in wasted resources or additional costs amending the defect

Over-processing:

It happens when more work is done than is required

Motion:

Needless motion effects in unnecessary spending of time and resources

Inventory:

It is same as to that of overproduction and consequences in the necessity for added handling, space, people, and paperwork to accomplish extra product

Waiting:

When labors and equipment are awaiting for material and parts, these resources are being misused

Additive manufacturing may influence an important number of these classifications. For example, additive manufacturing may substantially decrease the essential for large inventory, which is a significant cost in manufacturing. In 2011, there was an average of $208 billion or the equal of 14 % of yearly revenue seized in inventory for medium and high-tech manufacturing with an assessed cost of $52 billion or 3 % of revenue. Decreasing inventory liberates up capital and decreases expenditures. The possible savings and profits of additive manufacturing as well as its costs provided below.

POORLY-STRUCTURED COSTS

Various costs are hidden in the supply chain, which is a system that transfers products from supplier to customer. Additive manufacturing may possibly, have important influences on the design and size of this system, decreasing its related costs.

Inventory and Transportation:

At the start of 2011, there were $537 billion in inventories in the fabrication industry, which was equivalent to 10 % of that year's revenue. The resources consumed creating and storing these products could have been used somewhere else if the need for inventory were condensed. Suppliers often agonize from high inventory and delivery costs. Additive manufacturing delivers the capability to manufacture parts on request. For example, in the spare parts industry, a specific kind of part is rarely ordered; nonetheless, when one is ordered, it is desirable quite speedily, as idle machinery and equipment waiting for parts is costly. Conventional production technologies make it too costly and need too much time to yield parts on demand. The consequence is an important quantity of inventory of rarely ordered parts. This inventory is tangled up capital for products that are idle. They inhabit physical space, buildings, and land while demanding rent, utility costs, insurance, and taxes. In the meantime the products are worsening and becoming obsolete. Being able to produce these parts on demand by means of additive manufacturing decreases the necessity for upholding big inventory and excludes the associated costs.

Transportation:

Additive manufacturing permits for the production of several parts concurrently in the same build, making it likely to produce an entire product. Conventional manufacturing frequently comprises production of parts at multiple localities, where an inventory of each part might be kept. The parts are transported to a location where they are assembled into a product, as showed in Figure 10.1. Additive manufacturing has the possibility to substitute some of these stages for some products, as this procedure might permit for the production of the whole assembly. This would decrease the necessity to uphold big inventories for individual part of one product. It also decreases the transfer of parts manufactured at variable places and decreases the necessity for just-in-time distribution.

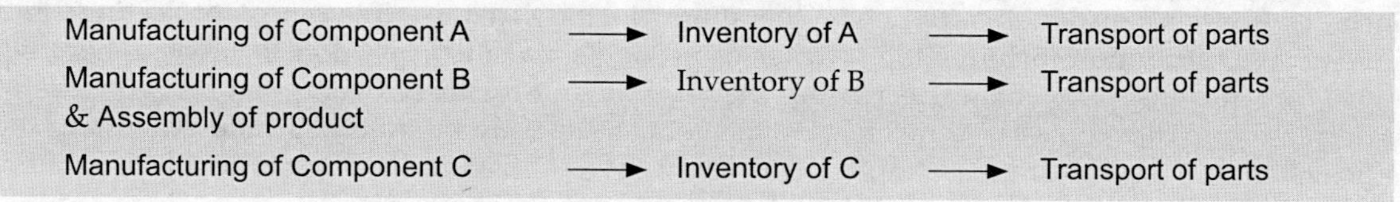

Fig.10.1: Example of Traditional Manufacturing Flow

CONSUMER'S PROXIMITY TO PRODUCTION

There are four substitutions that have been suggested for the diffusion of additive manufacturing:

The first is where a substantial amount of consumers buy additive manufacturing systems or 3D printers and create products themselves.

The second is a copy shop circumstances, where people yield their designs to a service benefactor that manufactures the parts.

The third situation encompasses additive manufacturing being implemented by the commercial manufacturing industry, altering the technology of design and production.

The fourth situation, because additive manufacturing can yield an end product in one build, there is restricted exposure to hazardous circumstances, and there is little hazardous waste, there is the possibility to carry out production nearer to the consumer for some products (i.e., distributed fabrication). For instance, presently, a more distant physical area may order automotive parts on request, which may take numerous days to be transported. Additive manufacturing might permit some of these parts or products to be created close the point of use or even onsite. Further, localized production joined with basic procedures may start to distort the line between manufacturers, wholesalers, and retailers as each could possibly produce products in their facilities.

By comparing the operational cost of consolidated additive manufacturing production and distributed production, where production is in close locality to the consumer. The examination can study the production of spare parts for the air-cooling ducts of the environmental control system for the US F-18 Super Hornet fighter jet, which is a well-documented case where additive manufacturing has by now been applied. The anticipated total cost per year for consolidated production was \$1.0 million and \$1.8 million for distributed production.

- Inventory obsolescence cost,
- initial inventory production costs,

- inventory carrying costs, and
- spare parts transportation costs
- are all reduced for dispersed production; though, substantial upsurges in employees costs and the original investment in additive manufacturing machines make it more costly than consolidated production.

Augmented automation and reduced machine costs are desirable for this condition to be cost effective. It is also significant to comprehend how to study the manufacture of a relatively simple component with little assembly. One possible advantage of additive manufacturing might be to yield an assembled product rather than individual components. It is prudent to examine spare parts in the aircraft industry, concurs that, currently, on demand consolidated production of spare parts is the most likely method to flourish; however, if additive manufacturing grows into a extensively adopted procedure, the distributed method becomes more feasible.

SUPPLY CHAIN MANAGEMENT

The supply chain contains purchasing, operations, distribution, and incorporation. Purchasing includes sourcing product suppliers. Operations include require planning, projecting, and inventory. Distribution includes the transfer of products and incorporation includes generating a well-organized supply chain. Decreasing the requirement for these events can consequence in a decrease in costs. Some large businesses and retailers mainly be indebted their success to the efficient administration of their supply chain. They have utilized technology to revolutionize the means they track inventory and restock shelves ensuing in cheaper costs.

Walmart, USA and around the world, for example, cut links in the supply chain, making the link between their stores and the manufacturers the shortest. It also started vender accomplished inventory (VMI), where manufacturers were accountable for managing their products in Walmart's warehouses. It has progressed its communiqué, and teamwork partnership. The management of the supply chain can be the influence that drives a company to market leadership. Additive manufacturing may have momentous effects on the manufacturing supply chain, decreasing the necessity for supply chain management. This technology has the possibility to bring manufacturers nearer to consumers, decreasing the links in the supply chain.

VULNERABILITY TO SUPPLY DISRUPTION

If additive manufacturing decreases the amount of links in the supply chain and carries out production nearer to consumers, it will end up in a reduction in the vulnerability to unforeseen catastrophes and disturbances. Every factory and warehouse in the supply chain for a product is a possible point where a catastrophe or disturbance can halt or delay the production and supply of a product.

A lesser supply chain with fewer links means there are fewer points for possible disturbance. Additionally, if production is transported closer to consumers it will effect in more decentralized production where many locations are creating a few products rather than a few locations producing numerous products.

Disturbances in the supply chain might effect in local impacts rather than regional or national impacts. Under conventional manufacturing, material resource benefactors transport to the manufacturers of parts and components, who might transport parts and components to each other and then to an assembly plant. From there the assembled product is brought to a retailer or distributer. A disturbance at any of the points in manufacturing or assembly may result in a disturbance of distributions to all the retailers or distributers if there is not redundancy in the system. Additive manufacturing with localized production does not have the same weakness.

First, there may not be any assembly of parts or components.

Second, a disruption to manufacturing does not impact all of the retailers and distributers.

Example of Traditional Supply Chain Compared to the Supply Chain for 3D Additive Manufacturing with Localized Production

Material Costs

With geometric liberty, additive manufacturing permits products to be made utilizing less material while upholding the essential achievements. Products can be created at the level of performance desired rather than considerably surpassing the essential performance level because of limitations in traditional manufacturing. Currently, however, the price of materials for additive manufacturing can often exceed those of traditional manufacturing.

Metal Material Costs

Metal and plastic are the primary materials used for 3D Printing technology. Presently, the cost of material for additive manufacturing can be high when equated to conventional manufacturing. It has been established that the material costs for a designated metal part made from aluminum alloys was $2.59 per part for conventional manufacturing and $25.81 per part for additive manufacturing utilizing selective laser sintering; thus, the additive manufacturing material was approximately ten times more expensive.

Other investigation on metal parts approves that material costs are a main cost incentive for this technology, predominantly parts made of stainless steel. For this material, four cost features are diverse and the production quantity is a fewer than for the base instance. This example delivers understanding into classifying the major costs of additive manufacturing. The **first** cost feature that is diverse is the construct rate, which is the speed at which the additive manufacturing system works. In this instance, it is calculated in cubic centimeters per hour. The **second** feature that is diverse is the machine use calculated as the amount of hours per year that the machine is activated. The **third** feature is the material cost and the last feature is the machine capital costs, which contain items connected to housing, utilizing, and upholding the additive manufacturing system. Among other things, this contains energy costs, machine purchase, and related labor costs to run the system.

The essential model has a construct rate of 6.3 cm/hr, and a consumption of 4500 h/yr, a material cost of $89, while a machine investment cost of $500, 000. On average, the machine costs percentage is 62.9 % of the cost estimates. This cost was the major cost even when construct rate was more than triplicated and other issues were constant. This cost was biggest in all but one case, where material costs were augmented to $600/kg. The second biggest cost is the materials, which, normally, averaged for 18.0% of the costs; nevertheless, it is vital to consider that this cost is possibly to shrink as more suppliers enter the field. Post processing, groundwork, oven heating, and construct process fix were around 8.4%, 5.4%, 3.3%, and 1.9%, respectively.

The material costs for additive manufacturing are substantial; nonetheless, technologies can often be paired, where two technologies are embraced together with each other and the advantages are superior than if they were embraced separately. One example is computer aided design CAD and computer aided manufacturing CAM, as together are required to be employed for the other to be prized.

Additive manufacturing and the raw materials that are utilized may be a state where they are balancing. All additive manufacturing necessitates raw materials; this may generate a response. Cumulative implementation of additive manufacturing may cause a discount in raw material cost over economies of scale. The decreased cost in raw material might then spread additional embrace of additive manufacturing. There may also be economies of scale in raw material costs if specific materials develop more common instead of an overabundance of dissimilar materials.

Plastic Material Costs

Associating the costs of fabricating a lamp holder utilizing injection molding associated to the additive manufacturing procedure of selective laser sintering utilizing of two dissimilar machines:

EOS **SLS** P730 and

EOS **SLS** P390.

A substantial share of the cost for injection molding is the mold itself, which contributes for between 84.6% and 97.7% of the cost. For additive manufacturing, the main costs are the machine cost per part, which varies as low as 58.7% to as much as 65.9% of the cost, and the material cost per part, which is varies as low as 29.1% to as much as 30.4% of the cost. The P730 is cost realistic for production volumes of 73,000 or less while the P390 is cost realistic for 87,000 or less.

Additive manufacturing costs of a polymer part are designated for two parts, a lever and a cover, employing stereolithography, fused deposition modeling, and laser sintering. A cost account of the lever material represented 25% of the cost for stereolithography, 39% for fused deposition modeling, and 74% for laser sintering.

Machine Cost

Additional to material costs, machine cost is one of the most substantial costs associated with in additive manufacturing. The regular selling price of a comprehensive additive manufacturing system was $73,220 in 2011. Even though, the price is increased from $62,570 in 2010, the price has dropped for most years proceeding to this point. Between 2001 and 2011, the price reduced 51 % after adjusting for inflation. Although the tendencies in machine costs are usually descending, great changes continue between the costs for polymer-based systems and metal-based systems, and the considerable progress in sales of low-cost, polymer-based systems throughout this time has intensely subjective the normal selling price of additive manufacturing systems. For metal material cost presented that machine costs reached from 23% to 75% of a metal part. The cost variance between the dissimilar kinds of additive manufacturing machinery was substantial reaching between $0.1 million characteristically for polymer systems and $1.0 million usually for metal systems. Obviously, deduce that the amount might have reduced over time; though, the machine cost approximations reached from 45% to 78% of the cost of a metal part. The machine cost per part was between 59% and 66% of the cost of a plastic part.

Build Envelope and Envelope Utilization

The dimensions of the construct envelope and the exploitation of this envelop both have an influence on the cost of an additive manufactured product. The dimensions of the construct envelope have two influences. **First**, products can only be constructed to the size of the construct envelope, which means that it might not be likely to construct some products utilizing additive manufacturing technologies deprived of increasing the construct envelope. **The second** influence of the construct envelope is associated to employ the total volume of construct capacity. A significant effectiveness factor lies in the capability to expend the available construct space. For example, the influence of volume utilization on energy using six different machines, and four different materials (titanium, stainless steel, and two kinds of polymers) are examined. The complete construct situation, where the construct envelope is fully employed, uses less energy per kilogram deposited than one single part being created for all six different machines. The EOSINT P 390 has the major construct volume and has the major variance in energy consumption between a single part and full construct.

Build Time

Build time is an important element with respect to approximating the cost of additive manufacturing and the amount of software packages is obtainable for approximating build time. There have a tendency to be two methods to approximating build time:

1. detailed analysis and
2. parametric analysis.

Detailed analysis uses information about the inside workings of a system, while parametric analysis uses information on procedure time and features such as layer thickness. Build time approximations incline to be precise to the system and material being used, while this is a significant influence in the cost of additive manufacturing.

Energy Consumption

Several cost studies for additive manufacturing, encompassed an investigation of energy consumption, as it added less than one percent to the final cost. Energy consumption, though, is a significant issue in bearing in mind the cost of additive manufacturing associated with other means of manufacturing, particularly in terms of investigating the costs from beginning to end. Energy investigations on additive manufacturing, nevertheless, incline to emphasize merely on the energy utilized in material manipulation and by the additive manufacturing system itself.

Metal

The investigation of energy use amid a number of machines displayed metal deposition to traditional tool and die manufacturing. The investigation recognizes that energy intake is determined by the solid-to-cavity volume ratio. At low ratios, the additive manufacturing process of direct metal deposition reduces energy, though at high ratios computer numeric controlled milling reduces energy consumption.

Other Investigations incline to concentrate on correctly forecasting energy consumption and reducing energy consumption for additive manufacturing. Envelope consumption and construct orientation are amid the subjects for dropping energy consumption. Scientists observed the influence of part orientation for three systems: Stratasys FDM 3000, 3D Systems Thermojet, and EOS EOSINT M250 Xtended. They studied locations for a single part. Due to the change in the position of the part, the energy consumed could upsurge between 75% and 160% contingent upon the system.

Plastic Material

Energy consumption was considered in the production of nylon parts utilizing selective laser sintering and likened these outcomes to that of injection molding. The investigation involved a minor construct of 50 parts and a full build of 150 parts. The parts were made by injection molded both with the energy spent for the production of the mold and without the mold. The small construct for selective laser sintering has utilized smaller amount of energy than the small construct, for injection molding, counting the energy for the mold. Nonetheless, the energy for the full construct was around 69% higher. For the full build, approximately 60% of the energy was utilized in nylon production and 37% was utilized in part manufacture for selective laser sintering.

The investigation of energy consumption of selective laser sintering was investigated utilizing nylon material constructing two "full chamber build" of prosthetic parts. They recognized the elements that are the main consumers of energy: chamber heaters (37 %), stepper motors for piston control (26%), roller drives (16%), and the laser (16%).

Labor inclines to be a minor share of the additive manufacturing cost. Labor might contain take away the finished product or replenishing the raw material amid other effects. Labor approximation is at 2% of the cost. Some extra labor is constructed into the other costs such as the material cost and machine cost, as these items also necessitate labor to produce.

Product Enhancements and Quality

Whereas the emphasis is on the costs of additive manufacturing, it is significant to consider that there are product augmentations and quality variances due to using this technology. There is more engineering design freedom with additive manufacturing and it generates more flexibility; nevertheless, there are restrictions, as some designs necessitate support structures and ways for drive away heat in production. Though, difficulty does not increase the cost of production as it does with conventional subtractive methods. With the exemption of the design cost, each product created can be customized at little or no expense. There is important requirement for custom products in the medical sector for developing joint implants, dental work, and hearing aids amid other matters. There is also the option of customers designing their individual products or adapting them. One concern with additive manufacturing, yet, is quality assurance. Presently, there is a requirement for standard methods to appraise and ensure precision, surface finish, and characteristic detail to accomplish desired part quality.

COST MODELS AND COMPARISONS

Two Main Factors to Additive Manufacturing Cost Modeling

Additive Manufacturing

The cost of additive manufactured parts is calculated. The average cost per part was obtained, and stated as follows:

1. the system yields a single type of part for one year,
2. it utilizes maximum volumes, and
3. the machine runs for 90 % of the time.

The analysis contains labor, material, and machine costs. Other aspects such as power consumption and space rental were well-thought-out but funded less than one percent of the costs; consequently, they were not considered in the results.

The average part cost is calculated by dividing the total cost by the total number of parts manufactured in a year. Costs can be distributed into machine costs, labor costs, and material costs. The costs are calculated for two parts, a lever and a cover, utilizing stereolithography, fused deposition modeling, and laser sintering. A cost distributed illustrates that in this investigation laser sintering was the most economical additive manufacturing procedure for this product. Machine cost was the foremost contributing cost influence for stereolithography and fused deposition modeling whereas the material cost was the main contributor for laser sintering.

The annual machine cost per part where the machine totally depreciates after eight years; that is the sum of depreciation cost per year, calculated as machine and auxiliary equipment divided by 8, and machine upkeep cost per year divided by production volume. The outcome is a machine cost per part that is not changing over time.

The cost of additive manufactured parts is calculated utilizing a created activity cost model, where each cost is related with a specific activity. They yield the similar lever that was produced utilizing selective laser sintering. In their model, the entire cost of a construct (C), is the sum of raw material costs and indirect costs. The raw material costs are the price (Pmaterial), measured in dollar per kilogram, multiplied by the mass in kg (M). The indirect costs are calculated as the entire construct time (T) multiplied by a cost rate (Pindirect). The entire cost of a constuct is then characterized as:

$$C = (Pmaterial * M) + (Pindirect * T)$$

The cost per part is calculated as the whole cost of a construct (C) divided by the quantity of parts in the build. The time and material utilized are the chief variables in the costing model. It was presumed that the machine operated 100 hours/week for 50 weeks/year (57 % consumption). The projected indirect cost per hour is obtained. Their cost model and the total costs are also obtained.

There are three different times that are calculated

1. "time to laser scan the section and its border in order to sinter;"
2. "time to add layers of powder;" and
3. "time to heat the bed before scanning and to cool down slowly after scanning, adding layers of powder or just waiting time to reach the correct temperature."

The sum of these times is the construct time (□).

At 1600 parts, the cost of the lever is projected at \$2.76 per part, and \$2.20 for laser sintering. The unexploited material was recycled. In this investigation, the per-unit cost was \$1.86. An assessment of the costs is made.

Several of the cost examinations accept a condition where one part is created frequently; though, one of the paybacks of additive manufacturing is the capability to yield diverse components concurrently. Consequently, a "smart mix" of components in the same construct might accomplish cheaper costs. In a sole part production, the per-part cost for a construct is the whole cost divided by the quantity of parts; though, the cost for dissimilar parts being made concurrently is more complex.

The first method is based on parts volume where

$$CostPi = (VPi / VB) * Cost_B$$

Where

$CostPi$ = cost of part i

VPi = volume of part i

VB = volume of the entire construct

$Cost_B = \Sigma(\text{indirect\_cost} / \text{working\_time})(t_{xy} + t_x + t_{HC}) + (\text{direct\_cost} / \text{mass\_unit})\, m_B$

mB = mass of the prearranged production proportional to the object volumes, and the time to manufacturing the whole construct

txy = time to laser-scan the section and its boundary to sinter powder

tz = time to add layers of powder

tHC = time to heat the bed before scanning and to cool down after scanning and adding layers of powder

i = an index going from one to the number of parts in the construct

$costB$ also equals C from above, which is the entire cost of a construct.

The second method is based on the cost of constructing a sole part and is signified as the following:

$$Cost_{pi} = \gamma_i * (CostB / n_j)$$

Where $\gamma_i = (Cost_{pi} * + n_i) / \Sigma\, (Cost_{pj} * j * n_j)$

Also, i is the index of the part being calculated, j is the index for all parts fabricated in the same bed, *ni* is the quantity of parts denoted with i, and *Costpi* * is the cost of a sole part i projected utilizing the earlier equation for C.

The third process is founded on the cost of a part construct in high-volume. It is comparable to the second way only the cost variables in γ_i are calculated utilzing a high quantity of parts instead of a solo part. It is signified as the following:

$$Cost_{pi} = \gamma_i^{\infty} * Cost_B / ni$$

Where

$$\gamma_i^{\infty} = Cost_{pi}^{\infty} + n_i / \Sigma_j (Cost_{pj}^{\infty} * ni)$$

Where $Cost_{pj}^{\infty}$ is a assumed quantity, which leads to infinity, of fabricated parts *i*

Additive Manufacturing Cost Advantage

Several of the cost investigations observe costs such as material and machine costs; though, many of the profits may be concealed in inventory and supply chain costs. For example, a dollar capitalized in automotive assembly takes:

10.9 days - to return in revenue.

7.9 days - spend in material inventory,

19.8 hours - pending to be utilized.

20.6 hours - employ in production and down time when the factory is shut down.

1.3 days - expended in finished goods inventory.

Furthermore, of the whole time utilized, only 8% is expended in real production. Conferring to notions from lean manufacturing, inventory and waiting, which resemble 92% of the automotive assembly time, are two of seven classes of waste. This is just the assembly of an automobile. The production of the engine parts, steering, suspension, power train, body, and others often happen independently and also have inventories of their own. Furthermore, all of these parts are transferred among localities.

The typical shipment of fabricated transportation equipment travels 801 miles. For the US, this amounts to 45.3 billion ton-miles of transportation equipment being trasported yearly. Since additive manufacturing is able, in some cases now and perhaps more in the future, construct an entire assembly in one build, it decreases the necessity for some of the transference and inventory costs, ensuing in influences through the supply chain.

It is significant to recognize that the capability to yield more intricate assemblies, such as those in an automobile, is still emerging and includes some assumption about future abilities. In addition to constructing whole or partial assemblies, there is also the possibility of decreasing the magnitude of the supply chain over distributed fabrication. Consequently, in order to comprehend the cost variance among additive manufacturing and other procedures, it is essential to study the costs from raw material transportation to production and through the sale of the final product. This might be characterized as:

$$C_{AM} = (MI_{R,} + MI_{M,}) + (P_{E,} + P_{R,} + P_{M,}) + (FGI_{E,} + FGIR,M + FGI_{M,AM}) + WT_{AM} + RT_{AM} + T_{AM}$$

Where C_{AM} = Cost of making an additive manufactured product

M_I = Cost of material inventory for refining raw materials (*R*) and for manufacturing (*M*) for additive manufacturing (*AM*)

P = Cost of the procedure of material extraction (E), refining raw materials (R), and manufacturing (M), including managerial costs, machine costs, and other pertinent costs for additive manufacturing (AM)

FGI = Cost of finished goods inventory for material extraction (E), refining raw materials (R), and manufacturing (M) for additive manufacturing (AM)

WT_{AM} = Cost of wholesale trade for additive manufacturing (AM)

RT_{AM} = Cost of retail trade for additive manufacturing (AM)

T_{AM} = Transportation cost through the supply chain for an additive manufactured Product (AM)

This could be associated to the cost of traditional manufacturing, which could be characterized as the following:

$$CT_{rad} = (MI_{R,Trad} + MI_{I,Trad} + MI_{A,Trad}) + (P_{E,Trad} + P_{R,Trad} + P_{I,Trad} + P_{A,Trad}) + (FGI_{E,Trad} + FGI_{R,Trad} + FGI_{I,Trad} + FGI_{A,Trad}) + WT_{Trad} + RT_{Trad} + T_{Trad}$$

Where $CTrad$ = Cost of manufacturing a product utilizing traditional procedures ($Trad$)

MI = Cost of material inventory for refining raw materials (R), manufacturing intermediate goods (I), and assembly (A) for traditional manufacturing ($Trad$)

P = Cost of the procedure of material extraction (E), refining raw materials (R), producing intermediate goods (I), and assembly (A), counting managerial costs, machine costs, and other pertinent costs for traditional manufacturing ($Trad$)

FGI = Cost of finished goods inventory for material extraction (E), refining raw materials (R), producing intermediate goods (I), and assembly (A) for traditional manufacturing ($Trad$)

$WTTrad$ = Cost of wholesale trade for traditional manufacturing ($Trad$)

$RTTrad$ = Cost of retail trade for traditional manufacturing ($Trad$)

$TTrad$ = Transportation costs through the supply chain for a product made utilizing traditional manufacturing ($Trad$)

Presently, there is a greater consideration about the cost of the additive manufacturing procedure cost (PAM) than there is for the other costs of additive manufacturing. Moreover, most cost investigations study a sole part or component; nonetheless, it is in the final product whereas additive manufacturing might have substantial cost savings.

Traditional manufacturing necessitates abundant transitional products that are transported and assembled, while additive manufacturing can accomplish similar final product with less component parts and many components constructed either concurrently or in the same site. For instance, consider the future option of an whole jet engine housing being fabricated in one construct utilizing additive manufacturing with respect to an engine housing that has parts fabricated and shipped for assembly from different sites with each site having its own factory, material inventory, finished goods inventory, administrative staff, and transportation infrastructure amid other matters. Furthermore, the jet engine housing might be fabricated utilizing not as much of material, achieve added capably, and last longer since the design is not restricted to the approaches employed in conventional manufacturing; nevertheless, numerous of these advantages would not be recognized in the earlier stated cost model. To cease these advantages one would require including an alpha and omega analysis.

ADDITIVE MANUFACTURING TOTAL ADVANTAGE

At the business level, the aim is to make the most of revenue; though, at the social level there are many shareholders to ponder and diverse costs and reimbursements. At this level, one might reflect the purpose to curtail resource use and exploit utility. Dollar values are expressed by many influences such as shortage, rules, and training costs among other things that influence how capitals are competently assigned. The distribution of resources is a significant matter; nevertheless, considering the societal effect of additive manufacturing necessitates segregating resource distribution subjects from resource exploitation subjects. The features of production are, characteristically, well-thought-out to be land, i.e., natural resources, labor, capital, and entrepreneurship; nevertheless, capital embraces machinery and tools, which themselves are made of land and labor. Moreover, a main component in the production of all goods and services is time, as exemplified in numerous operations management discussions. Consequently, one might reflect the greatest basic fundamentals of production to be land, labor, human capital, entrepreneurship, and time. The human investment and entrepreneurship exploited in producing additive manufactured goods are vital.

The residual items land, labor, and time establish the main cost basics for production. It is significant to recognize that there is a compromise between time and labor, calculated in labor hours per hour, as showed in Figure 10.2. For instance, it requires one hundred people less time to construct a house than it requires for one person to construct the house. It is also significant to consider that there is too a compromise between time/labor and land, i.e., natural resources, as showed in Figure 10.3. For instance, a machine can decrease both the time and the quantity of people desired for production, but uses additional energy. The triangular plane in the figure signifies likely amalgamations of land, labor, and time required for yielding a fabricated good. Scanning anywhere along this tringle plane is merely an modification of resource utilization. A company can exploit revenue by either shifting resources or by decreasing the resources desired for production. Scanning along the plane of the triangle in Figure 10.3 may consequence in a additional effective distribution of resources for a firm and for society; though, it does not decrease the mixture of resources desired for production.

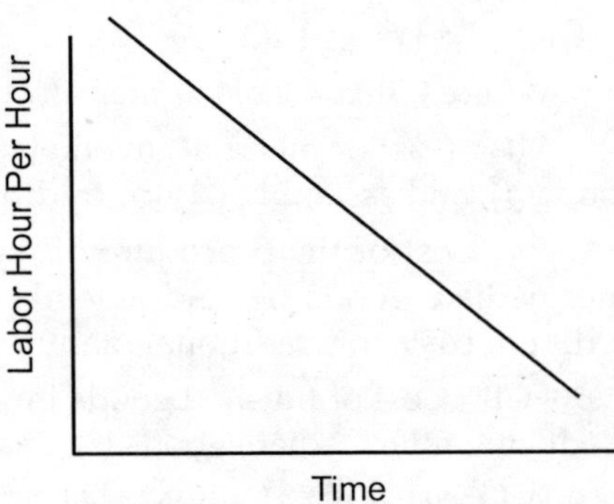

Fig.10.2: Time and Labor Needed to Produce a 3D Print Product

Consequently, once exploring the cost and benefits of a product or procedure from a social viewpoint, it develops ostensible that one wants to measure land, labor, and time required for production in order to comprehend if there has been a decrease in the combination of resources required to yield a fabricated good. If additive manufacturing consequences in a decrease in the resources required for production, then that triangular plane will move toward the origin as showed in Figure 10.3.

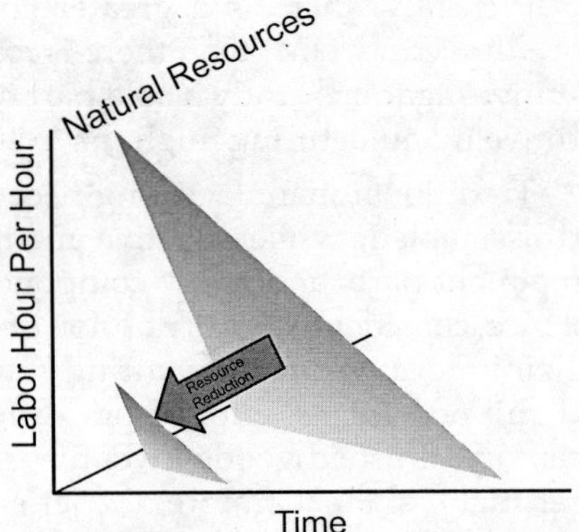

Fig.10.3: Time, Labor, and Natural Resources Needed to Produce a 3D Printed Product

Besides production, fabricated goods are yielded to aid a well thought of determination. For instance, automobiles carrying objects and people; cell phones enable communiqué; and monitors exhibit information. Each item made is intended for some intention. In the procedure of satisfying this goal additional resources are used in the form of land, labor, and time.

Furthermore, a product with a short-lifetime consequences more resources being consumed to replicate the product. In addition, the removal of the old product may outcome in disbursing additional resources. Additive manufactured products may deliver product improvements, new capabilities, or a lengthy useful life. The entire benefit of an additive manufactured good is the variance in the use of land, labor, and time used on production, utilization, and disposal shared with the utility added from the product compared to that of conventional fabrication approaches. This can be signified as the following:

$$TA_L = (L_{AM,P} + L_{AM,U} + L_{AM,D}) - (L_{T,P} + L_{T,U} + L_{T,D})$$
$$TA_{LB} = (LB_{AM,P} + LB_{AM,U} + LB_{AM,D}) - (LB_{T,P} + LB_{T,U} + LB_{T,D})$$
$$TA_T = (T_{AM,P} + T_{AM,U} + T_{AM,D}) - (T_{T,P} + T_{T,U} + T_{T,D})$$
$$TA_U = U(P_{AM}) - U(P_T)$$

TA = The total advantage of additive manufacturing compared to traditional methods for Land (L), labor (LB), time (T), and utility of the product (U).

L = The land or natural resources needed using additive manufacturing processes (AM) or traditional methods (T) for production (P), utilization (U), and disposal (D) of the product

LB = The labor hours per hour needed using additive manufacturing processes (AM) or traditional methods (T) for production (P), utilization (U), and disposal (D) of the product

T = The time needed using additive manufacturing processes (AM) or traditional methods (T) for production (P), utilization (U), and disposal (D) of the product

$U(P_{AM})$ = The utility of a product manufactured using additive manufacturing processes, including the utility gained from increased abilities, enhancements, and useful life.

(P_T) = The utility of a product manufactured using traditional processes, counting the utility increased from increased capabilities, improvements, and useful life. In this circumstance production includes material extraction, material refining, manufacturing, and transportation amongst other things. It is vital to consider that these items must be carefully taken into account when measuring the total benefit of additive manufacturing. An added challenge is that land, labor, time, and utility are calculated in different units, making them problematic to compare. An additive manufactured product might need more labor but reduce the natural resources required. In this case, there is a balance.

IMPLEMENTATION AND ADOPTION OF ADDITIVE MANUFACTURING

Additive manufacturing is considerably dissimilar from conventional ways; therefore, defining when and how to yield gain of the advantages of additive manufacturing is a challenge trial. Moreover, the fabricating industry is concerned with enhancing production employing conventional ways. Recognizing products that profit from enlarged difficulty or being created in nearer to consumers, or considering the influence on inventory is difficult and tough as it influences issues that are hard to quantify.

Additive Manufacturing and Firm Capabilities

With the intention of generating products and services, a corporation requires resources, recognized procedures, and abilities. Resources comprise natural resources, labor, and other items required for production. A corporation must have available resources to yield goods and services. The corporation must also have procedures in place that convert resources into products and services. Two corporations may have the similar resources and procedures in place; though, their products may not be equal due to quality, precise functions, or cost of the product or service. This variance is contributed to the competences of the corporation; that is, abilities are the corporation's capability to yield a good or service successfully.

Controllability is the corporation's capability to effectively regulate its procedures. Its prime purpose is to accomplish competence that reduces cost and exploits maximum accuracy and productivity. Flexibility is the corporation's capability to transact with internal and external ambiguities. It contains responding to fluctuating conditions whereas satisfying the influences of time, cost, or performance. Therefore, there is a compromise between controllability and flexibility; that is, in the short-term, a corporation selects amalgamating of flexibility and controllability, surrendering one for the other.

In the long term, a corporation can incorporate and upsurge both flexibility and controllability over technology or knowledge progression amid other things. Moreover to the objects of proficiencies, there are classes of abilities or a series of capabilities, which contain basic abilities, procedures-level abilities, system-level abilities, and achievements. The fundamental abilities embrace complete knowledge and skill of a corporation and its employees, counting their engineering expertise, safety talents, and work ethics among other things. Procedures-level abilities contain discrete functions such as assembly, welding, and other discrete activities. System-level abilities contain bringing talents together to convert resources into goods and services. Lastly, the performance is frequently quantified by revenues, profits or customer gratifications amid other things.

Accepting a new technology, such as additive manufacturing can have substantial effects on a corporation's abilities. In some occurrences, the per-unit cost can be complex for additive manufacturing than for conventional subtractive approaches. The outcome is that a corporation sacrifices controllability for flexibility; therefore, it may be acceptable for those corporations that pursue a high flexibility situation to implement additive manufacturing. In some occurrences, nevertheless, additive manufacturing can definitely shake controllability. Additive manufacturing can decrease costs for products that have intricate designs that are costly to manufacture utilizing conventional methods. As the price of material and systems reduced for additive manufacturing, the controllability associated with this technology will upsurge, causing it to be acceptable to more corporations.

Furthermore to the compromise among flexibility and controllability, additive manufacturing can also directly influence a corporation's chain of ability, counting the fundamental, process-level, and system-level abilities. At the fundamental level, additive manufacturing needs new information, methods, and designs. These new information areas can be costly and challenging to gain. At the process-level, a corporation that implements additive manufacturing is forsaking several of its present separate functions to implement a completely new production technique. Previous functions might have mandated substantial capital venture in order to entirely develop. Numbers of corporation may be anxious in forsaking these abilities for a new procedure, which itself may necessitate substantial capital venture to fully develop.

As a final point, additive manufacturing can influence the system-level competence, as it is not only a procedure that affects the production of discrete parts, but also the assembly of the parts. All of these variations can make it expensive and risky for a business to implement additive manufacturing technologies and can consequence in decreasing the rate at which this technology is adopted.

Additive Manufacturing Technology is Cost Effectiveness

The additive manufacturing costs disclose that this technology is cost effective for manufacturing small lots with continued centralized manufacturing; nonetheless, with improved mechanization distributed production may be cost effective. Integrating the influence of the production of sole parts and studies its equivalent supply chain effects such as inventory and transportation costs beside decreased risk to supply disturbance will deliver precise valuation to Additive Manufacturing application.

Presently, investigation also discloses that material costs establish a foremost amount of the cost of a product created by means of additive manufacturing. Technologies can frequently be harmonizing, where two technologies are implemented together with each other and the benefits are superior if they were adopted separately. Cumulative implementation of additive manufacturing may ensure a decrease in raw material cost through economies of scale. The decreased cost in raw material might then spread more embracing of additive manufacturing. There may also be economies of scale in raw material costs if specific materials developed more to become common rather than a excess of different materials. The additive manufacturing system is also a substantial cost aspect; nonetheless, this cost has repeatedly reduced. Between 2001 and 2018 the average price decreased 55% after adjusting for inflation.

A number of aspects obscure reducing the cost of additive manufacturing, including construct orientation, envelope exploitation, construct time, energy consumption, product design, and labor. The modest orientation of the part in the construct chamber can consequence in as much as 160 % upsurge in the energy consumed. Furthermore, entirely exploiting the construct chamber decreases the per-unit cost considerably. Each of these concerns must be well-thought-out in the cost of additive manufacturing, making it hard and complex to curtail costs. These matters, expectedly, slow the implementation of this technology, as it necessitates advanced knowledge.

Additive manufacturing not only has consequences for the costs of production, but also the operation of the final product. This technology permits for the fabrication of products that might not have been likely employing conventional subtractive techniques. These products may have novel capabilities, prolonged useful life, or decrease the time, labor, or natural resources required to utilize these products. For instance, automobiles might be made less weight, decreasing fuel costs or combustion engines might be designed to decrease cooling requirements.

For this motive, there is a necessity to track the land, i.e., natural resources, labor, and time consumed on production, operation, and discarding along with the utility added from new designs. The trouble in quantifying these items, possibly, decelerates the acceptance of additive manufacturing.

THE CONCLUSION OF THE MATTER

The conclusion of the matter is this "Additive Manufacturing Technology is Cost Effective." It is here to stay! However, nothing is new under the sun.

"All things are full of labor; man cannot utter it: the eye is not satisfied with seeing, nor the ear filled with hearing. The thing that hath been, it is that which shall be; and that which is done is that which shall be done: and there is no new thing under the sun".

King Solomon Ecclesiastes 1:8-9

REFERENCES

1. Kodama H., Automatic Method for Fabricating a Three Dimensional Plastic Model with Photo Hardening Polymer, Rev Sci Instrum, pp.1770-1773, 1981.
2. Tabusca Silvia, The Internet access as a fundamental right, Journal of Information Systems and Operations Management, Vol.4, No.2/2010, Universitary Publishing House, Bucuresti, ISSN 1843-4711.

3. Tabusca A., Electronic Online Advertising – The "Economic Crisis" is just the pretext of the fall, Journal of Information Systems and Operations Management, Vol. 6, No.2/2012, pp. 410- 418, Universitary Publishing House, Bucuresti, ISSN 1843-4711.
4. Evans B., Practical 3D Printers: The Science and Art of 3D Printing, Apress Publisher, 2012.
5. Soloman, Sabrie "Sensors Handbook, 2nd Edition," McGraw-Hill Publishing Co., NY – ISBN 978-0-07-160570-0
6. Soloman, Sabrie "Sensors and Control Systems in Manufacturing 2nd Edition," McGraw-Hill Publishing Co., NY – ISBN 0-07-059626-3
7. Soloman, Sabrie "Affordable Automation," McGraw-Hill Publishing Co., NY – ISBN 0-07-059633-6
8. Soloman, Sabrie "Sensors Handbook, 1st Edition," McGraw-Hill Publishing Co., NY – ISBN 0-07-059630-1
9. Soloman, Sabrie "Sensors and Control Systems in Manufacturing 1st Edition," McGraw-Hill Publishing Co., NY – ISBN 978-07-10572-4

Appendix

Certified Materials

STEREOLITHOGRAPHY

Material Certified Printers

Accura® 25 ProJet® 6000 HD, ProJet® 7000 HD, ProX® 800, ProX® 950
Accura® 48HTR ProJet® 6000 HD, ProJet® 7000 HD, ProX® 800, ProX® 950
Accura® 55 ProX® 800, ProX® 950
Accura® 60 ProX® 800, ProX® 950
Accura® ABS Black ProJet® 6000 HD, ProJet® 7000 HD, ProX® 800, ProX® 950
Accura® ABS White ProX® 800, ProX® 950
Accura® Amethyst Viper SLA (legacy printer)
Accura® Bluestone ProX® 800
Accura® CastPro™ ProX® 800, ProX® 950
Accura® CastPro™ Free ProX® 800, ProX® 950
Accura® CeraMAX Composite ProX® 800
Accura® ClearVue™ ProJet® 6000 HD, ProJet® 7000 HD, ProX® 800, ProX® 950
Accura® ClearVue™ Free ProX® 800, ProX® 950
Accura® HPC ProX® 800
Accura® PEAK ProX® 800, ProX® 950
Accura® Phoenix ProJet® 6000 HD, ProJet® 7000 HD, ProX® 800, ProX® 950
Accura® PP White ProX® 800, ProX® 950
Accura® Sapphire ProJet® 6000 HD, ProJet® 7000 HD
Accura® SL 5530 ProX® 800, ProX® 950
Accura® Xtreme ProJet® 6000 HD, ProJet® 7000 HD, ProX® 800, ProX® 950
Accura® Xtreme White 200 ProJet® 6000 HD, ProJet® 7000 HD, ProX® 800, ProX® 950
Accura® e-Stone ProX® 800

CJP Technology

Material Certified Printers

VisiJet® PXL ProJet® CJP 260Plus, ProJet® CJP 360, ProJet® CJP 460Plus, ProJet® CJP 660Pro, ProJet® CJP 860Pro

DENTAL

Material Certified Printers

NextDent Cast NextDent 5100
NextDent C&B MFH NextDent 5100
NextDent Denture 3D+ NextDent 5100
NextDent Gingiva Mask NextDent 5100
NextDent Model 1.0 Oker NextDent 5100
NextDent Model 2.0 NextDent 5100
NextDent Ortho IBT NextDent 5100
NextDent Ortho Rigid NextDent 5100
NextDent SG NextDent 5100
NextDent Tray NextDent 5100
NextDent Try-In NextDent 5100
VisiJet® M3 Dentcast ProJet® MJP 3600 Dental
VisiJet® M3 Pearlstone ProJet® MJP 3600 Dental
VisiJet® M3 Stoneplast ProJet® MJP 3600 Dental
VisiJet® M2R-TN ProJet® MJP 2500 Plus
Accura® e-Stone ProX® 800

ENTRY-LEVEL INDUSTRIAL

Material Certified Printers

FabPro™ Elastic Black FabPro® 1000
FabPro™ Flexible BLK FabPro® 1000
FabPro™ JewelCast GRN FabPro® 1000
FabPro™ Proto GRY FabPro® 1000
FabPro™ Tough BLK FabPro® 1000

Material Certified Printers

Figure 4™ ELAST-BLK 10 Figure 4® Modular, Figure 4® Standalone, Figure 4® **Production**
Figure 4™ FLEX-BLK 10 Figure 4® Modular, Figure 4® Standalone, Figure 4® Production
Figure 4™ HI TEMP 300-AMB Figure 4® Modular, Figure 4® Standalone, Figure 4® **Production**
Figure 4™ JCAST-GRN 10 Figure 4® Standalone, Figure 4® Production
Figure 4™ MED-AMB 10 Figure 4® Modular, Figure 4® Standalone, Figure 4® Production
3D SYSTEMS | MATERIALS BUYERS GUIDE | AUGUST 2019 43
Figure 4™ MED-WHT 10 Figure 4® Standalone
Figure 4™ TOUGH-BLK 20 Figure 4® Modular, Figure 4® Standalone, Figure 4®
Figure 4™ TOUGH-GRY 10 Figure 4® Modular, Figure 4® Standalone, Figure 4®
Figure 4™ TOUGH-GRY 15 Figure 4® Modular, Figure 4® Standalone, Figure 4®

Production

MJP

Material Certified Printers

VisiJet® Armor (M2G-CL) ProJet® MJP 2500 Plus

VisiJet® CE-BK ProJet® MJP 5600

VisiJet® CE-NT ProJet® MJP 5600

VisiJet® CR-BK ProJet® MJP 5600

VisiJet® CR-CL 200 ProJet® MJP 5600

VisiJet® CR-WT 200 ProJet® MJP 5600

VisiJet® M2 CAST ProJet® MJP 2500W

VisiJet® M2 ICast ProJet® MJP 2500 IC

VisiJet® M2 EBK ProJet® MJP 2500 Plus

VisiJet® M2 ENT ProJet® MJP 2500 Plus

VisiJet® M2R-BK ProJet® MJP 2500, ProJet® MJP 2500 Plus

VisiJet® M2R-CL ProJet® MJP 2500 Plus

VisiJet® M2R-GRY ProJet® MJP 2500 Plus

VisiJet® M2R-TN ProJet® MJP 2500 Plus

VisiJet® M2R-WT ProJet® MJP 2500, ProJet® MJP 2500 Plus

VisiJet® M3 Black ProJet® MJP 3600 Series

VisiJet® M3 CAST ProJet® MJP 3600W

VisiJet® M3 Crystal ProJet® MJP 3600 Series

VisiJet® M3 Hi-Cast ProJet® MJP 3600W

VisiJet® M3 Navy ProJet® MJP 3600 Series

VisiJet® M3 Procast ProJet® MJP 3600 Series

VisiJet® M3 Proplast ProJet® MJP 3600 Series

VisiJet® M3 Techplast ProJet® MJP 3600 Series

VisiJet® M3-X ProJet® MJP 3600 Series

VisiJet® ProFlex (M2G-DUR) ProJet® MJP 2500, ProJet® MJP 2500 Plus

SELECTIVE LASER SINTERING

Material Certified Printers

CastForm® PS sPro™ 60 HD-HS

DuraForm® EX Black sPro™ 60 HD-HS, sPro™ 140, sPro™ 230

DuraForm® EX Natural sPro™ 60 HD-HS, sPro™ 140, sPro™ 230

DuraForm® Flex sPro™ 60 HD-HS

DuraForm® FR1200 sPro™ 60 HD-HS

3D SYSTEMS | MATERIALS BUYERS GUIDE | AUGUST 2019 44

Appendix

DuraForm® GF sPro™ 60 HD-HS, sPro™ 140, sPro™ 230

DuraForm® HST Composite sPro™ 60 HD-HS, sPro™ 140, sPro™ 230

DuraForm® PA sPro™ 60 HD-HS, sPro™ 140, sPro™ 230
DuraForm® TPU Elastomer sPro™ 60 HD-HS
DuraForm® ProX® AF ProX® SLS 6100
DuraForm® ProX® EX BLK ProX® SLS 6100
DuraForm® ProX® EX NAT ProX® SLS 6100
DuraForm® ProX® FR1200 ProX® SLS 6100
DuraForm® ProX® GF ProX® SLS 6100
DuraForm® ProX® HST Composite ProX® SLS 6100
DuraForm® ProX® PA ProX® SLS 6100